5th International Symposium of Sustainable Landscape Development (The 5th ISSLD)

IOP Conference Series: Earth and Environmental Science
Volume 879

Bogor, Indonesia
22 – 24 September 2020

ISBN: 978-1-7138-4861-5
ISSN: 1755-1307

Printed from e-media with permission by:

Curran Associates, Inc.
57 Morehouse Lane
Red Hook, NY 12571

Some format issues inherent in the e-media version may also appear in this print version.

Institute of Physics
Dirac House, Temple Back
Bristol BS1 6BE UK

Phone: 44 1 17 929 7481
Fax: 44 1 17 920 0979

techtracking@iop.org

Additional copies of this publication are available from:

Curran Associates, Inc.
57 Morehouse Lane
Red Hook, NY 12571 USA
Phone: 845-758-0400
Fax: 845-758-2633
Email: curran@proceedings.com
Web: www.proceedings.com

TABLE OF CONTENTS

IOP Conf. Series: Earth and Environmental Science **879** (2021) 011001 doi:10.1088/1755-1315/879/1/011001

Preface

The 5th International Symposium of Sustainable Landscape Development (The 5th ISSLD) is a continuation of the previous symposium, The 4th ISSLD, which was conducted in 2019 in Bogor city. The main theme of the 5th ISSLD is "Landscape Transformation," which aims to address the various aspects of landscape transformation considering sustainable landscape development.

The landscape is a dynamic system that evolves. Landscape transformation can occur as a result of natural phenomena and human-made processes as well. The pandemic of Covid-19 is one example of how a catastrophic transform the lives of humans and the landscape. As a result of the Covid-19 pandemic, humans are known as rulers on earth with all their habits, having such limited movements that other landscape elements are also affected. The new normal habits make humans more disciplined in maintaining sanitation, thus impacting the cleanliness of the landscape. Restrictions on human movement have reduced human intervention in animal life, vegetation, and landscapes, thus providing opportunities for nature to heal. These things are happening around us, where the landscape is transforming during the COVID-19 pandemic.

According to the topic above, the 5th ISSLD features four sub-themes:

1. The role of man in the landscape transformation
2. The sustainable landscape transformation
3. The ecological approach for landscape development
4. Landscape Horticulture

The landscape transformation and research were presented in this symposium and discussed in panel sessions. The results of the discussion are the identification of Landscape transformation physically and recommendation to the impact of the landscape transformation, as follows:

- Increasing the area of public green open space and improving the quality of the landscape, which was previously a landscape with low use such as wasteland and ex-mining are, into a public park that is useful for ecological and social sustainability
- Conservation of animals and vegetation, such as ethnomedicine from the forest, protecting animals from roadkill, birds movements, and birds behavior, to conserve animals as essential elements of a landscape
- Improvement of landscape quality to anticipate changes caused by nature such as flooding, land subsidence, and soil erosion. These are also related to human involvement, especially in the field of policy and economy
- Presenting a productive garden in an edible landscape / edible garden, which is a response to changes in human culture in a landscape
- Improve the park/landscape quality that positively affects the health and comfort of people in the landscape.
- Increase community participation in landscape management and nature tourism to increase local people's income
- Restoring polluted landscapes through Biological Purification Systems
- Analyze human behavior in the landscape and their knowledge of ecology in a micro-middle-macro landscape
- Planning for public transportation routes, u-turns, bottlenecks, tourist routes that connect natural tourist destinations, cultural and historical tourism destinations, as well as other potential destinations in a landscape and fixing transportation problems such as congestion
- Use of Drone technology for more effective participatory mapping.

- Approaches to landscape design, planning, and management related to ecological and socio-cultural aspects are needed to deal with landscape changes to create better landscapes and increase environmental and human interactions in the landscape.

The recommendations above can be used as input in the implementation of landscape transformation research so that the definition of landscape transformation and its impact on humans and the landscape can be formulated in more detail.

Due to the Covid-19 pandemic not being over, we hold the 5th ISSLD of 2020 online. Initially, the organizers were skeptical about holding the symposium online. However, it turns out that this provides wider opportunities for participants from various countries and continents to participate. The total participants were from eight countries and three continents. The invited speakers come from Chile, the USA, Lebanon, and Indonesia. The speakers, presenters, and general participants represent some institutions and universities from Indonesia and abroad, such as IFLA, IALE, LIPI, Forest Research Development Agency, Ministry of Forestry Republik of Indonesia, University of Zagreb Croatia, Warsaw University of Life Science Poland, Chiba University Japan, University of Philippine Diliman, and universities in Indonesia, including the University of Gadjah Mada, University of Brawijaya, University of Trisakti, University of Pendidikan Indonesia, University of Indraprastha PGRI, Satya Wacana Christian University and IPB University.

In conducting the 5th ISSLD, the Department of Landscape Architecture, Faculty of Agriculture, IPB University cooperates with the Indonesian Society of Landscape Architecture IALI and the Indonesian Education Association of Landscape Architecture APALI. We gratitude to these two prominent landscape architecture institutions in Indonesia for their partnership and support.

Last but not least, We would like to thank you very much and high appreciation to all committee members who already worked hard and spent their time preparing this event.

Chief of Editor

Dr Akhmad Arifin Hadi, SP, MA

Invited Spakers

Ricardo Riveros Celis, *IFLA America, Chile*
Salma Samaha, *IFLA Edication and Academic Affair, Lebanon*
Robert Scheller, *IALE, USA*
Nizar Nasrullah, *IPB Univeristy, Indonesia*

Editorial Committees

Akhmad Arifin Hadi, *IPB University, Indonesia*
Beata Joanna Gawryszewska, *Warsar University of Life Sciences, Poland*
Iva Rechner Dika, *University of Zagreb Faculty of Agriculture, Croatia*
Ricardo Riveros Celis, *IFLA America, Chile*
Afra D.N Makalew, *IPB University, Indonesia*
Alinda F.M Zain, *IPB University, Indonesia*
Syartinilia, *IPB University, Indonesia*
Kaswanto, *IPB University, Indonesia*
Pingkan Nuryanti, *IPB University, Indonesia*

Scientific Committees

Afra DN Makalew, *IPB University, Indonesiaa*
Akhmad Arifin Hadi, *IPB University, Indonesia*
Alinda FM Zain, *IPB University, Indonesia*
Kaswanto, *IPB University, Indonesia*
Mohammad Zaini Dahlan, *ITB, Indonesia*
Nizar Nasrullah, *IPB University, Indonesia*
Pingkan Nuryanti, *IPB University, Indonesia*
Prita Indah Pratiwi, *IPB University, Indonesia*
Qodarian Pramukanto, *IPB University, Indonesia*
Sri Shindi Indira, *University of Pembangunan Panca Budi, Indonesia*
Syartinilia, *IPB University, Indonesia*

Steering Committees

Sugiyanta, *IPB Unversity, Indonesia*
Andi Gunawan, *IPB Unversity, Indonesia*
Aris Munandar, *IPB Unversity, Indonesia*
Bambang Sulistyantara, *IPB Unversity, Indonesia*
Daisy Radnawati, *APALI, Indonesia*
Dian Heri Sofian, *IALI, Indonesia*
Hadi Susilo Arifin, *IPB Unversity, Indonesia*
Nurhayati, *IPB Unversity, Indonesia*

Organizing Committees

Akhmad Arifin Hadi, *IPB Unversity, Indonesia*
Afra DN Makalew, *IPB Unversity, Indonesia*
Alfurqon Nur Aziz, *IPB Unversity, Indonesia*
Alinda FM Zain, *IPB Unversity, Indonesia*
Andika Yudianto, *IPB Unversity, Indonesia*
Avissa Ayuningdias Prasetya, *IPB Unversity, Indonesia*
Debby Wiramiharja, *IPB Unversity, Indonesia*
Dedy Guswandi, *IALI, Indonesia*
Dewi Rezalini Anwar, *IPB Unversity, Indonesia*
Fauzi Agustiawan, *IPB Unversity, Indonesia*
Fitriyah Nurul Hidayati Utami, *IPB Unversity, Indonesia*
Gita Mukti Rahayu, *IPB Unversity, Indonesia*
Indung Sitti Fatimah, *IPB Unversity, Indonesia*
Irma Normawati, *IPB Unversity, Indonesia*
Kartika Wulandari, *IALI, Indonesia*
Kaswanto, *IPB Unversity, Indonesia*
Kristianto Prasetyo, *IPB Unversity, Indonesia*
Lasta Musfra, *IPB Unversity, Indonesia*
Melana Efendi, *IPB Unversity, Indonesia*
Mohammad Fakhri Mashar, *IPB Unversity, Indonesia*
Mohammad Zaini Dahlan, *ITB, Indonesia*
Neviera Khairunisa Pratiwi, *IPB Unversity, Indonesia*
Nizar Nasrullah, *IPB Unversity, Indonesia*
Pingkan Nuryanti, *IPB Unversity, Indonesia*
Qodarian Pramukanto, *IPB Unversity, Indonesia*

IOP Conf. Series: Earth and Environmental Science **879** (2021) 011001 doi:10.1088/1755-1315/879/1/011001

Rachman Andra Wijaya, *IALI, Indonesia*
Restia Juniarti, *IPB Unversity, Indonesia*
Risdayatri Aulia, *IPB Unversity, Indonesia*
Smaradinda Harbowo Putri, *IPB Unversity, Indonesia*
Suranta Ginting, *IPB Unversity, Indonesia*
Sutoro, *IPB Unversity, Indonesia*
Syartinilia, *IPB Unversity, Indonesia*
Tati Budiarti, *IPB Unversity, Indonesia*
Vera Dian Damayanti, *IPB Unversity, Indonesia*

Assistant Editors

Gita Mukti Rahayu
Irma Normawati
Marwah Jadwa Izzati

Host

Department of Landscape Architecture, Faculty of Agriculture, IPB University, Indonesia.

Indonesian Society of Landscape Architecture IALI, Indonesia

The Indonesian Education Association of Landscape Architecture APALI, Indonesia

IOP Conf. Series: Earth and Environmental Science **879** (2021) 011001 doi:10.1088/1755-1315/879/1/011001

Figure 1. Online symposium of The 5th ISSLD

Peer Review Declaration Form Guidance

On the following page you will find the declaration form.

- Please answer each question.
- You should submit the form along with the rest of your submission files.
- The deadline is the submission date written in your publishing agreement.

All conference organisers/editors are required to declare details about their peer review.

We will published the information you provide as part of your proceedings.

Peer review declaration

The 5th International Symposium of Sustainable Landscape Development (The 5th ISSLD) was held on 22-23 September 2020 in Bogor, Indonesia.

All papers published in this proceeding have been peer-reviewed through a process managed by the Editor by involving reviewers from academics and professionals in the field of landscape architecture.

- **Type of peer review: Double-blind**

All papers were uploaded through the organizer's website http://arl.faperta.ipb.ac.id/symposium. All papers passed a basic check of writing format, including size, margins, font type and size, manuscript structure, compliance with submission instructions, quality of English grammar, and plagiarism checks. Reviewers provided opportunities to Authors for correction of discrepancies in the writing format and English grammar and plagiarism. Papers that still did not meet the minimum 25% of plagiarism checks were rejected.

We used a double-blind system for peer-review while keeping the identity of reviewers and authors anonymous. One external reviewer and one review editor staff member reviewed each paper using professional and scientific standards. The review process took approximately 30 days. The reviewers decided to accept or reject the paper along with the reasons. Editor sent forms of correction, acceptance, and rejection to the appropriate authors.

- **Submission management system:**

The submission was made through the ISSLD website management system at the address http://arl.faperta.ipb.ac.id/symposium developed by the ISSLD Operator. The submission, review, and acceptance systems were carried out through the website. Management of the website also includes:

1. Managing access of reviewers and editors to papers;

2. Automatic registration for ISSLD

3. Papers submission;

4. List of ISSLD every year;

4. Program book schedule of events, materials for Keynote and Invited Speakers, the title of the presentation, and presenter's name.

5. Communication via email between the program committee, reviewers, and authors;

6. Preparation and Publication of conference programs;

7. All authors who submit papers are assumed to have accepted the terms of Publication

License;

8. All requirements for papers, templates, and sample papers can be accessed for Author or aspiring writer.

- **Number of submissions received:** 65
- **Number of submissions sent for review:** 46
- **Number of submissions accepted:** 39
- **Acceptance Rate (Number of Submissions Accepted / Number of Submissions Received X 100):** 84,78
- **The average number of reviews per paper:** 2
- **Total number of reviewers involved:** 14
- **Any additional info on the review process:** All papers have passed the English language check using Grammarly and plagiarism check using Turnitin
- **Contact person for queries:**
Name : Akhmad Arifin Hadi
Affiliation: IPB University
Email : seminar_arl@apps.ipb.ac.id / arifin_hadi@apps.ipb.ac.id

IOP Conf. Series: Earth and Environmental Science **879** (2021) 012001 doi:10.1088/1755-1315/879/1/012001

Institutional model for ecotourism development in the *Gunung Pongkor* post-mining area

A Kusumoarto[1], A Gunawan[2], Machfud[3] and A Hikmat[4]

[1]Lecturer in Architecture Study Program, University of Indraprasta PGRI, Jl. Raya Nangka No. 58C, RT.5/RW.5, Tanjung Barat, Kecamatan Jagakarsa, Kota Jakarta Selatan, Daerah Khusus Ibukota Jakarta 12530
[2]Lecturer in Landscape Architecture Department, IPB University, Jl. Raya Darmaga, Bogor, Indonesia 16680
[3]Lecturer in AgroIndustrial Technology Department, IPB University, Jl. Raya Darmaga, Bogor, Indonesia 16680.
[4]Lecturer in Forest Resources Conservation and Ecotourism Department, IPB University, Jl. Raya Darmaga, Bogor, Indonesia 16680

Email: andrianto.kusumoarto@unindra.ac.id

Abstract. The *Gunung Pongkor* post-mining area has some resources potential for ecotourism destination. The development of area's ecosystem in the *Gunung Pongkor* post-mining area needs to protect the ecosystem and prevent activities that do damage within the area. In this case, it is necessary to make institutional model for ecotourism development in the *Gunung Pongkor* post-mining area. The objectives of this study are 1) to identify the elements in the development structure variable; 2) to analyze objective variables, needs variables, constraint variables, and actor variables; 3) to create institutional model for ecotourism development in the *Gunung Pongkor* post-mining area. This study uses descriptive qualitative methods with Interpretive Structural Modelling (ISM) analysis. The elements analyzed are 1) the objectives to be achieved in the development of the ecotourism areas, 2) the needs in the development of the ecotourism areas, 3) the obstacles that need to be overcome, and 4) the actors who play a role in the development of the ecotourism areas. The key element of the goal for ecotourism development in the *Gunung Pongkor* post-mining area is to provide education to the local community about environmental management and culture. In the development process, a master plan and legality of the area as ecotourism are needed. Some of the main obstacles in the development of this area are the absence of a change in the status of a forest area to an ecotourism area and a lack of knowledge about the conservation of natural resources and the environment. The key actors needed in developing this area are Regional Planning, Research, and Development Agency of Bogor Regency Regional Government, Tourism and Culture Office of Bogor Regency Regional Government, Environmental Office of Bogor Regency Regional Government. The diversity of the existing potential is utilized to the fullest it resolves problem that occur in this area in achieving an ecotourism area based on landscape characters in the *Gunung Pongkor* post-mining through collaborative ecotourism development.
Keywords: ecotourism planning and design, interpretive structural modeling, landscape development and management, mining reclamation.

1. Introduction

The *Gunung Pongkor* post-mining area has ecotourism objects potential, landscape character potential of lowland forest, visual aesthetics of the landscape character potential, and potential area as a habitat for flora and fauna [1]; [2]; [3]. In this area, there is also mine buildings and tailings storage facilities that could potentially be heritage materials. Some areas are used by the community for social and economic activities to become an attraction for activities and culture. Areas that are developed into ecotourism areas are in an excellent category if they have biophysical potential, cultural potential, visual aesthetic potential, historical heritage buildings potential, and biodiversity potential [4];[5];[6]. Ecotourism activities can provide economic benefits for the community and local government [7].

The development of an ecotourism area in the *Gunung Pongkor* post-mining area needs to protect the ecosystem, prevent damage in the area, prevent land-use change, protect ecotourism activities, organize and protect the facilities of ecotourism objects, and protect the visual quality of natural resources. Two of the six principles put forward by [8] that must be met in the implementation of ecotourism are related to the sustainability of the destination, namely: (1) as much as possible eliminating the negative impact of the presence of visitors on the ecotourism destination environment and residents; (2) participate in helping maximize the early and long-term participation of local communities in the decision-making process concerning the implementation of ecotourism. Sustainable ecotourism development policies need to carry out: (1) institutional arrangements; (2) setting policies to protect biodiversity in the context of ecotourism sustainability; (3) policies in protecting local uniqueness and participation of local communities; (4) policies in the protection and development of local culture; and (5) policies in the use of ecotourism resources that can provide economic benefits for local communities in particular and increase income for local governments in general [9]. The development of an ecotourism area in the *Gunung Pongkor* post-mining area requires institutional model based on expert perceptions. The objectives of this study are 1) to identify the elements in the development structure variable; 2) to analyze objective variables, needs variables, constraint variables, and actor variables; 3) to create institutional model for ecotourism development in the *Gunung Pongkor* post-mining area.

2. Methodology

The study was conducted in the Gold Mining Business Unit of *Pongkor* (GMBUP), at 9266297 North Latitude - 9263636 South Latitude and 673097 West Longitude - 674746 East Longitude UTM (Universal Transverse Mercator). The study was conducted from January 2018 to December 2018.

The method used in this research is descriptive qualitative with interpretive structural modeling (ISM) analysis. The elements analyzed are 1) the objectives to be achieved in the development of the ecotourism areas, 2) the needs in the development of the ecotourism areas, 3) the obstacles that need to be overcome, and 4) the actors who play a role in the development of the ecotourism areas.

2.1. Research Data Collection

Data collection methods consisted of literature studies, surveys, and interviews with experts. Primary data is obtained directly from interviews with stakeholders and experts. Secondary data were obtained through literature studies.

2.2. Data Analysis

This method is carried out in two stages. The first stage identifies the key stakeholders. The second stage, determining several key factors obtained from need analysis, which is reviewed through expert discussions and based on the theory of the ecotourism system.

Initial stakeholder information is obtained through the snowball method where stakeholder recommend other stakeholder as respondents [10]. Identification of needs and aspirations of each stakeholder according to the priority scale is carried out after identifying the stakeholders [11]. The identification of stakeholders is done by identifying community leaders and managers of the *Gunung Pongkor* post-mining area who play a role in the development of the ecotourism in the *Gunung Pongkor* post-mining area. The needs and aspirations of each stakeholder are then discussed together with experts to find critical stakeholders. Expert determination is based on considerations and criteria put forward by [12] due to differences of opinion [13]. Each selected expert has knowledge in the fields of ecotourism, ecotourism economics, social and institutional sciences, natural resources and environmental management, and regional planning. The main activities carried out are as follows: 1) to identify and map key stakeholders and their roles related to regional development, 2) identify the relationships among stakeholders, 3) to identify the elements related to ISM analysis, 4) to identify the institutions related to research.

The ISM analysis stage is the determination of contextual relationships, which are then converted into mathematical relationships. The relationship between elements fulfills reflexive and transitive properties [14]. in the process of transforming the contextual relationship (structural self-interaction matrix) into a mathematical relationship in the form of a reachability matrix with complete rules (Table 1) [15]. The determination of sub-elements is carried out through focus group discussions (FGD) of stakeholders involved in area management until a compromise is reached (Table 2).

Table 1. Transformation of Contextual Relationships between Elements Into Mathematical Relationships

Symbols	Definitions	SSIM
V	if sub-element *i* has a contextual relation to sub-element *j*, but sub-element *j* has no contextual relation with to sub-element *i*	$e_{ij} = 1$ and $e_{ij} = 0$
A	if sub-element *i* has no contextual relation to sub-element *j*, but sub-element *j* has a contextual relation to sub-element *i*	$e_{ij} = 0$ and $e_{ij} = 1$
X	if sub-element *i* has a contextual relation to sub-element *j*, and vice versa	$e_{ij} = 1$ and $e_{ij} = 1$
O	if sub-element i has no contextual relation to sub-element *j*, and vice versa	$e_{ij} = 0$ and $e_{ij} = 0$

Table 2. The Elements of the Ecotourism Development System

The elements	The contextual relationships
The Objective	The one objective sub-element contributes to the achievement of another objective sub element.
The Need	The one sub-element of needs supports the fulfillment of other sub elements of needs.
The Constraint	The one constraint sub-element causes another constraint sub element.
The Actor	The one actor sub-element needs the support of another actor sub element.

2.2.1. Analysis of Interpretive Structural Modeling (ISM)
The ISM analysis with the ISM-VAXO method goes through the following stages: 1) preparation of the Structured Self-Interaction Matrix (SSIM), in modeling with ISM, opinions or assessments of contextual relationships of elements are expressed in the symbols of V, A, X, O; 2) transformation of the SSIM matrix into a reachability matrix (RM) where the contextual relationship between sub-elements in the form of a matrix whose cells are in the form of letters (VAXO) is transformed into a binary number

reachability matrix; 3) testing and transformation of the reachability matrix, namely the reachability matrix, if the Boolean operation meets the reflexive and transitive requirements, if not, an adjustment is made by performing a recursive multiplication operation so that a closed matrix condition (causal looping) is formed.

2.2.2. The Classification of Sub-element

The sub-element classification is determined by the level of each sub-element with the following rules: 1) create a reachability matrix after going through the transitivity stage; 2) determine the level of each sub-element by first determining the reachability (horizontally for the sub element with value 1); 3) determine the antecedent (vertically for the sub-element with value 1); 4) specify the intersection (same value); 5) if the reachability and intersection values are the same, then the level of these sub-elements can be determined; 6) remove the sub-elements whose levels are known; 7) Repeat the procedure until all the sub-elements are leveled.

3. Result

3.1. The Objectives to be Achieved in the Development of the Ecotourism Areas in the Gunung Pongkor Post-Mining Area

Providing education to the local community about environmental management and about the culture are vital factors as the highest driving factor in the development of ecotourism (Figure 1a). Conserving living natural resources and their ecosystems and improving the community's welfare has the power as a high driving factor, but has a high enough dependency. Protecting against potential parks has a high dependence and has a high enough driving force. Increasing the active role and independence of the community has a fairly weak dependency and has a strong enough driving force. Protecting vital national objects has a high dependence and has a high enough driving force. Exploiting natural resource potential has a high dependency and has low driving factors (Figure 1b).

(a)

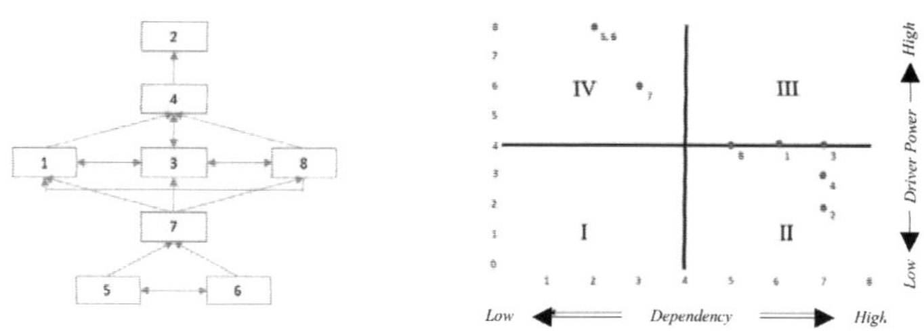

(b)

where :
1: conservation of living natural resources and their ecosystems; 2: exploitation of natural potential; 3: protection against potential parks; 4: protection of national vital objects; 5: provide education to the local community about environmental management; 6: provide education to the local community about the culture; 7: increasing the active role and independence of the community in ecotourism activities; 8: improve community's welfare; I: autonomous barrier; II: dependent barrier; III: linkage barrier; IV: independent (driver) barrier

Figure 1 The Hierarchy of Objectives for Ecotourism Development in the *Gunung Pongkor* Post-Mining Area (a); the Driving Factors and the Dependency Factors of Each Objective (b)

3.2. The Needs in the Development of the Ecotourism Areas in the Gunung Pongkor Post-Mining Area

Critical factors of need are the existence of a master plan document and legality for the ecotourism area (Figure 2a). Both of these needs have the strength as a highly motivating factor for the fulfillment of other needs. Besides, it also encourages the formation of the need for ecotourism area management. The need to improve accessibility to ecotourism areas, the need for increased skills and human resource capacity for ecotourism management, and the need for facilities and infrastructure are three needs that have strength as a high driving factor that encourages the fulfillment of investors needs. Investors' needs for ecotourism development in this area are highly dependent on the fulfillment of other needs (Figure 2b).

(a)

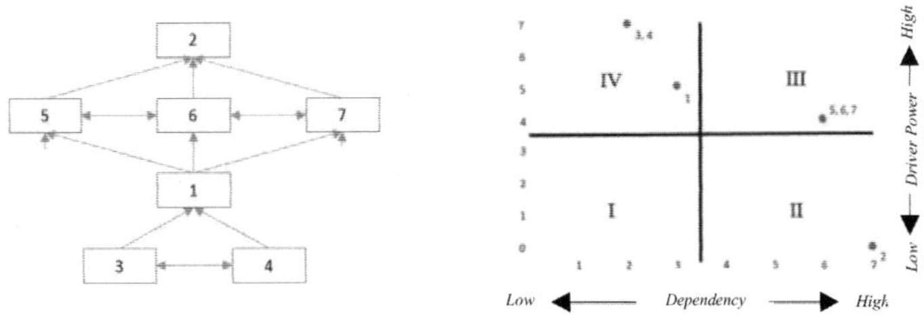

(b)

where :
1: there is a management institution; 2: there are investors; 3: there is a master plan document; 4: the existence of legality as an ecotourism area; 5: improving accessibility to ecotourism areas; 6: increasing skills and human resource capacity for ecotourism management; 7: the need for facilities and infrastructure; I: autonomous barrier; II: dependent barrier; III: linkage barrier; IV: independent (driver) barrier

Figure 2 The Hierarchy of Needs for Ecotourism Development in the *Gunung Pongkor* Post-Mining Area (a); the Driving Factor and the Dependency Factor of Each Need (b)

3.3. The Constraints in the Development of the Ecotourism Areas in the Gunung Pongkor Post-Mining Area

The critical constraint factors are the absence of a change in the forest area's status to an ecotourism area and a lack of knowledge about the conservation of natural resources and the environment (Figure 3a). Both of these constraints have the highest dependencies and have the lowest strengths as driving factors. These obstacles are due to the limited human resources who have knowledge, skills, and professionalism in ecotourism management. The fundamental thing is that there is no change in the forest area's status to an ecotourism area and there is a lack of knowledge about the conservation of natural resources and the environment (Figure 3b).

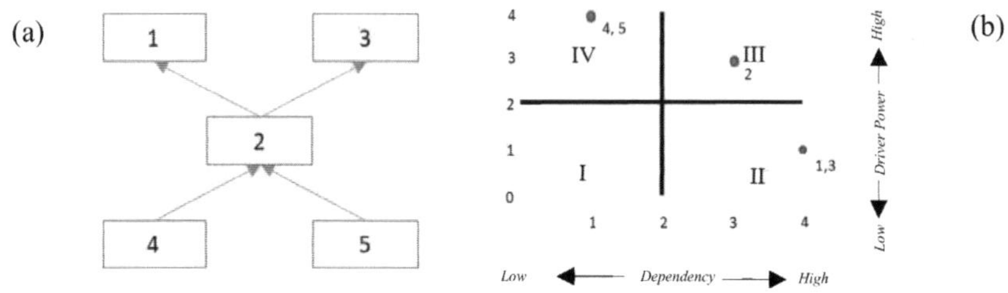

where :
1: there has been no change in the business of ex-illegal miners to the tourism business; 2: limited human resources who have knowledge, skills, and professionalism in ecotourism management; 3: lack of adequate facilities and infrastructure; 4: there has been no change in the forest area's status to an ecotourism area; 5: lack of knowledge about the conservation of natural resources and the environment; I: autonomous barrier; II: dependent barrier; III: linkage barrier; IV: independent (driver) barrier

Figure 3 The Hierarchy of Constraints for Ecotourism Development in the *Gunung Pongkor* Post-Mining Area (a); the Driving Factor and the Dependency Factor of Each Constraint (b)

3.4. The Actors in the Development of the Ecotourism Areas in Gunung Pongkor Post-Mining Area

The key actors involved in developing the ecotourism area in the *Gunung Pongkor* post-mining area are Regional Planning, Research, and Development Agency (*Bappelitbangda*), Tourism and Culture Office (*Dinas Pariwisata dan Kebudayaan*), and Environmental Office (*Dinas Lingkungan Hidup*), Bogor Regency Regional Government. (Figure 4a). Regional Planning, Research, and Development Agency of Bogor Regency Regional Government has the strength as a highest driving factor and has a high dependence on other actors (middle level). Tourism and Culture Office and Environmental Office of Bogor Regency Regional Government are the second institutions that have the power as a driving factor but have high dependencies (Figure 4b).

The Regional Leadership Consultative Forum (*Muspida*) of Bogor Regency Regional Government has a high driving force and has a high dependency. The Ministry of Environment and Forestry has a high dependency and power as a high driving factor. *PT. Aneka Tambang Tbk*, has a high dependence and has strength as a high driving factor. This company also supports to higher education institutions to carry out planning, research, development, and empowerment of communities around the area.

(a)

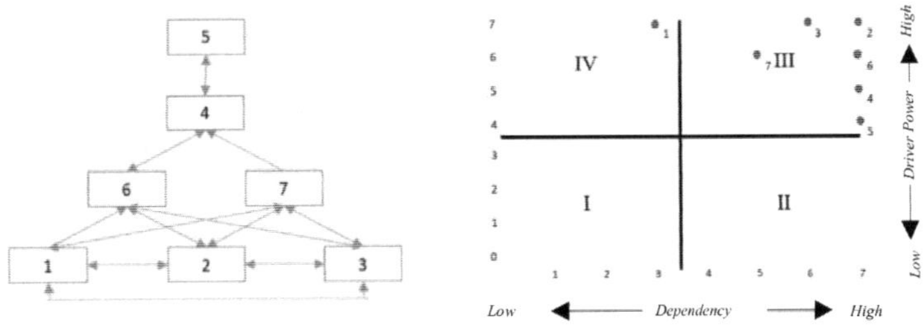

(b)

where :
1: Regional Planning, Research, and Development Agency of Bogor Regency Regional Government; 2: Tourism and Culture Office of Bogor Regency Regional Government; 3: Environmental Office of Bogor Regency Regional Government; 4: *PT. Aneka Tambang Tbk.*; 5: Higher Education Institutions; 6: Ministry of Environment and Forestry; 7: The Regional Leadership Consultative Forum of Bogor Regency Regional Government; I: autonomous barrier; II: dependent barrier; III: linkage barrier; IV: independent (driver) barrier

Figure 4 The Hierarchy of Actors for Ecotourism Development in the *Gunung Pongkor* Post-Mining Area (a); the Driving Factors and Dependency Factors of Each Actor (b)

3.5. The Institutional Model for Ecotourism Development in Gunung Pongkor Post-Mining Area

The development of an ecotourism area to achieve the desired goals requires institutional model in the form of "Collaborative Ecotourism Development". The seven actors, as described above, have contributed by their respective objectives with the principles of the protection and development of sustainable ecotourism in the *Gunung Pongkor* post-mining area. The actors collaborate in meeting the needs and overcoming obstacles for ecotourism development. The institutional model, as proposed, does not take over the role of *PT. Aneka Tambang, Tbk*. The institutional model aims to protect and manage the area so that the area can be characterized, sustainable, safe, and comfortable in the use of limited activities [16].

4. Conclusion

In the process of developing the area into an ecotourism area, the aim of providing education to the local community about environmental management and the culture is a critical element. In the development process, a master plan is needed and the area's legality as an ecotourism area. There are several obstacles in developing this area, especially the absence of a change in status from a forest area to an ecotourism area and a lack of knowledge about the conservation of natural resources and the environment. The key actors needed in developing this area are the Regional Planning, Research, and Development Agency of Bogor Regency Regional Government, Tourism and Culture Office of Bogor Regency Regional Government, and the Environmental Office of Bogor Regency Bogor Government. The diversity of the existing potential is utilized to the maximum extent possible to achieve an ecotourism area based on landscape characters in the *Gunung Pongkor* post-mining area through Collaborative Ecotourism Development.

Reference

[1] Kusumoarto A, Gunawan A, Machfud & Hikmat A 2017 Landscape Character of Pongkor Mining Ecotourism Area in : Kaswanto RL, Mugnisjah WQ, Arifin HS, Ismail NA, & Kobayashi T, editor *2nd International Symposium for Sustainable Landscape Development*; 2016 November 9-10; Bogor, Indonesia Bogor (ID): IOP Publ., pp 012028. Doi: 10.1088/1755-1315/91/1/012028.

[2] Kusumoarto A, Gunawan A, Machfud & Hikmat A 2019 Visual Aesthetic Analysis of Post-Mining Area For Ecotourism Destination AES Bioflux, 11 (3): 133-158.

[3] Kusumoarto A, Gunawan A, Machfud & Hikmat A 2020 The Biodiversity of Flora and Fauna in the Re-vegetation Area in The Post-Mining Area of Pongkor 4th International Symposium for Sustainable Landscape Development; 2019 October 10; Bogor, Indonesia Bogor (ID): IOP Publ., pp 012035. Doi: 10.1088/1755-1315/501/1/012035.
[4] Avenzora R 2008 Ekoturisme : Evaluasi Tentang Konsep Avenzora R, editor NAD-Nias (ID): BRR NAD-Nias.
[5] Gunn C A 1994 Tourism Planning : Basics, Concepts, and Cases Third Edition (London (UK): Taylor & Francis Ltd).
[6] Kusumoarto A & Ramadhan R 2016 Ecotourism and Suitability Evaluation of the Mount Salak Resort II in the Halimun-Salak National Park In Radzi et al. (ed). Heritage, Culture, and Society London (UK): Taylor & Francis Group pp 831.
[7] Jeyacheya J & Hampton MP 2020 Wishful thinking or wise policy? Theorising teorism-led inclusive growth: Supply chains and host communities World Development, 131: 1-12 Doi: 10.1016/j.worlddev.2020.104960.
[8] Fennel D & Eagles P F J 1990 Ecotourism in Costa Rica : A Conceptual Framework Journal of Park and Recreation Administration, 8(1): 3-34.
[9] Fennel D A & Dowling RK 2003 Ecotourism Policy and Planning London (UK): CABI Publ.
[10] Dunn W 2003 Methods of The Second Type : Coping with The Wilderness of Conventional Policy Analysis Policy Studies Review, 7:720-737.
[11] Wondirad A, Tolkach D, & King B 2020 Stakeholder collaboration as a major factor for sustainable ecotourism development in developing countries Tourism Management, 78: 1-21 Doi: 10.1016/j.tourman.2019.104024.
[12] Eriyatno & Sofyan F 2007 Riset Kebijakan Metodologi Penelitian Untuk Pasca Sarjana Bogor (ID): Institut Pertanian Bogor Press.
[13] Motlagh EY, Hajjarian M, Zadeh OH, Alijanpour A 2020 The difference of expert opinion on the forest-based ecotourism development in developed countries and Iran Land Use Policy, 94: 1-9 Doi: 10.1016/j.landusepol.2020.104549.
[14] Machfud 2001 Rekayasa Model Penunjang Keputusan Kelompok dengan Fuzzy Logic untuk Sistem Pengembangan Agroindustri Minyak Atsiri [disertasi] Bogor (ID): Institut Pertanian Bogor.
[15] Saxena JP, Sushil, Vrat P 1992 Hierarchy and classification of program plant elements using interpretive structural modeling: a case study of energy conservation in the Indian cement industry Syst.Pract, 5(6): 651-670.
[16] Rios CA, Amorocho R, Villarreal CA, Mantilla W, Velandia FA, Castellanos OM, Munoz SI, Atuesta DA, Jerez JH, Acevedo O, Vargas M, Caballero VM, Goso CA, Briggs A 2020 Chicamocha Canyon Geopark project: A novel strategy for the socio-economic development of Santander (Colombia) through geoeducation, geotourism and geoconservation International Journal of Geoheritage and Parks, 8: 96-122 Doi: 10.1016/j/ijgeop.2020.05.002.

GIS-based land suitability analysis for potential urban development sites in Diffun, Quirino, Philippines

E E A Sia[1], N Navarra[2] and J D Villa Juan[2]

[1] College of Architecture, University of the Philippines-Diliman, Quezon City, Philippines

[2] Supervising Faculty, College of Architecture, University of the Philippines-Diliman, Quezon City, Philippines

E-mail: easia1@up.edu.ph; edric.sia@gmail.com

Abstract. As the world moves towards urbanization, the demand for land for urban land use increases. If left unmanaged, this might result in the wastage of land that could have been used for more productive purposes, such as agriculture. Other developments can also pose risks to the environment and to those who inhabit it. Therefore, land suitability analyses must be carried out before proceeding to urban planning. This study produced an urban suitability analysis of the municipality of Diffun in Quirino Province as a basis for future urban planning. Conducting this analysis will be very timely, considering that the revision of the municipality's Comprehensive Land Use Plan (CLUP) will start in 2021. In this study, two methods of doing the analysis were integrated: Boolean overlay and weighted overlay. Using the gathered spatial data, a total of 18 factors and restrictions were produced for the analysis. Restriction layers used in the Boolean overlay method include existing built-up areas, erosion risk, flooding susceptibility, protected agricultural lands, other protected lands, Quirino Protected Landscape (QPL), slope, waterway buffers, and ecotourism site buffers. Meanwhile, the weighted overlay analysis used the following factors: elevation, land-use type, distance to ecotourism sites, distance to local roads, distance to national and provincial roads, proximity to the downtown, distance to protected agricultural lands, distance to waterways, and slope. The weights of the factors were calculated using a pairwise matrix. The study concluded that only 0.32% of the total land area of the municipality is most suitable for urban land-use, while 87.36% is considered not suitable for the development of built-up areas.
Keywords: land suitability analysis, weighted overlay, boolean overlay

1. Introduction

Throughout human history, the increase in population has brought about the development of built-up areas. With the Philippines having one of the highest population growth rates in ASEAN [1], urban areas in the country are expected to continue expanding. This expansion subjects agricultural land to urban pressures; such an example would be nearby provinces surrounding Metro Manila like Cavite [2]. If left unchecked, this might result in urban sprawl, eventually taking up too much prime agricultural land. Haphazard expansion can also mean that development may happen in areas that are unsuitable for built-up spaces. These areas may be vulnerable to different natural hazards that pose risks to those residing in them. Proper analysis for planning and enforcement must be done to avoid development that can negatively affect the environment and inhabitants of the areas.

Several studies have demonstrated the ability of GIS, and specifically the weighted overlay approach in being flexible enough in performing various kinds of spatial analyses. For example, the method of analyzing [3] as an improvement to the existing site evaluation technique used by the government, while [4] did an analysis of their study area for suitable areas for small farm reservoirs. Urban settlement analysis [5] and analysis for ecotourism [6] may seem vastly different at first, as one is more focused on developing built-up areas and the other is focused on preserving the natural environment. Nevertheless, the spatial factors that were used had a lot of similarities. These show that determining site suitability is mainly dependent on the relationship of the spaces adjacent to each other. The weighted overlay approach used by the studies was able to reflect the spatial relationships with the use of weights. Each study utilized various methods of determining the values. Riad [7] and Buncag, Santos, and Magpantay [8] gave weights based on previous literature, Şatir [9] assigned equal values to all factors. On the other hand, Kuru and Terzi [10] obtained the weights by surveying urban planners. Meanwhile, the others used the Analytic Hierarchy Process (AHP) in doing so [5, 6, 11]. Alongside the weighted overlay approach, Riad et al. [7] also integrated the Boolean overlay method into their study. The Boolean overlay method was used as an alternative way of determining suitable sites for artificial groundwater recharge.

Studies done in the Philippines that explore different methods on land suitability analyses for urban development seem to be limited, and most are focused on agriculture or siting facilities. [4, 8, 11] With the continued increase in population, the development of areas is sure to follow. Urbanization, as a phenomenon, inevitably needs to be managed through appropriate analysis and proper planning.

Diffun is a municipality located in the landlocked Province of Quirino, adjacent to the municipality of Cordon and Santiago City in Isabela Province. As it is nearer to the Pan-Philippine Highway, Diffun serves as the gateway to the province. The municipality prides itself as Quirino's largest producer of rice and corn—with around 70% of the total households of the municipality depending on agriculture as a source of livelihood. [12] Despite being heavily reliant on agriculture, it also serves as the province's center for trade and other economic activities. [13] Given its strengths in these areas, the municipality envisions itself to becoming a "prime agro-ecotourism center" in the province of Quirino.

According to the Local Government of Diffun [12], the municipality is experiencing positive economic spillover effects from its neighboring city, Santiago. As a result, population growth could surge as people migrate into the municipality. As established in the previous paragraphs, the development of built-up areas is brought about by population growth. Expansion of built-up areas without doing proper land suitability analyses can impede the municipality's progress in achieving its long-term vision. Using GIS, the suitability of areas for urbanization within the municipality could be determined, which can then help in urban planning for the years to come.

1.1. Study area

Diffun (Figure 1) is a municipality located in the Province of Quirino in Region II (Cagayan Valley Region). It is located along the northeast edge of the province, bordering Cordon and Santiago City. Other municipalities within the Province of Quirino adjacent to Diffun are the Municipalities of Sagu day and Cabarroguis. Diffun is the second most urbanized municipality in Quirino, with its neighboring town, Cabarroguis, coming in first. [13] The downtown (locally known as población) of the municipality comprises four small administrative divisions, otherwise known as barangays: Andres Bonifacio, Rizal, Aurora West, and Aurora East.

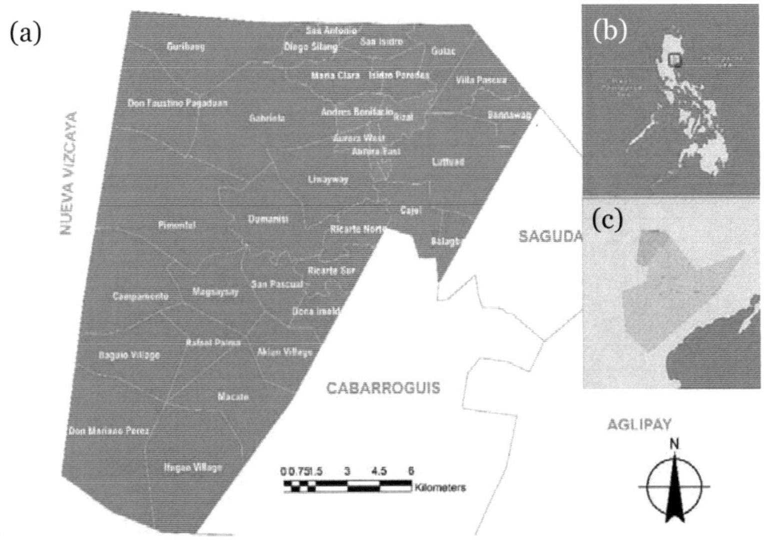

Figure 1 (a) The municipality of Diffun, with its urban barangays (downtown) colored in blue; (b) its location in the Philippines; and (c) its location in Quirino Province

2. Materials and methods

2.1. Spatial data

The spatial data used in the analysis were collected from Community Forestry Foundation-Quirino, Incorporated (CFFQI), a foundation under the government's jurisdiction. A digital elevation model (DEM) was obtained from the United States Geological Survey (USGS). Below (Table 1) is a summary of all spatial data gathered before preprocessing.

Table 1 List of gathered data and their sources

No.	Data	File Format	Scale	Source
1	Digital Elevation Model (DEM)	Raster (TIFF)	30m	USGS
2	Waterways	Shapefile (Line)	-	CFFQI
3	Ecotourism sites	Raster (JPG)	-	CFFQI
4	Erosion hazard	Shapefile (Polygon)	-	CFFQI
5	Flood susceptibility	Shapefile (Polygon)	-	CFFQI
6	Land-use 2015	Shapefile (Polygon)	-	CFFQI
7	Local Roads	Shapefile (Line)	-	CFFQI
8	National and Provincial Roads	Shapefile (Line)	-	CFFQI
9	Protection Agriculture	Shapefile (Polygon)	-	CFFQI
10	Quirino Protected Landscape (QPL)	Raster (JPG)	-	CFFQI

2.2. Methods

The study utilized GIS methods, namely Boolean overlay and weighted overlay, in determining urban land suitability. To make the spatial data gathered useful, these were preprocessed and then processed into different thematic maps. ArcMap 10.2.2 with the Spatial Analyst extension was used in the preprocessing of spatial data and in performing the analyses. For the weighted overlay method, a pairwise matrix was created using Microsoft Excel was used to calculate the weights needed. To make the final suitability maps, the outputs of both methods, Boolean and weighted overlay, were combined through the

Raster Calculator tool. Figure 2 shows the steps that were undertaken to determine urban land suitability in Diffun.

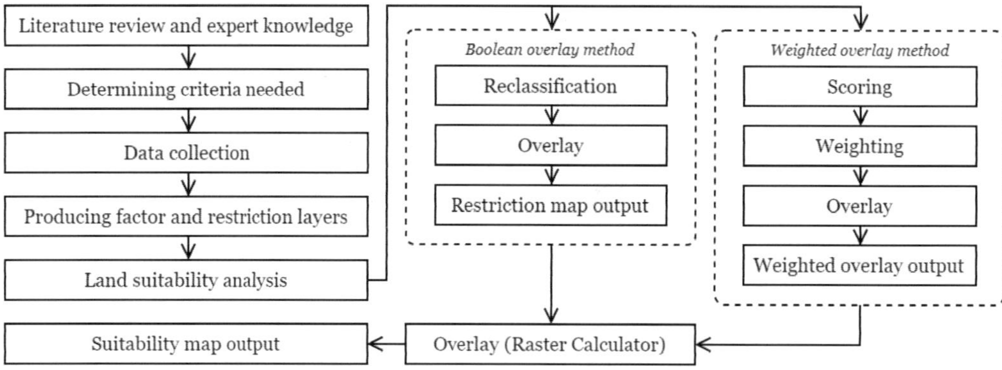

Figure 2 The workflow of the study

2.3. Boolean overlay

The Boolean overlay method is a type of suitability analysis that uses Boolean algebra. Boolean algebra is a branch of algebra that uses only the truth values, true or false, represented as 1 and 0, respectively. In this study, the Boolean overlay method was used to create a restriction map, which determines areas barred from being developed for urban purposes. Thematic maps that represent restrictions to urban development were derived from Philippine laws, related literature, and environmental hazards that are present in the municipality [12, 14-20].

Using the tools in ArcMap's Spatial Analyst toolbox, the raster thematic maps were reclassified into having two classes, with either a value of 1 or 0, depending on different conditions (Table 2). A value of 0 means an area is unsuitable for urban development, while 1 allows development.

The Boolean overlay operation was then done using the Raster Calculator. In place of the Boolean AND operation, overlaying was done by multiplying the restriction layers using the Raster Calculator.

Table 2 Layers used in the Boolean overlay method

No.	Restriction Layer	Value	
		1 (Suitable)	0 (Unsuitable)
1	Agricultural protection	Not considered as protected agricultural areas	Protected agricultural lands
2	Built-up areas	No existing built-up spaces	Existing built-up areas
3	Erosion risk	Minor risk of erosion	Prone to severe erosion
4	Ecotourism sites	Outside the vicinity of ecotourism sites	Ecotourism sites and areas within a 1 km buffer around these sites
5	Flood risk	Less risk of flooding	Highly susceptible to flood
6	Protected land	Alienable and disposable (A&D) land	Forest land
7	Quirino Protected Landscape (QPL)	Areas outside the QPL	Areas within the QPL
8	Slope	<18% slope	>18% slope
9	Waterways	Areas outside waterways	Water networks and areas within a 100-meter ecological buffer

2.4. Weighted overlay

A common and straightforward approach to weighted overlay is by doing weighted linear combination (WLC), also called weighted linear summation. [21, 9] It involves the standardization of criteria scores and the weighing of factors. Factors used for the weighted overlay analysis were determined by comparing the factors used in other studies and the availability of data. [5, 6, 9, 22-25] The data in each factor must be standardized to a specified range or scale because it will not make sense to calculate using different measurement scales. [21] In this study, scores of 1-10 were used to standardize the input values, with higher scores representing "high suitability." For instance, areas near waterways and are within the riparian buffer are unsuitable for development and thus are assigned a value of 1. Scoring was based on other literature and observations of the site. The reclassify tool was used to assign new cell values to the input data. The following conditions were considered to determine if an area is most suitable for development: proximity to ecotourism sites, road networks, protected agricultural land, the downtown and water systems, slope, elevation, and existing land use. In the layer maps, darker shades represent higher values or high suitability. Figure 3 below shows the reclassified factor maps for weighted overlay analysis.

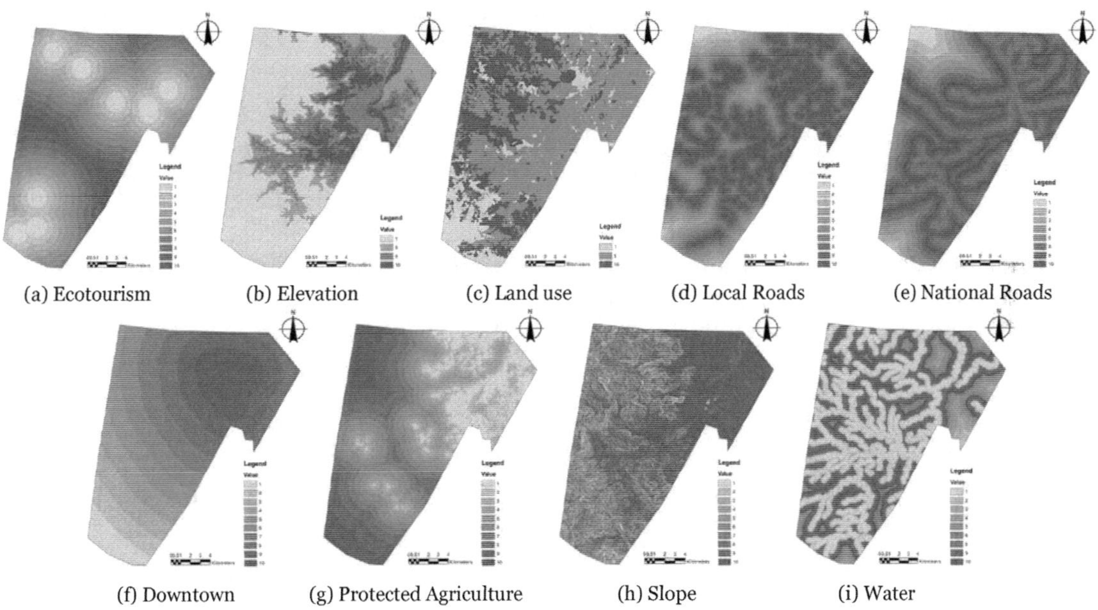

Figure 3 Factor maps used in the weighted overlay method

For the calculation of weights, the initial plan was to gather inputs from government officials, urban planners, and locals through questionnaires and field interviews. However, due to the COVID-19 and lockdowns imposed in the country during the time this study was undertaken, it was difficult to establish proper communication with the stakeholders. Thus, the study relied on the study area's existing conditions, the municipality's vision, and related literature [5, 6, 11-13].

Derived from the Analytic Hierarchy Process (AHP) [21], using a pairwise comparison matrix is a popular method in multi-criteria decision making (MCDM). [5] Therefore, it is used to determine the relative importance of the nine factors. In this study, Saaty's comparison scale [26] was used for pairwise matrices. As seen in Table 3, priority was given to the preservation of ecotourism sites and prime agricultural land as a reflection of the municipality's vision. Preservation of these areas is essential in the determination of potential urban development sites. This resulted in both being the top two factors with the highest level of importance with 24.72% and 29.52%, respectively.

To produce the weighted overlay output, the factors and their corresponding weights were then inputted into ArcMap's weighted overlay tool.

Table 3 The pairwise matrix used in calculating the weight of each factor

	Factors (F)	Factor *j*									Weights (%)
		F1	F2	F3	F4	F5	F6	F7	F8	F9	
Factor *i*	Ecotourism Sites (F1)	1	5	5	9	5	9	1/3	3	7	24.72%
	Elevation (F2)	1/5	1	1	3	2	3	1/3	1/5	3	7.65%
	Land Use (F3)	1/5	1	1	3	1/3	1	1/3	1/9	1/2	4.57%
	Local Roads (F4)	1/9	1/3	1/3	1	1/3	1/2	1/9	1/9	1/3	1.99%
	National Roads (F5)	1/5	1/2	3	3	1	1	1/5	1/3	3	6.44%
	Downtown (F6)	1/9	1/3	1	2	1	1	1/9	1/3	1	3.72%
	Protected Agricultural Land (F7)	3	3	3	9	5	9	1	4	5	29.52%
	Slope (F8)	1/3	5	9	9	3	3	1/4	1	3	16.96%
	Waterways (F9)	1/7	1/3	2	3	1/3	1	1/5	1/3	1	4.42%

3. Results and discussion

3.1. Restriction map (Boolean overlay output)

The Boolean overlay method was able to produce a restriction map (Figure 4a), defining areas that are unsuitable or prohibited from being developed. According to the output, around 87% (≈26,345.89 ha) of the municipality's land area is restricted for development, leaving only 13% (≈3,936.74 ha) available for possible expansion. The bulk of the restricted areas come from protected areas as around 65% of Diffun is considered as forest land. These are the mountainous areas on the western side of the municipality. Adding other factors such as flood risk, ecotourism site buffers, erosion risk, existing built-up areas, prime agricultural land, slope, and waterways further decreased the area available for development.

Most of the areas available for development are on the eastern side of the municipality, located near the downtown barangays. These areas have a slope of not more than 18%, which also makes them suitable for paddy fields. Hence, most of the available areas for development seen in the restriction map are adjacent to prime agricultural lands.

3.2. Weighted overlay output

Distinct levels of suitability from least suitable to the most suitable areas were determined using the weighted overlay approach, unlike in Boolean overlay, where the suitability values are binary. In this study, the suitability scale of 1-10 was further simplified and categorized into five (5) classes: least suitable, not very suitable, moderately suitable, highly suitable, and most suitable. Areas with a final score of 9 to 10 are the most suitable areas for urban development, while areas with the values 1 to 2 are deemed least suitable.

According to the weighted overlay map (Figure 4b), only 8.09% (≈2,449.86 ha) of the municipality is deemed most suitable. These areas are mostly located in the western barangays of Pimentel and Magsaysay. Large portions of the municipality are highly and moderately suitable for urban development, making up 46.07% (≈13,951.21 ha) and 33.09% (≈10,020.52 ha) of the total land area respectively. Not very suitable areas take up 12.31% (≈3,727.79 ha) and are located near the downtown (Aurora West, Aurora East, Rizal, and Andres Bonifacio) where the suitability scores are generally low. Lastly, the least suitable areas only take up 0.44% (≈133.24 ha) of the municipality's total land area.

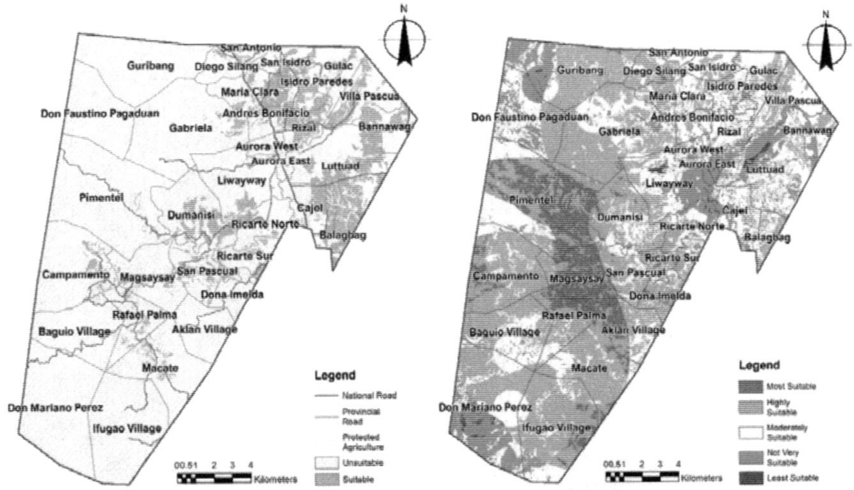

(a) Restriction map (b) Weighted overlay output

Figure 4 Maps produced using Boolean and weighted overlay

3.3. Comparison of the restriction map and the weighted overlay output

As seen in Figure 5, there are significant differences between the restriction map and the weighted overlay output. According to the weighted overlay output, many of the most suitable areas are on the western side of Diffun. However, according to the restriction map, most of these areas are unavailable for development. The reason is that most of the sites are located in or are adjacent to protected areas such as forest land. Meanwhile, barangays surrounding the downtown received low suitability scores in the weighted overlay output despite being available for development in the restriction map. The low suitability rating can be attributed to the sites' proximity to prime agricultural areas. Notwithstanding the differences, Figure 5 also shows that there are areas that satisfy the urban suitability criteria of both methods.

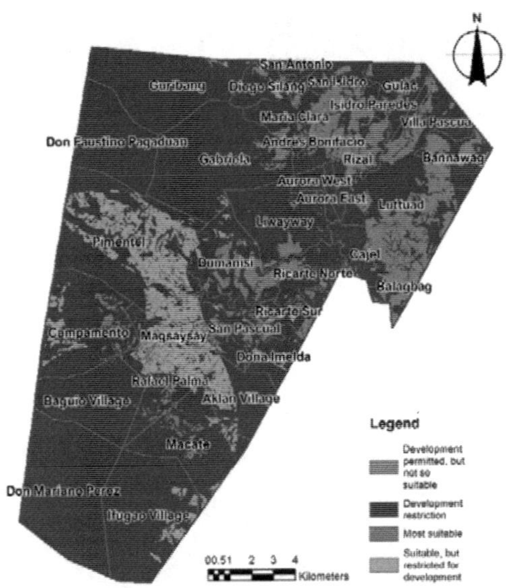

Figure 5 Comparison map

3.4. Urban suitability map

The final suitability map for the municipality of Diffun was created by combining the output maps of the Boolean overlay and weighted overlay analyses. The maps were consolidated into one with the use of the Raster Calculator tool in ArcMap. The final suitability map (Figure 6) shows the levels of suitability in different areas of the municipality with the added layer of development restrictions. It can be seen in the output map that areas near the urban core have medium suitability scores. Because of the added restriction map, the percentage of area covered by suitable areas were greatly diminished (Table 4). Areas that are considered most suitable now only consists of around 0.32% of the total land area of Diffun (≈96.90 ha) from the previously calculated 8.09% (≈2,449.86 ha) in the weighted overlay output. The area covered by highly suitable areas also decreased, now only making up 5.84% (≈1,768.51 ha). As seen in Table 5, Barangay Magsaysay has the largest area in the municipality that is most suitable for urban development. However, the area covered by these suitable sites is small relative to the size of the entire municipality.

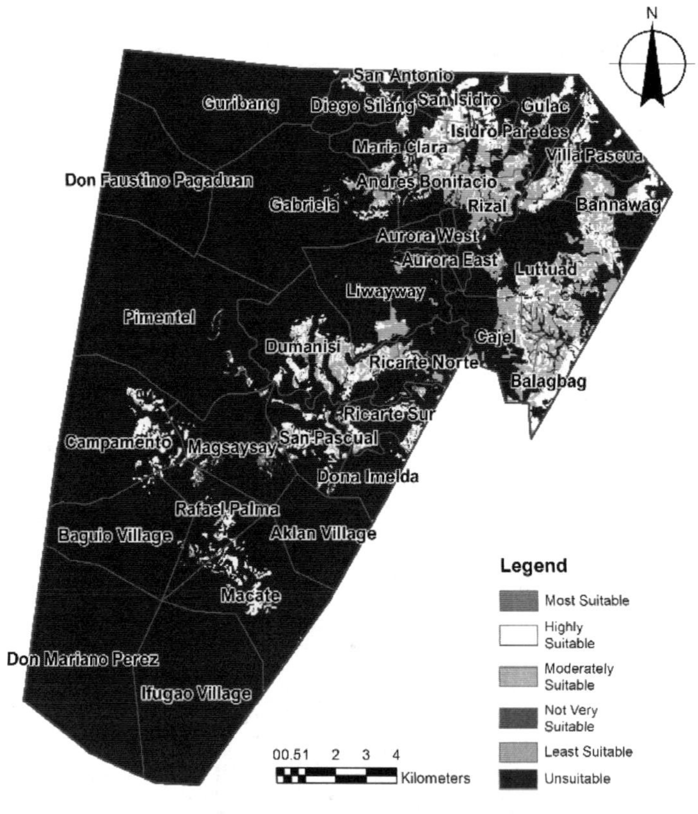

Figure 6 Final urban suitability map

Table 4 Percent area covered by each suitability level in the urban suitability map

Suitability Level	Approx. Area (ha)	Percent Area
Most Suitable	96.90	0.32%
Highly Suitable	1,768.51	5.84%
Moderately Suitable	1,907.82	6.30%
Not Very Suitable	54.51	0.18%
Least Suitable	-	0.00%
Unsuitable	26,454.95	87.36%

Table 5 Barangays with the most suitable areas for urban development

Barangay	Approx. Size of Most Suitable Area (ha)	% Area Relative to Diffun
Magsaysay	47.9274	0.1583%
Pimentel	18.1347	0.0599%
Rafael Palma	12.5833	0.0416%
Isidro Paredes	3.7935	0.0125%
San Isidro	3.5159	0.0116%
San Pascual	2.6832	0.0089%
Villa Pascua	2.2206	0.0073%
Gulac	1.2028	0.0040%
Bannawag	1.1103	0.0037%
Ricarte Sur	0.7402	0.0024%
Dumanisi	0.7402	0.0024%
Diego Silang	0.6477	0.0021%
Maria Clara	0.4626	0.0015%
San Antonio	0.1850	0.0006%
Luttuad	0.1850	0.0006%

4. Conclusion

As the population increases, the development of built-up areas is expected to follow. Diffun, being situated near important economic centers of the region, is foreseen to receive an influx of people from other places which can spur development. With proper planning, the adverse effects of urbanization can be mitigated. Conducting land suitability analyses is a part of the overall planning process. Without it, land use plans and policies may be more detrimental to the development of an area, hence, unsustainable.

Having used both Boolean and weighted overlay methods in the study, integrating the weighted overlay method into the former could provide more information about the suitability level of each area. The Boolean overlay method could only determine whether the land is suitable through binary values, 1 (true) or 0 (false). In contrast, in weighted overlay, an area can be more suitable than the other according to a specified numeric scale. Moreover, weighted overlay treats the factors with varying levels of importance, which is the same as in real-life decision-making processes. A total of 18 factors and restrictions were identified and used in the study in determining urban land-use suitability. After performing the analysis, it was found out that only 0.32% of the total land area of the municipality is most suitable for urban land-use, while 87.36% is considered not suitable for the development of built-up areas. The researcher recognizes that the study alone should not be the sole basis in planning as the dynamics of development is much more complex. The study can still, nonetheless, be a guide in carrying out future studies or analyses.

Acknowledgment

The authors would like to thank the University of the Philippines-Diliman for supporting this study. Likewise, the authors would like to extend their sincere gratitude to the provincial government of Quirino and the municipal government of Diffun for providing the relevant data needed for the completion of this research. We would also like to thank Dr. Brian Lee from the University of Kentucky for imparting his knowledge, which had helped in the initial development of the study.

References

[1] POPCOM Philippine Information Agency [Internet] Quezon City: PIA POPCOM: PH population growth slows down to 1.5%; 2019 Dec 29 [cited 2020 Mar 11] Available from:

https://pia.gov.ph/news/articles/1032129
[2] Malaque I R and Yokohari M 2007 Urbanization process and the changing agricultural landscape pattern in the urban fringe of Metro Manila, Philippines *Environ. Urban* **19**(1):191–206
[3] Lee B D, Adams J and Austin SD 2015 Landscape Tension Index. In: Mueller T, Sassenrath G, editors. GIS applications in agriculture Boca Raton FL: CRC Press p. 171-187
[4] Cacayan A O, Apdohan A G, Bocobo A E and Ruta J L 2019 Identifying suitable areas for small farm reservoir in Agusan del Norte using Geographic Information System *Int. Arch. Photogramm Remote Sens Spat Inf Sci - ISPRS Arch* **42**(4/W19):89–91
[5] Ruttoh D K 2015 GIS Approach in Spatial Suitability Analysis for Urban Residential Settlement: a Case Study of Nyeri Municipality [bachelor's project report on the Internet] Nyeri, Kenya: Dedan Kimathi University of Technology [cited 2020 Feb 10] Available from: https://www.academia.edu/25397980/GIS_APPROACH_IN_SPATIAL_SUITABILITY_ANALYSIS_FOR_URBAN_RESIDENTIAL_SETTLEMENT_A_CASE_STUDY_OF_NNYER_MUNICIPALITY_INSTITUTE_OF_GEOMATICS_GIS_AND_REMOTE_SENSISE
[6] Bunruamkaew K and Murayama Y 2011 Site suitability evaluation for ecotourism using GIS & AHP: A case study of surat Thani Province, Thailand *Procedia - Soc Behav Sci.* **21**:269–78
[7] Riad P H S, Billib M, Hassan A A, Salam M A and El Din M N 2011 Application of the overlay weighted model and boolean logic to determine the best locations for artificial recharge of groundwater *J Urban Environ Eng.* **5**(2):57–66
[8] Buncag M J J, Santos L M and Magpantay A 2019 Suitability Analysis for Sanitary Landfill Site in the Province of Ifugao, Philippines *Int. J Sci. Manag. Stud.* [Internet] **2**(6):85–94 Available from: https://www.researchgate.net/publication/338342473_Suitability_Analysis_for_Sanitary_Landfill_Site_in_the_Province_of_Ifugao_Philippines
[9] Şatir O 2016 Mapping the Land-Use Suitability for Urban Sprawl Using Remote Sensing and GIS Under Different Scenarios In: Sustainable Urbanization [Internet] InTech p. 116–24 Available from: http://www.intechopen.com/books/sustainable-urbanization/mapping-the-land-use-suitability-for-urban-sprawl-using-remote-sensing-and-gis-under-different-scena
[10] Kuru A and Terzi F 2018 Determination of New Development Area in Kırklareli by GIS Based Weighted Overlay Analysis *Int. J. Environ. Geoinformatics* [Internet] **5**(3):244–59 Available from: http://dergipark.gov.tr/doi/10.30897/ijegeo.427330
[11] Paquit J C and Rojo M J A 2018 Assessing suitable sites for Falcata (Paraserianthes falcataria Nielsen) plantation in Bukidnon, Philippines using GIS Int J Biosci [Internet] Feb 28;**12**(2):317–25. Available from: http://www.innspub.net/download/?target=wp-content/uploads/2018/03/IJB-Vol-12-No-2-p-317-325.pdf_19722
[12] Local Government of Diffun 2012 Comprehensive Land Use Plan Diffun, Quirino
[13] Provincial Government of Quirino 2012 Provincial Development and Physical Framework Plan Cabarroguis, Quirino
[14] *The Agriculture and Fisheries Modernization Act of 1997*
[15] *Department of Agrarian Reform Administrative Order No. 01, s. 2002*
[16] *Revised Forestry Code of the Philippines 1975*
[17] *Proclamation No. 548, s. 2004*
[18] *The Water Code of the Philippines 1976*
[19] Housing and Land Use Regulatory Board [Internet] 2013 Quezon City: HLURB CLUP Guidebook: A Guide to Comprehensive Land Use Plan Preparation: Vol 1 The Planning Process [cited 2020 Mar 11] Available from: http://hlurb.gov.ph/services/local-government-unit/clup-guidebook/
[20] Environmental Law Institute 2003 Conservation Thresholds for Land Use Planners [Internet] Washington DC: Environmental Law Institute [cited 2020 Mar 11] Available from: https://www.eli.org/sites/default/files/eli-pubs/d13-04.pdf
[21] Geographic Information Technology Training Alliance [Internet] GITTA Suitability analysis; [cited 2020 Apr 1] Available from: http://www.gitta.info/Suitability/en/html/index.html
[22] Chakma S 2014 Analysis of Urban Development Suitability In: Dhaka Megacity [Internet] Dordrecht: Springer Netherlands p. 147–61 Available from: http://link.springer.com/10.1007/978-94-007-6735-5_8
[23] Jain K and Subbaiah Y V 2007 Site suitability analysis for urban development using GIS *J Appl Sci.* 7(18):2576–83.
[24] Parry J A, Ganaie S A and Sultan Bhat M 2018 GIS based land suitability analysis using AHP model for urban services planning in Srinagar and Jammu urban centers of J&K, India *J Urban Manag.* [Internet] 7(2):46–56. Available from: https://doi.org/10.1016/j.jum.2018.05.002
[25] Liu R, Zhang K, Zhang Z and Borthwick A G L 2014 Land-use suitability analysis for urban development in Beijing *J Environ. Manage* [Internet] Dec;**145**(December):170–9. Available from: http://dx.doi.org/10.1016/j.jenvman.2014.06.020
[26] Saaty T L 1980 Analytic hierarchy process Maidenhead, England: McGraw Hill Higher Education

Distance analysis of U-turn round facilities for sustainable urban development (case study of Ahmad Yani Road, Tirtayasa and Cilegon road)

B Hidayat[1] and F A A Saputri[2]
[1]Lecturer, Polytechnic of Indonesian Land Transport, Ministry of Transportation, Jl. Raya Setu No.89 Cibitung Bekasi 17520, Indonesia,
Email: aim.harso@gmail.com
[2]Student, Polytechnic of Indonesian Land Transport, Ministry of Transportation, Jl. Raya Setu No.89 Cibitung Bekasi 17520, Indonesia,
Email: firdaauliya239@yahoo.com

Abstract. Cilegon City is an industrial city with internal and internal movements occurring on one access road stretching from Jalan Raya Cilegon, Jalan Ahmad Yani, and Jalan Sultan Ageng Tirtayasa. This arterial road section has a total road length of 5 km that passes through the Cilegon City area, which is the center of activity. Efficient use of traffic space by arranging the traffic space efficiency is used the distance between U-turns so that delays and queues are low. As long this road network has the highest volume of 2,399 pcu/h with an average speed of 32 km/h and reaching a VCR of 0.8 has a median number of 13-point openings, the number of uncontrolled U-turns causes many conflicts traffic that will cause traffic disruption when a vehicle makes a U-turn. The purpose of this study is to improve the performance of the road network in the Area of the City of Cilegon by knowing the best distance between U-Turns. Simulation analysis uses Vissim software to produce the ideal distance for each opening in the Cilegon City arterial road, which is then utilized in the arrange of the U-turn facility itself. The method used is the calculates of existing traffic performance and simulation to determine the ideal distance between U-Turn facilities. The simulation results with the ideal Vissim model of the distance between the turning facilities were 650 meters with an average delay of 17.56 seconds, a network speed of 33.07 km/h, a total distance of 28,453.34 km, and the total travel time to be 847.30 hours.
Keywords: distance, Network Performance, U-Turn facility

1. Introduction

The city of Cilegon, as one of the cities in Banten, which is the Java Island entrance has quite heavy traffic and an urban area accessible only to one main road. This city is on the northwestern tip of Java Island, on the edge of the Sunda Strait. It is an industrial city connected to the Jakarta – Merak toll road.

The city of Cilegon only has direct access, which extends from Jl. Raya Cilegon, Jl. Ahmad Yani and Jl. SA Tirtayasa is 5 kilometers long, resulting in the heavy traffic of the three roads where the heaviest traffic reaches 150 pcu-minute/km on Jl. Ahmad Yani and the highest vehicle volume of 2399 pcu/hour. The existing road conditions (Jl. Raya Cilegon, JL. Ahmad Yani, and Jl. SA Tirtayasa) where it gets denser during peak hours, especially in the morning where the VC *Ratio* reaches 0.8 with a speed of 32 km / hour[1]

This is also influenced by the fact that the road is a shopping center and shopping center, most of which do not have parking space and makes traffic space narrow due to the large number of vehicles parked on the road's side, so that vehicle space is blocked. However, this median's existence is also followed by the availability of a median opening (*U-Turn*) as a vehicle turning facility. Uncontrolled turnback facilities will cause traffic delays when a vehicle makes a, where the average delay at each *U-turn* reaches 10 seconds with a total of 13 *U*- Turns, the distance of each turning back facility (*U -Turn*) not more than 400 meters.

Besides, many turn-around facilities also can to cause conflicts between vehicles and vehicles and pedestrians. With a road length of 5 km and a median number of openings (*U-Turn*) of 13 points as evidenced by a travel time of up to 20 minutes, it is necessary to research to determine the ideal distance and arrangement of the turn-around facility based on traffic performance.

2. Literature View

2.1 U-Turn Facility

The median is the part of the road that is impassable by a vehicle with an elongated shape parallel to the road, located on the axis/center of the road, intended to separate the opposite traffic flow. The median may be in the form of an elevated median, a lowered median, or a flat median.

The median section or U-Turn facility is planned to accommodate the vehicle in order to perform a U-turn movement a divided road type and can accommodate to cut and turn right. In the Pedoman Perencanaan Putaran Balik (U-Turn) [2], U-Turn is the vehicle's traffic movement to turn back or turn 180°.

The median section for turning can be done at the following locations:

1) Locations between intersections to accommodate turning movements are not provided at the intersection.
2) Locations near the intersection to accommodate turning movement will affect the constant movement and turning movement at the intersection. Turning movement can be planned at a location with quite wide median near street approaches that have few openings.
3) Locations where there are critical public activities such as hospitals or other activities related to road activities. Openings for this purpose are required on the streets with access control and on divided roads with low traffic volume.
4) Locations on non-controlled road, is access where a turning facility at an optimum distance is provided to serve the development of the edge area and minimize the pressure for the return facility in front of it. The 400-800 meter gap between openings is considered sufficient in some cases.

The minimum distance between the opening and median opening width[3] is described in the Table 1.

Table 1 Minimum Distance Between Openings and Openings Size

Road Function	**Luar Kota**		**Urban**		
	Openings Distance (km)	Openings Space (m)	Openings Distance (km)		Opening Space (m)
			Pinggir Kota	Dalam Kota	
Arteri	5	7	2,5	0,5	4
Collector	3	4	1,0	0,3	4

Source : Pedoman Konstruksi dan Bangunan : Perencanaan Median Jalan (2004)

Planning for the location of the return round facility must consider several aspects of road and traffic geometric planning, including:

1) Street function
2) Street classification

3) Median width
4) Traffic lane width
5) Traffic volume per lane
6) Number of vehicles making turns per minute

The rotating facility must be equipped [4] with:

1) Retarding lane in approach lane entry;
2) The adequate turning radius for all types of vehicles according to road use class; and
3) Speed lane to join the mainline.

U-turns are permitted at locations with sufficient width for vehicles to make a turn without any violation/damage to the pavement's exterior. U-turns should not be permitted in continuous traffic as they may have impact on traffic operations, including reduced speed and the possibility of accidents.

2.2 Vissim application program

VISSIM is one of the transportation applications that can display microscopic simulations based on time and behavior developed for urban traffic models. This program can be used to analyze traffic operations under the confines of road line configurations, traffic composition, traffic signals, etc. So that this application can help to simulate various transportation engineering alternatives and the most effective level of planning. Not only about the road network, but also for intersections, public transportation, and pedestrians.

In simple terms, modeling using VISSIM is divided into five stages:

1) Identify the scope of the area to be modeled.
2) Data collection
3) *Network coding*
4) *Error checking*
5) Model calibration and validation

The data requirements to build a model using VISSIM are:

1) Geometric data
2) *Data traffic*
3) Vehicle characteristics

The data requirements to build a model using VISSIM to be used are geometric data, traffic volume, vehicle proportions, vehicle routes, and APILL / intersection priority data.

3. Research Methodology

In this case, the research method can be interpreted as a series of actions to obtain information in data with predetermined goals and benefits. This research is a type of comparative research, which is a study that compares several simulation results to the distance between U-turn points to obtain the best distance with the lowest delay.

The initial activity of the research was to identify the problem of the number of turn-around facilities that affect traffic performance along the observed road sections, to collect traffic data, and to perform simulations with several alternative U-Turn distances. The data collected includes traffic volume, delays and road inventory. In this study, the authors compared several delay conditions with various alternative distances between U-Turns.

This activity is a stage of observation and monitoring of conditions in the field to then formulate problems that will be used as research material. Observations were made in the City of Cilegon, precisely on median urban arterial roads where there are many turn-around facilities that have the potential to cause a decrease in the performance of the primary arterial roads and increase traffic congestion.

This stage is the process of identifying problems that occur in Cilegon City to obtain problem formulations that are more focused on the study location that will be used as research material, in this case relating to the effect of turning facilities on traffic on primary arterial roads.

This study requires two types of data, namely secondary data and primary data as the basis for solving problems that have been formulated previously.

Secondary data is data that has been compiled and obtained from several agencies related to the data required in this study. Secondary data required are as follows.

1) Road network map
2) Road geometric data
3) General Data of Cilegon City Transportation Patterns 2019
4) Laws and regulations
5) Land Use Map

Observations were made in the City of Cilegon, precisely on median urban arterial roads where there are many turn-around facilities that have the potential to cause a decrease in the performance of the primary arterial roads and increase traffic congestion.

Primary Data Collection

In this study, primary data was obtained through field surveys along Ahmad Yani, Tirtayasa and Raya Cilegon roads which were carried out for 16 hours starting at 06.00 - 21.00 to collect the following data.

1) Inventory survey *U-Turn* to determine the condition of the existing infrastructure, the location of the turn-around facility as the object of research.inventory data *U-turn* obtained directly from the field includes the length of the *U-turn*, the width of the *U-turn*, the distance of each opening, the number of openings and the direction system and land use conditions.
2) Classified traffic volume enumeration survey which is intended to determine the level of traffic density on thesection *U-Turn* based on the classified traffic volume and the direction of traffic flow and the type of vehicle in a certain time unit with direct enumeration in the field. The purpose of conducting this survey is to obtain traffic volume data and to find out the peak hour period at each point of the survey location.
3) A queue survey was conducted to determine how long and how many vehicles were delayed in making a U-turn. The queue survey at the *U-Turn is* carried out directly in the field by measuring the average queue length and the number of vehicles at the busiest hour or the busiest traffic.
4) Delay Survey Delay survey is conducted by calculating the delayed time when making a U-turn. The delay survey is carried out directly in the field by calculating the time of the vehicle from before turning until the vehicle is free and calculating the time the other vehicle is delayed behind the car that makes auntil the car can make a *U-turn*.

4. Analysis ad Problem Solving

4.1 Analysis of Existing Road Performance

Analysis of the performance of each road section for structuring and the effect of the distance between U-Turn facilities on arterial roads in Cilegon City was done by dividing the arterial roads in the city of Cilegon into 13 segments based on the U-Turn facility's location with details (Table 2).

Table 2 Round Facility Point

No	Round Facility Point		Length Segmen (m)
1	U-Turn 1	Junction PCI	165
2	U-Turn 2	Junction Teuku Umar Road(local)	170
3	U-Turn 3	In front of Restaurant (GB)	300
4	U-Turn 4	In front of Outlet Celcius	577
5	U-Turn 5	In front of Bank BNI Syariah	867
6	U-Turn 6	Junction KH Wasyid Road	277
7	U-Turn 7	In front of Yamaha	465
8	U-Turn 8	In front of Matahari Store	619
9	U-Turn 9	In front of Mayor's house	246
10	U-Turn 10	In front of Agung Mosque	120
11	U-Turn 11	In front of Sate Cilegon	257
12	U-Turn 12	Junction Antasari	157
13	U-Turn 13	In front of Honda	355

a) Road Section Capacity
Jl. Raya Cilegon to Jl. SA Tirtayasa has the same capacity, namely 2732,4 pcu/h, because these roads have the same geometric roads, road types, and side obstacles, namely commercial shopping areas (Figure 1).

b) Traffic Volume
The traffic volume of Jalan Raya Cilegon – SA Tirtayasa in the research study area was obtained from the enumeration survey that was converted into pcu/h. The road segment with the most enormous traffic volume is Jalan Ahmad Yani 3 2399 pcu/h, and the smallest volume is Jalan Raya Cilegon 1 with a volume of 1478 pcu/h

c) VC Ratio
According to the calculation of the VC Ratio, it can be seen the level service of the roads. The calculation of the VC Ratio is the ratio between the total volume and the capacity of the sections. The highest ratio is Jalan Ahmad Yani 3 with V/C Ratio 0.88 . As for the lowest rate of VC Ratio is Jalan Raya Cilegon 1 as much as 0.54

d) Speed
The road speed in Jalan Raya Cilegon – SA Tirtayasa was obtained from the MCO Survey (Moving Car Observer). The road segment with the highest average speed is Jalan Raya Cilegon 2 42 km/h. Meanwhile the road that has the lowest average speed is Jalan Ahmad Yani 2, which is 31 km/h.

e) Road Service Level
The lowest level of service is Jalan Ahmad Yani 3 with V/C Ratio of 0.88 at 32 km/h. The top-level of service was Jalan Raya Cilegon 1 with a V/C Ratio of 0.54 at 40 km/h.

Figure 1 Conditions in The Field

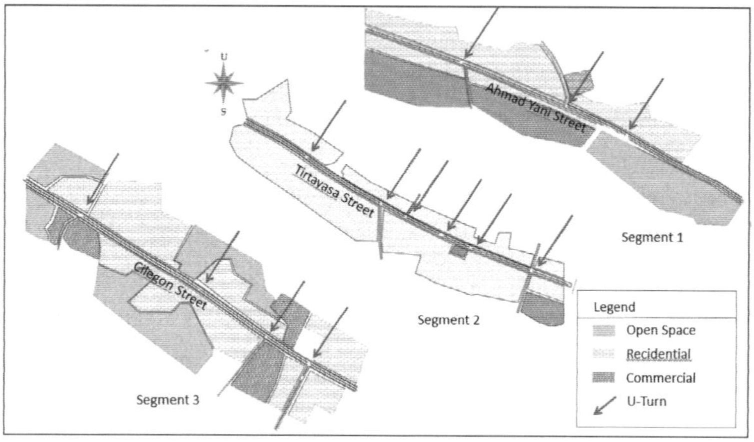

Figure 2 U-turn Preview

4.2 Transportation Modeling Using Vissim Software

a) Zoning
Zoning is dividing or splitting of an area or a region into several parts, according to function and purpose. This zoning is used to facilitate the travel distribution process to the next stage (Table 3).

Table 3 Zoning

No	Zone	Access
1	1	Jl. Raya Cilegon 2
2	2	Jl. Pondok Cilegon Indah
3	3	Jl. Teuku Umar
4	4	Jl. Imam Bonjol
5	5	Jl. KH Wasyid
6	6	Jl. Ishak
7	7	Jl. Panjaitan
8	8	Jl. Bojonegara 1
9	9	Jl. Sastradikarta
10	10	Jl. Pangeran Antasari
11	11	Jl. SA Tirtayasa

b) Proportion of Modal Choice

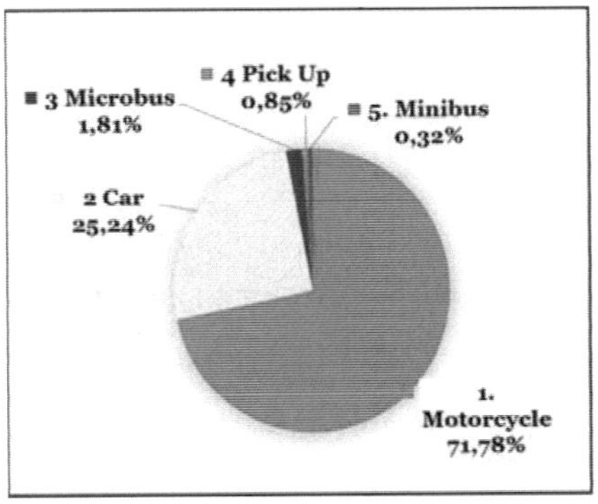

Figure 3 Proportion of Modal Choice

From the diagram above it can be seen that the mode most travels on arterial road is 71.78% motorbikes, while the least mode are medium busses, small trucks, and UM at 0.2% (Figure 3).

c) Travel Loads
By utilizing Vissim software to assess the micro-flow traffic based on the information provided (input) in road networks and traffic requests. This software will self-manage traffic in the road network according to its origin and destination by using routes based on the shortest travel time. The output of traffic charges with Vissim software demonstrates the kind of traffic that can be used as a basis for traffic management. The charges were done by using the volume and speed of each road segment.

d) Validation of the Road Network Model
After the road network model is built, validation is done by using the Chi-Square Test (Table 4). Data validation itself is a testing process of models with surveys to assess whether the network model created represents conditions in the field.

Table 4 Validation

I. HYPOTHESES Ho: Model with Aligned Survey H1: Model with the survey is not aligned		
II. Value Confidence Level	95%	0,05
III. Degrees of Freedom	(v) = (k-1) =	26
IV. So, the Chi Square Value Table	(X^2 table) =	38,8851
V. Calculating X^2 count = 32,225		
VI. Decision Rule :	Ho is accepted if X^2 count <	38,88514
	H1 is accepted if X^2 count >	38,88514
VII. Decision : Ho ACCEPTED		

Based on the calculations, X2 count = 32.22, then X2 count < 38.89 (X2 table) so Ho is accepted. In conclusion, the model results are in line with the observation results, it can be used because it represents conditions in the field.

e) Existing Network Performace
Along the Jalan Raya Cilegon - Jalan SA Tirtayasa as a road network currently has the following performance (Table 5).

Table 5 Existing Network Performance

Parameter	Network Performance
Delay Average (s)	19,81
Network Speed (km/h)	31,28
Total Travel Distance (km)	28.188
Total Time Travel (h)	901,04

4.3 Simulation
The U-turn facility simulation applies the distance between the U-Turn, which is calculated from the end of the median to the next median with a distance per 500 m, 550 m, 600 m, 650 m, 700 m, and 800 m without paying attention to any intersections to find the ideal distance with the best road network performance. This arrangement was made by utilizing Vissim software, with changes in distance between U-turn facilities so that after structuring it can be known the performance of the road network after structuring (Table 6).

Table 6 All of Alternative Network Performance

Parameter	Existing	Distance between U-Turn					
		Alternative 1	Alternative 2	Alternative 3	Alternative 4	Alternative 5	Alternative 6
		500 M	550 M	600 M	650 M	700 M	800 M
Delay Average (s)	19,81	18,45	18,35	18,01	18,63	19,05	19,84
Network Speed (km/h)	31,28	32,35	32,32	32,35	32,3	32,36	32,2
Total Travel Distance (km)	28.188	27.493,82	27.403,69	27.376,22	27.734,94	27.467,43	28.044,83
Total Time Travel (h)	901,04	849	847	846	858	848	870

Of the six proposed alternatives, it can be seen that only Alternative 3 shows an increased performance in traffic where the average delay decreases to 18.01 seconds, the network speed increases to 32.35 km/h, the total travel distance becomes 27,376.22 km, and the total travel time is 846.29 hours. Therefore, from several simulation analysis namely Alternative 1 to Alternative 6, it can be concluded that the ideal distance along CBD arteries in Cilegon City is 600 meters.

The suggestion to improve the performance of the road sections by applying the distance between U-Turns is to make changes to the ideal distance approach obtained and taking into account the land use conditions with the distance between the *U-Turn facilities* below (Table 7).

Table 7 The Simulation Recommendation Round Facility Point

No.	Round Facility				Recommendation distance between u-turn
	Location before		Location after		
1	Junction PCI	U-Turn 1	Junction PCI	U-Turn 1	165
2	Junction Jl. Teuku Umar (lokal)	U-Turn 2	In front of Restaurant (GB)	U-Turn 2	609
3	In front of Restaurant (GB)	U-Turn 3	In front of Toko Celcius	U-Turn 3	534
4	In front of *Celcius* Outlet	U-Turn 4	In front of SPBU Ahmad Yani	U-Turn 4	600
5	In front of Bank *BNI Syariah*	U-Turn 5	In front of Ramayana	U-Turn 5	632
6	Junction KH Wasyid	U-Turn 6	In front of Mie Kangkung	U-Turn 6	642
7	In front of ATPM MC 1	U-Turn 7	In front of Sate Cilegon	U-Turn 7	789
8	In front of Mall	U-Turn 8	In front of Honda	U-Turn 8	636
9	In front of Mayor's House	U-Turn 9			
10	In front of Agung Mosque	U-Turn 10			
11	In front of Warung Sate	U-Turn 11			
12	Junction Antasari	U-Turn 12			
13	In front of ATPM Mc 2	U-Turn 13			

This arrangement is done by utilizing Vissim software, with changes in the distance between the U-turn facilities obtained from ideal distances except for the UT1 PCI intersection, due to its being too close to the public corner Simpang APILL Serdang. It will be applied only to vehicles entering Jalan Pondok Cilegon Indah, not allowed to make a U-Turn, existing provisions along with land use so after the structuring it can be seen the performance of final road network after structuring (Figure 4).

From the table, it is seen that the performance gas increased from the existing conditions, namely the average delay to 17,56 seconds, network speed increased to 33,07 km/h, and total travel time to 847,30 hours.

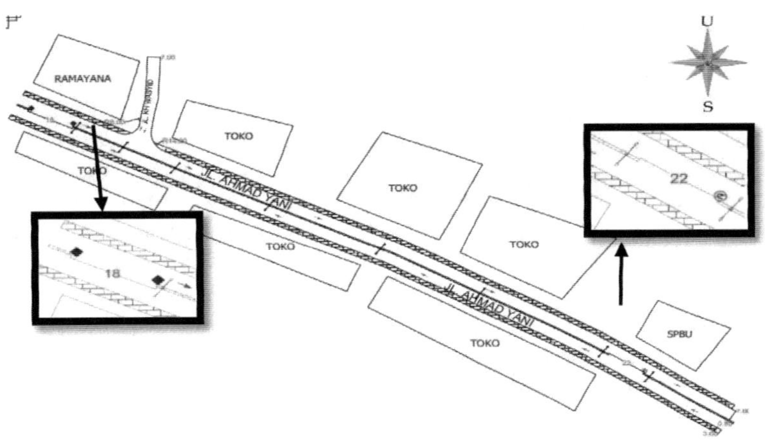

Figure 4 Arrangement of U-Turn facility

5. Conclusion

The conclusions are drawn from the research result if the effect of distance between U-Turn facilities on CBD traffic in Cilegon city are as follows.

1) Modeling existing traffic conditions using Vissim software where existing road network on median arterial roads in Cilegon city includes an average delay of 19,81 seconds,

network speed of 31,28 km/h, total travel distance 28,188 km, and total travel time is 901,04 hours.
2) A simulation analysis was done using Vissim software an alternative experiment of the distance between turning facilities starting from 500 meters, 550 meters, 600 meters, 650 meters, 700 meters, and 800 meters. By looking for the lowest average delay, the ideal distance for a median arterial road in Cilegon city is 600 meters.
3) Arrangement of U-Turn facility using an ideal distance of 600 meters was done in addition to referring the ideal distance, it is also taking consideration of land use conditions around the location so that the distance between turning facilities is 165 meters, 609 meters, 534 meters, 600 meters, 632 meters, 642 meters, 789 meters, and 636 meters. After applying the Vissim model, the road network performance results show an increase where the average delay decreases to 17.56 seconds, the network speed increases to 33,07 km/h, the total travel distance to 28.453,34 km, and total time to 847,30 hours.

6. Suggestion

Some suggestions may be offered based on the result of the research include the following:
1) Alternative treatment applications related to the arrangement of U-Turn facilities needs to be done to improve the performance of the arterial road network in Cilegon city.
2) Based on traffic performance, to obtain the ideal distance between turn-around facilities on arterial roads in Cilegon City, it is recommended to use an ideal distance of 600 meters between U-Turn facilities accompanied by consideration of land use conditions around the location of the median opening.
3) It is necessary to provide 15 U-Turn signs at 8 U-Turn facility locations.
4) It is important to socialize with the public regarding U-Turn facility's arrangement so that the public knows the latest conditions. It is necessary to enforce the law by the police against people who violate the facility's terms.

References

[1] General Report on Job Training in Cilegon Cityn2019 ***General Pattern of Land Transportation for Cilegon City 2019***. Bekasi, Indonesian Land Transportation Polytechnic – STTD
[2] Directorate General of Highways 2005 *Guidelines for U-Turn Planning*. Jakarta (ID): Directorate General of Highways
[3] Department of Settlement and Infrastructure 2004 Temporary Area based on Construction and Building Guidelines: Road Median Planning
[4] Ministry of Public Works 2011 Regulation of the Minister of Public Works of the Republic of Indonesia Number 19 Year 2011 concerning Road Technical Requirements and Road Technical Planning Criteria. Jakarta (ID): Ministry of Public Works
[5] Kristy, C. Jotin & Kent Lall 2005 *Fundamentals of Transportation Engineering*. Third Edition Volume 1. Jakarta (ID): Erlangga. Salter, RJ 1989 *Traffic Engineering Examples and Problems*. Macmillan
[6] Directorate General of Highways 1997 *Indonesian Road Capacity Manual* (MKJI). Jakarta (ID): Directorate General of Highways
[7] FD Hobbs 1984 *Traffic Planning and Engineering*. Birmingham (UK): University of Birmingham
[8] Indonesian Ministry of Transportation 2015 Regulation of the Minister of Transportation of the Republic of Indonesia Number 96 of 2015 concerning Guidelines for the Implementation of Traffic Management and Engineering. Jakarta (ID): Ministry of Transportation
[9] Morlok, Edward K 1984 *Introduction to Engineering and Transportation Planning*. Jakarta (ID): Erlangga
[10] Salter, RJ 1989 *Traffic Engineering Examples and Problems*. Macmillan
[11] Tamin, Ofyzar Z 2008 Planning, Modeling, and Transportation Engineering. Bandung (ID): Publisher ITB

Planning tour transport connectivity in Tomohon City

I M Suraharta[1] And N K A Cahyani[2]
[1]Indonesian Land Transportation Polytechnic - STTD, Bekasi
[2]Indonesian Land Transportation Polytechnic - STTD, Bekasi

E-mail: isuraharta@ptdisttd.ac.id

Abstract. Tomohon City is one of the tourist destinations which is currently experiencing an increase in tourist visits. This impacts traffic density because tourists still use private vehicles and large buses originating from Manado. Therefore, transportation and accessibility and connectivity are needed to connect one area to another through planning tour transport connectivity. The method used in this research is field data collection and descriptive quantitative and qualitative analysis. After going through the data processing, the result is the type of vehicle used to plan of tour transport in Tomohon City is a small bus with 16 passengers. With the number of vehicles needed in the planning of tour transport, there are ten vehicles to accommodate the highest number of tourist visits on holidays with IDR. 400.000 / bus. The alternative route chosen is Tomohon Extreme Market -Lake Linow- Tomohon Prayer Hill.
Keywords: tourist, tour transport, accessibility, connectivity, route

1. Introduction

Tomohon City has a very rapid development of tourism. Based on data from tourist visits, the past five years obtained from the Department of Culture and Tourism of the City of Tomohon showed an increase in tourist visits, both foreign and domestic tourists. The high number of tourist visits impacts the increasingly high level of traffic density in the tourist destination [1]. Because to visit the tourist attraction, the tourists still use private vehicles (such as private cars and motorbikes) and large buses originating from Manado, which cause congestion on the access in and out of attractions. This is because the City of Tomohon does not yet have a Tourist Transport that connects one tourist attraction to another tourist object, taking the tourists to go to the destination. Also, road conditions are inadequate both in terms of geometric and width of the road body, especially in some tourist objects with a width of ± 7 m that are traversed by private vehicles and large tourism buses, which causes congestion on roads that have a width of fewer than 7 m and traffic jams in and out of tourist attractions [2].

As a tourist destination, the City of Tomohon requires transportation and accessibility, to support tourists to enjoy the attractions in the city of Tomohon and the development of Tomohon City as a Tourist City [1] [3] [4]. Also, the existence of these tourist transportation must connect one tourist attraction with other attractions, which can save time and money [5]. Thus we need a transportation plan that can overcome the problem that is by planning a transportation connectivity that can connect one tourist attraction with other attractions as well as an easy [5], smooth and comfortable transfer by paying attention to inadequate road conditions both in terms of geometric and width the road body so as not to cause traffic jams in and out of attractions. Based on this background, the authors are interested in raising the issue as a study entitled "Planning Tour Transport Connectivity in Tomohon City."

2. Methodology

The research locations were Tomohon Beriman Terminal, Tomohon Extreme Market Attraction, Lake Linow Tomohon, and Tomohon Prayer Hill in November 2019. These three tourist sites are part of the master plan of tourism development in Indonesia in 2010-2025 [6]. Data collection was carried out through field research by observing survey and interview objects and collecting secondary data obtained from the Tomohon City Transportation Agency [4]. Tomohon City Regional Planning and Development Agency, Tomohon City Culture and Tourism Office. Besides that, it is also supported by library research, and previous research relating to tourism transportation planning and connectivity. The method used in this study is the collection of field data and qualitative descriptive analysis and quantitative descriptive [3]. A descriptive analyzing model is a way of analysis by describing data that has been collected, including characteristics of tourists, vehicle operations, vehicle operating costs, tariffs [5] [7] [8].

3. Results and discussion

3.1. Analysis of tourist characteristics

3.1.1. Vehicles used by travelers

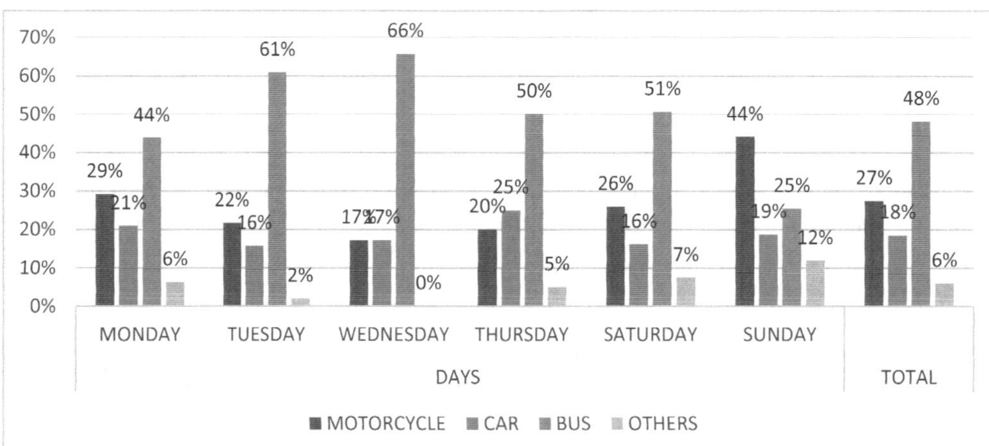

Source: Analysis results, 2020

Figure 1 To go to Tomohon City travel destinations

Based on the table and graph above (Figure 1), it can be seen that the vehicles used by tourists to Tomohon City attractions are on weekdays, most tourists use large buses and on holidays, 51 percent use large buses on Saturdays and 44 percent use motorbikes on Sundays.

3.1.2. Travel expenses

Based on the results of an interview survey that has been carried out, the highest tourist travel costs in visiting tourist objects in the City of Tomohon are more than IDR 75,000 with a percentage of 48%, and then the lowest travel costs are IDR 5,000 - 25,000 with a percentage of 27% (Figure 2).

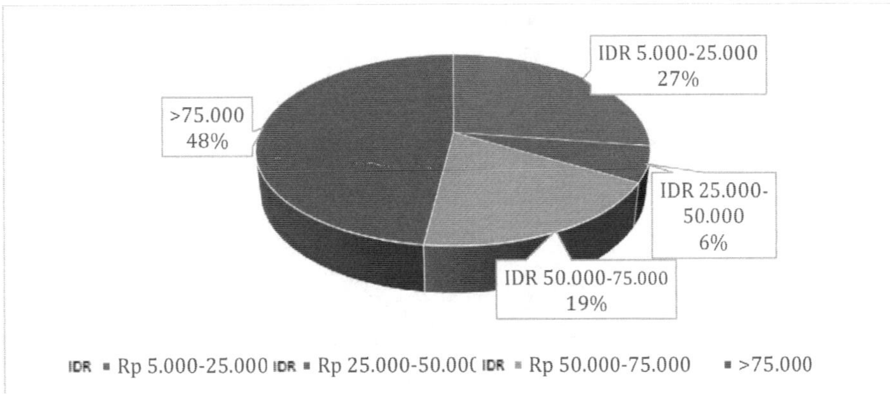

Source: Analysis results, 2020

Figure 2 Percentage of tourist travel costs in visiting Tomohon City tourism object

3.1.3. Tourist willingness to switch modes to use tour transport

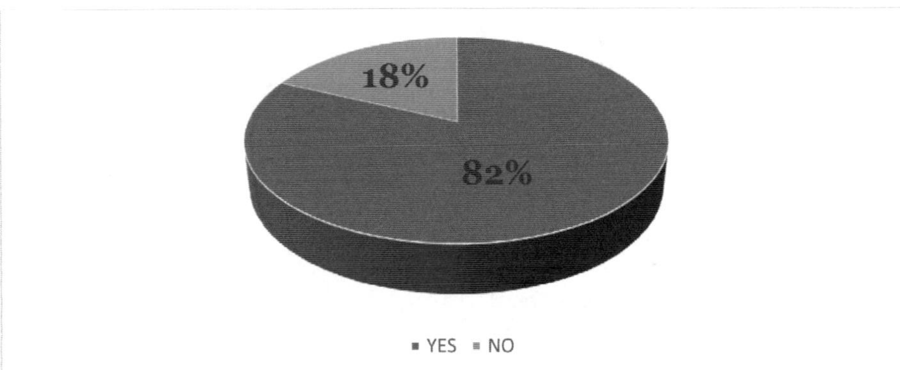

Source: Analysis results, 2020

Figure 3 Percentage of tourist willingness to switch mode using travel transport

After the interview, it is found that most tourists are willing to move to the mode of tourism transportation that will be planned with a percentage of 82% of tourists and for tourists who are not prepared for 18% (Figure 3).

3.1.4. Tourist destinations most attracted by travelers

As mentioned above, in Tomohon there are three tourist sites that are often visited by visitors, namely Extreme Market, Linow Lake and Tomohon Prayer Hill. Extreme market is a place to buy and sell extreme food such as dogs, bats, snakes and others. Linow Lake is a tourist destination that offers beautiful views around the Lake from above. While Tomohon Prayer Hill is a tourist area with religious purposes, especially for christians, but everyone, other than Christians can visit this tourist site. These 3 locations were chosen because these attractions are in one lane and are the most popular tourist attractions in Tomohon City.

From the analysis results conducted with an interview survey of tourists, 41% of tourists chose the Extreme Market as a tourist attraction most interested in visiting. Then in the second position is 35% of tourists choosing the Linow Lake tourist attraction after the Extreme Market and the next is Tomohon Prayer Hill which is 24% (Figure 4).

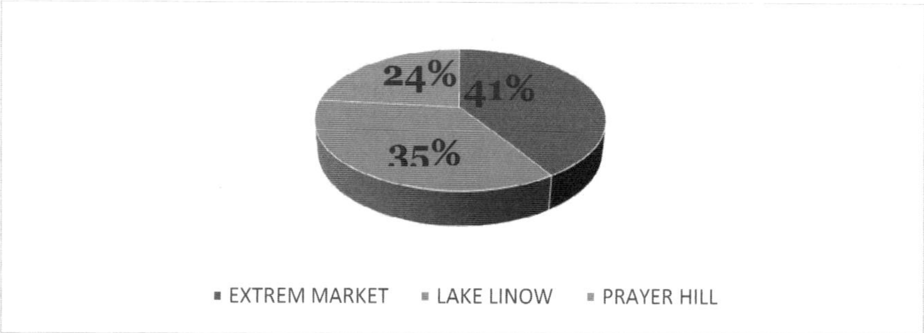

Source: Analysis results, 2020

Figure 4 Percentage of tourist attractions most attracted by travelers in Tomohon City

3.2. Transportation connectivity index

Transportation infrastructure connectivity index can be measured based on distance/length of route, travel time from origin to destination, and availability of public transportation modes [9].

3.2.1. Transportation connectivity used by travelers in Tomohon City mileage

Table 1 Distance between attractions in Tomohon City

No.	Origin - Destination	Distances (Km)
1	Extreme Market-Linow Lake	9
2	Linow Lake - Prayer Hill	13
3	Prayer Hill - Extreme Market	5

Source of data : Analysis results, 2020

Based on the above table, the total distance from the starting point to the endpoint is 22 km, and the distance from the starting point to returning to the starting point again is 27 km.

3.2.2. Traveling time

Table 2 Travel time between attractions in Tomohon City

No.	Origin - Destination	Minutes
1	Extreme Market-Linow Lake	25
2	Linow Lake - Prayer Hill	35
3	Prayer Hill - Extreme Market	15

Source: Analysis results, 2020

Based on the above table, the travel time from what is needed from the starting point to the endpoint is 60 minutes and the time required from the starting point to return to the starting point is 75 minutes.

3.2.3. Accessibility index

Based on an interview survey of tourists in Tomohon City, the tourists still use private vehicles and use large buses to go to tourist destinations in Tomohon City. Besides, tourists outside the island who do not use private vehicles must rent a vehicle to reach tourist attractions in the City of Tomohon. So the accessibility of tourists in reaching Tourism Objects in the City of Tomohon is still lacking due to the absence of tourist transports that deliver tourists to be able to reach attractions in the City of Tomohon in a connected way.

3.3. Actual and potential demand

The actual demand is the number of tourists who are visiting the Lake Linow tourist attraction and the Tomohon Prayer Hill tourist attraction obtained from one day at the busiest hour. Meanwhile, potential demand is obtained from the number of tourists who can come back using tourist transportation, which will be planned to accommodate the number of tourists on weekdays and holidays (Table 3).

Table 3 Actual and Potential Demand (Person/day)

No	Days	Actual Demand	Potential Demand
1	Monday	55	68
2	Tuesday	58	77
3	Wednesday	38	49
4	Thursday	44	59
5	Saturday	101	140
6	Monday	69	106
TOTAL		365	499

Source: Analysis results, 2020

3.4. Route choice

The next step is to determine the best travel route after determining the origin and destination of the travel. The determination of transportation routes to tourist destinations in Tomohon

City is precisely important due to the lack of accessibility of transportation infrastructure in Tumohon City [7] [8]. The path that will be taken to get to the tourist destinations in Tomohon City by transport. Tomohon Lake Linow, then Tomohon Prayer Hill, and finally back to Believers Terminal, is the first route. In choosing the route, there are three criteria: the shortest route length, the dense land use of prospective tourist destination passengers, short travel times. The map of transport routes to tourist destinations in Tomohon City is as shown in Figure 5.

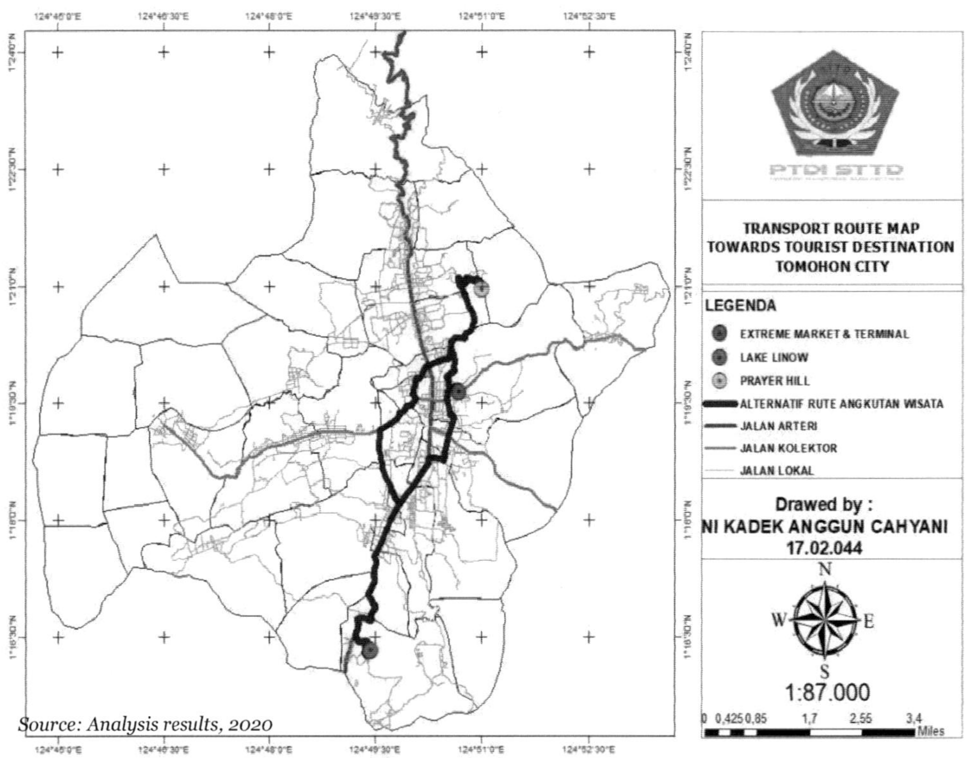

Source: Analysis results, 2020

Figure 5 Transport route map towards tourist destinations Tomohon City

3.5. Operations planning

3.5.1. Type of vehicle

Determination of the type of tourist transport fleet uses MINIBUS vehicles with a capacity of 16 passenger seats, assuming that the vehicles have small dimensions to match the characteristics of land use and road conditions at the tourist sites [4].

3.5.2. Service indicators

3.5.2.1 Load factor

The planned vehicle loading factor is 95% of the number of tourists in the highest volume hours with a vehicle capacity of 16 seats.

3.5.2.2 Traveling time

The time required for tourist transportation to get to the tourist location can be determined by calculation using the formula:

$$WT = \frac{PR}{KR} \times 60 \quad \text{.......(1)}$$

= 44 minutes

Information:

WT : travel time
PR : length of the route
KR : speed of plan

3.5.2.3 Travel transport circulation time

The circulation of tourist transportation is the travel time of tourist transportation from origin to destination and back again to the beginning [10].

$$CTABCA = (TAB + TBC + TCA) + (\sigma AB + \sigma BC + \sigma CA) + (TTA + TT + TTC) \quad \text{.......(2)}$$

= 62,1 minutes

Information:

CT ABCA : the time between circulation from A-B-C back to A
TAB : average travel time from A to B
TB : average travel time from B to C
TCA : average travel time from C to A
δ AB : travel time deviation from A to B
δ BC : deviation of travel time from B to C
δ CA : deviation of travel time from C to A
TTA : vehicle downtime in A
TTB : vehicle downtime in B
TTC : vehicle downtime in C

3.5.2.4 Time between vehicles (headway)

Here is the headway calculation formula [7] [10]:

$$H = \frac{60 \times C \times Lf}{P} \quad \text{.......(3)}$$

Headway on weekdays
$H = 15{,}72$ minutes, for actual demand
$H = 11{,}84$ minutes, for potential demand
Headway on weekend
$H = 9{,}03$ minutes, for actual demand
$H = 6{,}51$ minutes, for potential demand

Information:

H : the time between vehicles (minutes)
P : average number of passengers per hour in the most populous section (number of passengers obtained on average from 3 busy hours)

IOP Conf. Series: Earth and Environmental Science **879** (2021) 012004 doi:10.1088/1755-1315/879/1/012004

C : vehicle capacity
Lf : load factor, taken from 95% (in dynamic conditions)

3.5.2.5 Frequency
Frequency is the number of tourist transport trips on one route within one hour. The frequency obtained from the analysis results are as follows:

$$F = \frac{60}{H} \quad (4)$$

Frequency on weekdays
$F = 4\ vehicles/hour$, for actual demand
$F = 5\ vehicles/hour$, for potential demand

Frequency on weekend
$F = 7\ vehicles/hour$, for actual demand
$F = 9\ vehicles/hour$, for potential demand

Information:
F : Frequency ($vehicle/hour$)
H : Headway (minutes)

3.5.3. Number of vehicle needs
The number of fleets is determined by the vehicle's circulation time and the time between vehicles [8]. The number of circulation fleets required is calculated using the formula:

$$K = \frac{CT}{H \times fA} \quad (5)$$

Number of pessimistic vehicles
$$K = \frac{CT}{H \times fA} = \frac{62{,}1}{11{,}69 \times 100\%} = 5\ vehicles$$

Number of optimistic vehicles
$$K = \frac{CT}{H \times fA} = \frac{62{,}1}{6{,}38 \times 100\%} = 10\ vehicles$$

3.5.4. Vehicle operating cost analysis

Table 4 Recapitulation of transport transport operational costs

No	Recapitulation of cost/km	Price
1	**Direct Cost**	
	a. Depreciation	IDR4.394,90
	b. Capital Interest	IDR2.966,56
	c. Driver's Salary and Allowances	IDR2.996,52
	d. BBM	IDR1.044,44

No	Recapitulation of cost/km	Price
1	**Direct Cost**	
	e. Tire	IDR156,00
	f. Small Service	IDR279,00
	g. Large Service	IDR119,88
	h. Over Houl Engine	IDR55,00
	i. Over Houl Body	IDR599,30
	j. Terminal Retribution	IDR18,52
	k. STNK /Vehicle tax	IDR48,44
	l. Kir	IDR12,98
	m. Insurance	IDR49,44
2	**Indirect Cost**	
	a. Non-Bus Employee Salary Costs	
	b. Cost Management	IDR77,91
3	**AMOUNT**	**IDR12.818,91**

Source: Analysis results, 2020

From the recapitulation of the operational costs of the tourist transport vehicle, then: Vehicle operating costs / vehicles / km = direct costs + indirect costs = IDR12.818,91

3.5.5. Tourist transport cost (tariff) analysis

Tourist transport in this study, the tourist transport rates were measured taking into account the vehicle's operational costs, the operating cost and revenue balance point (BEP) and the expected profit rate (10 percent). The total cost is calculated for one trip from Extreme Market -Lake Linow to Prayer Hill, per passenger unit (IDR/pers) or per bus unit. The overall cost of a bus is a fare multiplied by the average number of passengers per bus for each passenger. The bus under consideration is a bus with a capacity of 16 seats. Here is the estimate of the tariff at issue.

BEP Cost of Extreme Market to Lake Linow = IDR. 7.590,15
BEP Cost of Lake Linow to Prayer Hill = IDR. 10.963,55
BEP Cost of Prayer Hill to Extreme Market = IDR. 4216,75

Tariff (Cost of Transport) = Cost (BEP) +(10% x Cost (BEP))

Cost of Extreme Market - Lake Linow = IDR. 8.349,17
Cost of Lake Linow-Prayer Hill = IDR. 12.059,91
Cost of Prayer Hill – Extreme Market = IDR. 4638,43
Total Cost/pers = IDR. 25.000
Total Cost/Bus = IDR. 400.000

Based on the calculation above, assuming the average number of passengers is 16 passengers per vehicle, the transportation cost from Extreme Market - Lake Linow,-Prayer Hill and return to Extreme Market is IDR. 25,000/passenger or IDR. 400,000/bus.

4. Conclusion

The average time taken by tourists to visit tourist destinations in Tomohon City from Tomohon Extreme Market-Lake Linow-Pray Hill of Tomohon- Tomohon Extreme Market in 60 minutes. Meanwhile, when planned tourist transport connectivity in Tomohon City with the calculation result is 44 minutes. Determination of alternative routes from interviews, namely the tourist objects that are most in-demand by tourists in Tomohon City, namely from Tomohon Extreme Market -Lake Linow Tomohon-Tomohon Prayer Hill-Extreme Market. The type of vehicle used is a small bus with 16 seats, with the number of vehicles needed is ten vehicles, and the operating time for tourist transportation is 9 hours. The operational cost of the vehicle/vehicle / km is IDR. 12,818.91, then the planned fare for tourist transportation in Tomohon City is IDR. 400,000 per bus.

The operation of tourism transportation in Tomohon City is expected to be held soon to help tourists reach the tourist objects with ease. The number of vehicles provided is 5 armadas for the initial stage, then you can add up to 10 fleets according to the calculations above to accommodate the number of tourists on holidays. To realize the operation of Tomohon City tourism transportation, the transportation and tourism department can fulfill the required fleet.

References

[1] Basuki I and Amos S 2015 *The Potential of Public Transportation in Tourism in the Special Region of Yogyakarta* Yogyakarta **15**(2):135–42
[2] Indonesia Transportation Department 2009 Constitution No. 10 Year 2009 about Tourism **4**(3)
[3] Widyastuti H, Marsoyo A and Setiawan B 2019 Analysis of Connectivity Between Coastal Tourism Destinations in the Special Region of Yogyakarta **14**(1):1–12
[4] Rismarini N, Munawar A and Priyanto S 2018 *Tourism Transport Planning as a Link for Accommodation and Tourist Destinations in Yogyakarta City* Yogyakarta 19-20
[5] Hartanto and Budi D 2014 Tourism Transport Planning in the Special Province of Yogyakarta (DIY) *Implement Sci.* **39**(1);1-15
[6] Government of Republic of Indonesia 2011 PP No.50 year 2011about The Master Plan for Tourism Development RI 2010-2025
[7] Rien T, Akhmadali and Kadarini S N 2002 *Pontianak City Tourism Transportation Planning* Pontianak 1-10
[8] Lestari D A, Suthanaya P A and Wedagama D M P 2017 *Planning of the Transportation Transportation Operational System in the City of Denpasar* Denpasar **5**(1):64–70
[9] Indonesia Transportation Department 2009 Constitution No. 22 Year 2009 4(3)
[10] Land Transportation Agency of Transportation Departement of Indonesia 2002 Technical Guidelines for the Implementation of Public Passenger Transport in Urban Areas in Fixed and Regular Routes (SK.687/AJ.206/DRJD/2002):2–69

Urban landscape for birdwatching activities

I Kurnia[1], H Arief[2], A Mardiastuti[2] and R Hermawan[2]
[1]Ecotourism Management and Environmental Service Study Program, Faculty of Forestry and Environmental, IPB University, Bogor, Indonesia
[2]Department of Forest Resources Conservation and Ecotourism, Faculty of Forestry and Environmental, IPB University, Bogor, Indonesia

Email: insan_ipb@yahoo.com

Abstract. Urban landscapes are usually dominated by built spaces and human-made vegetation, which different from natural landscapes. This difference will affect the composition of birds that can be found in the urban landscape. For birdwatchers, birds are the main object in birdwatching activities. The objectives of this paper were to analyze the feasibility of urban landscapes for birdwatching activities and find out the characteristics of urban landscapes favored by birdwatchers in four cities in Java (Bogor, Sukabumi, Bandung, and Surabaya). Birdwatchers' site preference and perception were surveyed through online questionnaires in February through May 2020 (n=1,247 respondents). Surveyed data revealed that birdwatching sites' size varied between 0.05 and 76.82 ha, mainly urban forests and city parks. Most of the urban landscape was habitat to various bird species typical of the urban landscape (e.g., Eurasian tree-sparrow, Cave swiftlet, Black-headed Bulbul). Raptors (e.g., Black-thighed Falconet) are found in several locations. Among the respondents, 25 % (n=309) had previously conducted birdwatching activities in urban landscapes, while 72.0% (n=808) expressed their interest in birdwatching in urban landscapes, indicating that urban landscapes was feasible for birdwatching. The most favorite locations were the Bogor Botanical Gardens, Darmaga Research Forest (both in Bogor), Merdeka Field Park, Cikundul Agrotourism Area (Sukabumi), Bandung Zoo, Babakan Siliwangi City Forest (Bandung), Bungkul Park, and Flora Park (Surabaya). Characteristics of sites favored by birdwatchers were a shady area, not noisy location, and any facilities for birdwatching. The diversity of bird species did not became the main reason.
Keywords: birdwatchers, city parks, recreational activities, urban forest

1. Background

Urban landscapes are changing natural ecosystems in such extreme times that their natural forms have changed or disappeared completely [1]. Urban landscapes are characterized by the dominance of awakening space and the vegetation grown by humans as green open space. The dominance of socio-economic components over biotic and abiotic components impacts the composition of birds as part of the ecosystem. Only some species can adapt and survive in the urban landscape, while others are lost [2]. Urban landscapes, therefore, contain a lower variety of birds and are composed of generalist species with broad environmental tolerance [3] [4]. Nevertheless, urban landscapes become so important as birds' global habitat that many urban development will become interested in the birds' interests as part of the ecosystem [5] [6].

The presence of birds in the urban landscape became crucial concerning the birdwatching recreational object. For birdwatchers, birds are the principal objects in birdwatching recreational activities. Birdwatching is bird observational recreation type in the wild by naked eye surveys using aids such as telescopes and binoculars, or only listening the bird songs. Birdwatching is part of nature tourism focused on the bird observation [7] to be enjoyed through watching or listening [8]. This recreational activity is carried out in a very varied way from bird observation, discovery listing, until new species discovery competition;

therefore, the experts specifically distinguish birdwatching with twitching [9] to create a birdwatcher specialization [10].

Green open spaces in the urban landscape have been exploited by communities as one of the centers for physical activities [11] [12] and provide recreational satisfaction, which is close to the residence [13]. Like a bird's primary habitat in the urban landscape, green open spaces' existence has potential and opportunity as a birdwatching site [14]. While birds are the main objects in birdwatching, many factors influence the birdwatchers' choice of a location, especially in urban landscapes that are different from natural landscapes. The objectives of this paper were analyze the feasibility of urban landscapes for birdwatching activities and find out the characteristics of urban landscapes favored by birdwatchers in four cities in Java (Bogor, Sukabumi, Bandung and Surabaya).

2. Methodology

The research was carried out from February to May 2020 in four cities in Java (Bogor, Bandung, Surabaya and Sukabumi). Area data is taken by processing maps from google maps with the software Google Earth Pro. Habitat data are described. The vegetation strata refer to [15] with modifications, namely tree (A), shrub (B), bush (C), ground cover (D), and water plant (E). Bird data were collected using the presence-absence for five times each location. Observations of the birds were carried out during morning (05.30–11.00) or evening (14.00–18.00).

Online questionnaires used for birdwatchers' site preference and perception which distributed via email, WhatsApp, and social media to 1500 people. The answers to the questionnaire are close-ended with a ranking scale of 1-4 (never at all), 2 (seldom), 3 (often), and 4 (always), or the answers agree/disagree. The ranking answers are made an average, while agree/disagree answers are made a percentage. The questionnaire contains questions regarding preferences for birdwatching (interest, priority, birds, habitat, and location) and perceptions about birdwatching in urban landscapes (conservation support, ability, and advantage). Data were analyzed using (i) a chi-square for interest and priority preference and (ii) an exploratory factor analysis (EFA) for bird and habitat preference. The significant value level used is 5% (<0.05). Data processing uses version 24 of IBM SPSS.

3. Result

3.1. Birdwatching sites

The size of birdwatching sites varied between 0.05 and 76.82 ha. The smallest area are Aspirasi Park Sukabumi and the largest are Bogor Botanical Garden. The majority location had four strata. Only three locatioan has five strata. The dominant habitat type is a forest and garden. The forest habitat is dominated by trees from stratum A, while garden is dominated by shrub (statum B) and bush (stratum C). The field is dominated by ground cover vegetation (stratum D) or open space.

The dominant habitat type is a garden and the others habitat are forests, fields, ponds and rivers. The garden habitat is dominated by non-trees or trees with low height (sratum B and C), while the field is dominated by ground cover vegetation (stratum E) or open space. The ponds that were found were only in Bogor, namely the Bogor Botanical Garden (artificial pond) and Darmaga Research Forest (a natural pond). River habitats are found in the Bogor Botanical Garden and Sempur with the same river, namely the Ciliwung River. Other locations with rivers are Cikundul Agro Area, Kibitay Forest City, Maluku Park, Bandung Zoo, and Teras Cikapundung Park. The ponds found were only in Bogor and Bandung, namely the Bogor Botanical Garden and Bandung Zoo, both an artificial pond. The other are Darmaga Research Forest a natural pond. The field paddy habitat are only found in Cikundul Agrotourism Area, meanwhile the beach habitat found were only in Kenjeran Lama Park.

The number of bird species varied between three and 53 species. The bird species richness was highest in Bogor Botanical Garden (53 species) followed by Dramaga Research Forest (36 species). While the bird species richness was lowest in Aspirasi Park (three spesies). The species of birds encountered are generalists in the urban landscape; only a few specialists were found. Common types of generalists include the Eurasian tree-sparrow (31 locations), Cave swiftlet (30 locations), Black-headed Bulbul (29 locations), Plaintive cuckoo (26 locations), and Olive-backed Tailorbird (25 locations). As many as 29 bird species were only found in only one location, indicating that this species is not a generalist urban landscape but a specialist species but can survive or only visit in urban landscapes (e.g., Great Egret, Grey-cheeked Green Pigeon, Lesser coucal, and Hill Blue Flycatcher).

Table 1 Birdwatching sites condition

	Birdwatching site	Area (ha)	Habitat type	Strata	Number of bird species
	Bogor				
1	Bogor Botanical Garden	76.82	Fo, Ga, Fi, Rv, Po	A-B-C-D-E	53
2	Dramaga Research Forest	70.11	Fo, Po	A-B-C-D	36
3	Sempur Park	2.53	Ga, Fi, Rv	A-B-C-D	16
4	Heulang Park	2.25	Ga, Fi	A-B-C-D	9
5	Ahmad Yani City Forest	1.58	Fo	A-B-C-D	8
6	Malabar Park	0.70	Ga	A-B-C-D	15
7	Kencana Park	0.64	Ga	A-B-C-D	10
8	Peranginan Park	0.39	Fo	A-B-C-D	6
	Sukabumi				
9	Cikundul Agrotourism Area	6.00	Fo, Fi, Rv, Fp	A-B-C-D-E	18
10	Kibitay City Forest	4.16	Fo, Rv	A-B-C-D	13
11	Kerkof City Forest	1.87	Fo	A-B-C-D	16
12	Merdeka Field	1.78	Fo, Fi	A-B-C-D	9
13	Sugema Park	0.82	Ga	A-B-C-D	5
14	Digital Park	0.58	Ga	A-B-C-D	11
15	Aspirasi Park	0.05	Fo	A-B	3
	Bandung				
16	Tegalega Park	17.23	Fo, Ga, Fi	A-B-C-D	11
17	Bandung Zoo	13.57	Fo, Ga, Fi, Rv, Po	A-B-C-D-E	21
18	Balai Kota Park	6.12	Fo, Ga	A-B-C-D	14
19	Maluku Park	5.48	Fo, Ga, Fi, Rv	A-B-C-D	19
20	Babakan Siliwangi City Forest	4.61	Fo, Ga	A-B-C-D	14
21	Teras Cikapundung Park	2.69	Fo, Ga, Rv	A-B-C-D	11
22	Lansia Park	1.85	Fo, Ga	A-B-C-D	13
23	Pramuka Park	1.42	Fo, Ga	A-B-C-D	10
24	Ganesha Park	1.40	Fo, Ga, Fi	A-B-C-D	13
25	Cibeunying Park	1.00	Fo, Ga	A-B-C-D	10
26	Pustaka Bunga Park	0.83	Fo, Ga	A-B-C-D	5
	Surabaya				
27	Harmoni Park	8.58	Fo, Ga	A-B-C-D	12
28	Pakal City Forest	7.27	Fo, Ga	A-B-C-D	18
29	Flora Park	3.98	Fo, Ga	A-B-C-D	16
30	Kenjeran Lama Park	2.75	Fo, Ga, Fi, Bc	A-B-C-D	10
31	Bungkul Park	1.68	Fo, Ga	A-B-C-D	10

Note: Habitat (Fo: forest, Ga: garden; Fi: field, Rv: river; Po: pond; Fp: field paddy; Bc: beach)

The raptor that was found also indicated that it was not a generalist species, because it was only found in few locations, namely Crested Goshawk and Black-thighed Falconet. Only a few species of birds made nests at the site, namely Black-crowned Night Heron (Bogor Botanical Garden and Bandung Zoo), Red-breasted Parakeet (Ganesha Park), Javan Munia

(Sugema Park), White-headed Munia (Digital Park), Scarlet-headed flowerpecker (Pakal Forest City), Black-headed Bulbul (Pakal Forest City and Flora Park), and Zebra doves made use of artificial nests (Harmoni Park).

3.2. Birdwatching preference

Preference for birdwatching is linked with (a) respondent interest, (b) priority of activities, (c) birds' uniqueness, (d) bird characteristics, (e) birdwatching activities, (f) bird habitat, and (f) birdwatching sites in each city. A total of 1,296 citizens responded to the survey, though not all questions were answered (full complete 96%; n=1,247). A 72% of the total respondent was interested for birdwatching in the urban landscape, and of the total respondents 25% had previously birdwatched in the urban landscape. The interest respondents in urban landscape did differ significantly (χ^2 = 241.70; df=1; P<0.05) and as well as birdwatching experience (χ^2 = 317.27; df=1; P<0.05). Most respondents chose a balanced activity between looking for birds and other leisure activities (e.g., jogging and sport) (Figure1).

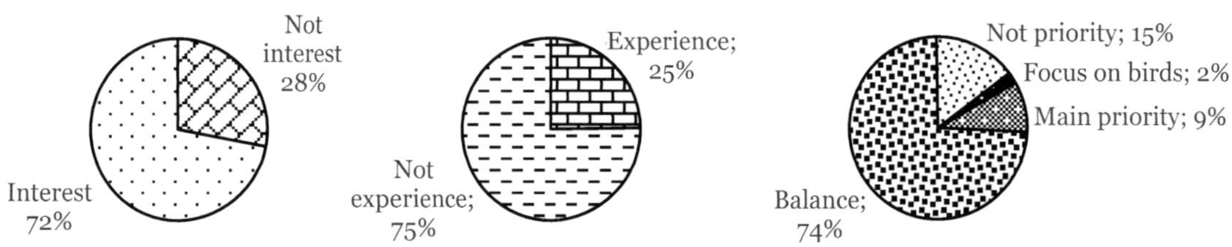

Figure 1 Percentage of respondents interest (left), respondent experience (center), and activities priority (right) for birdwatching in the urban landscape

The preference for the uniqueness of birds and the characteristics of birds to be seen presents only one factor. While the preference for birdwatching activities and the preference for birdwatching types of habitats make up two factors (Table 2). The first factor of birdwatching activities are the most severe activities in birdwatching (e.g., draw, song record, white the second factor of birdwatching activities are common activities during birdwatching (e.g., to see and to listen).

Table 2 Exploratory factor analysis of birdwatching preference

Preference	Cumulative initial eigenvalues	Variable of factor	EFA Indicator	Cronbach's α	Eigenvalues
Bird uniqueness	68.96%	All species, feather, song, behavior, habitat, distribution, new species	8	0.932	5.516
Bird characteristics	78.77%	Easy to see, easy to hear, many individuals	3	0.864	2.363
Birdwatching activities	73.47%	Draw, song record, video record, encounter record, discuss	5	0.844	4.836
		To see, to listen, to take a photo	3	0.844	1.042
Bird habitats	74.16%	City forest, city park, green roadside, river, pond, field paddy, farm, other green open space	8	0.938	6.621
		Backyard, housing, office ground	3	0.893	1.536

IOP Conf. Series: Earth and Environmental Science **879** (2021) 012005 doi:10.1088/1755-1315/879/1/012005

Generally, the most popular location is Bogor Botanical Garden and Dramaga Research Forest, both in Bogor. The other's location are Merdeka Field and Cikundul Agrotourism Area (Sukabumi), Bandung Zoo and Babakan Siliwangi City Forest (Bandung), and Bungkul Park and Flora Park (Surabaya). All locations that are most desirable in each city are those near the city center, excluding Dramaga Research Forest in Bogor. However, they have low bird diversity (e.g., Sempur Park, Merdeka Field, and Bungkul Park) (Figure 2).

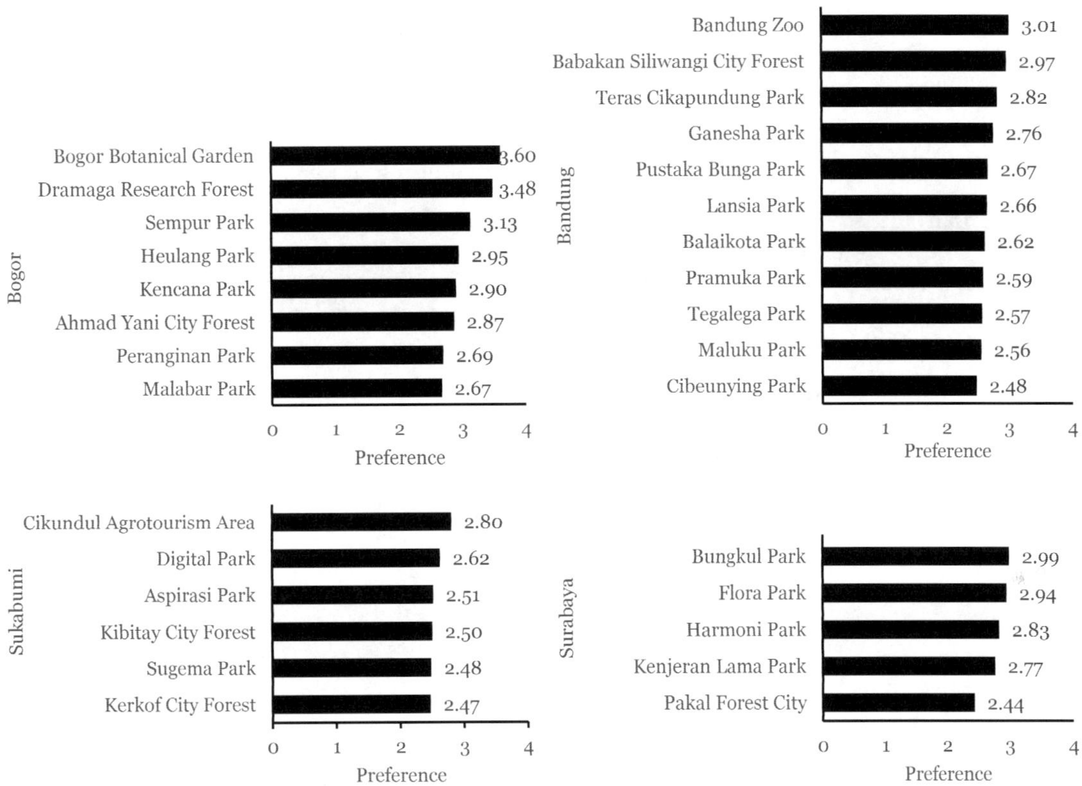

Figure 2 The birdwatching sites preference

The main reason of respondents for choosing the birdwatching sites is a shady area. These patterns are taking place all over the city. Bird diversity is not the main reason for choosing a preferred location for birdwatching. Even locations with high bird diversity get a poor choice. However, respondents gave a higher reason score for the birdwatching facilities (Figure 3).

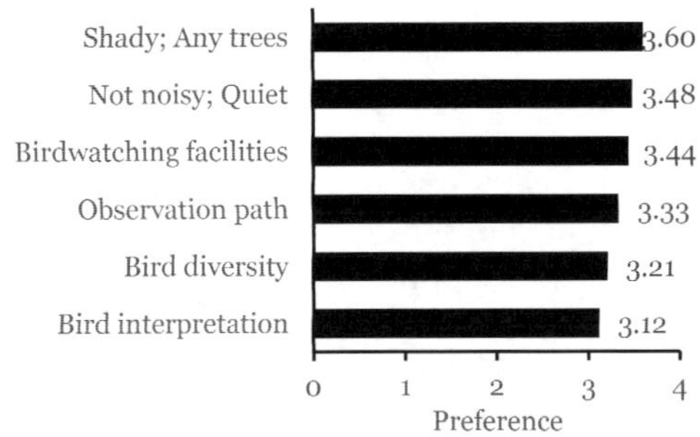

Figure 3 The reason choice of birdwatching sites preference

3.3. Birdwatching perception of the urban landscape
The birdwatching perception of the urban landscape is connected with (a) the birdwatching for conservation support, (b) the urban landscape's ability to birdwatching, and (c) the birdwatching advantage. Most of the responders responded to their support if birdwatching in the urban landscape can support conservation (Figure 4).

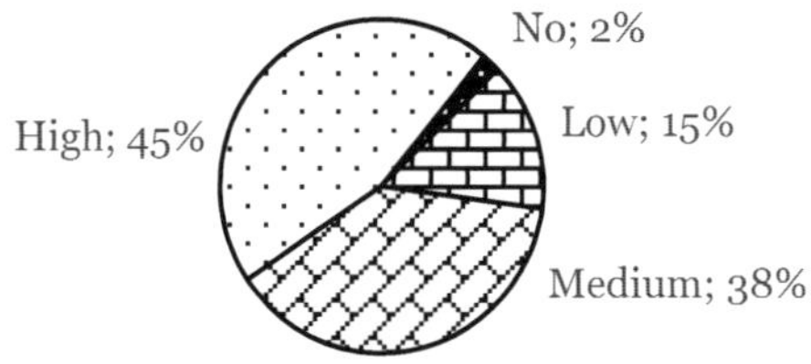

Figure 4 Percentage of support for birdwatching in urban landscape

The urban landscape is a positive reflection of the birdwatching landscape. The most dominant reason is accessibility ease, followed by reason as easy and cheap inexpensive recreation, this same value pattern for an entire city. The highest perception of birdwacthing gains in the urban landscape is that children and students are learn media and are then followed by a perception of green or green cities (Figure 5).

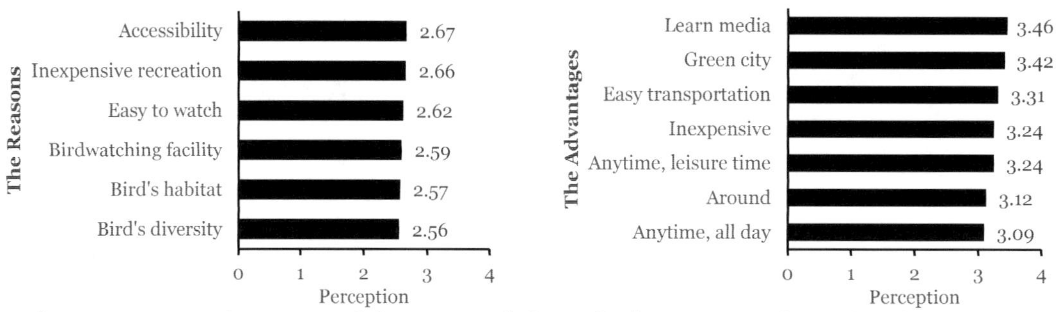

Figure 5 Perception score of the reason (left) and advantage (right) urban landscape utilization for birdwatching

4. Discussion
Birdwatching sites demonstrate the urban landscape's ability as a habitat for birds. Though it widens to influence bird attendance [16] [17], large is not the only factor affecting the presence of birds in a particular habitat, but many other factors that support the presence of birds, such as vegetation and landscape structure [18] [19] [20] [21] [22]. The range that the birdwatching is a nature trip to enjoy birds in the wild, making birds on the birdwatching sites vital as it is the main object of the birdwatching. Generally speaking, bird species encountered in urban landscapes are generalist species found in many habitats, though some specialists in urban landscape. Raptor presence on urban landscape became something of interest. The raptor, as the top predator in an ecosystem, is an indicator of biodiversity in an ecosystem [23] [24].

Respondent interest in birdwatching, which is shown by a higher percentage of interest than an uninterested one, indicates that birdwatching is a form of recreation that has an interest in the community. The respondent interest is similar to people in the USA for birdwatching and hunting [25]. The aesthetic value of birds provides the hope of satisfaction for society in recreation [26]. These conditions of bird's value are similar to those found by [8] about the positive perceive of Chicago's birds around their home to aesthetic aspects and the role of birds in the ecosystem. The same is true in Sweden that people provide a positively perceive of birds in open green space [27].

The birdwatching experience shows that the birdwatching has been done by people in Indonesia even though it is not as popular as recreational activities. Birdwatchers' experience shows that birdwatching has existed and continues to unfold as a hobby to this day [28] [29]. Bird's diversity research for birdwatching in the urban landscape is also conducted by [30] [31] [32]. This research implies that the urban landscape is also part of a bird's habitat that can be enjoyed for birdwatching [33].

The preference for birdwatching activities with a dominant value, not birdwatching, as a priority is very likely to occur because the respondent's main activity is outdoor recreation, so birdwatching is considered an additional activity while doing other recreational activities. Urban parks are generally used by the community for recreational activities [13]. Other activities outside of birdwatching are also common for birdwatchers when making birdwatching trips but are generally performed by beginner and intermediate birdwatcher groups, not expert birdwatchers [34]. Despite positive perceptions of birds and birdwatching, appreciation for birds may not yet fully emerge. Park visitor knowledge in Chile was not related to bird species richness in city parks, but they tended to only know about species that were abundant, widespread, and exotic [35].In general, almost no research has examined visitor preferences for bird choices and various aspects related to them. The formation of a preference factor for the uniqueness of birds and the nature of the birds they want to see shows that the respondents' pattern is similar for all variables. A preference for seeing as many bird species as possible suggests that most birdwatchers are generalists, while a minorities are specialists or twitchers. Generalists refer to birdwatchers who generally study entire groups of birds. In contrast, specialists are groups of birdwatchers who specifically target a specific species [10]. This condition is different from what found that birdwatchers in Australia are dominated by specialist groups, compared to generalists [36].

Birdwatching' activity preferences form two factors, general activity factors and specific activity factors. Activities that include general factor (seeing, listening, and taking photos) are everyday activities that stick to the definition of birdwatching [28], while other activities require more specific equipment. Equipment can provide a picture of strong or weak motivation. The birdwatcher's motivation for recreation reflected by an equipments and skills possessed, that high motivation is indicated by high quality equepment and expert skills [37]. The first factor is common for all birdwatchers, while the second factor is generally used by expert birdwatchers [34, 38]. The beginner birdwatchers can be distinguished from expert birdwatchers by the frequency and the motivation of birdwatching. The beginner birdwatcher less frequently than expert. The beginner birdwatcher has a variety of motivations, while expert birdwatcher, only focus on birds [38].

Birdwatching habitat preference forms a general green open space factor and a green open space factor around a residential or working building. The first factor has a higher value than the second factor. This condition illustrates; although, birdwatching is done in urban landscapes, respondents expect a more natural atmosphere than those around where they live or work. This condition is similar to what was found in Australia that natural habitat is more desirable than backyard [36]. Although sometimes, the type of habitat that is preferred is not related to the type of bird you are interested in seeing. Even though it gets a lower choice score, the backyard is a site that can be used for birdwatching every day [14].

Bird diversity are not a consideration for respondents in choosing a location, because, respondents' preference are shady location, followed by not noisy conditions and passing visitors. Conditions in open green space are generally more relaxed than other environments, attracting people to come [39]. Not noisy locations are often the reason people take advantage of green urban space [11]. Most of the respondents are beginner birdwatchers, so that there is no motivation to see birds as the main activity or even the only activity. The landscape quality, in this case, bird diversity, is not yet the main reason for visitor choice, similar to what was found in China in that the relationship between selected activities in urban landscapes [40].

Urban landscapes are perceived positively as a location for birdwatching. The most dominant reason is the ease of accessibility, followed by reasons as easy and cheap recreation. Birdwatching in urban landscapes will save a lot of expenses. This condition different with birdwatching in natural landscapes, which are generally high cost [28]. Both accessibility and distance influenced on visits to green open spaces [12] [41].

Birdwaching in urban landscapes is perceived as a learning medium, especially for children and students, in line with the research in England, which made birds an object of study related to school students' awareness [42]. There is a positive perception that birdwatching is related to environmentally friendly cities or green cities because it is related to the concept of green open space as an indicator of green city [43]. Understanding this perception is important because perceptions uses the basis for management. Many motivations and perceptions studies produce a good recommendations for park management [8] [29].

5. Conclusion

Possible to do birdwatching activities in urban landscapes because of the existence of birds as their primary resource. The species diversity found in each location was generally low; only a few locations had high species diversity of birds. The types of birds encountered are generally generalist. The majority of respondents are interested in and support birdwatching in urban landscapes. A popular location is desirable for its coolness and minimal disturbance, but does not require that it be rich in bird species.

6. Recommendation

It is necessary to build a green open space that can become a bird's habitat and preferred by birdwatchers with enrichment of feed vegetation and shady vegetation. The development of birdwatching in the urban landscape can be done to conserve biodiversity to form a sustainable city.

7. Acknowledgments

This research did not receive grant funding. Thanks are given to the respondents. Alfian Wiguna Hutama, Martino Batara, and Siti Fidhia Syahirah are help for data collection. Dr. Ruhyat Partasasmita shared and discuss bird's diversity of Bandung.

References

[1] Silva C P, García C E, Estay S A and Barbosa O 2015 Bird richness and abundance in response to urban form in a Latin American City: Valdivia, Chile *J. PLoS ONE* **10**(9): 1-16

[2] Martin P R and Bonier F 2018 Species interactions limit the occurrence of urban adapted birds in cities *PNAS* **115** (49): 11495-11504

[3] Bonier F, Martin P R and Wingfield J C 2007 Urban birds have broader environmental tolerance *Biol. Lett.* **3**: 670–673

[4] Palomino D and Carrascal L M 2006 Urban influence on birds at a regional scale: A case study with the avifauna of northern Madrid Province *Landsc. Urban Plan.* **77**: 276–290

[5] Jokimäki J, Suhonen J and Marja-Liisa J K 2018 Urban core areas are important for species conservation: A European-level analysis of breeding bird species *Landsc. Urban Plan.* **178**: 73–81

[6] Snep R P H, Kooijmans J L, Kwak R G M, Foppen R P B, Parsons H, Awasthy M, Sierdsema H L K, Marzluff J M, Fernandez-Juricic E, de Laet J, van Heezik Y M 2016 Urban bird conservation: presenting stakeholder-specific arguments for the development of bird-friendly cities *Urban Ecosyst.* **19**: 1535-1550

[7] Biggs D, Turpieb J, Fabriciusc C and Spenceley A 2011 The value of avitourism for conservation and job creation-An analysis from South Africa *Conserv. Soc.* **9**(1): 80-90

[8] Belaire J A, Westphal L M, Whelan C J and Minor E S 2015 Urban residents' perceptions of birds in the neighborhood: Biodiversity, cultural ecosystem services, and disservices *The Condor* **117**: 192-202

[9] Connell J 2009 Birdwatching, twitching and tourism: towards an Australian perspective *Aus. Geogr.* **40**(2): 203-217
[10] Hvenegaard G T 2002 Birder specialization differences in conservation involvement, demographics, and motivations *Hum. Dimens. Wildl.* 7: 21–36
[11] Burrows E, O'Mahony M and Geraghty D 2018 How urban parks offer opportunities for physical activity in Dublin, Ireland *Int. J. Environ. Res. Public Health* **15**: 1-11
[12] Cohen D A, McKenzie T L, Sehgal A, Williamson S, Golinelli D and Lurie N 2007 Contribution of public parks to physical activity *Am. J. Public Health* **97**(3): 509-514
[13] Zhu J, Lu Hm Zheng T, Rong Y, Wang C, Zhang W, Yan Y and Tang L 2020 Vitality of urban parks and its influencing factors from the perspective of recreational service supply, demand, and spatial links *Int. J. Environ. Res. Public Health* 17(1615): 1-17
[14] Cammack P J, Convery I and Prince H 2011 Gardens and birdwatching: recreation, environmental management and human–nature interaction in an everyday location *Area* **43**(3): 314-319
[15] Lembaga Ilmu Pengetahuan Indonesia 2013 3500 Plant Species of The Botanic Gardens of Indonesia (Jakarta: Lembaga Ilmu Pengetahuan Indonesia)
[16] Chamberlain D E, Gough S, Vaughan H, Vickery J A and Appleton GF 2007 Determinants of bird species richness in public green spaces *Bird Study* **54**(1): 87-97
[17] Kim J, Chae J and Koo T H 2007 Variation in bird diversity in relation to habitat size in the urban landscape of Seoul, South Korea *Acta Ornithol.* **42**(1): 39-44
[18] Leveau L M 2019 Primary productivity and habitat diversity predict bird species richness and composition along urban-rural gradients of central Argentina *Urban For. Urban Green.* **43**: 1-11
[19] Kaban A, Mardiastuti A and Prasetyo L B 2018 Landscape structure affects bird community in Bogor, West Java *Jurnal Penelitian Kehutanan Wallacea* 7 (2): 109-118
[20] Aronson M F J, Lepczyk C A, Evans K L, Goddard M A, Lerman S B, MacIvor J S, Nilon C H and Vargo T Biodiversity in the city: Key challenges for urban green space management *Front. Ecol. Environ.* **15** (4): 189–196
[21] Schutz C and Schulze C H 2015 Functional diversity of urban bird communities: Effects of landscape composition, green space area, and vegetation cover *Ecol. Evol.* **5**(22): 5230–5239
[22] Sulaiman S, Mohamad N H N and Idilfitri S 2013 Contribution of vegetation in urban parks as habitat for selective bird community *Procedia - Social and Behavioral Sciences* **85**: 267-281
[23] Tinajero R, Barragan F and Chapa-Vargas L 2017 Raptor functional diversity in scrubland-agricultural landscapes of Northern-Central-Mexican dryland environments *Trop. Conserv. Sci.* **10**: 1–18
[24] Roth T and Weber D 2008 Top predators as indicators for species richness? Prey species are just as useful *J. Appl. Ecol.* **45**(3): 987-991
[25] Wilkins E J, Cole N W, Miller H M, Schuster R M, Dayer A A, Duberstein J N, Fulton D C, Harshaw H W and Raedeke A H 2019 Rural-urban differences in hunting and birdwatching attitudes and participation intent *Hum. Dimens. Wildl.* **24**(6): 530-547
[26] Clucas B, Rabotyagov S and Marzluff J M 2015 How much is that birdie in my backyard? A cross-continental economic valuation of native urban songbirds *Urban Ecosyst.* **18:** 251-266
[27] Gunnarsson B, Knez I, Hedblom M and Sang A O 2017 Effects of biodiversity and environment-related attitude on perception of urban green space *Urban Ecosyst* **20**: 37–49
[28] Cordell H K and Herbert N G 2002 The popularity of birding is still growing *Birding* **34**: 54-61
[29] Sekercioglu C H 2002 Impact of birdwatching on human and bird communities *Environ. Conserv.* **29**(3): 282-289
[30] Agita A, Nurcahyani N, Kanedi M and Busman H 2020 Study of birds and plant types for birdwatching tourism development in The Liwa Botanical Garden, West Lampung *Jurnal Ilmiah Biologi Eksperimen dan Keanekaragaman Hayati* **7**(1): 6-10
[31] Janra M N 2019 Birding backyard: birdwatching in Andalas University *Earth and Environmental Science* 327: 1-12 *IOP Conf. Ser.: Earth Environ. Sci.* **327** 012025
[32] Hasibuan R S, Nitibaskara T U and Mahardika R 2018 Birdwatching Interpretation in Bogor Botanical Garden *Media Konservasi* **XXIII** (1):28-36
[33] Callaghan C T, Slater M, Major R E, Morrison M, Martin J M and Kingsford R 2018 Travelling birds generate eco-travelers: The economic potential of vagrant birdwatching *Hum. Dimens. Wildl* **28**(1): 71-82
[34] Maple L C, Eagle P F J and Rolfe H 2010 Birdwatchers' specialization characteristics and national park tourism planning *J. Ecotourism* **9**(3): 219-238
[35] Celis-Diez J L, Muñoz C E, Abades S, Marquet P A and Armesto J J 2017 Biocultural homogenization in urban settings: public knowledge of birds in city parks of Santiago, Chile *Sustainability* **9**(485): 1-14
[36] Steven R 2015 The Relationship between birders, avitourism, and avian conservation (Griffith: Griffith University)
[37] Ditton R B, Loomis D K and Choi S 1992 Recreation specialization: Re-conceptualization from a social worlds perspective *J. Leis. Res.* **24**(1): 33-51

[38] Scott D, Ditton R B, Stoll J R and Eubanks Jr. T L 2005 Measuring specialization among birders: utility of a self-classification measure *Hum. Dimens. Wildl.* **10**: 53-74
[39] Martini A, Biondi D and Batista A C 2018 Distance and intensity of microclimatic influence provided by urban forest typologies *Floresta e Ambiente* **25**(2): 1-12
[40] Wang H, Dai X, Wu J, Wu X and Nie X 2019 Influence of urban green open space on residents' physical activity in China *BMC Public Health* **19**: 2-12
[41] Zhang Y, Van den Berg A E, Van Dijk T and Weitkamp G 2017 Quality over quantity: contribution of urban green space to neighborhood satisfaction *Int. J. Environ. Res. Public Health* **14**: 1-10
[42] White R L, Eberstein K and Scott D M 2018 Birds in the playground: Evaluating the effectiveness of an urban environmental education project in enhancing school children's awareness, knowledge, and attitudes towards local wildlife *Journal PLoS ONE* ***13***(3): 0193993
[43] Pujiati A, Bowo P A and Nihayah D M 2018 The urban sustainability index in urban aglomeration *J. of Economics and Policy* **11**(2): 294-305

Landscape planning of historical tourism route of Siak Sultanate in Siak Sri Indrapura, Riau

M Z Islami[1], Nurhayati[2] and A Gunawan[2]
[1]Graduate Student of Landscape Architecture, IPB University, Indonesia
[2]Lecturer of Landscape Architecture Department, IPB University, Indonesia
E-mail: mzainulislami@gmail.com

Abstract. Siak Sri Indrapura City designates as the National Cultural Heritage regarding the Siak Sultanate Historic Area (1723-1945). Some heritages of Siak Sultanate have become historical tourism objects, but they cannot yet convey the Siak Sultanate history. Therefore, the historical tourism route plan is needed. This study aims to identify the use of historical landscapes in Siak Sri Indrapura; identifying historical tourism objects; analyze the potential of historical tourist objects, and; create a landscape plan for the historical tourism route of the Siak Sultanate. The analysis used in this study is the significance value of the historical landscape to assess the originality and uniqueness of the historic landscape, spatial analysis to obtain suitable land for historical tourism areas, and descriptive analysis to assess other aspects of the history of the Siak Sultanate. Part of the historical landscapes in Siak Sri Indrapura have become historical tourism areas, and there are 27 Siak Sultanate heritages found during the survey. Based on the analysis result, all the heritages and Siak Sri Indrapura areas are suitable to be developed into historical tourism areas. This research's output is the historical tourism route map and the estimated historical tourism activity table in the Siak Sultanate. The Siak Sultanate tourism area divides into four: The Beginning of Siak Sultanate, The Sultan's and Datuk Tomb, The Palace, and The Colonial. The separation of the area based on location considerations and historical links between historical sites. With the historical tourism route plan, tourists can quickly learn and understand each Siak Sultanate heritage objects' historical linkages.
Keywords: historical heritage, historical tourism, Siak Sultanate

1. Introduction

The development of the tourism area can be carried out in Indonesia because it has abundant natural, historical, and cultural wealth. Each area has tourism objects and attractions that are diverse and interesting to enjoy. The tourism sector provides many benefits for the community and local government. Through tourism, it can introduce culture, preserve nature and the environment, increase pride in the area, increase the love to preserve culture, create jobs and economic opportunities, create prosperity, and create good relations between tribes and nations [1]. In addition to providing benefits, tourism activities can also harm the area. The development of tourism that is not supported by proper planning and management of the area will cause a decrease the quality of the environment, followed by changes in the local community's culture [2].

Siak Regency has become a popular tourist destination for domestic and foreign tourists in Riau Province recently. Siak was the capital of the Malay-Siak Sultanate, so this area has much their historical heritage. In 2018, the Indonesia Minister of Education and Culture established the Siak Sultanate area as a National Cultural Heritage [3]. The area is currently also used as a historical tourism area. Since establishing the Siak Sultanate Area as a National Cultural Heritage area, the Regional Government has been intensively conducting renovations and improving cultural heritage objects' quality [4]. Unfortunately, regardless of that effort, the historical value among Cultural Heritage objects in Siak Sri Indrapura City has not been conveyed. This research needs to do so that tourists who visit to witness historical heritage objects understand these historical heritage objects' value and historical linkages. Therefore, this study aims to identifying and analyzing the potential of the Siak Sultanate area and plan the historical tourism route to make it easier to facilitate, provide information, and improve tourists' understanding of the historical value and link of the historical tourism objects.

2. Methods

The research conducts in Siak Sri Indrapura City, Siak Regency, Riau Province (Figure 1), for seven months, from December 2019 to June 2020.

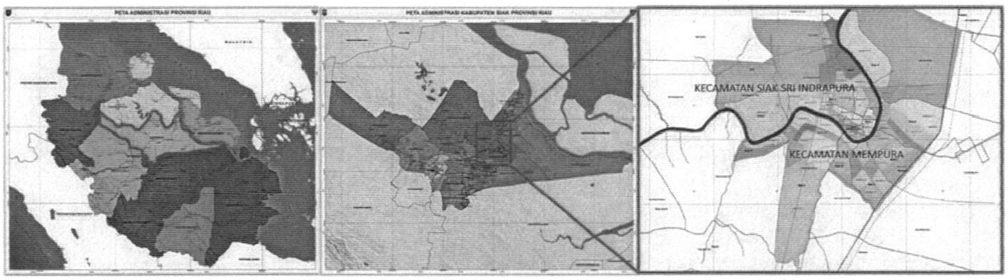

(a) Riau Province **(b) Siak Regency** **(c) Siak Sri Indrapura City**

Figure 1 Research location

This research uses descriptive method through historical tracing approach and consists of several stages: inventory, analysis, synthesis, concepts, and planning [5]. The inventory stage is to identify historical landscapes and historical tourism objects in Siak Sri Indrapura. The analysis stage is to analyze the potential attractiveness of the historical tourism object of the Siak Sultanate through the Historical Landscape Significant Value Analysis, Spatial Analysis, and Descriptive Analysis. Historical landscape significance value analysis assesses each historical landscape object based on criteria in the aspects of originality and uniqueness (Table 1) [6].

Table 1 Significance value assessment of historical landscape

No.	Criteria	Value		
		1 (Low)	2 (Moderate)	3 (High)
Originality				
1	Land use	> 50 % changing	25 - 50 % changing	< 25% changing
2	Building	There are changes in structure and building elements; does not represent the character and architectural style of the past; there is only one historical element with > 50 years	There are changes in structure and building elements, but represent the character and architectural style of the past; there are 2-5 historical elements with > 50 years	There are no changes in structure and building elements, represent the character and architectural style of the past; there are > 5 historical elements with > 50 years
Uniqueness				
1	Historical association	Landscape/element has no historical association	Landscape/element has a weak historical association	Landscape/element has a strong historical association
2	Integrity	Historical landscape element scattered in small amounts and inharmonious	Historical landscape element scattered in moderate amounts and harmonious	Historical landscape element scattered in large amounts and very harmonious
3	Different diversity	The landscape has only one historical element	The landscape has 2-5 historical elements	The landscape has > 5 historical elements
4	Aesthetic quality	The landscape element has no distinctive style	The landscape element has a distinctive style	The landscape elements have a distinctive style, including the details of the ornaments

The assessment was carried out through the scoring method to identify historical objects with high potential for historical tourism development. The scoring method calculates using the class interval formula [7]:

$$\text{Class Interval (CI)} = \frac{\text{Highest value (Hv)} - \text{Lowest value (Lv)}}{\text{Number of categories (n)}}$$

Category:

High	= Lv + 2CI + 1	to	Hv
Moderate	= Lv + CI + 1	to	Lv + 2CI
Low	= Lv	to	Lv + CI

Calculation of scoring method:

- Hv = 18
- Lv = 6
- n = 3

$$CI = \frac{18 - 6}{3} = \mathbf{4}$$

High = 6+2(4)+1 to 18
= **15 to 18**

Moderate = 6+4+1 to 6+2(4)
= **11 to 14**

Low = 6 to 6+4
= **6 to 10**

Spatial analysis is to obtain suitable land to develop into historical tourism routes by overlaying topography, climate, land use, circulation, and historical objects data. The descriptive analysis assesses other aspects of Siak Sultanate's history to support historical tourism routes development. The synthesis stage is to develop the potential of the historical area of the Siak Sultanate, which discovers during the analysis stage. The concept stage is to determine the basic concept of planning the historical tourism route of the Siak Sultanate. The planning stage was to produce a landscape plan for the historical tourism route of the Siak Sultanate, which presents in a historical tourism route map and an estimated table of distances, travel times, and historical tourism durations. The estimated distance and travel time calculation uses the Google Maps apps, while the tour's estimated duration base on the field survey results.

3. Result and discussion

3.1. General condition of research location

Siak Sri Indrapura is the capital of Siak Regency, Riau Province. The distance of Siak Sri Indrapura City from the provincial capital is around 120 km. Siak Sri Indrapura City locates in two sub-districts, namely Siak and Mempura. The Siak River separates both sub-districts, the north side is Siak Sub-district, and the south side is Mempura Sub-district. Siak Sub-district geographically located between 0°42'-0°57' North Latitude and 101°43'-102°17' East Longitude, while Mempura Sub-district located between 0°42'-0°57' North Latitude and 101°43'-102°14' East Longitude. Siak Sri Indrapura City's topographic conditions are flat with a slope of 0°-3° [8]. The average temperature is 25°C, and humidity is 88.9% [9]. Land cover in Siak Sri Indrapura dominates by green open space consisting of plantations and forests. Land use in Siak Regency for plantations is 200.648 ha (23,45%), and for forests is 191.757 ha (22,42%) [10].

3.2. History of the Siak Sultanate

The Siak Sri Indrapura Sultanate is an Islamic Malay Sultanate with control over large areas of the island of Sumatra, the Malay peninsula, to the Natuna Sea. The Siak Sultanate establish in 1723 AD by Raja Kecik, the son of Sultan Mahmud Syah II, the king of Johor Sultanate [11]. The historical timeline of the Siak Sultanate shows in Figure 2.

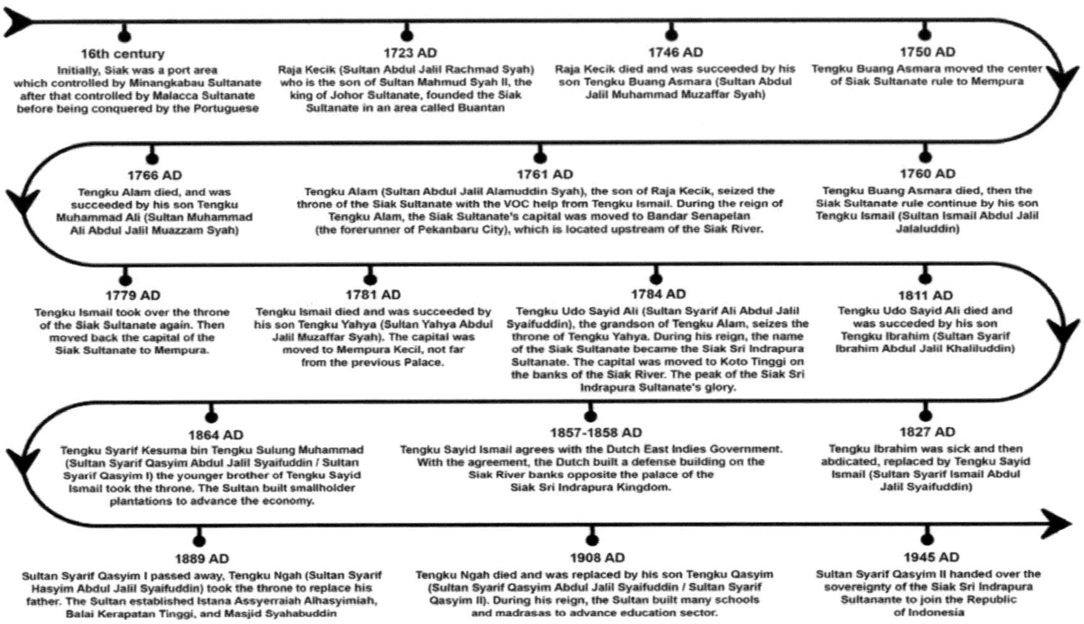

Figure 2 Historical timeline of Siak Sultanate

3.3. *Heritage of Siak Sultanate*

Siak Sri Indrapura City designates as a National Cultural Heritage Area regarding the Siak Sri Indrapura Sultanate rule center. Many historical heritages have been registered as cultural heritage objects and have legal protection from the district to the national level and scattered in several villages in Siak Sri Indrapura City. Besides, these objects have become historical tourist destinations in Siak Sri Indrapura City. The list of historical heritages of the Siak Sri Indrapura Sultanate is described in Table 2 [12].

Table 2 Heritage of Siak Sultanate

Historical Heritage	Legal Protection	Explanation
Siak Sri Indrapura Palace (*Istana Siak Sri Indrapura*)	Ministry of Culture and Tourism Decree: No. KM.13 / PW.007 / MKP / 2004	The Palace currently functions as a museum for Siak Sultanate heritages.
Syahabuddin Mosque (*Masjid Syahabuddin*)		The Sultan's mosque, still used as a place of worship for Muslims.
Koto Tinggi Tomb Complex (*Komplek Makam Koto Tinggi*)		Tomb of the sultans (Sultan Siak VII, VIII, IX, X, XI) and their families
Sultan Syarif Kasim II Tomb (*Makam Sultan Syarif Kasim II*)		Tomb of Sultan Siak XII
Siak Royal Audience Hall (*Balai Kerapatan Tinggi*)		Sultan Siak coronation place, meeting, and court; currently functions as a museum of Siak culture and history
Raja Kecik Tomb (*Makam Raja Kecik*)		Tomb of Sultan Siak I
VOC Barracks (*Tangsi Belanda*)		The resident office, arsenal and logistics, detention center, and barracks for Dutch East Indies troops; currently functioned as a library, office, museum, and exhibition space
Landraad Building (*Bangunan Landraad*)		Dutch East Indies government office
Controlleur Building (*Gedung Controlleur*)		Offices and official residences of Dutch East Indies officials
Siak Sultanate Palace Bride (*Jembatan Istana Kesultanan Siak*)	Ministry of Culture and Tourism Decree: No. PM.87 / PW.007 / MKP / 2011	Functions as part of the defense of the Siak Palace
Gunpowder Building (*Gudang Mesiu*)		Storage for cannon bullets and various war equipment
Hock Siu Kiong Temple (*Klenteng Hock Siu Kiong*)		A place of worship for the Chinese community
Tengku Buang Asmara Tomb (*Makam Tengku Buang Asmara*)		Tomb of Sultan Siak II

Historical Heritage	Legal Protection	Explanation
Syaithunissa' Madrasa (*Madrasah Syaithunissa'*)	Siak Regent Decree: No. 436 / HK / KPTS / 2017	The education place for women; currently functions as a display area for archives of Siak Regency history.
Taufikiyah Madrasa (*Madrasah Taufikiyah*)		The education place for men; currently functions as the secretariat of Religious Harmony Forum (*Forum Kerukunan Umat Beragama/FKUB*)
Hollandsch-Inlandsche School		The colonial primary school for noble children; currently functions as an elementary school for the Siak community children
Green Pond (*Kolam Hijau*)		The keris washing place and the heirlooms of the Siak Sultanate
Datuk Pesisir House (*Rumah Datuk Pesisir*)	Has been registered by the Tourism Department of Siak Regency	The residence of one of the advisers of the Sultan Siak; currently functions as a place to store Siak traditional weaving tools
Latifah Palace (*Istana Latifah*)		The sultan's resting place
Kato Ship (*Kapal Kato*)		Sultan Siak ships for cruising and visit the territories under his rule
Sultan's Well (*Sumur Sultan*)		Wells for ablution and sultan baths
Chinatown (*Kampung Pecinan*)		Settlements of immigrants from China
Sultan Ismail Tomb (*Makam Sultan Ismail*)		Tomb of Sultan Siak III
Tomb in the Banyan Tree (*Makam di dalam Pohon Beringin*)		The grave of scholars who teach Islam in the Siak area
Datuk Pesisir Tomb (*Makam Datuk Pesisir*)		Tomb of one of the four advisers of the Siak Sultanate
Datuk Kampar Tomb (*Makam Datuk Kampar*)		Tomb of one of the four advisers of the Siak Sultanate
Tuk Sesido Tomb (*Makam Tuk Sesido*)		Tomb of an architect in the Siak Sultanate

3.4. Historical landscape significance value analysis

The significance value of the historical landscape in Siak Sri Indrapura is assessed based on the originality and uniqueness of the Siak Sri Indrapura Sultanate's historic landscape. The originality assessment consists of two aspects: land use and building, and the uniqueness assessment consist of four aspects: historical associations, integrity, different diversity, and aesthetic quality. The significance value analysis result shows in Table 3.

Table 3 Analysis results of the significance-value of the Siak Sultanate historical landscape

No	Historical Heritage	Criteria						Total	Category
		Originality		Uniqueness					
		A1	A2	B1	B2	B3	B4		
1	Siak Sri Indrapura Palace	3	3	3	3	3	3	18	High
2	Syahabuddin Mosque	3	3	3	3	3	3	18	High
3	Koto Tinggi Tomb Complex	3	3	3	3	3	2	17	High
4	Sultan Syarif KasimII Tomb	3	3	3	3	3	3	18	High
5	Siak Royal Audience Hall	3	3	3	3	3	3	18	High
6	Raja Kecik Tomb	3	3	3	3	3	3	18	High
7	VOC Barracks	3	2	3	3	3	3	17	High
8	Landraad Building	3	2	3	2	3	2	15	High
9	Controlleur Building	3	2	3	2	3	2	15	High
10	Siak Sultanate Palace Bride	3	3	3	2	3	3	17	High
11	Gunpowder Building	2	3	3	2	3	1	14	High
12	Hock Siu Kiong Temple	3	3	3	3	3	3	18	High
13	Tengku Buang Asmara Tomb	3	3	3	2	3	2	16	High
14	Syaithunissa' Madrasa	2	2	3	2	2	2	13	Moderate
15	Taufikiyah Madrasa	2	2	3	2	2	2	13	Moderate
16	Hollandsch-Inlandsche School	2	2	2	2	2	2	12	Moderate
17	Green Pond	3	2	3	2	2	2	14	Moderate
18	Datuk Pesisir House	3	2	2	2	2	2	13	Moderate
19	Latifah Palace	3	3	3	3	2	2	16	High
20	Kato Ship	1	3	2	2	2	2	12	Moderate
21	Sultan's Well	3	3	3	3	2	1	15	High
22	Chinatown	2	2	2	2	2	2	12	Moderate
23	Sultan Ismail Tomb	3	2	2	2	2	2	13	Moderate
24	Tomb in the Banyan Tree	3	2	2	2	1	1	11	Moderate
25	Datuk Pesisir Tomb	3	2	2	2	2	2	13	Moderate
26	Datuk Kampar Tomb	3	2	2	2	2	2	13	Moderate
27	Tuk Sesido Tomb	3	2	2	2	2	2	13	Moderate

Note:

Originality assessment	Uniqueness assessment	Value	Category
A1 = Land Use	B1 = Historical association	1 = Low	High = 15 - 18

Originality assessment	Uniqueness assessment	Value	Category
A2 = Building	B2 = Integrity	2 = Moderate	Moderate = 11 - 14
	B3 = Different diversity	3 = High	Low = 6 - 10
	B4 = Aesthetic Quality		

Based on the analysis results above, it knows that the object or site: Siak Sri Indrapura Palace, Syahabuddin Mosque, Koto Tinggi Tomb Complex, Sultan Syarif Kasim II Tomb, Siak Royal Audience Hall, Raja Kecik Tomb, VOC Barracks, Landraad Building, Controlleur Building, Siak Sultanate Palace Bride, Gunpowder Building, Hock Siu Kiong Temple, Tengku Buang Asmara Tomb, Latifah Palace, and Sultan's Well have a high significance value of the historical landscape. In comparison, the object or site of Syaithunissa' Madrasa, Taufikiyah Madrasa, Hollandsch-Inlandsche School, Green Pond, Datuk Pesisir House, Kato Ship, Chinatown, Sultan Ismail Tomb, Tomb in the Banyan Tree, Datuk Pesisir Tomb, Datuk Kampar Tomb, and Tuk Sesido Tomb has a moderate significance value of the historical landscape. However, there are no historical objects or sites with a low historical landscape significance value on the study object.

3.5. Spatial analysis
Spatial analysis overlays topographical data, climate, land use, circulation, and historical objects. The aim is to obtain suitable areas for the historical tourism route planning of the Siak Sri Indrapura Sultanate. The results of the analysis shown in Figure 3. The picture shows that the blue area is suitable to be developed as a historical tourism path of the Siak Sri Indrapura Sultanate. While the green area is the area that is quite suitable, and the brown area is not suitable.

3.6. Descriptive analysis
Based on surveys, literature studies, and interviews about the Siak Sri Indrapura Sultanate, some historical facts are found and can use in planning historical tourism routes. The Siak Sri Indrapura Sultanate is one of the Malay Sultanates in Indonesia. The planning of the Siak Malay Sultanate area has a pattern and elements consisting of: (1) palace, (2) large open fields, (3) mosque, (4) river, (5) market, (6) nobleman housing, and (7) commoner's settlements [13]. Illustration of patterns and elements of the Siak Sultanate shown in Figure 4. The river is one of the main elements considered in selecting the location of the Malay Sultanate rule [14,15,16,17]. The market builds close to the river because it functions as a place for economic activities and transactions. Another essential element in the Malay Sultanate area is the mosque. Malay Sultanate is closely related to Islam, so the mosque was built close to the palace. In front of the Malay sultanate palace, there is an open field used as a gathering place. The settlement of nobles and ordinary people were built around the palace with a pattern, as illustrated. Considering the importance of rivers' role for the civilization of the Malay Sultanate, in planning the Historical Tourism Route of the Siak Sri Indrapura Sultanate, it is necessary to consider the Siak River as part of the planning.

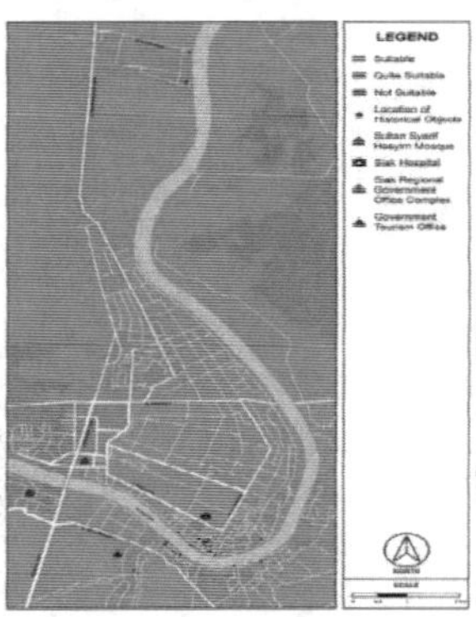

Figure 3 Spatial analysis result

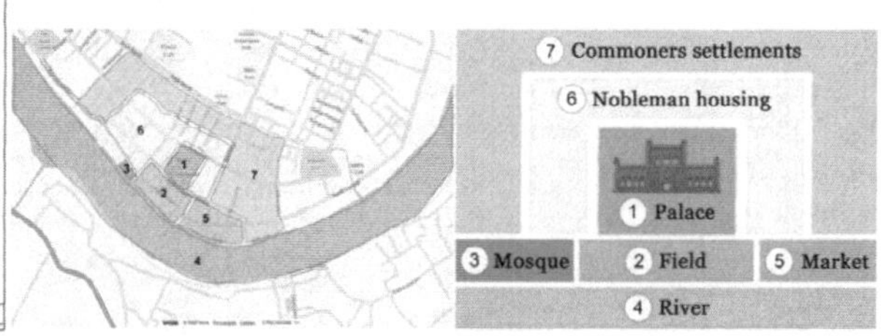

Figure 4 Pattern of Siak Sultanate area

The analysis results above show that the historical objects of the Siak Sri Indrapura Sultanate have high and medium historical landscape significance values. These historical objects can be a historical tourist attraction for the Siak Sultanate. Besides, Siak Sri Indrapura City also has potential areas to be developed into the historical tourism route, including the Siak River. The planning of the Siak Sri Indrapura Sultanate's historical tourism route will be carried out only in suitable areas based on spatial analysis results.

3.7. Historical tourism path planning

The plan's basic concept is to preserve historical values in the historical landscape of the Siak Sri Indrapura Sultanate, utilize the historical landscape as a historical tourism area, and introduce its history through the historical tourism route. Due to the historical sites and objects of the Siak Sultanate are scattered in several villages, the tourism route planning divided into several zones according to its location and history, which divided into four: The Beginning of Siak Sultanate Zone, The Sultan's and Datuk Zone, The Palace Zone, and The Colonial Zone (Figure 5).

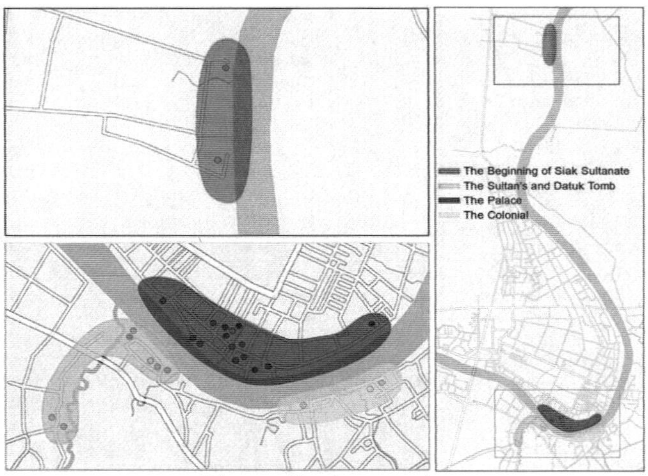

Figure 5 Tourism zone plan

Each zone's tourism area divides into three areas: the historical tourism area, service area, and welcome area. The historical tourism area is the area that has historical sites and objects in it. The service area supports tourism activities to provide information and facilities related to the historical tourism area. The welcome area is the area for welcoming tourists before entering the historical tourism area. The welcome area's determination is based on the most strategic location because it connects the circulation from outside to the tourism area. The activity and facilities plan is arranged based on each area's needs because they have different functions. The activity plan in each area is needed to determine the appropriate facilities to accommodate tourists' needs. Table 4 outlines the plan of activities and facilities for the Siak Sri Indrapura Sultanate's historical tourism area.

Table 4 Activities and facilities plan for historical tourism areas

Tourism Area	Area		
	Historical Tourism	Service	Welcome
The Beginning of Siak Sultanate	• Activities: seeing historical sites; studying the early history of the Siak Sri Indrapura Sultanate (the location of the Tomb of Raja Kecik was the first point for the establishment of the center of the Siak Sri Indrapura Sultanate); taking pictures; enjoying the scenery • Facilities: historical sites and objects: Raja Kecik Tomb and Green Pond; information	• Activities: enjoy the atmosphere and scenery, sightseeing, souvenir shopping, eating and drinking, worship, short breaks, sanitation • Facilities: Information center, interpretation board, shopping center, food court,	• Activities: purchase the entrance ticket, park the vehicle, get information about the tourism area, sanitation • Facilities: Entrance gate, ticket window, parking area, guard post, tour guide, information center,

Tourism Area	Area		
	Historical Tourism	**Service**	**Welcome**
	and interpretation boards; tourism transportation	tourism transportation, prayer room, picnic shelter, ATM, toilet	interpretation board, tourism transportation, toilet
The Sultan's and Datuk Tomb	• Activities: seeing historical sites; make a pilgrimage to the tombs of Sultan and Datuk (Datuk is a title given to the advisors of the Siak Sri Indrapura Sultanate); taking pictures; enjoying the scenery • Facilities: historical sites and objects: Tengku Buang Asmara Tomb, Sultan Ismail Tomb, Datuk Pesisir House, Datuk Pesisir Tomb, Datuk Kampar Tomb, Tuk Sesido Tomb, and Tomb in the Banyan Tree; information and interpretation boards; tourism transportation	• Activities: enjoy the atmosphere and scenery, sightseeing, souvenir shopping, eating and drinking, worship, short breaks, sanitation • Facilities: Information center, interpretation board, shopping center, food court, tourism transportation, prayer room, picnic shelter, ATM, toilet	• Activities: purchase the entrance ticket, park the vehicle, get information about the tourism area, sanitation • Facilities: Entrance gate, ticket window, parking area, guard post, tour guide, information center, interpretation board, tourism transportation, toilet
The Palace	• Activities: seeing historical sites; studying the history of the Siak Sri Indrapura Sultanate rule and the life of its people (this area is the area have the most historical heritage); taking pictures; enjoying the scenery • Facilities: historical sites and objects: Siak Royal Audience Hall, Sultan Syarif Kasim II Tomb, Syahabuddin Mosque, Syaithunisa' Madrasa, Siak Sri Indrapura Palace, Sultan's Well, Latifah palace, Kato Ship, Siak Sultanate Palace Bridge, Taufikiyah Madrasa, Gunpowder Building, Koto Tinggi Tomb Complex, China Town, Hock Siu Kiong Temple, and Hollandsch-Inlandsche School; information and interpretation boards; tourism transportation	• Activities: enjoy the atmosphere and scenery, sightseeing, souvenir shopping, eating and drinking, worship, short breaks, sanitation • Facilities: Information center, interpretation board, shopping center, food court, tourism transportation, prayer room, picnic shelter, ATM, toilet	• Activities: purchase the entrance ticket, park the vehicle, get information about the tourism area, sanitation • Facilities: Entrance gate, ticket window, parking area, guard post, tour guide, information center, interpretation board, tourism transportation, toilet
The Colonial	• Activities: seeing historical sites; studying the history of Dutch intervention in the Siak Sultanate (this area was the center of colonial rule in the Siak area); taking pictures; enjoying the scenery • Facilities: historical sites and objects: VOC Barracks, Landraad Building, and Controlleur Building; information and interpretation boards; tourism transportation	• Activities: enjoy the atmosphere and scenery, sightseeing, souvenir shopping, eating and drinking, worship, short breaks, sanitation • Facilities: Information center, interpretation board, shopping center, food court, tourism transportation, prayer room, picnic shelter, ATM, toilet	• Activities: purchase the entrance ticket, park the vehicle, get information about the tourism area, sanitation • Facilities: Entrance gate, ticket window, parking area, guard post, tour guide, information center, interpretation board, tourism transportation, toilet

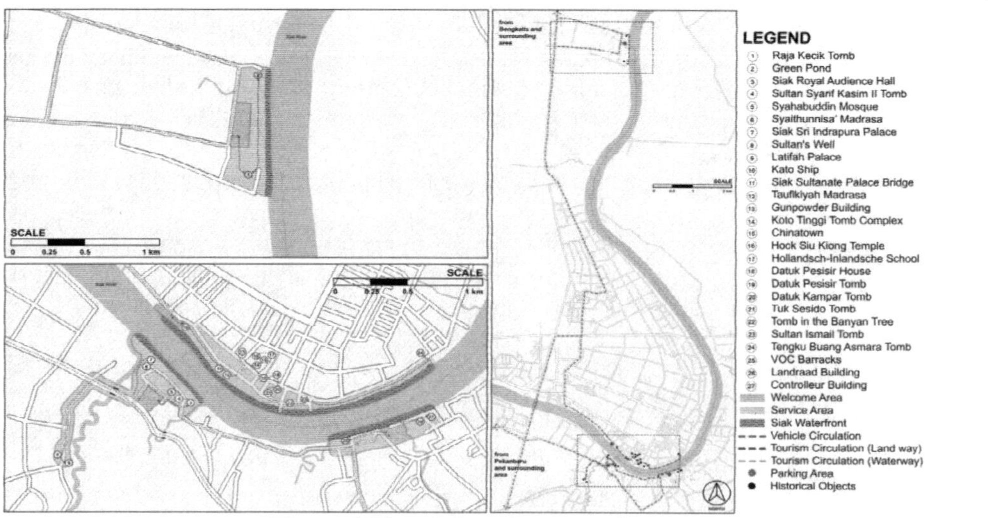

Figure 6 Historical tourism route plan

The historical tourism route plan for the Siak Sri Indrapura Sultanate shows in Figure 6. On the map, there are three types of circulation: vehicle circulation (red line), land way tourism circulation (blue line), and waterway tourism circulation (orange line). The limit of vehicle circulation only to the Welcome Area; after that, tourists can choose to go to tourism objects by foot or tourism transportation on the map's blue line. If tourists want to go to other tourism areas, namely the Sultan's and Datuk Tomb, the Palace, or the Colonial, they can use the waterway tourism transportation (orange line) and sail on the Siak River or with the private vehicle (red line). Table 5 shows the estimated distance, time, and duration of Siak Sultanate historical tourism.

Table 5 Estimation of the distance, time, and duration of the historical tourism

	Tourist attraction	Distance (km)	Estimated Travel Time (minutes)		Estimated Tourism Duration (minutes)
			With TT	Without TT	
	The Beginning of Siak Sultanate (Z1)				
	Welcome Area				
1	Raja Kecik Tomb	1	1	4	15
2	Green Pond	2	2	13	10
	The Sultan and Datuk Tomb (Z2)				
	Welcome Area				
3 4	Datuk Pesisir House Datuk Pesisir Tomb	1	1	4	20
5	Datuk Kampar Tomb	1	1	1	10
6 7	Tuk Sesido Tomb Tomb in the Banyan Tree	1	1	4	15
8 9	Sultan Ismail Tomb Tengku Buang Asmara Tomb	3	3	14	25
	The Palace (Z3)				
	Welcome Area				
10	Siak Royal Audience Hall	1	1	1	20
11 12	Sultan Syarif Kasim II Tomb Syahabuddin Mosque	1	1	4	20
13	Syaithunisa' Madrasa	2	2	3	15
14 15 16 17	Siak Sri Indrapura Palace Sultan's Well Latifah Palace Kato Ship	1	1	1	60
18	Siak Sultanate Palace Bridge	1	1	1	5
19	Taufikiyah Madrasa	1	1	1	15
20 21	Gunpowder Building Koto Tinggi Tomb Complex	1	1	1	20
22	Chinatown	1	1	1	20
23	Hock Siu Kiong Temple	1	1	4	15
24	Hollandsch-Inlandsche School	2	2	10	15
	The Colonial (Z4)				
	Welcome Area				
25	VOC Barracks	1	1	2	25
26 27	Landraad Building Controlleur Building	3	3	12	25

Note
TT: Tourism Transportation

Furthermore, the estimate of the distance and time required to move from one tourism area to another using a tourism vehicle or private vehicle shows in Table 6. Travel by tourism vehicle (the waterway) is more effective and efficient than using a private vehicle (vehicle circulation).

Table 6 *Estimated distance and time of travel by private vehicle and tourism vehicle*

		Z1	Z2	Z3	Z4
Z1	A	-	16.6 km (23 minutes)	14.5 km (22 minutes)	18.6 km (26 minutes)
	B	-	15 km (33 minutes)	15 km (33 minutes)	13.5 km (30 minutes)
Z2	A	16.6 km (23 minutes)	-	9.6 km (13 minutes)	2 km (3 minutes)
	B	15 km (33 minutes)	-	0.2 km (1 minutes)	1.5 km (3 minutes)
Z3	A	14.5 km (22 minutes)	9.6 km (13 minutes)	-	11.4 km (16 minutes)
	B	15 km (33 minutes)	0.2 km (1 minutes)	-	0.2 km (1 minutes)
Z4	A	18.6 km (26 minutes)	2 km (3 minutes)	11.4 km (16 minutes)	-
	B	13.5 km (30 minutes)	1.5 km (3 minutes)	0.2 km (1 minutes)	-

Note
Z1: The Beginning of Siak Sultanate
Z2: The Palace
Z3: The Sultan and Datuk Tomb
Z4: The Colonial
A: Land way (private transportation)
B: Waterway (tourism transportation)

Table 7 shows the estimated total duration of the tour needed for visiting all Siak Sri Indrapura Sultanate's historical tourism objects. The tour's total duration according to the recommended route if using the tourism transportation planned is about 410 minutes, while if the tourist does not use that facility, the total duration is about 483 minutes.

Table 7 Estimated total tour duration

Tourism Area	Estimated Tour Duration (minutes)	
	With TT	Without TT
The Beginning of Siak Sultanate	28	42
Travel Time*	33	23
The Sultan's and Datuk Tomb	76	93
Travel Time*	1	13
The Palace	217	232
Travel Time*	1	16
The Colonial	54	64
Total Tour Duration	**410**	**483**

Note
TT: Tourism Transportation
*Data from Table 6

4. Conclusion

Based on the identification results, part of the Siak Sri Indrapura Sultanate landscape has become a historical tourism area, and there are 27 historical objects recorded. The analysis results show that the whole historical objects of the Siak Sultanate; and the Siak Sri Indrapura City have the suitability to be developed into historical tourism areas. In the planning of historical tourism routes, the Siak Sri Indrapura historical tourism area divides into four: The Beginning of Siak Sultanate with the main object is Raja Kecik Tomb; The Sultan's and Datuk Tomb with the main object is Tengku Buang Asmara Tomb; The Palace with the main object is Siak Sri Indrapura Palace; The Colonial with the main object is VOC Barracks. The division of the area based on location considerations and historical links between historical sites.

References

[1] UNESCO 2009 Ecotourism: Operational basic guidance [Internet] [cited 2020 Jun 20] Available from: https://unesdoc.unesco.org/ark:/48223/pf0000185506
[2] Inskeep E 1991 Tourism planning: An integrated and sustainable development approach vnr tourism and commercial recreation series New York (US): Van Nostrand Reinhold 508 p
[3] Kemendikbud 2018 Decree of the minister of education and culture of the republic of indonesia number: 164/M/2018 [Internet] [cited 2020 Jun 20] Available from: https://kebudayaan.kemdikbud.go.id/bpcbsumbar/wp-content/uploads/sites/28/2019/03/164-M-2018-Kawasan-Cagar-Budaya-Pusat-Pemerintahan-Kesultanan-Siak-Sri-Indrapura-Kab.-Siak-Riau.pdf
[4] Firzal Y and Temas I 2018 Cultural significance of the place: The key factor in urban heritage programme IOP Conf. Series: Earth and Environmental Science 213 DOI:10.1088/1755-1315/213/1/012008
[5] Gold S M 1980 Recreation planning and design New York (US): McGraw Hill Book
[6] Harris C W and Dines N T 1988 Time-saver standards for landscape architecture: Design and construction data New York (US): McGraw-Hill Companies 800 p
[7] Sari D C and Munandar A 2019 Characteristic assessment and cultural landscape conservation of Urug Indigenous Village IOP Conf. Series: Earth and Environmental Science 501 DOI:10.1088/1755-1315/501/1/012045
[8] Siak Regency Spatial Plan 2002-2011
[9] Siak Regency Government Work Plan 2018
[10] Siak Regency Government Performance Accountability Report 2016
[11] Barnard T P 2001 Texts, Raja Ismail and violence: Siak and the transformation of Malay Identity in the eighteenth century *Journal of Southeast Asian Studies* **32**(03): 331-342
[12] Ministry of Education and Culture of the Republic of Indonesia 2019 List of Cultural Heritage of West Sumatra, Riau and Riau Islands Provinces [Internet] [cited 2020 Jun 20] Available from: https://kebudayaan.kemdikbud.go.id/bpcbsumbar/wp-content/uploads/sites/28/2019/07/PENETAPAN-CB-Di-Wilker-BPCB-Sumbar-OK.pdf
[13] Final Report on the Preparation of Building and Environmental Planning for Priority Heritage Areas of Siak Regency 2018
[14] Andy U F 2017 The influence of global trading network on the Malay Sultanates Central of Government structure and spatial configuration in West Kalimantan Langkau Betang **4** (1): 67-78
[15] Sumarno E, Karina N, Ginting J S and Handoko 2018 River and Malay Identity in East Sumatra in the XX Century Indonesian Historical Studies **2**(2): 83-88
[16] Putra B A 2006 Jambi Malay Settlement Patterns (Case Study of Tanjung Pasir Sekoja Area) [thesis] Semarang (ID): Diponegoro University
[17] Rijal M 2018 The configuration of residential area in urban structure of the palace in Siak Sri Indrapura – Riau IOP Conf. Series: Materials Science and Engineering 352 DOI:10.1088/1757-899X/352/1/012060

Ligtas Agtas: a sustainable and disaster-ready resettlement area for the Agta Community in Sitio Pulang Lupa

S R Ang[1] and H T Palarca[2]
[1] Student, College of Architecture, University of the Philippines - Diliman, Philippines
[2] Senior Lecturer, College of Architecture, University of the Philippines - Diliman, Philippines

E-mail: srang1@up.edu.ph

Abstract. Most indigenous peoples experience homelessness in the Philippines. This is the case for Mount Kalbo's indigenous tribe, the Agta Community. They face the challenges of dealing with mass wasting disasters potentially devastating their houses and compromising the lives of their families. It is with high hopes that they will be resettled to Sitio Pulang Lupa to rebuild a community, in which this research aims to propose and develop. The proposed resettlement's goal is to improve and sustain the Agtas' living conditions in Sitio Pulang Lupa by addressing the unstable slopes and preparing the community to cope with mass wasting disasters. Sitio Pulang Lupa is analyzed compared to Mount Kalbo regarding its advantages of keeping the resettlement sustainable and safe from mass wasting. The resettlement merges the Agtas' necessities to achieve the research's objectives, which are improved living conditions, cultural representation, and disaster preparedness into one landscape development in Sitio Pulang Lupa. The resettlement comprises of spaces and facilities that reflect their cultural practices and their needs. Disaster management protocols were also formulated by mapping safe routes, and refuge landscapes to equip the Agtas with preparedness in case disaster occurs. To sustain the vision for the Agtas, soil conditions are firstly addressed by using contour planting. It was used as slope protection and to over-all enhance the environmental quality of the Agtas' resettlement area. The resettlement as a whole shall be able to meet the research's goal of proposing a landscape development that mitigates the mass wasting threats experienced by the Agtas, and improves their living conditions. Having said that, Agtas shall finally feel safe and secured in their own homes.
Keywords: Culture, Disaster Preparedness, Mass Wasting, Resettlement, Sustainability

1. Introduction

Quirino province comprises of 80 per cent highlands that make it mountainous.[1] Entailed with its rich topography are threats of mass wasting. Mass wasting, commonly known as landslides, is a life threatening occurrence where soil, rocks, or sand naturally moves downward according to gravity through falling, sliding, or rolling. [2]

In addition to this concern, Nagtipunan, Quirino was recently hit by a magnitude 4.8 earthquake at the evening of 14th of November 2019.[3] Although there are reportedly no damages caused, the sudden jolt of the ground in Quirino may lead to the fragility of its soil and cause disasters in the future. It is an enormous threat to the people and communities that reside in the province, most especially if not equipped with knowledge on how to act or prevent such disasters from happening.

Mount Kalbo's Agta Community, one of the indigenous peoples in Quirino Province, faces these kinds of threats.[4] Indigenous peoples' welfare is often neglected with government support, which hinders them from building a safe and sustainable community.

They hope to resettle in Sitio Pulang Lupa, where the soil is more stable, and be able to develop their community towards unity and resiliency, in which this research's goal will focus.

1.1 Problem Setting
Having mentioned the situation of the Agtas, this research's main goal is to address the problem by formulating a functional and secure landscape development design for the resettlement of Agta Community. The resettlement's objective is to improve and sustain the Agtas' living conditions by addressing the unstable slopes and preparing the community to cope with mass wasting disasters. It is achieved by characterizing the soil and topography conditions in the resettlement site and making use of landscape architectural elements and natural slope protection measures to address these. In designing the landscape development plan, the Agta culture was considered, and procedures to mitigate threats of mass wasting was provided. Accessible and functional refuge landscapes that serve as an evacuation area for the Agta Community in times of mass wasting disasters will be included in the development. Altogether, Agtas' resettlement will be able to rebuild a community that is not only sustainable but enhances the environmental quality of the area as well.

2. Methodology
A research method design is formulated reflecting the processes conducted in this

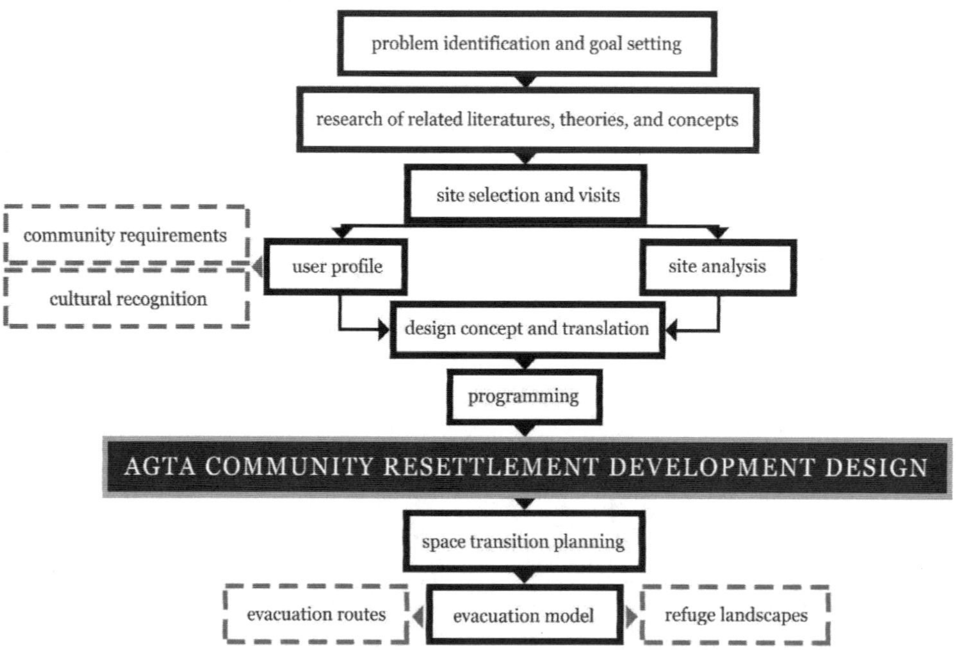

Figure 1 Methods Used to Develop The Resettlement of The Agta Community

research. This research is divided into two phases as reflected in Figure 1. The first phase is the development of the resettlement for the Agta Community, and second is the transitioning to an evacuation model in case of disasters, especially mass wasting.

As the research established the problems and goals, a site selection was conducted to identify the site that best embodies the objectives of the research. A site visit and interviews were then accomplished to acquire perspective to the situation of the previous site. With this, a site within the Agtas' ancestral domain was selected following criteria that highlights the convenience of the location to the Agtas. The chosen resettlement site was then compared to the previous site to justify that the chosen site is suitable for the resettlement of the Agta Community. Agtas' culture and lifestyle were also recognized in planning the development. With this, a design concept was developed that reflects the culture of the Agta Community. The development then transitions to the second phase where the development spaces shift into disaster-resilient spaces in case of disasters. Evacuation routes were mapped, and refuge landscapes or evacuation areas were determined. In the end, a landscape design will be proposed where it will secure the Agta Community from the risks of mass wasting disasters.

3. Findings

3.1 Geographic Information

The brangay where the Agta communities reside is the Barangay of Disimungal. Agta communities reside in different areas in Disimungal such as in Mount Kalbo, Masog Falls, Tili-tilan, Junuan Falls, and Sitio Pulang Lupa. Reflected in Figure 2 is the boundary of Disimungal where most land area is prone to moderate erosion due to the mountainous topography of Quirino.

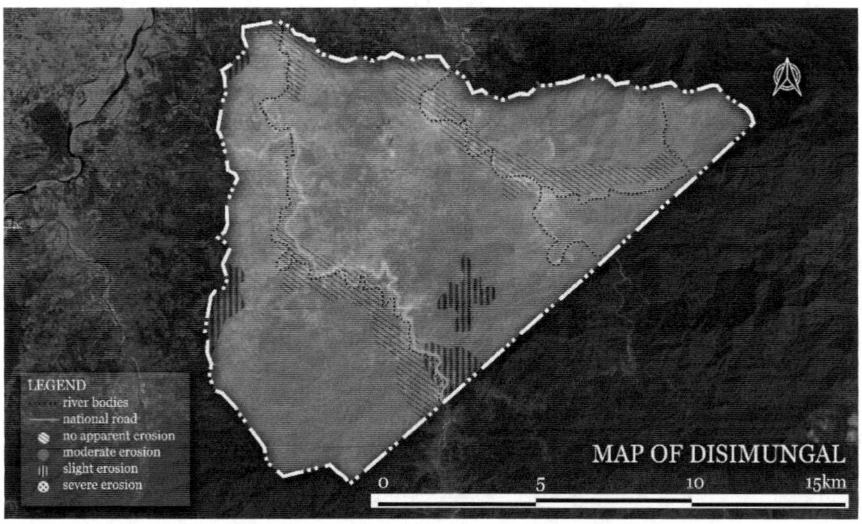

Figure 2 Map of Barangay Disimungal, Quirino Province

Some areas experience slight to severe erosion, while there are also areas with no apparent erosion. These areas with no erosion are usually those along with the river bodies because of the slight slope going to the river. One main river body runs at the west side of the national road, and this is where most Agta communities fish for their livelihood. Generally, the Agta communities are only accessible through the national road.

3.2 Cultural Information

Documenting and analyzing the following cultural practices and beliefs of the community is significant to apply these to achieve cultural preservation in the resettlement.

1) The Agtas practice communal activities due to limited space inside their houses because they are accustomed to doing chores outside with their neighbours.
2) The public has the rights towards the whole land included in their ancestral domains.
3) Agtas plant crops for sustenance, raw materials, and medicinal purposes.

4) Their swidden farmland is their primary source of income, food, and medicine. Mostly Agtas plant beans, banana, ginger, cassava, ube, and sweet potato.
5) Agtas through history relocate from places to places where they have access to rivers. They fish as their alternative source of food and income if their crops are dull.
6) They practice animal farming. Currently, they own carabaos, wild boars, and chicken, which they also raise, sell, or eat.
7) They sell their products as one of their sources of income. They also make a living from handmade products. One particular culture they are skilful at is weaving crafts.
8) As part of their culture as hunters, seeing a broad vantage point from their homes assures them security as they can see their surroundings from their very own homes.
9) There is no definite pattern of housing layout. Agtas' housing pattern is usually dictated by the slope in the site.

4. Analysis

4.1 Site Selection Analysis

The main drive behind this research is prompted by their safety concern in their homes because of mass wasting disasters. With this, it is necessary to select the most effective and safe site for the resettlement.[5] The selection shall analyze the following criteria:

1) Location - The location should have the necessities of the community for their everyday living. The land area should cater all needed infrastructures. It should allow a room for expansion to anticipate the growing population of the community.
2) Topography – The site that shall be selected shall not have brutal slopes that it will introduce mass wasting disasters.
3) Soil Characteristics and Condition - The soil should nurture vegetation. With this, the soil shall be stable and prevent possible mass wasting disasters.
4) Accessibility of Utility, Service Needs, and Transportation - The site should be accessible to modes of transportation, especially by emergency response vehicles.
5) Public Acceptance - The community should wholeheartedly accept and feel at home and secured at the site where they will be resettled.

The site selection was made among the different areas within the Agtas' ancestral domain in Barangay Disimungal, where an Agta community exists. These are Masog Falls, Tili-tilan, Junuan Falls, and Sitio Pulang Lupa.

By conducting a series of site visits, analyses were conducted on the mentioned sites and, an evaluation shall determine the ideal resettlement site for the Agtas by rating each set criteria from 1 to 5 with a corresponding percentage equivalent. With this, the result of the evaluation of the sites is reflected on the table below.

Table 1 Site Selection Evaluation

CRITERIA	%	SITES							
		Masog Falls		**Tili-tilan**		**Junuan Falls**		**Sitio Pulang Lupa**	
		Rate	**%**	**Rate**	**%**	**Rate**	**%**	**Rate**	**%**
Location	**25**	3	15	3	15	3	15	5	25
Topography	**20**	2	8	3	12	2	8	4	16
Soil	**20**	4	16	2	8	4	16	4	16
Accessibility	**15**	4	12	3	9	2	6	5	15
Acceptance	**20**	3	12	2	8	3	12	5	20
Total	**100**	**63**		**52**		**57**		**92**	

As seen on the table above, Sitio Pulang Lupa earned the highest rating as the Agtas' resettlement area. Its location provides areas for their livelihood such as swidden farming and fishing. Moreover, the land area is possible for expansion. Sitio Pulang Lupa has a slope that ranges from 20 per cent to 40 per cent. It requires minimal earthmoving interventions for it to be buildable by infrastructures and houses. The area is not highly susceptible to landslide than the other sites with steep slopes ranging from 60 per cent. From the name itself, Sitio Pulang (red) Lupa, the site's soil has a distinct characteristic making it red. It appears to be stable, buildable, and minimal soil erosion is experienced. Moreover, the area is accessible immediately through the national road. It has road paths inside the community beneficial for transport or emergency response vehicles. Most importantly, according to conducted interviews, Sitio Pulang Lupa is the chosen location for resettlement by the Agtas in Mount Kalbo. It is ideal because the location is resided by Agtas solely. They consider Sitio Pulang Lupa as the "core of the Agta community."

4.2 Comparative Micro-Site Assessment

To analyze better the need for the Agta Community to resettle to Sitio Pulang Lupa from their previous area in Mount Kalbo, site inventory and analysis are provided to determine the difference between the two sites. These will provide information such as threats and weaknesses of the previous area that may be prevented and solved for the proposed resettlement area in Sitio Pulang Lupa.

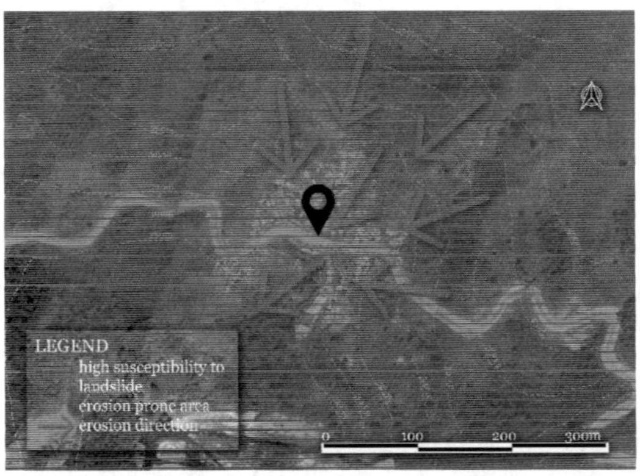

Figure 3 Hazard Analysis of Mount Kalbo

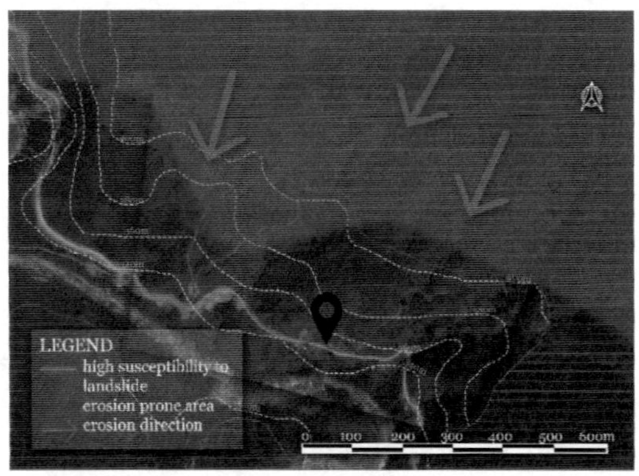

Figure 4 Hazard Analysis of Sitio Pulang Lupa

As seen in Figure 3, most areas in Mount Kalbo is prone to soil erosion, and the direction of the erosion is towards their houses. Because the majority of the site in Mount Kalbo has 31 to 50 per cent slope, the area is highly susceptible to landslides. On the other hand, the proposed resettlement area, Sitio Pulang Lupa has few areas prone to severe soil erosion. Most areas have 0 to 30 percent slope, which makes it more ideal and less hazardous.

5. Synthesis

To combine the findings and analyses in one development, their weaving culture will be used to produce a design for the resettlement of the Agta Community. *Habin*, or weft in English, are lateral threads used in weaving inserted over-and-under the warp. It is responsible for keeping intact all the materials used in the process to produce a new product.

Like a Habin that ties together its materials to serve its purpose in weaving, the Agta Community also does most of their activities together as a community. Their culture of working together, if channeled in this development, can enrich the people of Agta and strengthen them as a community.

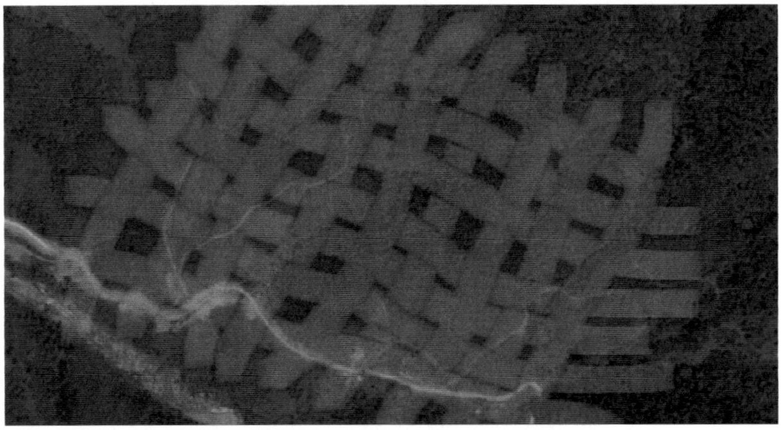

Figure 5 Weaving used as the Design Concept

The general form for the whole development, as seen in Figure 5, reflects a representation of interlocking threads in weaving that follows the slope of the site. The pathways also reflect the form by having vertical and horizontal connections within site.

5.1 Site Development Plan
By consolidating the data gathered from the site analysis, interviews and focus group discussions with the Agta community, specific spaces that shall cater to the community's needs were programmed in the site.

The spaces in Figure 6 reflect the resettlement plan of the Agta Community in Sitio Pulang Lupa. Housing is the primary concern of the Agtas. The Agta population in Mount Kalbo and Sitio Pulang Lupa combined comprises of 100 families. The usual house in the present Agta community in Mount Kalbo is smaller than a standard house and accommodates over three families in one roof. To address this inadequacy, the proposed housing provided 145 houses with 42 square meters to provide a 1:1 house to every family with enough space for outdoor activities as part of their culture. The resettlement also provides additional space for expansion, and these should have access to significant facilities in the community. Such spaces include a cultural area, where they can do their cultural activities and handmade activities such as weaving, and a wet and dry trade market. The development also fulfils the Agta community's wish of a school that will encourage their children to continue their studies. Moreover, a livestock farm, and a health centre was proposed. To further enrich their culture, the community is also provided with communal facilities, such as outdoor communal laundry areas, gymnasium and open communal spaces. Lastly, to support their primary source of livelihood, their swidden farmland is preserved for the community. To address soil stability in the area, contour farming will be applied along with their houses to serve as their garden. Water for the whole community will be provided through a water tank that connects to every communal faucet for every five households. The water in the tank is driven from a spring at the top of the mountain where the site is located.

IOP Conf. Series: Earth and Environmental Science **879** (2021) 012007 doi:10.1088/1755-1315/879/1/012007

Figure 6 Site Development Plan of the Resettlement Area in Sitio Pulang Lupa

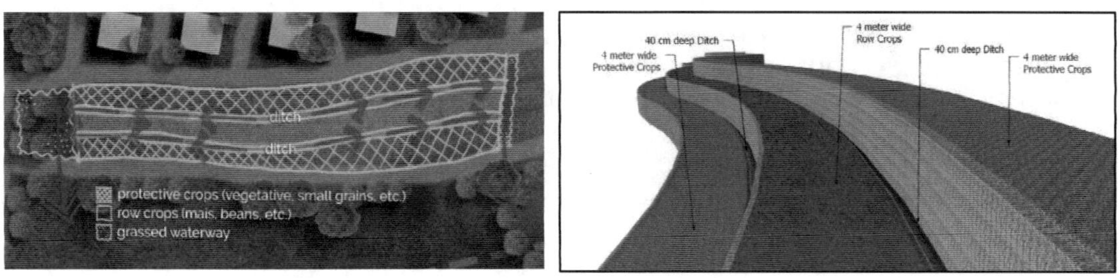

Figure 7 Blow-Up Details of Contour Planting

To address the slopes in Sitio Pulang Lupa, planting across the contour shall help mitigate soil erosion as seen in Figure 7. The contour planting shall follow a combination of strip-cropping and ditch and barrier system where strips of protective crops are alternated with row crops. Strips are placed with ditches in between that catches water. This heightens the effectivity of contour planting in decreasing water runoff speed.[6] Moreover, every home has access to the contour farm where they can plant edible plants or medicinal herbs. A list of suggested protective and row crops are listed in Table 2.

Table 2 Contour Planting Plant Examples

Protective or Vegetative Crops	
Scientific Name	Common Name
Brassica rapa subsp. chinensis	Pechay
Colocasia esculenta	Gabi
Gliricidia sepium	Madre de Cacao
Ipomoea batatas	Kamote
Leucaena leucocephala	Ipil - ipil
Saccharum officinarum	Sugarcane
Zingiber officinale	Ginger
Row Crops	
Scientific Name	Common Name
Chrysopogon zizanioides	Vetiver Grass
Fagopyrum esculentum	Buckwheat
Glycine max	Soybean
Lablab purpureus	Abitsuelas
Lactuca sativa	Lettuce
Pachyrhizus erosus	Singkamas
Phaseolus vulgaris	Common Beans
Psophocarpus tetragonolobus	Winged Bean
Vigna unguiculata	Sitaw
Zea mays	Mais

Plants reflected in Table 2 are also based on local plants and trees found in Sitio Pulang Lupa. Included also on the table are plants familiar to Agtas as plants for their food or medicine. Aside from these listed plants can serve as soil protection, Agtas can maximize the contour farm to practice their culture of planting.

5.2 Disaster Risk Reduction Management Plan – Transitioning and Simulation

To ensure the security of Agtas and promote their self-independence, a disaster management plan was also developed to prepare them from disasters in the future, especially mass wasting.

The spaces will transition into six stations for evacuation, as seen in Figure 8. The whole community will be divided into five clusters that will be provided with stations for the resident's safety, namely handling, emergency, storage, hygienic, and receiving stations. Residents within their cluster will be heading to their designated assembly area where refuge landscapes are located as reflected in Figure 9 (legends A, B, C, D, and E) in case of evacuation.

Routes towards refuge landscapes are mapped as seen in Figure 9. In case the emergency intensifies, all residents should be gathered at the central camp, refuge landscape B, to ensure everyone's safety. According to the Japan Institute for Fire Safety and Disaster Preparedness, the average speed of refugees during an evacuation is 1.1 meter per second.[7] This will measure the time to reach the refuge landscape from the farthest point of the connecting path walks. The routes are mapped in which no path will be overcrowded to lessen panic. Residents shall be educated by these evacuation protocols through training, seminars, and visual aids. It is assumed that with residents who are aware and equipped with the knowledge of their distance to the safe zone and how they shall respond to the emergency, they will be more alert and all individuals from different houses can reach the safe zone in time.

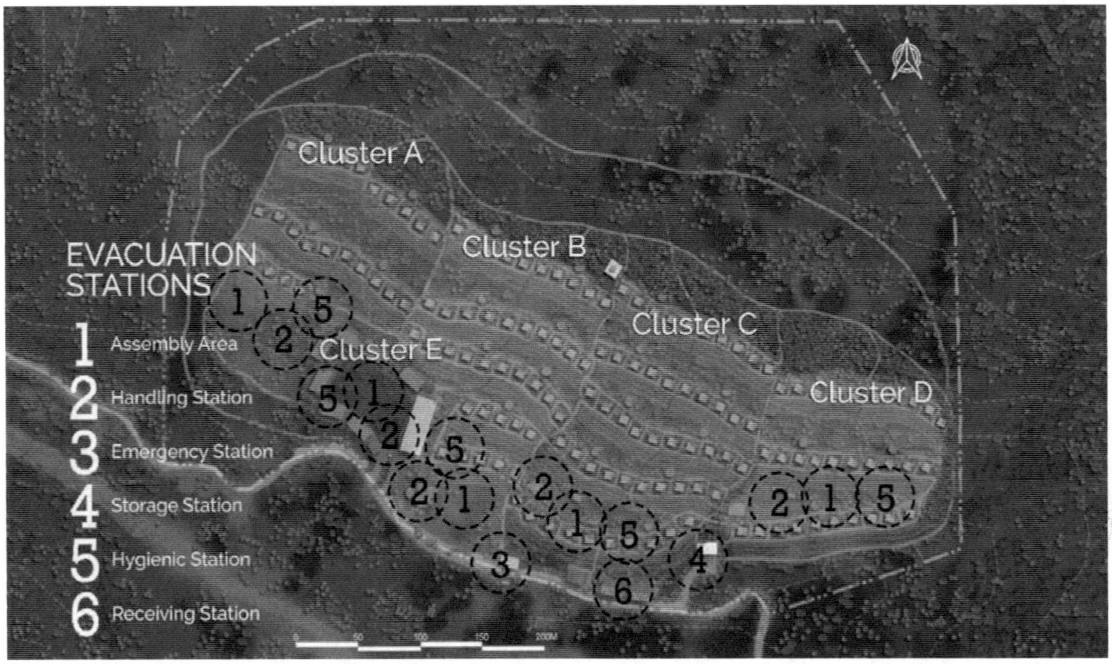

Figure 8 Transition Map of the Resettlement for Disaster Management

Figure 9 Simulation Map of the Resettlement for Disaster Management

Table 3 Evacuation Routes Distance and Time

Evacuation Route	*Total Seconds*	*Length in Meters*
Route A1	164 seconds	180 meters
Route A2	160 seconds	175 meters
Route A3	119 seconds	130 meters
Route A4	110 seconds	120 meters
Route A5	87 seconds	95 meters
Route B1	228 seconds	250 meters
Route B2	210 seconds	230 meters
Route B3	210 seconds	230 meters
Route B4	160 seconds	175 meters
Route B5	173 seconds	190 meters
Route B6	123 seconds	135 meters
Route B7	110 seconds	120 meters
Route B8	73 seconds	80 meters
Route C1	196 seconds	215 meters
Route C2	205 seconds	225 meters
Route C3	150 seconds	165 meters
Route C4	164 seconds	180 meters
Route C5	119 seconds	130 meters
Route C6	128 seconds	140 meters
Route C7	91 seconds	100 meters
Route C8	73 seconds	80 meters
Route D1	128 seconds	140 meters
Route D2	119 seconds	130 meters
Route D3	78 seconds	85 meters
Route D4	96 seconds	105 meters

As calculated in Table 3, the longest route distance to reach its designated refuge landscape is Route B1. Figure 9 shows that it is located at a high elevation of the site, and it measures 250 meters with 228 seconds to travel. Nevertheless, the most efficient way possible to reach the destination was determined through mapping a perpendicular route from the starting point to the refuge landscape.[7] With this, an individual taking any route can reach the safe zone in an estimated time of no less than 5 minutes. Each safe zone or refuge landscape can hold more than the Agtas' population in every cluster, which is necessary to provide every individual with space for their needs.

Table 4 Carrying Capacity of Refuge Landscapes

Evacuation Area	*Area*	*Carrying Capacity*
Refuge Landscape A	5 100 square meters	3 642 individuals
Refuge Landscape B (central camp)	4 870 square meters	3 478 individuals
Refuge Landscape C	2 670 square meters	1 907 individuals
Refuge Landscape D	3 200 square meters	2 285 individuals
Refuge Landscape E	2 590 square meters	1 850 individuals

Table 4 shows the area and occupant capacity of each refuge landscape based on the Fire Code of the Philippines. A less concentrated capacity will be 1.4 square meters per person. It shows that the area of the refuge landscapes will be enough to carry the whole population of 1,200 Agtas safely in case of disasters.

6. Conclusion

The Agta Community highly expects from their resettlement for the safety of their families and homes. By analyzing the soil and topography conditions in the sites selected, the research has arrived at Sitio Pulang Lupa as the optimal option. It can improve the living conditions of the whole community and create a well-prepared community from mass wasting disasters and a sustainable environment where they can grow their community.

The effect of a thorough and well-planned landscape development is evident in the results found in the methods used in this study. It is concluded that soil conditions contribute significantly in community development as a stable and nurtured ground creates a safe environment for the community. More importantly, landscape elements mostly take part in mitigating the effects of natural disasters such as mass wasting catastrophes. In addition to this, the impact of landscape architecture can channel the culture of the Agtas. The spaces proposed in the development is in line with their interests that they will be encouraged to stay permanently in this proposed resettlement and nurture their community.

Landscape architecture takes part in creating a safe and effective environment when disaster comes. By providing easily accessible and functional refuge landscapes, the Agtas are well-prepared and educated that their worry of safety and security they used to have from their previous settlement will no longer appear to mind.

With these landscape architectural interventions done, the Agtas will develop environmental and community sustainability that enriches future generations.

References

[1] Guia J 2012 Quirino: The forest heartland of Cagayan Valley [Internet]. *Vigattin Tourism.* [cited 2019 Sep 10]. Available from: https://www.vigattintourism.com/tourism/articles/Quirino-The-Forest-Heartland-Of-Cagayan-Valley
[2] Nelson S 2018 Mass movements and mass movement processes [Internet]. *Nat Disasters.* [cited 2020 Mar 4]. Available from: https://www.tulane.edu/~sanelson/Natural_Disasters/masswastproc.htm
[3] Parco B 2019 Mild quake hits Quirino [Internet]. *GMA News Online.* [cited 2019 Dec 12]. Available from: https://www.gmanetwork.com/news/news/regions/715375/mild-quake-hits-quirino/story/?utm_source=GMANews&utm_medium=Twitter
[4] The Agta tribe of the Sierra Madre [Internet]. *Ethnic Groups of the Philippines.* 2016. [cited 2020 Feb 20]. Available from: http://www.ethnicgroupsphilippines.com/2016/12/19/the-agta-tribe-of-the-sierra-madre/
[5] Arnuco D, Caasi J, Loon D and Sumang R 2013 Site selection criteria [Internet]. *Slideshare.* [cited 2020 Feb 13]. Available from: https://www.slideshare.net/dabdabeedabs/site-selection-criteria
[6] Leonard D 1986 Techniques for combatting soil erosion. In: *Soil, Crops and Fertilizer Use* [Internet]. 4th Editio. South Carolina. [cited 2020 Jan 18]. Available from: http://www.nzdl.org/gsdlmod?e=d-00000-00---off-0hdl--00-0----0-10-0---0---0direct-10---4-------0-1l--11-en-50---20-about---00-0-1-00-0--4----0-0-11-10-0utfZz-8-00&cl=CL1.16&d=HASH412cd503b5262205ac14c6.5&gt=1
[7] Ashar F, Amaratunga D and Haigh R 2017 Tsunami evacuation routes using network analysis: A case study in Padang. Procedia Eng [Internet]. 2018;212(2017):109–16. [cited 2019 Mar 29]. Available from: https://doi.org/10.1016/j.proeng.2018.01.015

Park users' perception and preference of public park in Bogor City post Large Scale Social Restriction (LSSR)

A F Fithriyah[1], N Nasrullah[2] and B Sulistyantara[2]

[1] Graduate Student, Department of Landscape Architecture, Faculty of Agriculture, IPB University, Bogor, Indonesia

[2] Graduate Student, Department of Landscape Architecture, Faculty of Agriculture, IPB University, Bogor, Indonesia

E-mail: ff.aliya@yahoo.com

Abstract. On 31st March 2020, Indonesia pulls out a large scale social restriction (LSSR) policy to depress the spread of Covid19. The regulation leads to the lay-off of schools, workplaces, worship places, and public facilities, including public parks. Despite the time, it is only a matter of time until the new normal and reopening of all public facilities. For that, public parks as public facilities should be prepared in a way before it is reopened so as not to worsen the situation or even create a new pandemic wave. It is an excellent opportunity to build better by applying health protocol for a healthier future, and transforming the future sustainable landscape. This research is studying two public parks in Bogor City, a satellite city of the megapolitan Jakarta area, which affected the most by Covid19, i.e.,, Sempur Park and Kencana Park. This research aims to explore users' perceptions and preferences of using public parks to plan health protocol for public parks in the new normal era. Data collected through an online questionnaire survey. As much as 192 responses are analyzed. Data and instruments are valid and reliable, with an overall Cronbach alpha value of 0.896. Results showed that the perception of park use pre-LSSR and post-LSSR was significantly different. The trend of park use was similar but at different levels. This indicates that the motivation and willingness to visit and use public parks is decreasing, therefore applying health protocol is a necessity in the future. This study also found that existing parks perceived to be quite appropriate in accordance with health protocols and need to be moderately modified. Another finding was that the considered most required health protocols application were procurement of handwashing facility, park benches placed at least 1 meter apart, and solitary park bench. It is elicited that despite the type of park, the visitors' preference was the same.

Keywords: health protocol, new normal, parks, perception, preference

1. Introduction

The Covid-19 pandemic is a global threat that affects health, economic, and social aspect. On 30th January, WHO declared the Covid-19 outbreak as a public health emergency of international [1]. With thousands of cases reported all over the world within one month, and still increasing, the declaration announced along with various recommendations regarding early detection of infection, isolating and treating infected patients, contact tracing, and social distancing measures that correspond to the level of risk in each country, with hope to slow down and hopefully limit the virus spread. To contain the COVID-19 epidemic, one of the non-pharmacological control measures is reducing the transmission rate in the population through (physical) social distancing [2]. Social distancing measures were initially introduced by the World Health Organization in response to the exponentially global coronavirus spread. The measures were general and were subject to governments' interpretations, so they may respond to the level of risk in each country or even local neighborhood. These interpretations range from minor physical distancing measures to lockdown. On 31st March 2020, Indonesia pulls out a large-scale social restriction (LSSR) policy to depress the spread of Covid19, which leads to the closures of schools, workplaces, worship places, and public facilities, including public parks. Social distancing is widely recognized as the most effective approach to depress COVID-19 cases [3][4], yet it has simultaneously disrupted the daily lives of entire humankind wherever it was implemented [5].

During these extraordinary circumstances of social distancing, urban nature offers resilience for maintaining well-being [5] by providing opportunities to escape from household confinement and enjoy well-being positivity [6], provide a connection with nature and the outside world [7], also maintain social relationships even with limited distance [8]. Cities' nature areas are proven repeatedly played a crucial security role in times of crisis [9][10]. One example is public park. Public park is a free accessed place which lets people be in touch with nature and able to do many various activities at the place. Parks offer a unique setting within the urban landscape, present opportunities for nature enjoyment, physical activity, social interaction, and escape [11]. Park visitation as a recreational activity is a form of nature-based tourism, by definition, is undertaken in a natural setting where either the recreational activity or visitor experience depends on, or is enhanced by, the natural environment [12]. Physical elements and activities at public open space significantly proven to offer many benefits to quality of life such as health, social interaction, and economic value and its quality relate to the usability depend on people needs and perception which if not meet, public open space becomes useless and unsuccessful [13].

Despite the time, it is only a matter of time until the new normal and reopening of all public facilities. For that, public parks as public facilities should be prepared in a way before it is reopened so as not to worsen the situation or even create a new pandemic wave, as the likelihood of new pandemic outbreaks is continuously on the rise, partly because humanity is continuing to urbanize [5]. Planners may see this as an excellent opportunity to build better by applying health protocol for a healthier future and as a transformation for the future sustainable landscape.

Perception provides an essential basis for the planning and management of green open spaces including parks without exception [2]. However, studies showed that human perception of the environment is subjective and personal. It differs from person to person [14]. Thus the benefits derived from urban green spaces and its objective properties are interpreted individually [15]. Attributes such as safety, aesthetics, amenities, maintenance, and proximity are essential for encouraging park use [16]. Therefore, the analysis of perceptual information on green spaces is necessary. This research aims to explore users' perceptions and preferences of using public parks in order to plan health protocol for public parks in the new normal era. Objectives of this study are: 1) discover distinction of public parks use in pre-LSSR and post-LSSR period, 2) identify users' perception of existing parks regarding health protocol, and 3) acquire preference of health protocol application in public parks.

2. Methods

The authors studied two parks in Bogor City, West Java, Indonesia, named Sempur Park and Kencana Park. Kencana Park is a general public park; meanwhile Sempur Park is sports thematic public park. Both are very well-known parks that many people visit from all over Bogor, also from outside the city, thus chosen for this study. Sempur Park is located in the center of Bogor City, next to Bogor Botanical Garden, while Kencana Park is located 350 meters north of Sempur Park. With 1,25 hectares of area, Sempur Park offers running tracks, grass fields, children's playground, reading corner, basketball court, rock climbing, skateboard arena, and culinary area. Kencana Park has a 4.795,56 m^2 area with a leafy and calm ambiance, which is excellent for relaxing and sightseeing. On the other hand, Kencana Park also offers a culinary area which serves much traditional foods of Bogor. Hence many visitors visit this park for its food. Activities in Sempur Park are mainly active, while activities in Kencana Park are generally passive.

Table 1 Question items perception and preference variables

Variables	Question Items	
	Sempur Park	**Kencana Park**
Perception of Park Use	To rest or relax. Connection with nature. Close to living quarters. To see the iconic structure. Work or work reasons. To socialize (get together, chat, meet new people). To taste various culinary delights. For sports or physical exercise. To see a festival or event. To shop. To find joy. Because of the view. To show it off on social media. Just looking around. For recreation with family. On the way to another place. For study or academic reasons. Celebration of an important day. Engage in hobbies.	
Perception of Existing Health Protocol in Public Parks	Hygiene facilities Pedestrian Park benches Grass fields Running track Basketball court Plaza in Expression park Sitting area in Expression park Skateboard arena. Woodhouse in Kaulinan park. Park benches in Kaulinan park Playground in Kaulinan park Canteen area	Hygiene facilities Pedestrian Park benches Spaciousness Vegetation arrangements Sitting area Plaza
Preference of Health Protocol Application for Public Parks	Procurement of hand washing facility Solitary park bench Park benches placed at least 1 meter apart Vegetation hedge as space/seat dividers Use of less dense vegetation to maximize sunlight One entrance - one exit Usable running track reduction	Procurement of hand washing facility Solitary park bench Park benches placed at least 1 meter apart Vegetation hedge as space/seat dividers Use of less dense vegetation to maximize sunlight One entrance - one exit Wider pedestrian (at least 2 meters)

This study was conducted through several steps namely preparation, data collection through a questionnaire survey, and statistical analysis. Before the analysis is carried out, a

IOP Conf. Series: Earth and Environmental Science **879** (2021) 012008 doi:10.1088/1755-1315/879/1/012008

series of tests are carried out on the data to ensure that the data obtained is good, robust, and sufficient for further statistical analysis. The analysis used in this research are descriptive analysis, correlation analysis using Spearman correlation test, and comparative analysis using Mann Whitney test. Mann Whitney and Spearman analysis were conducted to comprehend park users' perception and preference thoroughly.

The questionnaire survey was conducted in June-July 2020. The questionnaire was spread online due to the impossibility of face-to-face interviews regarding the pandemic. This potential bias should be minded concerning study results. Hence, the questionnaire was spread widely through various social media, messages, and mouth-to-mouth to high schools, workplaces communities, neighborhood communities, universities, and many other communities within Bogor city. They were thus reaching an expansive and diverse range of the city's society. Respondents were selected using a snowball sampling technique. Respondents are adult citizens of Bogor who have visited both parks and willing to fill the questionnaire. The questionnaire consists of demography variables (i.e.,, age, gender, domicile residence, marital status, educational background, occupation), visitation characteristics (i.e.,, distance, companion, frequency, length of stay, time, transportation), also perception and preference variables. Demography and visitation characteristics questions were asked in multiple-choice questions, while perception and preference questions were asked in 5-scale Likert. The same set of park use perception questions were asked for the pre-LSSR and post-LSSR periods to distinguish the differences. Questions of perception and preference variables are listed in Table 1.

Perception and preference about health protocol were asked through a questionnaire with representing photos. Question items in the perception variable are arranged based on the existing park facilities in each park. The results were assessed by the mean value of each question item and categorized into five assessments, as shown in Table 2. Question items in the preference variable asked were health protocols which possible to apply in the park, arranged according to the existing facilities in each park. Health protocol preferences asked in this study are in the respects of landscape architecture perspective, which are limited to applications in spatial planning and landscape elements i.e., hardscape and softscape elements.

Table 2 Categories of perception and action assessment

Mean Value	Rank	Action
$1{,}00 \leq M \leq 1{,}80$	Very inappropriate	Total modification
$1{,}80 < M \leq 2{,}60$	Inappropriate	Major modification
$2{,}60 < M \leq 3{,}40$	Quite appropriate	Moderate modification
$3{,}40 < M \leq 4{,}20$	Appropriate	Minor modification
$4{,}20 < M \leq 5{,}00$	Very appropriate	No modification

This survey managed to gather 231 responses in total with 192 completed responses, thus the margin of error level is 7,04%, and the confidence level is 95%. All respondents were Bogor citizens. The result showed that the research instrument is valid and reliable with a Cronbach alpha value of 0.896. Comparison analysis was conducted using the Mann Whitney test with a threshold of Asymp. Sig < 0.05. Correlation analysis was conducted to explore the correlation between perception and preference variables with demography and visitation characteristics variables using Spearman analysis with a threshold of Sig. < 0.05.

3. Results and discussions

3.1. Demography and visitation characteristics

Of 192 respondents, as many as 87 respondents were male (45.3%) and 105 people were female (54.7%). A total of 13 people were under 20 years old (6.8%), 95 people aged 20-29 years (49.5%), 29 people aged 30-39 years (15.1%), 27 people aged 40-49 years (14.1%), 26 people aged 50-59 years (13.5%), and 1% aged 60 years or more. Most of respondents were married and had children (52.6%), while 35.9% were unmarried, 9.9% married but did not have children, and others for 1.6%. Respondents' educational background were varying. As much as 2.1% graduated from junior high school / equivalent, 12.5% were high school graduates / equivalent, more than half of the respondents (59.4%) had diploma or bachelor degree, 21.4% were master graduates, and 4.7% of respondents had doctoral degree. Most of the respondents were private employees (37%). Others were students (15.6%), civil servants (14.6%), self-employed (13.5%), housewives (10.9%), educational staff (3.1%), freelance workers or not working (2.1%), and others (3.1%). The demographics of respondents can be seen in Figure 1. It can be concluded that visitors are generally women aged 20-29 who are married and have children, work as private employee with a diploma or bachelor degree.

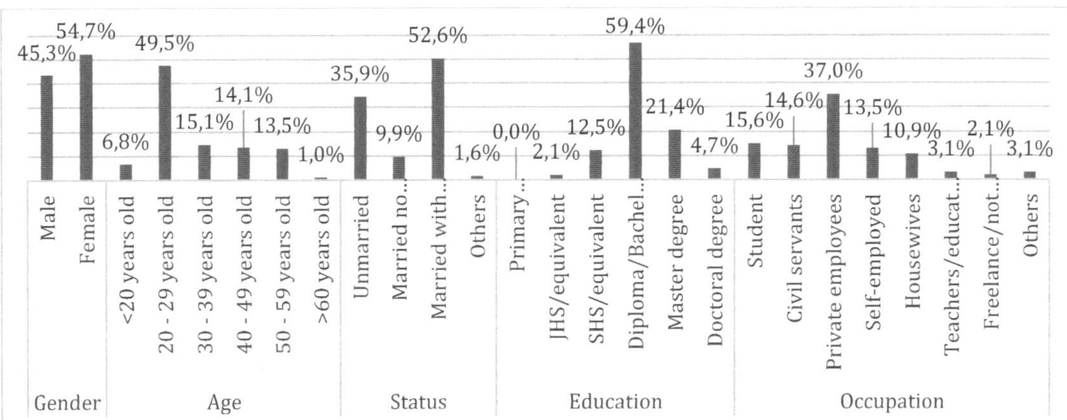

Figure 1 Respondents' demography

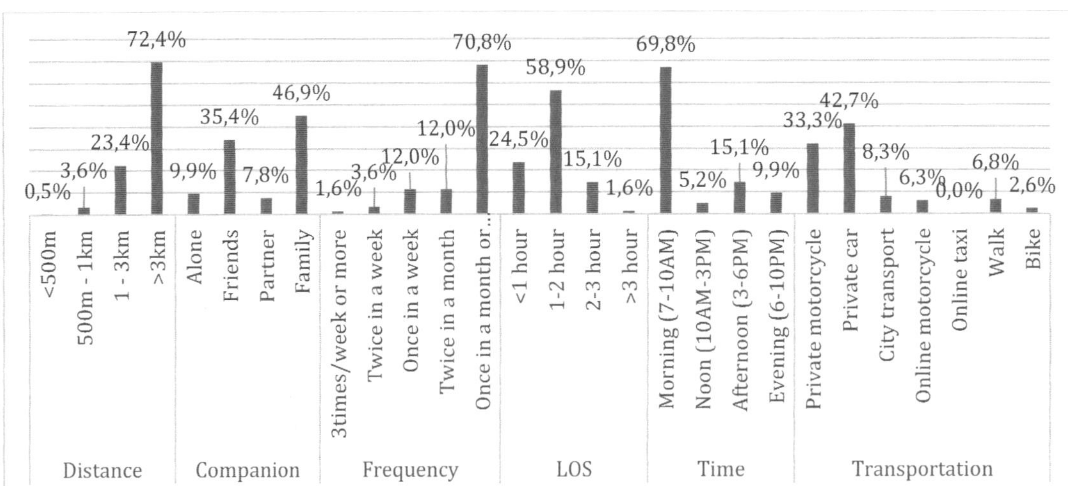

Figure 2 Visitation characteristics of Sempur Park

In general, visitors of Sempur Park were domiciled in areas more than 3 km from the park. Visitors usually came by a private car with their family in the morning, between 07.00 - 10.00 AM. Visitors regularly visited Sempur Park once a month or less and spent a 1-2 hours in the park. Visitation characteristics of Sempur Park is shown in Figure 2. Similar to Sempur Park, visitors of Kencana Park were also live in areas more than 3 kilometers from Kencana Park, usually visited the park by driving private car along with the family in the morning, between 07.00 – 10.00 AM once a month or less. However, the LOS of Kencana Park is shorter. Visitors mostly spent less than an hour in this park. This is probably because the size of Kencana Park is smaller than Sempur Park, and because Sempur Park offers more complete facilities than Kencana Park such as sport facilities and children playground. Visitation characteristics of Kencana Park is shown in Figure 3.

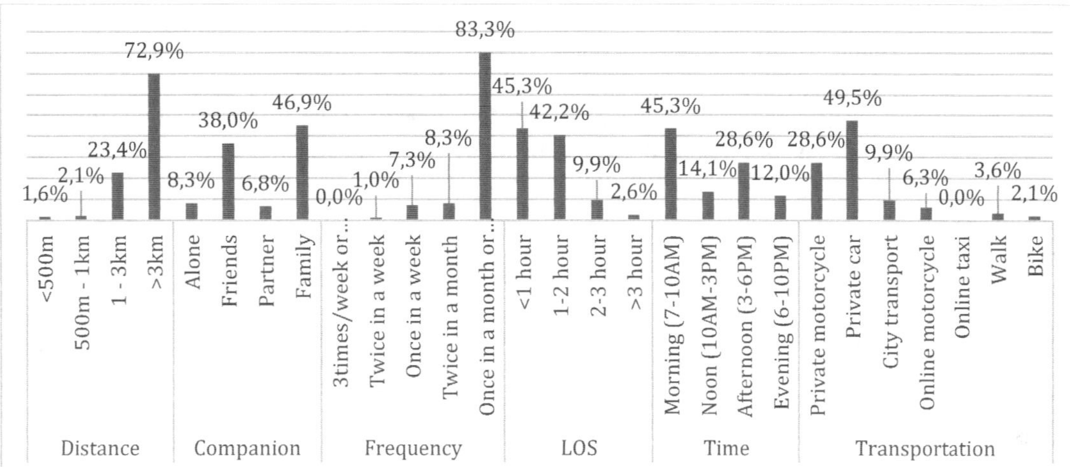

Figure 3 Visitation characteristics of Kencana Park

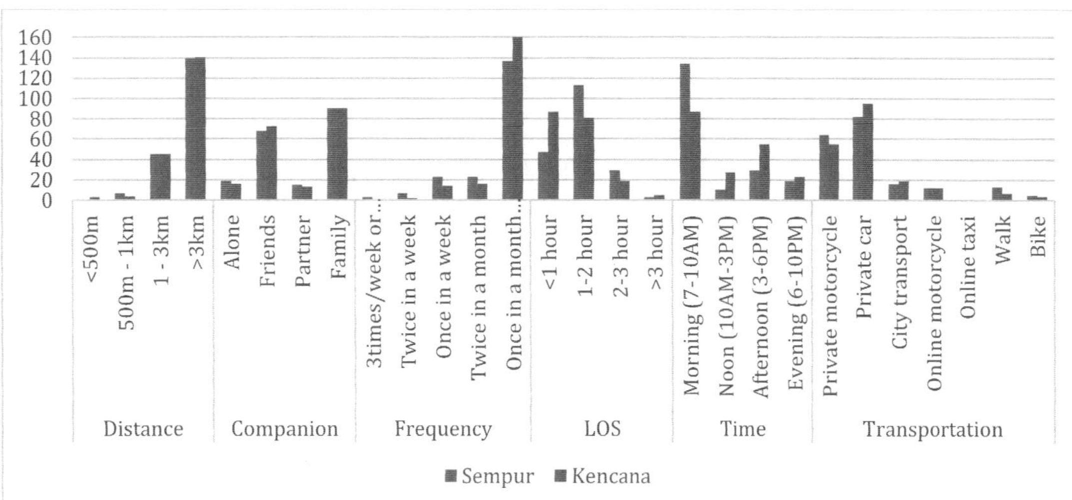

Figure 4 Visitation characteristics of Sempur and Kencana Park.

Further, this study would like to distinguish visitation characteristics of both parks. Figure 4 shows the comparison between the two parks. Mann Whitney analysis was conducted with results shown in Table 3. Results showed than visitation characteristics between Sempur Park and Kencana Park were significantly different in frequency, LOS, and time visit. This may be due to different function of the two parks. Kencana Park is general park while Sempur Park is sports thematic park. Sempur Park often used for jogging therefore morning intensity is high. Moreover, usually flea market is held every weekend morning, thus many visitors came.

3.2. Perception of park use

Pre-LSSR, users usually visited Sempur Park for physical exercise (M = 4.12 SD = 0.819), to found joy (M = 3.73 SD = 0.868), and gastronomic reasons (M = 3.71 SD = 1.043). After the LSSR, the user's motivations to visit Sempur Park were for physical exercise (M = 3.69 SD = 1.041), to connect with nature (M = 3.31 SD = 1.114), and to enjoy the scenery (M = 3.24 SD = 1.067). For Kencana Park, users visited Kencana Parkpre-LSSR mainly for gastronomic reasons (M = 3.92 SD = 0.950), to rest or relax (M = 3.44 SD = 1.091), and found joy (M = 3.44 SD = 1.037). Post-LSSR, users stated that they might visit Kencana Park for gastronomic reasons (M = 3.53 SD = 1.121), to find joy (M = 3.10 SD = 1.153), and to rest or relax (M = 3.04 SD = 1.186). The Mann Whitney test was conducted to determine differences in perceptions of park use pre-LSSR and post-LSSR. The results showed that the perception of public park use in general differed that significantly ($p = 0.003$ $\alpha = 0.05$). Comparison test was also carried out for each park, namely Sempur Park and Kencana Park. The study showed that perceptions of the use of Sempur Park before and after LSSR ($p = 0.000$ $\alpha = 0.05$) and Kencana Park before and after LSSR ($p = 0.027$ $\alpha = 0.05$) also differed significantly, as shown in Table 4. This proves that Covid19 changes perception of public parks use. It was apparent the trends for pre-LSSR and post-LSSR were similar, but at different levels. This indicates that the people's motivation and willingness to visit and use public parks is decreasing, despite the prohibition of using public places by the government, therefore applying health protocol on the design is a necessity.

Table 3 Comparison analysis of visitation characteristics

	Distance	Companion	Frequency*	LOS*	Time Visit*	Transporta tion
Mann-Whitney U	18322,0	18384,5	15998,0	14696,0	14367,0	18165,5
Wilcoxon W	36850,0	36912,5	34526,0	33224,0	32895,0	36693,5
Z	-0,130	-0,047	-3,046	-3,778	-4,188	-0,263
Asymp. Sig. (2-tailed)	0,896	0,962	0,002	0,000	0,000	0,793

Grouping Variable: Park
*Significantly different at 95% level

Table 4 Comparison analysis of park use perception pre-LSSR and post-LSSR

	Sempur Park	Kencana Park	Public Park
Mann-Whitney U	14500,000	16022,500	15169,500
Wilcoxon W	33028,000	34550,500	33697,500
Z	-3,616	-2,216	-3,000
Asymp. Sig. (2-tailed) / (p)	0,000	0,027	0,003

a. Grouping Variable: Pre_Post

Spearman rho's correlation coefficient was conducted to identify factors that affect perception, by assess the relationship between perception variables, pre-LSSR and post-

LSSR, with demographic variables and visitation characteristic variables. Results shown in Table 5. The results indicated that prior to LSSR, the older the visitor is, married, the longer the LOS, the lower the frequency of visits to the park, and the male gender, had higher motivation and a better perception of park use. While after the LSSR, the older the visitors, the closer home range, the lower the frequency of visits to the park, and the male gender is more likely to visit public parks in LSSR period. This is due to the decreasing people's motivation to visit public places, and probably correlates with the LSSR regulation which lowers people's mobility thus only people in the neighborhood will visit, either to stop by or only pass by.

3.3. Perceptions about suitability of health protocols in public parks

In addition to understand the user's perceptions of using public parks post-LSSR, this study also analyzes user's perception of the suitability of existing facilities in the park regarding Covid-19 health protocol. The results showed that visitors perceived the existing Kencana Parkfacilities (M = 3.283 SD = 0.733) as a whole to be slightly better than Sempur Park (M = 3.121 SD = 0.683), although both parks considered in the same category, which is quite appropriate in accordance with health protocols and necessarily modified moderately. Area setting of Kencana Park was more detached while Sempur Park was more open thus allows crowding. The vegetation arrangements (M = 3.54 SD = 0.824) and pedestrians (M = 3.44 SD = 0.872) in Kencana Parkwere deemed appropriate and only minor modification needed to meet health protocols. Meanwhile, cleanliness facilities (M = 3.32 SD = 0.992), park benches (M = 3.20 SD = 1.026), spaciousness (M = 3.18 SD = 0.978), sitting area (M = 3.15 SD = 1.013), and the plaza (M = 3.14 SD = 0.979) is considered quite suitable and needs moderate modification. There were no elements that are considered inappropriate and very inappropriate in Kencana Park.

Table 5 Factors affect perception of park use

	PRE-LSSR			POST-LSSR		
	Correlation Coefficient (r)	Sig. (2-tailed) / (p)	N	Correlation Coefficient (r)	Sig. (2-tailed) / (p)	N
Park	-,157**	0,002	384	-0,083	0,103	384
Age	,186**	0,000	384	,246**	0,000	384
Gender	-,107*	0,035	384	-,106*	0,038	384
Status	,126*	0,013	384	0,061	0,235	384
Education	0,055	0,282	384	0,084	0,100	384
Occupation	-,109*	0,033	384	-,103*	0,043	384
Distance	0,073	0,153	384	,107*	0,036	384
Companion	-0,035	0,494	384	-0,083	0,105	384
Frequency	-,119*	0,019	384	-,120*	0,019	384
LOS	,161**	0,002	384	0,065	0,204	384
Time visit	-0,089	0,082	384	-0,058	0,254	384
Transport	0,086	0,092	384	0,057	0,267	384
Pre	,186**	0,000	384	-,107*	0,035	384
Post	,246**	0,000	384	-,106*	0,038	384

**. Correlation is significant at 0.01 level (2-tailed).
*. Correlation is significant at 0.05 level (2-tailed).

On the other hand, for Sempur Park, visitors perceived that only the grass field (M = 3.65 SD = 0.849) and the running track (M = 3.62 SD = 0.872) were compliant with the health protocol so that only minor modification needed. Meanwhile, pedestrians (M = 3.38 SD = 0.913), basketball court area (M = 3.26 SD = 0.828), hygiene facilities (M = 3.20 SD = 1.004), plazas in expression parks (M = 3.12 SD = 0.887), skateboard arena (M = 3.07 SD = 0.857), sitting area in expression park (M = 3.01 SD = 0.963), park benches in grass field (M = 2.99 SD = 1.021) , park benches in Kaulinan Park (M = 2.92 SD = 0.965), playground in Kaulinan Park (M = 2.87 SD = 0.968), wood house in Kaulinan Park (M = 2.84 SD = 1.003), and canteen area (M = 2.65 SD = 0.965) is categorized as quite appropriate and needs moderate modification.

3.4. Preference of health protocol application in public parks

After acquire user's perceptions of the existing park facilities suitability regarding health protocol, this study also tried to identify what kind of health protocols visitors preferred and desired. This study discovered that preference for health protocol application in both parks has a similar trend. This suggests that despite the type of the park, the preference is the same. Results showed that procurement of hand washing facility is the most preferred (Sempur Park: M = 4.40 SD = 1.211, Kencana Park: M = 4.46 SD = 1.033), followed by park benches placed at least 1 meter apart (Sempur Park: M = 4.20 SD = 1.172, Kencana Park: M = 4.24 SD = 1,017), and solitary park bench (Sempur Park: M = 4.07 SD = 1.180, Kencana Park: M = 4, 16 SD = 1.062). Comparison test was carried out to confirm this similar preference trend between Sempur Park and Kencana Park (α = 0.05). The results are presented in Table 6. The test confirmed the hypothesis that whilst the type of park is different, the preferences are the same.

Table 6 Factors affect perception of park use

	Item 1	Item 2	Item 3	Item 4	Item 5	Item 6	Item 7
Mann-Whitney U	18208,5	17993,5	18123,5	17620,5	16424,5	18089,5	11526,5
Wilcoxon W	36736,5	36521,5	36651,5	36148,5	34952,5	36617,5	30054,5
Z	-0,257	-0,435	-0,313	-0,773	-1,913	-0,334	-6,563
Asymp. Sig. (2-tailed)	0,797	0,663	0,755	0,439	0,056	0,738	0,000

a. Grouping Variable: Park

4. Conclusion

This study adds to the knowledge of perception and preference study of landscape architecture. It also provides subsidiary comprehension for the future sustainable landscape, especially regarding health protocol, which is a currently an essential issue due to the Covid-19 pandemic. This study is useful as direction and consideration in city landscape development. This study found that the park use perception of Sempur Park and Kencana Park between pre-LSSR and post-LSSR have similar trends but differed significantly. This indicates that the people's motivation and willingness to visit and use public parks is decreasing, therefore applying health protocol on the park redesign is a necessity. The results indicated that before LSSR, the older the visitor is, married, the longer the LOS, the lower the frequency of visits to the park, and the male gender had higher motivation and a better perception of park use. While after the LSSR, the older the visitors, the closer the home range, the lower the frequency of visits to the park, and the male gender is more likely to visit public parks in the LSSR period. This study also tried to convey visitors' perception about the suitability of existing park with health protocol. It is discovered that visitors perceived Kencana Park to be slightly better than Sempur Park, although both parks considered to be quite appropriate in accordance with health protocols and necessarily modified moderately. It is also elicited that visitors' preference is the same for any kind of parks and the considered most required health protocols application are procurement of handwashing facility, park benches placed at least 1 meter apart, and a solitary park bench. It is suggested to provide these elements in every park as the primary standard required health protocol for public parks.

References

[1] Salama A M and Brown J 2020 Coronavirus questions that will not go away : interrogating urban and socio-spatial implications of COVID-19 measures [version 1 ; peer review : 3 approved]
[2] Gao S, Rao J, Kang Y, Liang Y and Kruse J 2019 Mapping county-level mobility pattern changes in the United States in response to COVID-19
[3] Tian H, Liu Y, Li Y, Wu C, Chen B, Kraemer M U G et al 2020 An investigation of transmission control measures during the first 50 days of the COVID-19 epidemic in China **642**(May):638–42
[4] Wilder-Smith A and Freedman D O 2020 Isolation , quarantine , social distancing and community containment : pivotal role for old-style public health measures in the novel coronavirus (2019-nCoV) outbreak 1–4
[5] Samuelsson K, Barthel S, Colding J, Macassa G and Giusti M 2020 Urban nature as a source of resilience during social distancing amidst the coronavirus pandemic
[6] Markevych I, Schoierer J, Hartig T, Chudnovsky A, Hystad P, Dzhambov A M et al 2017 Exploring pathways linking greenspace to health : Theoretical and methodological guidance *Environ Res* [Internet] **158**(February):301–17. Available from: http://dx.doi.org/10.1016/j.envres.2017.06.028
[7] Weimann H, Björk J and Håkansson C 2019 Experiences of the Urban Green Local Environment as a Factor for Well-Being among Adults : An Exploratory Qualitative Study in Southern Sweden
[8] Jennings V and Bamkole O 2019 The Relationship between Social Cohesion and Urban Green Space : An Avenue for Health Promotion
[9] Barthel S, Isendahl C, Vis B N, Drescher A, Evans D L and Timmeren A Van 2019 Global urbanization and food production in direct competition for land : Leverage places to mitigate impacts on SDG2 and on the Earth System
[10] Colding J and Barthel S 2013 The potential of ' Urban Green Commons ' in the resilience building of cities *Ecol Econ* [Internet] **86**:156–66 Available from: http://dx.doi.org/10.1016/j.ecolecon.2012.10.016
[11] Hayward D G and Weitzer W H 1984 The public's image of urban parks: past amenity, present ambivalance, uncertain future **8**:243–68
[12] Hewer M, Scott D and Fenech A 2016 Seasonal weather sensitivity , temperature thresholds , and climate change impacts for park visitation 6688(April)
[13] Nasution A D and Zahrah W 2014 Community Perception on Public Open Space and Quality of Life in Medan, Indonesia *Procedia - Soc Behav Sci* [Internet] **153**:585–94 Available from: http://dx.doi.org/10.1016/j.sbspro.2014.10.091
[14] Hernández-morcillo M, Plieninger T and Bieling C 2013 An empirical review of cultural ecosystem service indicators *Ecol Indic* [Internet] **29**:434–44 Available from: http://dx.doi.org/10.1016/j.ecolind.2013.01.013
[15] Kothencz G and Blaschke T 2017 Urban parks: Visitors' perceptions versus spatial indicators *Land use policy* [Internet] **64**:233–44 Available from: http://dx.doi.org/10.1016/j.landusepol.2017.02.012
[16] Mccormack G R, Rock M, Toohey A M and Hignell D 2010 Health & Place Characteristics of urban parks associated with park use and physical activity : A review of qualitative research *Health Place* [Internet] **16**(4):712–26. Available from: http://dx.doi.org/10.1016/j.healthplace.2010.03.003

Waterbirds roosting on gas drilling platforms in Mahakam Delta, East Kalimantan: the Unexpected Impact of Landscape Transformation

A Mardiastuti[1], Y A Mulyani[1], T Sutrisna[1], Y Hidayat[1], A S Widodo[2], I A R Agustin[2], and E Santosa[2]
[1]Department of Forest Resources Conservation and Ecotourism,
Faculty of Forestry and Environment, IPB University, Bogor, Indonesia
[2]PT. Pertamina Hulu Mahakam, Balikpapan, Indonesia

Email: aniipb@indo.net.id; ani_mardiastuti@ipb.ac.id

Abstract. Since 2014 some offshore gas platforms operated by Pertamina Hulu Mahakam near Mahakam Delta has been used by hundreds of egrets as roosting sites, causing nuisance to the platform and hampering the duty of workers/operator. The objective of this paper was to analyse the situation of the waterbirds' nuisance, analyse the possible causes, and provide some recommendations. Field visits were conducted in August 2019 and February 2020 to three unmanned platforms (Ax, D, C) that suffered the most from the bird roost. Birds that infested the platforms were identified as great egret and intermediate egret, both are very similar in appearance and ecological requirements. The maximum number of egrets visiting C-Platform (the worst platforms) was 671 birds. These birds roost at night in the platform and leave the platform in the morning to forage for fish, eel, and shrimps in the wetlands of Mahakam Delta, about 2-4 km from the platform. The root cause of the infestation was believed to be the landscape transformation in the Mahakam Delta. Mangrove forest has been diminished and heavily converted into shrimp-ponds. The egrets' foraging areas became ubiquitous, while the roosting/nesting site greatly decreased, forcing the egrets to roost on gas platforms. In the short-medium term, platforms need to be managed as bird tolerated-zone and bird-free zone (using deterrent), while in the long term, land-use in Delta Mahakam need to be restored.
Keywords: egrets, mangrove, Nypa, Pertamina Hulu Mahakam, wetlands

1. Introduction

Seabirds that used offshore rigs for roosting has been reported in some sites outside Indonesia, for example in Bering Sea [1], North Sea [2] and North-west Atlantic [3]. However, published report on non-marine birds roosting on rigs could not be found. In Indonesia there has been no scientific report related to birds (both seabirds or non-seabirds) that use offshore rigs as part of their habitat.

Offshore riggs (or categorized as 'swamp rigs') managed by Pertamina Hulu Mahakam (PHM) in the eastern side of Mahakam Delta in West Kalimantan (Indonesia) has been reported being used by many birds - suspected to be egrets - for roosting. PHM is a subsidiary of Pertamina, a state-owned oil and gas company. PHM operates several gas and oil platforms in the Mahakam Delta areas.

Egrets that have been roosting on the gas platform was noted by the PHM Management since early 2014. The birds were considered nuisance to the PHM Management due to various reasons. The accumulation of egrets' faeces on the platform could cause damage to the sensitive apparatus. Fallen feathers and regurgitated food (shrimp, fishes, eels, other macroinvertebrates) might clogged the sewer. For the workers (operators) who occasionally need to inspect the platform, the faeces would be slippery when raining and might impacted the safety of the workers. There was also a health concern that might linked to the decomposition of faeces and regurgitated food, although there has been no previous report about human disease due to microorganism in bird's faeces.

There has been no scientific report on birds that use gas or oil rigs in Indonesia. The phenomena of egrets roost on human-made gas platform is quite unique and need to be observed on a scientific basis. The objective of this study was to document the occurrence of the birds nesting on gas drilling platform, to analyse the situation, and discovered the possible causes.

2. Methods

The study area was located in the Mahakam Delta in East Kalimantan, Indonesia, specifically on and surrounding offshore gas platforms (Ax, D, and C), managed by PHM, located about 1.7km from the nearest mainland (**Figure 1**). Population and the daily habit of egrets as nuisance were examined on or near the gas platforms. Bird species that have been roosting in the gas platforms were directly observed. Counting was conducted only to C-Platform, which has been suffered the most from bird nuisance. C-Platform was an unmanned small platform, consisted of a two-deck main platform and two trunklines that branched out from the main platform.

Bird population was estimated by counting all birds (i.e. census) when the birds flying from the mainland to the gas platforms in the evening (5:30-7pm, 5-min interval). The three-day counting was conducted by four observers (two pairs), taken from a boat or the tip of the gas platforms without disturbing the birds' flight behavior and flight path. Each pair of observers counted birds in different directions, avoiding sight overlap. Bird species was identified during flight, based on morphological character and size. Binoculars (Nikon Action EX; 8x40) were used to improve sighting of bird flight and when roosting on the platform. Visit during the night when birds were roosting was prohibited, following standard operational procedure and due to security reason. Morning count when the birds leave the Platform was not conducted due to technical reason.

Habitat components were also analyzed, both for feeding sites in the mainland of Mahakam Delta and roosting sites on the gas platforms. For birds' feeding sites, the habitat components were analyzed based on several boat trips to the Mahakam Delta to overview the condition of the nearby feeding sites. The available land-use map was also studied to aid interpretation about landscape transformation that has been going on in the Mahakam Delta.

Habitat components of the gas platforms were analyzed based on visits to three gas platforms that suffered the most from bird nuisance, during the day when the birds were not present on the gas platforms. The analysis included identification of specific sites and constructions used for bird roosting. Land-use type surrounding the C-Platform was analyzed from Google Earth map, within 5-km radius, as the main potential foraging sites. A larger foraging areas in the mainland was analyzed based on the available maps and published papers related to the land use system in the terrestrial areas of Mahakam Delta. Interviews with PHM Management Team, mainly to the Health, Safety & Environment Division, were also conducted to provide a clearer understanding about the situation. Some necessary recommendations were generated from the analysis.

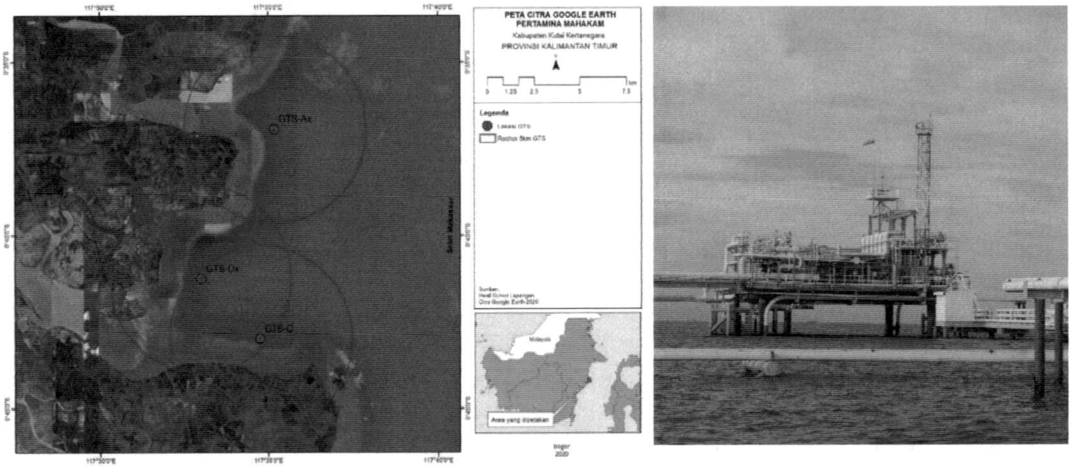

Figure 1 Google Earth map of the gas platforms as the study sites at the east side of the Mahakam Delta, East Kalimantan (left; circle denoted 5-km radius from C-Platform), and C-Platform as the main focus of the study (right).

3. Results

3.1. Daily habit and population of egrets roosting on C-Platform

The birds arrived from the mainland to the C-Platform in late evening, spent the night on the platform, and leave the platform very early in the morning, before the sun rise. The peak of arrival time occurred on 6:45pm during the observation days. The bird species were most probably consisted of two species, namely great egret (*Egretta/Areda alba*) and intermediate egret (*Egretta intermedia*). During the flight, these species mixed together and difficult to distinguished one and another.

The maximum number of bird count was 671 individuals, occurred in day two (**Figure 2**). Underestimated bias might still happen due the poor lighting during the end of the census period (7pm). There was a dim light at the top of the platform, which could navigate some late comers, if any. However, the number of uncounted birds was considered low, judging based on the trend of the timing of the birds' arrival. Birds selected the central part of the platform first (mainly the top part of the large piping), or trunk lines. Hand railings (on the main decks and along walkways) as well as small piping also were used intensively. They seemed to keep distance to each other on their perched spot, and some fighting over a specific perch site might happen.

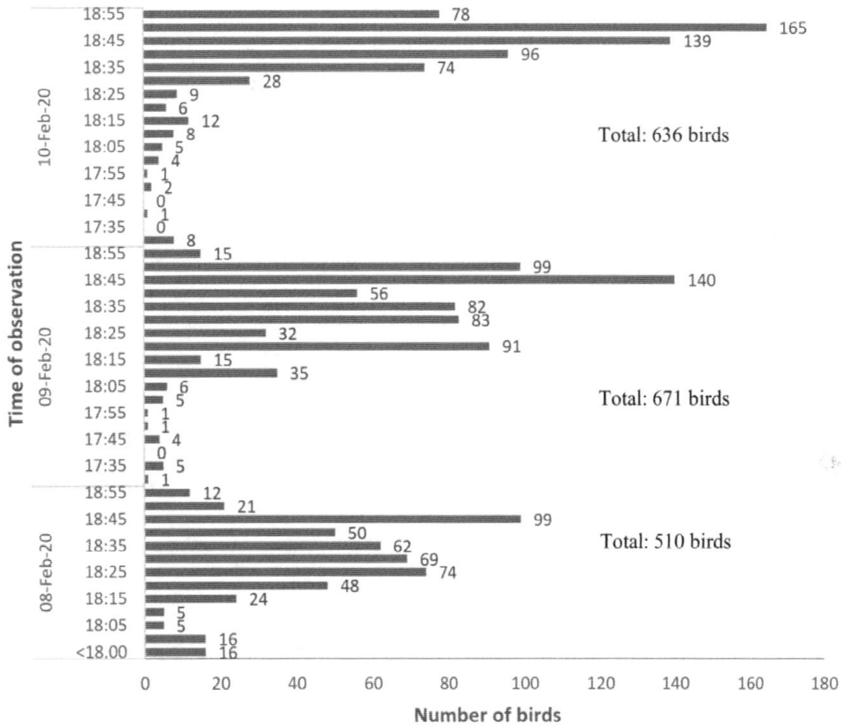

Figure 2 Estimation of the number of egrets that use C-Platform as roosting site during the night, presented in 5-min interval; average egret population was 603 birds

3.2. Analysis of habitat components in the mainland and gas platforms

Observation during bird arrival in the evening showed that the C-Platform (as well as other platforms in the nearby area) were able to provide a roosting site for the birds, due to the abundant availability of piping, railing, and other human-made construction for perching. Other factors might also contributed, including short distance to the mainland and the lack of roosting site on the mainland of Mahakam Delta (**Table 1**).

Land-use type within the 5-km radius from the C-Platform was confirmed to be dominated by shrimp ponds (**Table 2**). Visit to the surrounding areas in the Mahakam Delta by a boat showed that there was almost no mangrove tree left due to the intensive clearing for the shrimp pond. Nypa palm (*Nypa fruticans*) was still can be found in a narrow strip along the water periphery. The shrimp pond was managed in a non-intensive operation and almost all shrimp ponds were unattended. Land-use map in the larger area of the Mahakam Delta confirmed that most of the delta region has already converted into vast areas of shrimp ponds (**Figure 3**).

Table 1 Habitat component analysis of the mainland and C-Platform

Sites	Advantages for Birds	Limiting Factor for Birds
Mainland as feeding sites for birds during the day	• Vast feeding/foraging areas • Abundant of food from shrimp/fish ponds (shrimps, eels, fishes, other invertebrates) • Relatively safe from anthropogenic disturbances • Sufficient cover from climatic factors (heat, rain) during the day	• Lack of perching/roosting site for resting during the nights (i.e. trees to perch) • Lack of breeding sites during breeding season (i.e. trees to lay nests and eggs)
Gas platforms as perching/roosting sites for birds during the night	• Availability of the various constructions (e.g. hand-railing, piping, trunk lines) that resembled perching sites • The occurrence of the deck construction that resembled tree canopy layers • Short distance from the mainland to the offshore gas platforms • Warm temperature during the night • Provide protection from the strong sea wind and rain • Safe from anthropogenic disturbance	• Un-natural; human-made; (could be not a limiting factor because egrets apparently were able to adapt to this condition) • Lack of breeding sites during breeding season (i.e. construction resembled trees to lay nests and eggs)

Table 2 Land-use type within the radius of 5km from C-Platform

Landscape type	Land-use type	Area (ha)	Percentage	
			To the total area	To land area
Marine area				
	Sea	4886.14	62.02	-
	Estuary	0.00	0.00	-
	Tidal area	1850.55	23.58	-
Land area				
	Shrimp pond	751.03	9.57	66.48
	Vegetated area	357.74	4.56	31.66
	Housing	21.00	0.27	1.86

IOP Conf. Series: Earth and Environmental Science **879** (2021) 012009 doi:10.1088/1755-1315/879/1/012009

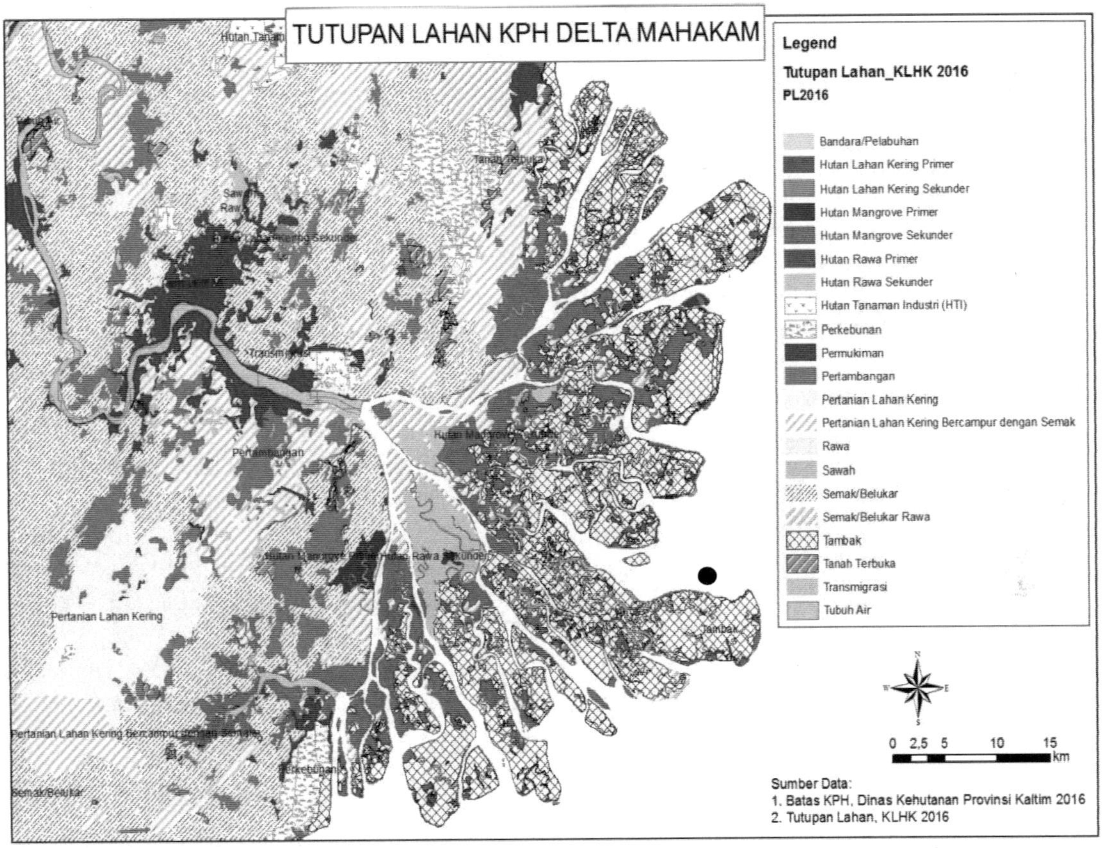

Figure 3 Land-use map of the Mahakam Delta, showing a very intensive shrimp ponds in most areas (criss-cross shading); black dot is the location of the C-Platform
Source of map: Kesatuan Pemangkuan Hutan (KPH) Delta Mahakam

4. Discussion

Food, water and cover are known to be habitat components for wildlife [4]. For egrets, cover consists of sheltered area during the day (trees are the best), trees to roost during the night, as well as trees for breeding (i.e. to lay eggs and to care of chicks). In Pulau Rambut Wildlife Reserve, a small island in the Jakarta Bay area, about 3 km off the coast [5], egrets also behave similar to birds that roosting on the PHM's gas platform. In the late afternoon, the egrets flew back to the island to roost, and then leave the island to the mainland of Java early morning to forage for food. The wildlife reserve is a natural island, and mangroves as roosting and nesting habitat was available.

Landscape transformation in the Mahakam Delta surely has a major impact on wildlife that inhabit and depend on the landscape. The landscape of Mahakam Delta that used to be covered by mangrove has been transformed into shrimp ponds. Analysis of mangrove deforestation in the Mahakam Delta by using high fidelity images [6] suggested that a total of 21,000 ha of mangrove land had been converted to shrimp ponds during 2000-2010. The deforested lands increased sharply, from 47% in 200 to 75% in 2010.

Shrimp ponds are excellent for egrets as feeding or foraging areas. A study on South Kalimantan [7] suggested that the food of egrets were fish, frogs, crabs, molluscs, mud worms, and other shallow-water vertebrates and invertebrates. With the vast expansion of the shrimp pond in the Mahakam Delta to replace mangroves, food for the egrets were abundant. In addition, visit to the surrounding areas has confirmed that the shrimp-ponds were mostly unattended by the fishermen during the day, creating a more convenient condition for the egrets to forage for food.

Mangroves are needed by egrets for roosting and breeding. This bird species, however, has been known to be very adaptive and even able to roost and breed successfully on non-mangrove trees, for example Melaleuca leucadenron and Acacia mangium, as reported occurred in the industrial area in West Java [8]. Nypa palm that was still scattered in the study area was not appropriate for roosting sites due to the absence of perching sites. Unfortunately, there has been no report on the breeding sites of the egrets or other waterbirds nearby.

The lack (and almost absent) of mangroves eventually drove the egrets to discover alterative roosting sites near the foraging areas. The alternative sites that was the gas platforms, located offshore, within a flight distance from the land. Study on egret's flight in the USA [9] had showed that the round-trip flight distances of egrets averaged 16.1±3.2 km (range 6-49 km). Furthermore, the optimum foraging distance were less than 1 km distance [10]. As the closest distance from C-Platform to the nearest mainland was 1.7km, basically most of the foraging areas in the Delta Mahakam were within the reach to the egrets, although not within the optimum flight range.

Flight velocities of the egrets in the USA was known to be 10.8-14.1 m/sec (Maccarone et al. 2012), and thus it will take about 5min for the egrets to fly from the C-Platform to the nearest shore. Under strong sea wind, the flight trip might take longer, but still manageable by the egrets.

5. Conclusion and Recommendation

Wildlife species can easily react when landscape was transformed or has been transforming into other land-use types. Although bio-ecological knowledge has been greatly gathered and expanded, some impact of the landscape transformation on a certain species is less predictable and could be unexpected. The fact that the platform has been used by egrets was unexpected and was not foreseen beforehand. The case of egrets roosting on gas platforms could happened because the birds' natural habitat was not able to provide necessary components for their survival due to landscape transformation. The ultimate cause of this unfortunate event was the landscape change in the egrets' foraging habitat.

Restoring the current habitat into its original mangrove-dominated habitat is the best solution. However, this solution is very difficult to be implemented, due to the already complicated land ownerships of various stakeholders and local communities, and thus require a long-plan collaboration of multi-stakeholders. While waiting to implement the long-term plan, platforms need to be managed as a bird friendly roosting sites. Some less-sensitive parts of the platform can be manage as a 'bird tolerated-zone', while other are managed as a 'bird-free zone' by using appropriate deterrent.

References

[1] Baird PH 1990 *The Condor* **92** 168-771
[2] Tasker ML, Jones PH, Blake BF, Dixon TJ, Wallis AW 1986 *Ringing & Migration* **7** 7-14 doi: 10.1080/03078698.1986.9673873
[3] Wiese FK, Montevecchi WA, Davoren GK, Huetmann F, Diamond AW, Linke J 2001 *Marine Pollution Bulletin* **42** 1285-1290
[4] Moen WH, Aaron N, Freeman 1973 *Wildlife Ecology: An Analytical Approach* (USA: San Francisco)
[5] Mardiastuti A 1992 Habitat and Nest-Site Characteristics of Waterbirds in Pulau Rambut Nature Reserve, Jakarta Bay, Indonesia PhD dissertation (Michigan (USA):

Michigan State University)
[6] Rahman AF, Dragoni D, Didan K, Barreto-Munoz A, Hutabarat JA 2013 *Remote Sensing of Environment* **130** 96–107 doi: 10.1016/j.rse.2012.11.014
[7] Jannah M and Mahruddin H 2019 Prosiding Seminar Nasional Lingkungan Lahan Basah **4** 555-561
[8] Mardiastuti A, Mulyani YA, Suratno 2018 *IOP Conf. Series: Earth and Environmental Science* **197** 012024 doi: 10.1088/1755-1315/197/1/012024
[9] Maccarone AD, Brzorad, John N, Stone, Heather M 2012 *Waterbirds* **35** 394-401 doi: 10.1675/063.035.0304
[10] Kelly JP, Stralberg D, Etienne K, McCaustland M 2008 *Wetlands* **28** 257–275

Assessing the impact of land cover changes on land surface temperature and the relation to urban heat island in Makassar City, South Sulawesi

A S Liong[1], N Nasrullah[2] and B Sulistyantara[2]

[1] Department of Landscape Architecture, Faculty of Agriculture, IPB University, Jl. Meranti, Kampus IPB Dramaga, Bogor 16680, Indonesia

[2] Department of Landscape Architecture, Faculty of Agriculture, IPB University, Jl. Meranti, Kampus IPB Dramaga, Bogor 16680, Indonesia

E-mail: nizar_nasrullah@apps.ipb.ac.id

Abstract. Makassar City, the capital of South Sulawesi Province, is the largest metropolitan city in the eastern part of Indonesia, with a population development rate of 1.19% in 2019. An increase in population impacts city development and results in land use and land cover changes. Changes in land use and land cover pattern bring impact to Land Surface Temperature (LST). This study examines land cover's influence on land surface temperature in Makassar City using multi-temporal satellite data. Land cover and LST data were extracted using Landsat 7 and Landsat 8 over the period of 1999, 2009, and 2019. The result shows that the highest increase in land cover changed was a built-up area of 13.1%, and vegetation decreased by 8.6%. The change in average LST value in the last 20 years was 0.39°C with the highest LST distribution areas was in 30-32°C and 32-34°C classes. The result of LST analysis in 2019 shows that the Urban Heat Island phenomenon has occurred in Makassar in the downtown area and several areas with the densely built-up area. With an overview of the UHI phenomenon in Makassar, the government is expected to raise public awareness of this phenomenon so that preventive actions can be taken, so the effects of UHI do not spread more widely.

Keywords: GIS, land surface temperature, urban heat island

1. Introduction

Makassar City, the capital of South Sulawesi Province, is the largest metropolitan city in the eastern part of Indonesia, with a population development rate of 1.19% in 2019 [1]. An increase in population in urban areas has an impact on city development by modifying the biological and physical surface characteristics and can lead to climate change, land degradation, and changes in biological diversity and ecosystem services, and this development impact associated with land use and land cover changes [2]. Urban development in Makassar city results in a land conversion from green open spaces and open fields to build-up areas to fulfil urban communities' needs.

Transformation in land use and land cover was an essential driving factor for the Land Surface Temperature (LST) increase in the city [3]. The uneven and centralized development in the downtown areas in Makassar City has a possibility to bring adverse impact like Urban Heat Island (UHI) phenomenon. Investigating the spatial distribution of LST is essential to understanding the progression of land use and land cover in the city and its relation and shows the spatial pattern of UHI [4] so that prevention can be done and the UHI phenomenon doesn't spread wider in urban areas.

This study examines the influence of land cover changes on land surface temperature in Makassar City by using multi-temporal satellite data. Land cover and Land Surface Temperature data are extracted using Landsat 7 ETM+ and Landsat 8 OLI over the period of 1999, 2009, and 2019. The year's selection in this study was based on the latest data, namely 2019, and was pulled back for 20 years with a time interval of 10 years. This study's results are expected to be the basis of determining policies to prevent or decrease the phenomenon of urban heat island that occurs in Makassar City.

2. Data and Methods

Three main steps were taken to assess the influence of land cover changes on LST in Makassar City. First of all, classifying land cover maps of Makassar City in three periods of time in 1999, 2009, and 2019. Second, identifying spatial distribution LST of Makassar City also in three time periods. Third, assessing the impact of land cover changes on LST in Makassar City. To classifying and identifying the land cover and LST of Makassar. Primary data used was Digital Satellite Landsat 7 ETM+ and Landsat 8 OLI path 114-row 64 from United States Geological Survey (USGS) website. The Landsat 7 satellite images were acquisition on 20 September 1999 and 17 October 2009, and Landsat 8 satellite image was acquisition on 19 September 2019 when the season is the dry and relatively clear atmospheric condition under 30% cloud cover. Makassar City administrative boundary as the study area was obtained from the Development Planning Agency (BAPPEDA) of Makassar City.

2.1. Land cover map classification

The land cover analysis was carried out using ERDAS Imagine 2014 with a supervised classification method. To classify land cover classes, an image classification sub-menu was used with the maximum likelihood classification. The first step of this analysis was layer stack by combining three bands to make a natural colour; in this study, bands 4, 3, and 2 were combined to get the results. The next step is pan sharpening using haze reduction to sharpen the objects' showing in visual analysis. Then cut the study area based on the administrative boundary map of Makassar City using a subset. The next step was to create a training sample area in polygon shape in each class that will be classified. Training sample areas should be spread and evenly distributed to each classification class. The five types of land cover identified are (1) cloud, (2) build-up areas (commercial buildings, office, settlements, etc.), (3) water body (river and lake), (4) bare ground, and (5) vegetation (forest, green open spaces, and land that has sparse, medium, or high densities vegetation). After that, the accuracy test is carried out using a reference point generated by a random sample point with the final result accuracy of not less than 75%.

2.2. Land surface temperature map distribution

To analyze LST, the converted thermal image data from satellite images were used. The first step is to define the subset of the study area. Second, converting the Digital Number into Spectral Radiance (L_λ). In Landsat 7, ETM+ band six is used, and in Landsat 8 OLI, it is used band 10 and 11 converted from raw image to Top of Atmospheric (ToA) spectral radiance [5]. The equation to be used is as follows:

$$L_\lambda = \frac{L_{max} - L_{min}}{QCal_{max} - QCal_{min}}\ x\ (\mathbf{QCal - QCal_{min}}) + \mathbf{L_{min}}$$

Where: L_λ is Top of Atmospheric Spectral Radiance ($W/m^2.sr.\mu m$); L_{min} is spectral radiance scaled to QCALMIN ($W/m^2.sr.\mu m$); L_{max} is spectral radiance scaled to QCALMAX ($W/m^2.sr.\mu m$); $QCal_{min}$ is a minimum quantized calibrated pixel value in DN = 1;

$QCal_{max}$ is a maximum quantized calibrated pixel value in DN = 255, and QCal is quantized calibrated pixel value in DN.

Third, Thermal band was taken to estimate Brightness Temperature (BT) in Kelvin by converting spectral radiance to BT for every thermal band in Landsat 7 and Landsat 8 [5] by using the following formula:

$$BT_n = \frac{K2}{ln\left(\frac{K1}{L_\lambda} + 1\right)}$$

Where: BT is ToA brightness temperature (°K); L_λ is ToA spectral radiance ($W/m^2.sr.\mu m$); K1 is Band-specific thermal conversion constant from the metadata, and K2 is Band-specific thermal conversion constant form the metadata.

And then, for Landsat 8, the combination of BT band ten and band 11 generates mean with the following formula:

$$BT = \frac{BT_{10} + BT_{11}}{2}$$

Furthermore, to convert BT to LST, it's important to calculate land surface emissivity (e) value. To get e value, Sobrino and Raissouni proposed a methodology by using a normalized difference vegetation index (NDVI) estimation algorithm [6]. To estimate the value of NDVI using the near-infrared band and red band, with the following formula:

$$NDVI = \frac{(NIR - RED)}{(NIR + RED)}$$

Where: Near-infrared (NIR) represented by band 4 in Landsat 7 and band 5 in Landsat 8; and Red represented by band 3 in Landsat 7 and band 4 in Landsat 8.

The proportion of Vegetation (Pv) can be calculated by using this formula:

$$Pv = \left[\frac{NDVI - NDVI_{min}}{NDVI_{max} - NDVI_{min}}\right]^2$$

After getting the Pv value, the land surface emissivity (e) value is then calculated using the following formula:

$$e = m\,Pv + n$$

Where: m = 0.004, Pv is proportion of vegetation, and n = 0.986

The final step to estimate the LST, using the following formula [7] and the temperature acquired in Kelvin, was converted to a Celsius scale by subtracting 273.15.

$$LST = \frac{BT}{[1 + (\frac{\lambda * BT}{a})\, ln(e)]}$$

Where: LST is Land Surface Temperature; BT is ToA brightness temperature (°K); λ is wavelength of emitted radiance (Landsat 7 band 6 = 11.45 and Landsat 8 band 10 = 10.8; band 11 = 12); a = 14388 μmK is constant obtained from h*c/σ (where h is the Plank's constant = $6.626*10^{-34}$ Js; c is the velocity of light = $2.998*10^{8}$ ms^{-1}; and σ is Boltzmann constant = $1.38*10^{-23}$ J K^{-1}); and e is land surface emissivity.

3. Result and Discussion

3.1. Land cover changes

The results of a land cover classification in Makassar can be seen in Figure 1. Figure 1a shows that in 1999 there were many built-up areas and bare grounds cover, that is 6.040 Ha (34.4%) and 5.187 Ha (29.6%), respectively. Meanwhile, vegetation area covering 4.984.4 Ha (28.4%) of the total area of Makassar, and water body 854.2 Ha (4.9%), which is Jeneberang watershed and Tallo river. The presence of clouds in Landsat satellite images requires the clouds cover class in the land cover classification because Indonesia has no seasonal cloud-free moment offering the opportunity to capture cloud-free Landsat satellite image [8].

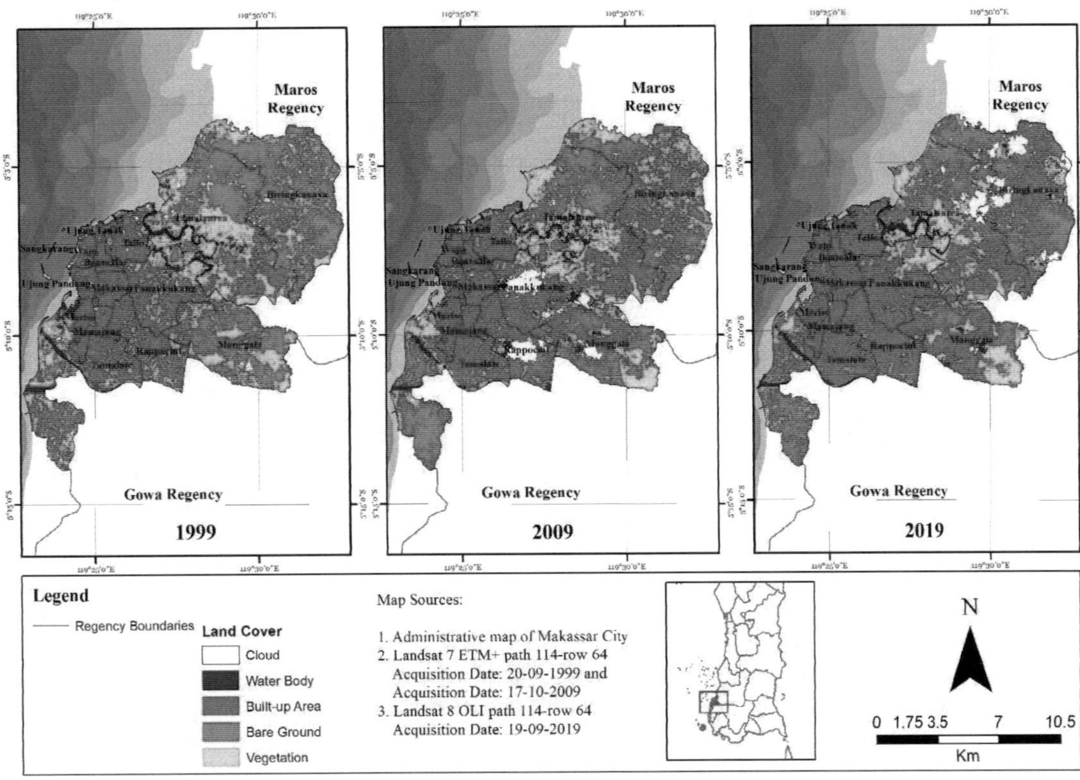

Figure 1 Land cover map of Makassar City in (a) 1999, (b) 2009, and (c) 2019

In 2009 (Figure 1b), bare grounds in the north of Makassar have begun to decrease and turn into built-up and vegetation. Still, in the downtown area, the built-up areas have increased and become 8540.3 Ha, or 48.9% of the total area of Makassar. Vegetation has decreased to 24.8%, while the water bodies have not experienced drastic changes. Changes in water bodies can be due to the conversion of water bodies into the built-up area or vegetation on river boundaries detected as vegetated land cover. In the 1999 to 2009 period, built-up areas increased by 14.5% and mostly in the north and downtown area. Bare grounds, vegetation, and water bodies have decreased by 12%, 3.6%, and 0.2%, respectively.

In 2019, Figure 1c clearly showed that the built-up areas in the downtown area have increased; on the contrary, vegetation has decreased. The total Built-up area is 47.5%, water bodies 4%, bare ground 23.7%, and vegetation covering 19.8% of the total area in Makassar. Also, cloud cover has 5% of the total study area.

Based on land cover changes in Table 1, from 2009 to 2019, built-up areas, vegetation, and water bodies have decreased, while the cover of clouds and bare grounds has increased. Although the built-up areas in the downtown area increase, total built-up areas in 2019 are not consistent increase because of the increase and existence of clouds cover. So the built-up areas decreased by 1.4% in the Landsat satellite images because the actual land cover under clouds couldn't be detected accurately. Meanwhile, the increase in bare grounds by 6.1% could come from the land cover of vegetation, which turned into bare grounds or dry vegetation due to drought and then classified as bare grounds.

Table 1 Land cover changes of Makassar in 1999, 2009 and 2019

Land Cover Classifications	Areas						Land Cover Changes					
	1999		2009		2019		1999-2009		2009-2019		1999-2019	
	Ha	%	Ha	%	Ha	%	Ha	%	Ha	%	Ha	%
Cloud	486.7	2.8	707.3	4.0	887.4	5.0	+220.6	+1.3	+180.1	+1.0	+400.7	+2.3
Water body	854.2	4.9	810.4	4.6	700.2	4.0	-43.8	-0.2	-110.2	-0.7	-154.0	-0.9
Built-up area	6040.0	34.4	8540.3	48.9	8371.8	47.5	+2500.3	+14.5	-168.5	-1.4	+2331.8	+13.1
Bare ground	5187.2	29.6	3075.0	17.6	4177.1	23.7	-2112.2	-12.0	+1102.1	+6.1	-1010.2	-5.9
Vegetation	4984.4	28.4	4337.6	24.8	3491.9	19.8	-646.8	-3.6	-845.6	-5.0	-1492.5	-8.6

In 20 years, land cover changes that occurred in Makassar based on the results of land cover classification analysis, built-up areas have increased by 13.1%, and water bodies, bare grounds, and vegetation each have decreased by 0.9%, 5.9%, and 8.6%. After obtaining the results of the land cover map classification, the accuracy-test was carried out. The accuracy-test results with an overall accuracy of the land cover classification map in 1999 it was 78.81%, for 2009 it was 77.48%, and for 2019 it was 81.46%.

3.2. Land surface temperature

Based on the analysis result of the Land Surface Temperature distribution of Makassar City in three periods of time, Table 2 shows the average increased temperature from 1999 to 2019. Still, in 2009 it tends to be lower than in 1999 and 2019. The lower mean temperature in 2009 is expected to occur because the Landsat satellite image's acquisition date has entered the beginning of the rainy season, so the recorded temperature is lower than the others. The LST is likewise influenced by the cloud cover pattern from Landsat satellite images [9] so that the presence of cloud cover affects the average value of LST in 2009. Changes in the average value of LST in 20 years in Makassar is 0.39°C.

Table 2 The land surface temperature of Makassar in 1999, 2009 and 2019

Land Surface Temperature	1999 (°C)	2009 (°C)	2019 (°C)
Minimum	20.64	14.93	15.93
Maximum	38.79	36.82	39.92
Mean	30.11	29.10	30.50

Land surface temperature distribution in Makassar can be seen in Figure 2. Table 3 shows the classifications of LST distribution classes and contains the area per class changes in 11 distribution classes. Distribution of LST in 1999 shows that the LST classes that have the most extensive area are in the range of 30-32°C with an area of 5.394.60 Ha and the sub-district that have the highest LST are in Biringkanaya and sub-districts located in downtown areas such as Bontoala, Ujung Tanah, Wajo and Mariso with temperature ranges in the 32-

34°C and >34°C classes. Simultaneously, the lowest temperature is in the 22-24°C range covering three sub-districts such as Tamalanrea, Tallo dan Panakukang.

In 2009, the lowest LST was around 14-22°C, classified as clouds land cover, and the highest LST was in the class >34°C. The LST class had the most area still in the 30-32°C range increased by 578.97 Ha from 1999, while LST class 32-34°C and >34°C has decreased by 1.375,47 Ha dan 885.69 Ha respectively. Meanwhile, the 2019 LST analysis (Figure 2c) shows temperatures in the range of 14-22°C, classified as clouds land cover on the land cover classification map (Figure 1c). Distribution of LST classes shows, LST in classes 30-32°C, 32-34°C, and >34°C respectively have increased to 3.402,45 Ha, 2.717,64 dan 540,72 Ha since 2009; while the distribution area of LST classes under 30°C was decreased.

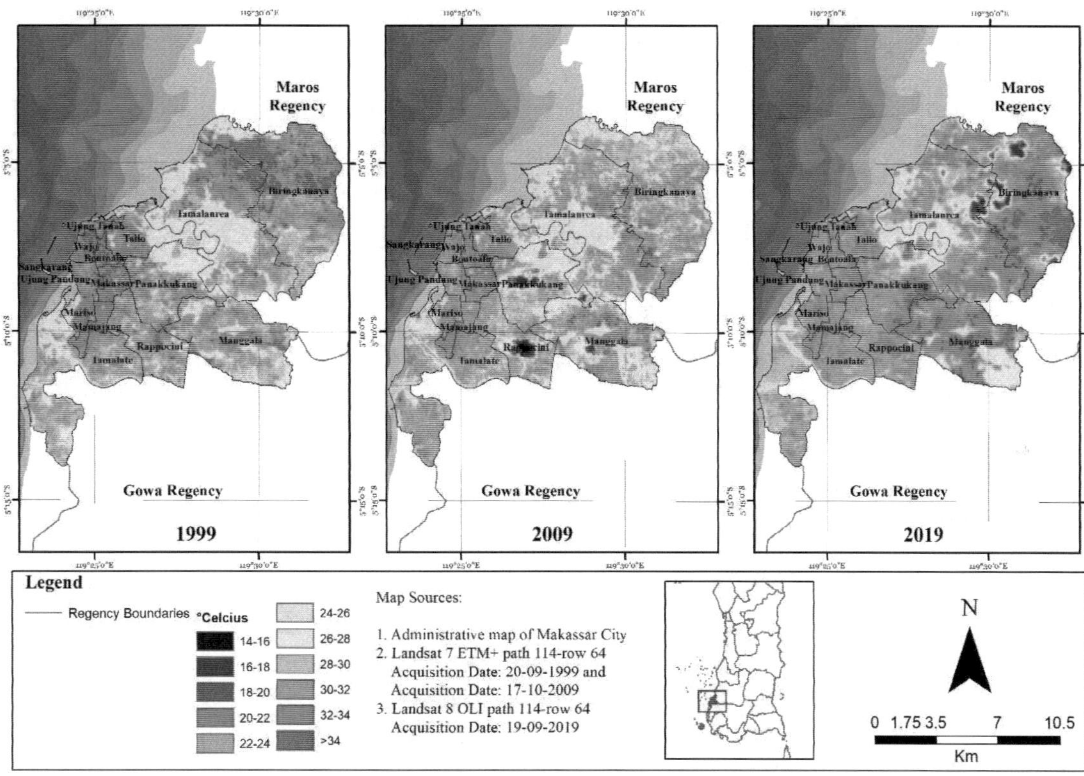

Figure 2 Land surface temperature (°C) of Makassar City in (a) 1999, (b) 2009, and (c) 2019

Table 3 Land surface temperature classes distribution of Makassar in 1999, 2009 and 2019

Land Surface Temperature (°C)	Area (Ha)			Changes in Area (Ha)		
	1999	2009	2019	1999-2009	2009-2019	1999-2019
14-16	0.00	33.75	0.00	+33.75	-33.75	+0.00
16-18	0.00	67.14	98.64	+67.14	+31.50	+98.64
18-20	0.00	78.84	101.79	+78.84	+22.95	+101.79
20-22	0.99	75.24	88.02	+74.25	+12.78	+87.03

Land Surface Temperature (°C)	Area (Ha)			Changes in Area (Ha)		
	1999	2009	2019	1999-2009	2009-2019	1999-2019
22-24	39.96	446.76	107.46	+406.80	-339.30	+67.50
24-26	1671.03	2207.25	596.79	+536.22	-1610.46	-1074.24
26-28	2215.17	2275.20	1817.01	+60.03	-458.19	-398.16
28-30	3811.14	4236.30	2982.06	+425.16	-1254.24	-829.08
30-32	5394.60	5973.57	9376.02	+578.97	+3402.45	+3981.42
32-34	3382.92	2007.45	4725.09	-1375.47	+2717.64	+1342.17
>34	954.72	69.03	609.75	-885.69	+540.72	-344.97

Within 20 years, the area of LST distribution in Makassar has experienced a relatively large increase, specifically in LST 30-32°C and 32-34°C classes, respectively experiencing an increase in the area of 3.981,42 Ha dan 1.342,17 Ha. Meanwhile, LST >34°C decreased by 344,97 Ha, and LST classes below the 30°C also experienced a wide decrease in area.

Based on the LST distribution analysis in Makassar in 2019, it can be seen that in the downtown area and several areas with dense built-up areas forming a pattern like an island with the highest temperature than the surrounding area. Xiong et al. found that built-up areas and thickly populated areas were firmly related to high-temperature anomalies [10]. With the LST analysis results, it can be illustrated that the UHI phenomenon has occurred in Makassar, so prevention action needs to be taking, so its effect doesn't spread more widely. The UHI impact in urban areas can be mitigated by making changes in construction structure, material determination, and land use ratio in urban areas to increase the albedo value and cool the surrounding environment [11]. To make a better proportion of land use in urban areas to reduce UHI effects, the use of greens, trees, and vegetation is significantly useful to decrease the UHI effects by evaporation and shade mechanisms, which can ultimately cool the surroundings [12].

3.3. Relation between land cover changes, land surface temperature, and urban heat island

According to land cover results and land surface temperature analysis from Landsat satellite images, temperature changes based on the land cover classes were obtained (Table 4). In the last 20 years, there has been an increase in LST in Makassar on four land covers: bare grounds, built-up areas, vegetation, and water bodies. The highest increase occurred in water bodies by 1.477°C, then in vegetations increased by 0.984°C, in bare grounds also increased by 0.242°C and in built-up areas was 0.023°C.

Although the highest temperature changes occurred in water bodies and vegetations, the average temperature of both land cover was lowest in three years in this study. Meanwhile, the built-up areas and bare grounds have a relatively high temperature than the other two land covers. These results are consistent with the research results of Adulkongkaew et al. and Feizizadeh et al. that bare grounds and built-up areas had a positive correlation pattern with LST and overall have higher temperatures than water bodies and vegetation areas [13,14].

Table 4 Average land surface temperature changes based on land cover classes

Land cover classes	LST Average per Year (°C)			LST Changes (°C)		
	1999	2009	2019	1999-2009	2009-2019	1999-2019
Cloud	30.444	23.608	25.546	-6.836	+1.938	-4.898
Water body	25.393	25.141	26.869	-0.252	+1.728	+1.477
Build-up area	31.452	30.393	31.475	-1.059	+1.081	+0.023
Bare ground	31.562	31.059	31.803	-0.503	+0.745	+0.242
Vegetation	27.774	26.981	28.758	-0.793	+1.777	+0.984

Changes in vegetation land cover by 8.6% and an increase in built-up area by 13.1% resulted in an increase in the average LST in Makassar for 20 years by 0.39°C. The significant difference in LST occurred in different land cover types. Built-up area positively affects LST because the built-up area holds the maximum LST compared to the other land covers [15]. Moreover, a decrease in vegetation land cover also increased LST distribution in the 30-34°C range and decreased LST area over the 24-30°C range. Vegetation was the way to bring down the LST, and the more vegetated land cover, the lower LST will be [13].

Urban heat island phenomenon in Makassar City in 2019 can be seen in Figure 2c. From this figure shows isotherm lines form the heat islands in the city. Isotherm lines can be seen in several subdistricts in downtown areas such as Makassar, Bontoala, and Panakkukang. This phenomenon mostly found in areas with a high density of built-up areas when comparing to Figure 1c. Urban heat island is dominant in areas with high building density due to the structures and materials used, such as asphalt, concrete, cement, etc. Those materials increase solar energy absorption and increase the temperature in urban areas and then affect the surrounding environment and form the isotherm line [16,17].

4. Conclusion

The highest increased land cover change in Makassar City in 1999 to 2019 was the built-up area, which increased by 13.1%, while the land cover that was decreased the most was vegetation by 8.6%. The change in average LST value in the last 20 years in Makassar was 0.39°C with the highest LST distribution areas was in 30-32°C and 32-34°C classes. The result of LST analysis in 2019, it can shows that the Urban Heat Island phenomenon has occurred in Makassar in the downtown area and several areas with the densely built-up area. With an overview of the UHI phenomenon in Makassar, the government is expected to raise public awareness of this phenomenon so that preventive actions can be taken, so the effects of UHI do not spread more widely.

References

[1] BPS-Statistics of Makassar 2020 *Makassar City in Figures* (Makassar: BPS-Statistics of Makassar)
[2] Rindfuss R R, Walsh S J, Turner B L, Fox J and Mishra V 2004 Developing a science of land change: Challenges and methodological issues *Proc. Natl. Acad. Sci. U. S. A.* **101** 13976–81
[3] Jiang J and Tian G 2010 Analysis of the impact of Land use/Land cover change on Land Surface Temperature with Remote Sensing *Procedia Environ. Sci.* **2** 571–5
[4] Tran D X, Pla F, Latorre-Carmona P, Myint S W, Caetano M and Kieu H V 2017 Characterizing the relationship between land use land cover change and land surface temperature *ISPRS J. Photogramm. Remote Sens.* **124** 119–32
[5] USGS 2018 *LANDSAT 7 (L7) Data Users Handbook*
[6] Sobrino J A and Raissouni N 2000 Toward remote sensing methods for land cover dynamic monitoring: Application to Morocco *Int. J. Remote Sens.* **21** 353–66
[7] Artis D A and Carnahan W H 1982 Survey of emissivity variability in thermography of urban areas *Remote Sens. Environ.* **12** 313–29
[8] Broich M, Hansen M, Stolle F, Potapov P, Margono B A and Adusei B 2011 Remotely sensed forest cover loss shows high spatial and temporal variation across Sumatera and Kalimantan, Indonesia 2000-2008

Environ. Res. Lett. **6**

[9] Li S, Qin Z, Du W, Fan J, Zhao S and Rozenstein O 2019 Linking cloud cover patterns to land surface temperature change in Landsat 8 images *Int. Geosci. Remote Sens. Symp.* 7270–3

[10] Xiong Y, Huang S, Chen F, Ye H, Wang C and Zhu C 2012 The Impacts of Rapid Urbanization on the Thermal Environment: A Remote Sensing Study of Guangzhou, South China *Remote Sens.* **4** 2033–56

[11] Nastran M, Kobal M and Eler K 2019 Urban heat islands in relation to green land use in European cities *Urban For. Urban Green.* **37** 33–41

[12] AboElata A A A 2017 Study the Vegetation as Urban Strategy to Mitigate Urban Heat Island in Mega City Cairo *Procedia Environ. Sci.* **37** 386–95

[13] Feizizadeh B, Blaschke T, Nazmfar H, Akbari E and Kohbanani H R 2013 Monitoring land surface temperature relationship to land use/land cover from satellite imagery in Maraqeh County, Iran *J. Environ. Plan. Manag.* **56** 1290–315

[14] Adulkongkaew T, Satapanajaru T, Charoenhirunyingyos S and Singhirunnusorn W 2020 Effect of land cover composition and building configuration on land surface temperature in an urban-sprawl city, case study in Bangkok Metropolitan Area, Thailand *Heliyon* **6** e04485

[15] Pal S and Ziaul S 2017 Detection of land use and land cover change and land surface temperature in English Bazar urban centre *Egypt. J. Remote Sens. Sp. Sci.* **20** 125–45

[16] Bokaie M, Zarkesh M K, Arasteh P D and Hosseini A 2016 Assessment of Urban Heat Island based on the relationship between land surface temperature and Land Use/ Land Cover in Tehran *Sustain. Cities Soc.* **23** 94–104

[17] Priyadarsini R 2009 Urban heat island and its impact on building energy consumption *Adv. Build. Energy Res.* **3** 261–70

The landscape design of Karawang Islamic center based on Islamic garden

D Nugrahatama[1] and F N H Utami[2]

[1] Student, Department of Landscape Architecture, Faculty of Agriculture, IPB University, Indonesia

[2] Lecturer, Department of Landscape Architecture, Faculty of Agriculture, IPB University, Indonesia

E-mail: [1] davin.nt@gmail.com; [2] fitry_nurul@apps.ipb.ac.id

Abstract. Karawang is the capital as well as the town center of Karawang Regency. As the capital and the town center, it requires public spaces with functional and aesthetic values that reflect and accommodate the town's social characteristics. As a city where most of the population is Muslim, it requires adequate space for worship and public space. The local government can develop the worship and public space from the existing Islamic Center in the township. Islamic Center Karawang is an Islamic Center that has long been built and is located in the township of Karawang. The Islamic Center Karawang consists of a collection of supporting buildings in various fields and the landscape itself. The landscape at the Islamic Center Karawang requires a good landscape design to accommodate visitors' needs and has a good aesthetic value. This study aims to identify the landscape's character at the Islamic Center Karawang, analyze the potential and constraints on the Islamic Center Karawang landscape, and design the landscape of the Islamic Center Karawang. The concept of Islamic garden is adopted from depicting the beauty of heaven in the Quran to a landscape character. The method used in this research is the method of field activities, which includes direct site surveys, interviews with stakeholders and visitors, and filling out questionnaires to visitors. This study's results are in concept diagrams, schematic designs, and Islamic Center Karawang designs. Concept diagrams include basic concepts, design concepts, and development concepts. The schematic design consists of a block plan and a phased scenario. The design output of the Islamic Center Karawang includes site plan, planting plan, section, detail, and design illustrations. The result is the design of a garden that can be enjoyed optimally by the five senses of humans. It can influence humans and definite sharia rules or boundaries regarding elements, characters, and activities in an Islamic garden.
Keywords: cultural landscape, landscape planning, tourism, weaving tradition

1. Introduction

Karawang is a city located in West Java Province, Indonesia. It has approximately 1,737.30 km2, with 2,125,234 people and a density of 1,223 people per km2. The population is getting increasing. Hence it makes the density of the population in Karawang is getting higher as well. The majority of the population is Muslim. As a city where most of the population is Muslim, it requires adequate space for worship and public space.

The Islamic Center Karawang is a place of worship and an important place that reflects Islamic development in the Karawang Regency and has various functions. The landscape at the Karawang Islamic Center. Karawang requires a good landscape design to accommodate the needs of visitors. The current Karawang Islamic Center landscape problems are untidy arrangement, hot climate, not optimal in accommodating visitors' needs, and the lack of landscape elements on the site.

This study aims to identify the landscape's character at the Karawang Islamic Center, analyze the potential and constraints on the Islamic Center Karawang landscape, and design the Karawang Islamic Center landscape. This study's benefit is as an input of the Karawang Islamic Center landscape design and as a reference of Islamic center landscape design with garden-based Islamic landscape design.

2. Methods

2.1. Study site and period

This research was conducted in the Karawang Islamic Center, in the Karawang Regency, West Java (Figure 1), from October 2019 until March 2020.

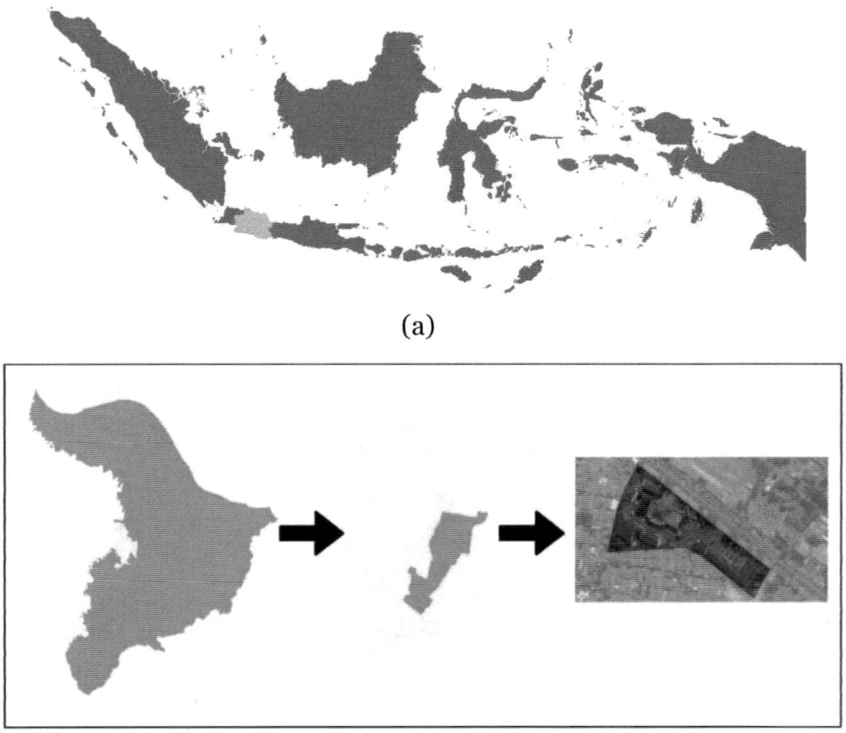

(a)

(b) (c) (d)

Figure 1 Research site
(a) Map of West Java, (b) Map of West Karawang Sub-District, (c) Map of Karangpawitan Village, (d) Karawang Islamic Center Complex

2.2. Research methods

The method used in this design research [1] are: 1. Site inventory (Table 1): (a) site observations; (b) interview; (c) study literature from book/journals of reference and articles; 2. Analysis: Analyzing the inventory data results is done through various methods, such as descriptive, qualitative, spatial, and tabulation methods; 3. Synthesis: The potential and amenities can be developed further in their utilization, while the solutions of the constraints and danger signals can be found; 4. Concept: (a) concept of space; (b) concept of circulation, (c) concept of facilities, and utilities; (d) concept of vegetation; 5. Design: (a) block plan; (b) site plan; (c) section drawing; (d) illustrations; (e) detailed drawings.

Table 1 Data used in the study

No	Aspects and type of data	Form of Data	Data Source	Obtain
1.	Physical and biophysical			
	a. Location and site boundary	Map	Primary	Inventory
	b. Accessibility and circulation	Map	Primary	Inventory
	c. Type of soil	Map/description	Secondary	Study Literature
	d. Hydrology	Map/description	Secondary	Study Literature
	e. Topography	Map/description	Secondary	Study Literature
	f. Climate	Map/description	Secondary	Study Literature
	g. Vegetation and animals	Map/description	Primary	Inventory
	h. Visual	Map/description	Primary	Inventory
2.	Socio-cultural			
	a. Social circumstances	Tabular/description	Primary	Interview
	b. Users	Tabular/description	Primary	Interview/ questionnaire
	c. Activity and intensity	Tabular/description	Primary	Interview/ questionnaire
3.	History			
	a. Historical aspects	Tabular/description	Secondary	Interview
4.	Institutional			
	a. Institutional structure	Tabular/description	Secondary	Interview

3. Result and discussion

3.1. Physical and biophysical aspect

3.1.1. Location and site boundary
The Islamic Center Karawang complex is located on General Ahmad Yani street No. 10, Karangpawitan Village, West Karawang District, Karawang Regency, West Java. It is located in a strategic location in Karawang City's heart and takes a vital role in disseminating Islam in Karawang Regency. Physically, the Islamic Center Karawang complex is bordered by Jendral Ahmad Yani Street and RA Kartini Street. It is connected by Karangpawitan Field, Hotel Omega, and Karawang Public Health Center to the north. In the east, it is bordered by PT Jasa Raharja Karawang and commercial areas. In the south, it is bordered by community settlements, commercial areas, and Permata Ruby Hotel. In The west, it is bordered by commercial spaces and community settlements.

3.1.2. Accessibility and circulation
The Islamic Center Karawang Complex Access can be reached by using public transportation, private vehicle, or on foot. The Karawang Islamic Center complex is facing forward the main road with a provincial road status, Jenderal Ahmad Yani Street, Karangpawitan. The road consists of two lanes which plant bound as a road media. There is a pedestrian lane on the road's outer side with 6m width on each street. There are two accesses to the Karawang Islamic Center Complex, the main gate for vehicles and another gate for pedestrians.

3.1.3. Type of soil
The Karawang Islamic Center complex is located in West Karawang District, and the type of soil in the complex refers to the land data in West Karawang District. Karawang Regency has 3 types of soil, alluvial land is in the north, Mediterranean Land is in the east, and latosol

land is in the southern part of Karawang Regency. The type of soil in West Karawang District is an alluvial soil type [2]. Alluvial soil is formed from river sediment. Alluvial soil/sediment is formed from river mud that settles in the lowlands with fertile soil properties and is suitable for agricultural land. Alluvial soils are classified as young soils that are formed from fine sediment from river flow. Alluvial soil can be used as agricultural land because of its relatively high nutrient content. Alluvial soils also have a dense soil structure and are classified as clay or sandy loam [3].

3.1.4. Hydrology
Based on direct field observations, the hydrological system for water systems and drainage channels has no problems. The Runoff, which comes from rainwater, is channeled into a closed drainage channel in the form of a sizeable concrete-covered trench outside the Islamic Center of Karawang, then the flow will be directed to an irrigation channel called "Saluran Induk Tarum Utara Cabang Barat" (The Main Channel of West Branch North Tarum) in Karawang, and the channel will flow towards the Citarum River.

3.1.5. Topography
The Islamic Center Karawang Complex is located in West Java's lowlands with elevations mostly below 100 m above sea level. West Karawang District is included in the slope class of 0-2% (Table 2). The land's slope in the Karawang Islamic Center Complex is flat between 0-2% slope [2].

Table 2 The slope of Karawang regency

Slope	Sub-distric	Large
0-2%	Pakisjaya, Batujaya, Tirtajaya, Cibuaya, Pedes, Jayakerta, Lemahabang, Cilamaya Kulon, Cilamaya Wetan, Kotabaru, Banyusari, Jatisari, Tirtamulya, Rengasdengklok, Kutawaluya, Tempuran, Rawamerta, Karawang Barat, Karawang Timur, Majalaya, Telagasari, Purwasari, Cikampek, Sebagian Telukjambe Barat, Telukjambe Timur, Sebagian Klari, Sebagian Tegalwaru, Sebagian Pangkalan, Sebagian Ciampel.	80,44%
2-15%	Sebagian Ciampel, Sebagian Tegalwaru, Sebagian Pangkalan, Sebagian Klari, Sebagian Telukjambe, Sebagian Cikampek.	8,93%
15-40%	Sebagian Tegalwaru, Sebagian Cikampek, Sebagian Klari, Sebagian Telukjambe Barat, Sebagian Pangkalan.	8,22%
>40%	Sebagian Telukjambe Barat, Sebagian Pangkalan, Sebagian Tegalwaru	2,95%

Source: (Central Bureau of Statistics of Karawang Regency 2019)

3.1.6. Climate
Based on the user's questionnaire data, the most dominant deficiency in the Islamic Center Karawang Complex Landscape is the hot climate. The Central Statistics Agency of Karawang Regency is a lowland with an average air temperature of 27° C, air pressure of 0.01 millibar, 66% solar irradiation and 80% relative humidity, and wind velocity between 30-35 km/hour, the average length of blowing is 5-7 hours [2].

The most considerable rainfall occurs in February with 304 $mm^{3,}$ and the lowest rainfall occurs in August, which is 1 mm^3. The highest number of rainy days appears in February, which was about 16 days. Whereas Rainfall conditions in West Karawang District, the most considerable rainfall occurred in February, which was 279 mm^3 and the lowest rainfall occurred in July - September, which was 0 mm^3 [2].

3.1.7. Vegetation and animals
The Karawang Islamic Center Complex vegetation is planted intentionally by humans, in random patterns, clumping together, following garden patterns, and lined up. Vegetation in

the Karawang Islamic Center Complex is planted for various purposes, such as shade, display or beauty, and road guidance. The types of vegetation planted in the Islamic Center Karawang Complex also vary from growth, such as ground cover, shrubs, and trees. The existence of a variety of vegetation in the Karawang Islamic Center Complex attracts animals to live there. Some of the animals found in the Karawang Islamic Center Complex are various kinds of insects and birds.

3.1.8. Visual
The viewpoint of the visual aspect of the Karawang Islamic Center Complex is magnificent of Al-Jihad Karawang Mosque with an octagonal star as its distinctive architectural design. Based on direct field observations, the Karawang Islamic Center Complex Landscape on the north is the main gate of the vehicle, the Karawang Regency Regional Library Building on the west, and the Karawang Islamic Center Building on the eastern part. At the same time, in the south is the old building, the P2TP2A, and Wisma Haji Building. While the outside view of the Islamic Center Karawang Complex Landscape, in the north is the Karangpawitan Field, Omega Hotel, and the commercial area; in the south and west. On the east side is the PT. Raharja Karawang services.

3.2. Social and cultural aspects
The population growth in The Karawang Regency has the same number as Karawang District. The highest population density in Karawang Regency occurs in West Karawang and Klari District. The people religion in Karawang Regency is mostly Islam [2].

The Karawang Islamic Center Management Agency has activity programs, as follows:

1. Education and training sector
    - The Tahfidz Quran training and education
    - Islamic Center Karawang workshop with the Muslim community
2. Dakwah sector
    - The PHBI – Tabligh Akbar
    - Fajr prayer congregation program
3. Socio-cultural sector
    - The week of The Karawang community culture
4. Economics and business sector
    - Food court Islamic Center Karawang Unit
5. Public relations sector
    - Consolidation between elements of society and Islamic mass organizations
    - Journalism Training
6. Facilities and infrastructure sector
    - Cleaning, arranging and renovating the entire area of Islamic Center Karawang
    - Inventory of assets in the Islamic Center Karawang

3.3. Historical aspects
The interview was conducted with Ustadz Endang Muliana, the elder of the Al-Jihad Karawang Mosque and vice chairman of the Al-Jihad Mosque's DKM for the 2017-2022 period. He explained that the land used as the Islamic Center Karawang Complex was rice fields that the government took over. The first building in the Karawang Islamic Center Complex is Masjid Al Jihad Karawang, then that other buildings have been built. The construction of additional buildings took place gradually at different times (Table 3).

Table 3 The identity of existing buildings in the Karawang Islamic Center Complex

Name of Building	Year of Construction	Function
Karawang *Islamic Centre* Building	The early 2000s	• There are rooms used by BAZNAS, SATGAS, Diskominfo Karawang, etc. • A place to hold Islamic events • Office of the Karawang Islamic Center Management Agency
Al Jihad Karawang Mosque	• 1960an: 1st construction • 2010: 2nd construction	• Place for prayer, dzikir, reciting the Quran, preach • A place to hold Islamic events • The DKM Office of Masjid Al Jihad Karawang
Region Library of the District Karawang	2010	• A place read books • Where to find book references • Book collection place
Multipurpose Hall Building	1991	• A place to hold events, such as weddings, circumcisions, seminars, etc.
Wisma Haji Building	1994	• Temporary residence for prospective pilgrims
Child Empowerment and Protection Office Building	-	• Office for 'Women's Empowerment and Child Protection Office
P2TP2A Building	-	• Head office for Integrated Service for Women and Children Empowerment
MUI Building	The early 2000s	• Management office of MUI Karawang Islamic Center branch • HMI office of Karawang Islamic Center

3.4. Institutional aspects
Based on the 'Regent's Decree Number: 978/Kep.899-Huk/2017 concerning the Management of the Islamic Center Management Agency for the Karawang Regency service period 2017-2022, the Karawang Islamic Center's institutional structure is submitted to the Karawang Islamic Center Management Agency which is responsible to the Regional Government Karawang Regency.

4. Analysis and Synthesis

4.1. Physical and biophysical aspect analysis

4.1.1. Location and site boundary
The Karawang Islamic Center Complex has approximately 1.59 hectares or 15,957.21 m2 with an open space area of 10,549,703 hectares or 66.11% of the total built-up area of 5,407,503 hectares or 33.89% of the whole area. The open space area larger than the built space in the Karawang Islamic Center Complex should support the existing building spaces, such as being a garden around the building, but this function is not optimal yet. The research questionnaire data to the visitors from different districts to the Karawang Islamic Center Complex visited it as a stopover place for performing prayers only.

4.1.2. Accessibility and circulation
Based on the regulation of the Minister of Public Works Number 03/PRT/M/2014 concerning Guidelines for Planning, Provision, and Utilization of Pedestrian Network Infrastructure and Facilities in Urban Areas, that the main road/access conditions on Jl. Jenderal A. Yani is classified as a good one because it has complete road attributes by

IOP Conf. Series: Earth and Environmental Science **879** (2021) 012011 doi:10.1088/1755-1315/879/1/012011

pedestrian facility standards. The circulation analysis map is presented in Figure 2.

4.1.3. Type of soil

If it is viewed from the main nutrient source characteristics, alluvial lowland soils generally have a reasonably high mineral content of feldspars [4]. The high mineral content of feldspars affects the productivity of paddy soil because the soil will contain high reserves of Ca and K nutrients. The soil fertility level is maintained in the long term. Based on alluvial lowland soils' chemical characteristics, soils generally have relatively good chemical properties because the formation of the ground occurs in basins or plains, which are areas for material accumulation and deposition. The way to overcome micronutrient deficiencies, both primary and secondary, which is not too obvious, can be done by fertilizing it with fertilization recommendations and applying agricultural lime based on soil test results.

4.1.4. Hydrology

Based on direct field observations, the hydrological system for water systems and drainage channels has no problems. There are no obstacles in the hydrological system and drainage channels on site, such as clogging, water pollution, etc. The hydrological analysis map is presented in Figure 3.

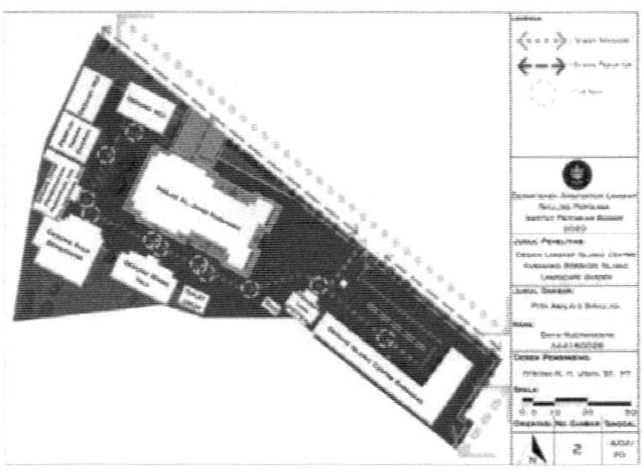

Figure 2 Map of analysis circulation

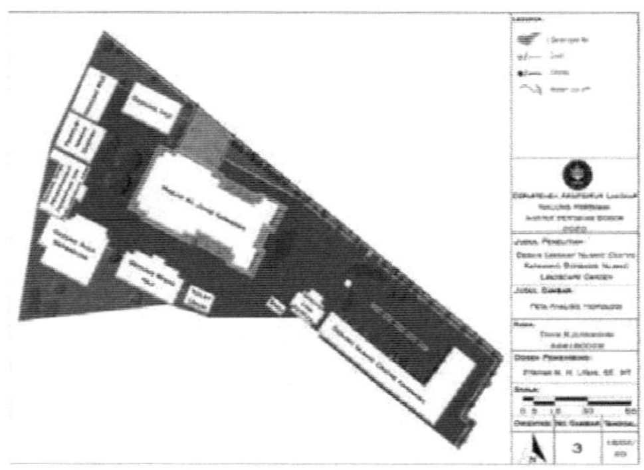

Figure 3 Map of analysis hydrology

4.1.5. Topography

The existing Islamic Center Karawang Complex state is included in the flat topography category of around 0-2%. A land slope of 1-5% is an ideal land slope category for various spatial functions without the need to face problems in shaping the land face (grading) [5]. However, this land slope category can be related to the hydrological system on the site, and the visual aspect, significantly if the soil type in this land slope category cannot absorb water properly and can cause monotony. One of the efforts that can be made to reduce visual redundancy on a flat site is by changing the atmosphere spatially. It can be formed by using colors on landscape elements, strengthening focal point characters, and adding other components, such as vegetation, walls, and buildings.

4.1.6. Climate

A microclimate is a site-specific climate created from topography, vegetation, openness to the wind, shadow patterns caused by buildings and trees, height from sea level, and site correlation to a large area of water [6]. The research site's temperature is included in the optimal warm comfort class [7]. Karawang Regency has a wind speed class classified as class 4 with moderate wind properties and can fly the paper flying and move the twigs and small branches [8]. The effort to overcome the microclimate problem increases shade vegetation, creates a pergola or gazebo, and provides a water element that can give a dark atmosphere, control the microclimate, and become comfortable for visitors or users.

4.1.7. Vegetation and animals
Analysis of vegetation conditions [5] in the Islamic Center Karawang Complex has the process of vegetation consists of three main things, structural uses, environmental functions (environmental services), and visual functions (visual services). The same size and shape vegetation will create low visual interest compared to various sizes and shapes [5]. Based on direct observations in the field, the animals found in the Islamic Center Karawang Complex are multiple insects and birds. To increase the diversity of animals in the Islamic Center Karawang Complex is by expanding the variety of vegetation to attract animals to come in it. The types of plants in the Quran related to Islamic gardens. There is some consideration the types of plant which can live and be used in the site, such as [9] [10]:

1. As mentioned in the Quran in the Islamic garden
2. Visually resemble what the Quran says in the Islamic garden (habitus, exciting part)
3. Supporting to function and aesthetics
   The plants are *Acalypha macrophylla, Alpinia purpurata, Bougainvillea sp., Cheilocostus globosus, Ficus benjamina, Ficus carica, Ficus elastica sylvie, Ficus longifolia, Ficus lyrata, Ficus nitida, Gomphrena globosa, Heliconia rostrata. azedarach, Phoenix canariensis, Phoenix sylvestris, Punica granatum, Quisqualis indica, Senna pendula, Spathiphyllum sp., Vitis vinifera, and Ziziphus mauritiana.*

4.1.8. Visual
A view must be exploited or analyzed because it is a potential site [1]. A picture that could be potential on the site is called a good view. One of the potential good views in the Islamic Center Karawang Complex that can be developed is an octagonal star, the unique/distinctive architectural design of the Al-Jihad Karawang Mosque, the center of the Karawang Islamic Center Complex. The octagonal star is an example of a typical Islamic geometrical pattern and can be developed into the design concept of this study with a geometric line formation. The octagonal star is a pattern resulting from the evolution of Islamic Geometrical Patterns (Figure 4) [11]. The visual analysis map is presented in Figure 5.

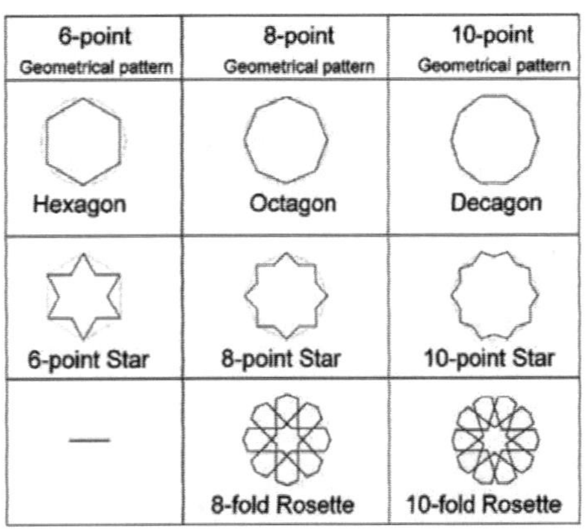

Figure 4 Evolution Islamic geometrical patterns

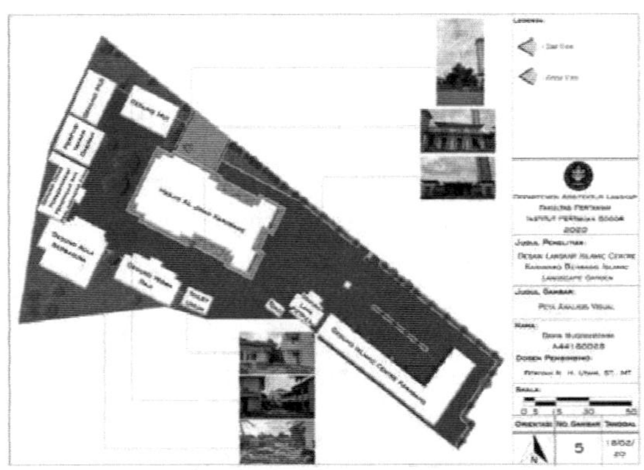

Figure 5 Map of analysis visual

4.2. Social and cultural analysis
The research questionnaire results identified that most visitors/respondents know the information related to the Islamic Center Karawang Complex due to the location of the

Islamic Complex. Center Karawang is on the edge of the provincial road Jl. Jend A. Yani and is in the middle of Karawang City. Meanwhile, the length of visits by visitors/respondents who visited the Karawang Islamic Center Complex was 30-60 minutes (40%), 15-30 minutes (30%),> 60 minutes (16.67%), and for < 15 minutes (13.33%).

5. Concept

5.1. Basic concept

The basic concept for Karawang Islamic Center landscape design is "Oasis Garden." This basic concept is obtained based on the problems and potential of the site. The presence of an "Oasis Garden" is a contrasting thing in the hot environment. As a result, "Oasis Garden" is like a heavenly garden in the world that creates comfort and beauty of the human environment; people call it "Paradise" [12]. An example of the basic concept of "Oasis Garden" is applied in the garden of Shazdeh Garden, Iran (Figure 6). The idea of philosophy in the Islamic Center Karawang Complex's landscape design is "The Phase of Human Life." This philosophy was born from the Oasis Garden, which has a meaningful life for humans and other living things in the arid desert. Human life is divided into three (3) phases, the stage in the womb towards birth, the world life, and death.

Figure 6 The basic concept of "Oasis Garden"

5.2. Design concept

The design concept used is a geometric line that develops an octagonal star as the distinctive design of the Al-Jihad Karawang Mosque. The geometric line design concept originates from this octagonal star found in hot climates like Arabia. Arab countries and other Islamic countries have a distinctive Islamic ornament art called "Arabesque". Geometric lines or straight lines are the characteristic formations of Arabesque. Besides, the octagonal star is a pattern resulting from the evolution of Islamic Geometrical Patterns and then transformed into the design concept (Figure 7) [11].

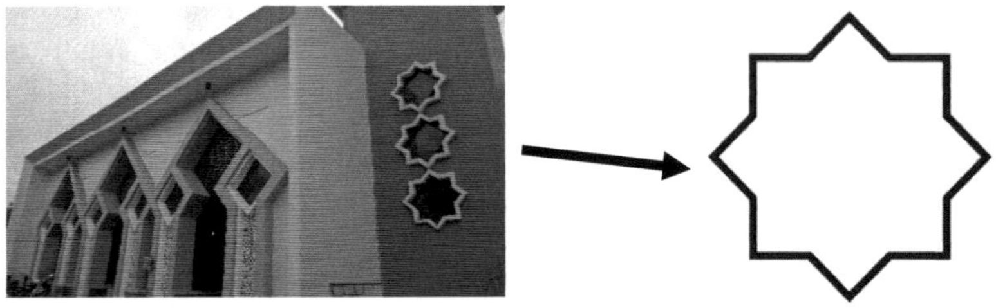

Figure 7 Transformation of design concept formation

5.3. Design
Several products are produced at the design stage, including the concept of space, circulation, facilities and utilities, vegetation, block plans, site plans or drawings of site plans, cutouts, illustrations or perspectives, and detailed drawings.

6. Conclusions and suggestions

6.1. Conclusion
The Karawang Islamic Center Complex is also an essential function in developing Islam in Karawang Regency, which has a majority Muslim population. It is a strategic location in the heart of Karawang City, on Jl. General Ahmad Yani No. 10, Kelurahan Karangpawitan, West Karawang District, Karawang Regency makes the Islamic Center Karawang Complex a critical building that can support public activities image of Karawang Regency name. There are various landscape characters in the complex landscape of the Islamic Center Karawang. There are also different kinds of potentials and obstacles in the Islamic Center Karawang Complex that must be exploited and repaired.

6.2. Suggestion
This study can be a consideration and reference in designing the landscape of the Islamic Center Karawang Complex based on an Islamic garden, and can be implemented by the local government of Karawang Regency. However, the number of pools in this design and considering visitor behavior will require extra maintenance so that cleanliness is maintained and has aesthetic value and good comfort value for visitors.

References
[1] Simonds J O and Starke B W 2006 Landscape architecture McGraw-Hill Book Company New York
[2] Central Bureau of Statistics Karawang Regency 2019 Karawang regency in numbers
[3] Soepraptohardjo M 1976 Soil types in Indonesia series 3 C Soil Classification *Soil Mapping Training* 1976-1977 Soil Research Institute, Bogor 16
[4] Subardja D and Buurman P 1980 A toposequence of Latosols on Volcanic rock in the Bogor-Jakarta area Pp 2545
[5] Booth N K 1983 Basic element of landscape architectural design Waveland Press Inc. Illionist
[6] Laurie M 1986 Introduction to landscape architecture (Translation) Intermatra 2nd Printing Bandung
[7] Talarosha B 2005 Creating thermal comfort in buildings *J. of Industrial Engineering Systems* Vol **6** 3
[8] Kartasapoetra 2004 Cultivation of medicinal plants (Jakarta: PT Bineka Karya)
[9] Text F 2019 50 plants mentioned in the Holy Quran at Islamic garden
[10] El-Seedi H R, Khalifa S A M, Yosri N, Khatib A, Chen L, Saeed A, Efferth T and Verpoorte R 2019 Plants mentioned in the Islamic Scriptures (Holy Qur'ân and Ahadith): Traditional uses and medicinal importance in contemporary times *J. of Ethnopharmacology* **243** 112007 https://doi.org/10.1016/j.jep.2019.112007
[11] Abdullahi Y and Embi M R 2013 Evolution of Islamic geometric patterns *Frontiers of Architectural Research* Vol **2**(2) 243–251 https://doi.org/10.1016/j.foar.2013.03.002
[12] Pramukanto Q 1992 Oasis garden: typical Islamic garden *Azan Magazine* **10** 122-127

Attachment

Siteplan

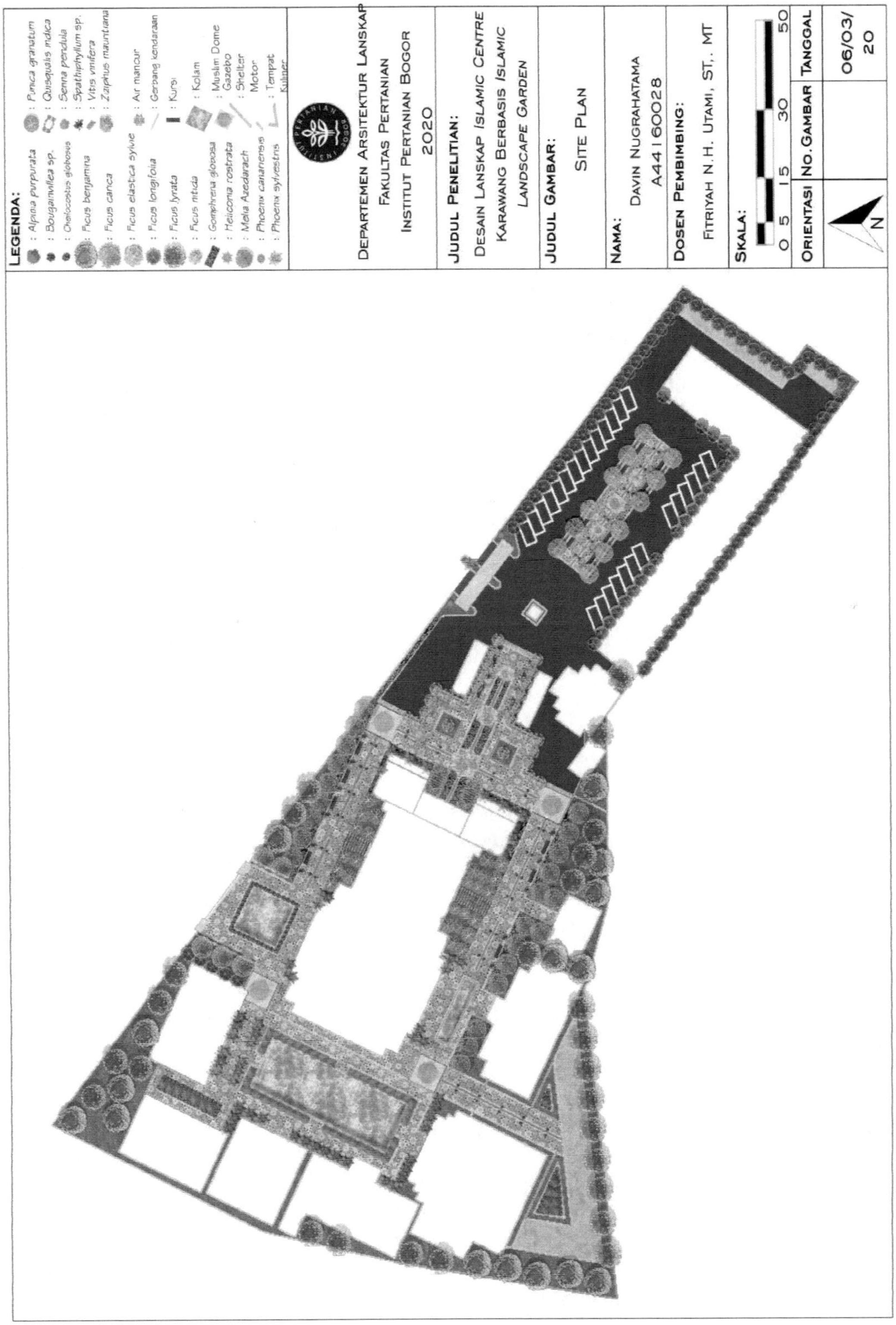

Selokan Mataram in Yogyakarta as a cultural landscape: Heritage values and pressures

D H Rahmi[1] **and B Setiawan**[2]
[1]Universitas Gadjah Mada, Yogyakarta, Indonesia
[2]Universitas Gadjah Mada, Yogyakarta, Indonesia

E-mail: dwitahr@ugm.ac.id

Abstract. Selokan Mataram, a 31.2 km irrigation canal linking Opak and Progo Rivers in the Province of Yogyakarta, initiated and built by Sultan HB IX during the Japanese Occupation, has essential historical values and cultural heritage. It was built to provide water for agricultural activities in the area. At the moment, however, the Selokan Mataram is under the pressure of urbanization and rapid development. Such pressures are threatening its heritage values. This paper aims to examine Selokan Mataram from a cultural landscape perspective, examine the values of its cultural heritage, and document the pressures, threats, and efforts that the government has undertaken to preserve it. This research uses a case study approach, where data and information are obtained through secondary data, field observation, and interviews with multiple sources. The research argues that Selokan Mataram is a 'cultural landscape' that is important to Yogyakarta. It has and fulfills important values to be registered as a cultural heritage. At the same time, as one crucial cultural landscape, Selokan Mataram and the surrounding area is threatened by the rapid development of the city where the local government is not fully able to manage and control it. This research recommends that Selokan Mataram should be declared as a cultural heritage area, so that efforts can be further supported and strengthened to preserve it.
Keywords*:* Selokan Mataram, cultural landscape, conservation, pressure

1. Introduction

1.1. Cultural landscape and heritage values

A cultural landscape is interpreted as the process and products of the relationship between natural and cultural heritage in space and time units, and a complex phenomenon with tangible and intangible identity [1] [2]. The world cultural landscape predicate is inscribed by UNESCO (United Nations Educational and Cultural Organization) to the areas or regions around the world that can show the *outstanding universal value* of the relationship between nature and culture. Values related to universality, uniqueness, and representation of the particular cultural landscape. This outstanding universal value can be fulfilled if the area has authenticity and integrity values[3,4]. Authenticity is the originality, which is the value of the specificity of the cultural landscape form. At the same time, integrity lies in the general values of traditional values that can still withstand the various change threats.

At the locality, however, there are so many cultural landscapes that are significant and have essential values to the locals.There are perhaps do not meeting the criteria of *outstanding universal value* adopted by UNESCO, but important part of the history and identity of specific socio-cultural group[5]. Such local cultural landscapes are then an integral part of the living culture of a given society and therefore should be conserved and developed [6]. Because all localities have unique relations between nature and socio-cultural activities of the people, it is important to understand the specific historical values of such relations [7]. In many cases, only the locals clearly understand the history and meaning of a certain cultural landscape [5].

Cultural landscapes consist of tangible physical patterns and elements, and importantly, reflect intangible values and associations [8]. Cultural landscapes entail natural and human-made components of the environment and how they have changed overtimes [9]. It is essential, therefore, that studies on cultural landscape understand both elements and how they dynamically interchanges. Further, because cultural landscapes are an expression of long and complex relationships between tangible and intangible elements, the conservation and development of such cultural landscapes depends very much on the local communities' participation and management. In this context, the stewardship is therefore becoming an important issue in cultural landscape conservation–the stewardship of these landscape happens in different ways, with different models for decision-making and management [5]. Studying cultural landscape conservation should understand the existing planning and management system of a given society or administrative boundaries. Effective protection and management of living cultural landscapes require a coordinated and multilevel system to address the complexity [6].

1.2. Selokan Mataram: overview

Selokan Mataram or Mataram Canal is a 31.2 km irrigation channel built-in 1942, from the initiative of Sultan Hamengku Buwono IX (Sultan HB IX), the Sultan of Yogyakarta Hadiningrat Kingdom. Selokan Mataram was built based on two primary considerations. First, consideration of usability or functional, namely that sewer will be able to irrigate thousands of hectares of land in the Yogyakarta area. This will guarantee food production and the welfare of many farmers in this province. Second, socio-political reasons, namely as a tactic of Sultan HB IX, so that his people were not asked by the Japanese colonial Government to serve as 'the forced labor' program that was initiated by the Japanese colonial Government at that time.

The two ideas and objectives of Selokan Mataram development initiated by Sultan HB IX are important to know and understand, because it is an integral part of the history of Yogyakarta region, which later gained status as the Special Region or Province of Yogyakarta. In 2012, the specific historical position and role of such region or province finally gained a Special Region status. This is strengthened by Law No. 13 of 2012, on the Special Region of Yogyakarta. This law specifically provides several special affairs in DIY, namely: 1) procedures for filling positions, duties, and authority of governors and vice governors; 2) structure of local government institutions; 3) culture; 4) land; and 5) spatial planning. These five aspects are undoubtedly exciting and important because it is seen as an essential factor in realizing DIY as a region with its cultural peculiarities.

Concerning Selokan Mataram, the specialty of this region is relevant, because with its history and function, Selokan Mataram can be a 'leverage' and an important factor to realize this region's ideals as a distinctive, unique, and prosperous area. From this perspective, Selokan Mataram should be viewed and treated as a 'cultural landscape' with significant historical value for the region and its people. Selokan Mataram is a cultural landscape because it is a unity of natural and cultural elements that dynamically interact with each other [8] [9].

1.3. Yogyakarta City: historic city under transformation

As the center of the development of Yogyakarta, and an education and tourism city, the city of Yogyakarta is growing very rapidly and physically has crossed the boundaries of the administrative area of Yogyakarta, which is narrow, only an area of 46 Km2. Currently, the development of the Yogyakarta urban area has reached four times the size. This urban area of Yogyakarta, or referred to as Yogyakarta Urban Area (KPY), covers areas in the Sleman Regency on the Northside and Bantul Regency on the Southside. Within KPY itself, there are still exist some rural areas and 'desa-kota'-a combination of urban and rural elements. In terms of population, KPY is an urban agglomeration area with a population of 1 million people, double the number in the administrative area of the city Yogyakarta.

With the rapidly increasing level of urbanization and the rapid growth of KPY, the pressures and threats to the environment and agricultural activities in the KPY area of Yogyakarta as well as the Sleman Regency are increasing. Directly and indirectly, the pressures and threats of urban development also occur along Selokan Mataram, which does cross the KPY area; some areas are even experiencing rapid growth acceleration. More importantly, accelerating the development of KPY is also triggered by various government policies and programs, which are often at odds with environmental and agricultural policies and programs. On the other hand, development control systems and mechanisms in Sleman, especially those related to land use control, have also not been fully effective. This led to efforts to preserve Mataram Sewer as a cultural landscape also experienced obstacles [10] [11] [12] [13].

1.4. Objective of the paper

The paper aims to examine Selokan Mataram from a cultural landscape perspective, the values of its cultural heritage, and document the pressures, threats, and efforts that the government has undertaken to preserve it. It is hoped that the paper would then be able to provide some practical recommendations for ensuring the conservation of Selokan Mataram as a historic cultural landscape of the region.

2. Methods

This paper is based on a case study research that explores Selokan Mataram, from cultural heritage perspectives. Selokan Mataram was chosen as a case study as it is a unique case, from its historical development and the present condition. It represents a unique cultural landscape commonly neglected because of its changing roles in present-day situation. This research is also a reminder for the policymakers and community on how important is such a historic cultural landscape which now is becoming everyday people's activities.

Primary data was collected from the field observation, focus group discussion (FGD), and in-depth interviews. Field observation was conducted several times from March to August 2020, to get an understanding and collect data on the condition of land use, farmers' activity in agriculture, and area development progress.Taking photos and notes was taken during the field survey. Based on the result of the field observation and the in-depth interviews was also done to get more understanding and information on the problems and management of Selokan Mataram.

Qualitative analysis was used by explanation building, constructing and evaluating the organized data and information into a narration. Narrative explanation based on the information from field observation is used to interpret the meaning and roles of Selokan Mataram for the local community. Whereas the result of field observation and in-depth interviews was analyzed by using explanation building to understand the pressures faced by the rice terrace, and the management system implemented in the area.

3. Result and discussion

3.1. *Selokan Mataram: brief history*

Selokan Mataram, initiated by Sultan HB XI during the Japanese colonial era, was based on the argument of the importance of optimizing existing land for agriculture, especially rice. Inspired by the early idea of Sultan Agung (the Sultan of Yogyakarta Kingdom in 16 century), Sultan HB IX believes that if the abundant water source of the Progo River on the West side of Yogyakarta region can be taken and used to irrigate dry land in the region of Yogyakarta, then it will be very useful for the people of Yogyakarta, especially the farmers. Technically, efforts to utilize the water source of the Progo River were then implemented by the construction of the canal, which connects the Progo river and the Opak River on the Eastern side of Yogyakarta. This 31.2 Km irrigation channel or sewer becomes an irrigation channel that flows or supplies water on its Southside.

Behind these very practical and rational reasons, the idea of Selokan Mataram construction has another political reason that is no less important, namely to avoid the demand of the Japanese Occupation which is requiring thousands of volunteers to be forcibly employed in various regions of Indonesia for the benefit of the Japanese invaders. By submitting the proposed construction of Mataram sewers, Sultan HB XI could argue with the Japanese colonial Government not to send its citizens as forced laborers in commercial areas of Japan. This reason is important to note because it describes the political nuances of the struggle at that time.

From the technical side, Selokan Mataram is a regular irrigation channel, with a length of 31.2 Km, with a width ranging from 2 meters to 8 meters, depending on the area and its topography. This canal flowed the water from the Progo river to the East and was poured into the Opak River in Prambanan region on the Eastside. Along this channel, there are floodgates and dividing channels to its Southside. Such a canal is so crucial as it serves irrigation for at least 10,000 hectares of land in this region.

3.2. *Present condition and function*

Today, Selokan Mataram is still fully functioning, although the water debit tends to decrease or fluctuate. Based on data from the Sleman Local Government, the area of land served by Selokan Mataram covers at least 10,000 hectares of agricultural land, especially rice fields. The area of agricultural land, especially the rice fields that can be served but the water supply from this irrigation channel allows farmers to grow rice three times a year with good results. Besides, the canal also supplay water for domestic uses to the Southern part of the region.

Another development that occurs in the Mataram sewer area is the use of water that develops, not only for agriculture but also for fisheries. The development of this fishery activity began exist since the early 1990s and until now continues to grow to about 1,000 hectares. In some cases and locations, there has been a conflict of water benefit between agriculture and fisheries. In some locations, this conflict was tried to be resolved by developing a model of 'minapadi' which is a combination of rice farming and fisheries that is quite successful.

3.3. *Pressures and changes*

With the growing population in Yogyakarta province and the urban area of Yogyakarta, economic activities and housing development and other functions along with the Mataram sewers. Housing and urban development also changed the system of regional infrastructure that also affected the Mataram sewer.

IOP Conf. Series: Earth and Environmental Science **879** (2021) 012012 doi:10.1088/1755-1315/879/1/012012

The representation of pressure and change was realized in the change of spatial/land use, which generally changed from natural rural land to urban form. This pattern of land use along Selokan Mataram can be categorized into three classifications, namely: natural/rural (30%); 2) semi-urban (30%); and 3) urban (40%). Each has characters like in Table 1 below.

Table 1 Classification of land use in Selokan Mataram (2020)

No.	Classification	Characteristics	Pictures
1	Natural/Rural (30%)	Built-up area: 10-20%; Building density: 10-20%; Population density: 10-50 jiwa/Ha; Number of building floor: 1; Architecture: Traditional/rural style; Dominant occupation: farmers	
2	Sub-urban (30%)	Built up area: 20-50%; Building density: 20-50%; Population density: 50-100 jiwa/Ha; Number of building floor: 1-2 Architecture: mixed between traditional and modern; Dominant occupation: farners and labor	
3	Urban (40%)	Builtup area: 50-90%; Building density: 60-90%; Population density: 100-200 jiwa/Ha; Number of building floor: 2-6 Architeture: modern Dominant occupation: Office staffs	

Source: Field observation and analysis, 2020

In terms of development and change, there has been a significant change in land use classification in Selokan Mataram. As seen in Table 2 below, the classification of rural natural areas that in the 1940s dominated the entire Selokan Mataram area currently only reaches 30% of them. In contrast, the development of urban land classification reached 40% of 0% at the beginning of the construction of this sewer. This development will continue to occur given that the area is in a rapid urban development area, including the planned construction of toll roads that will pass through about 20% of the entire length of this sewer.

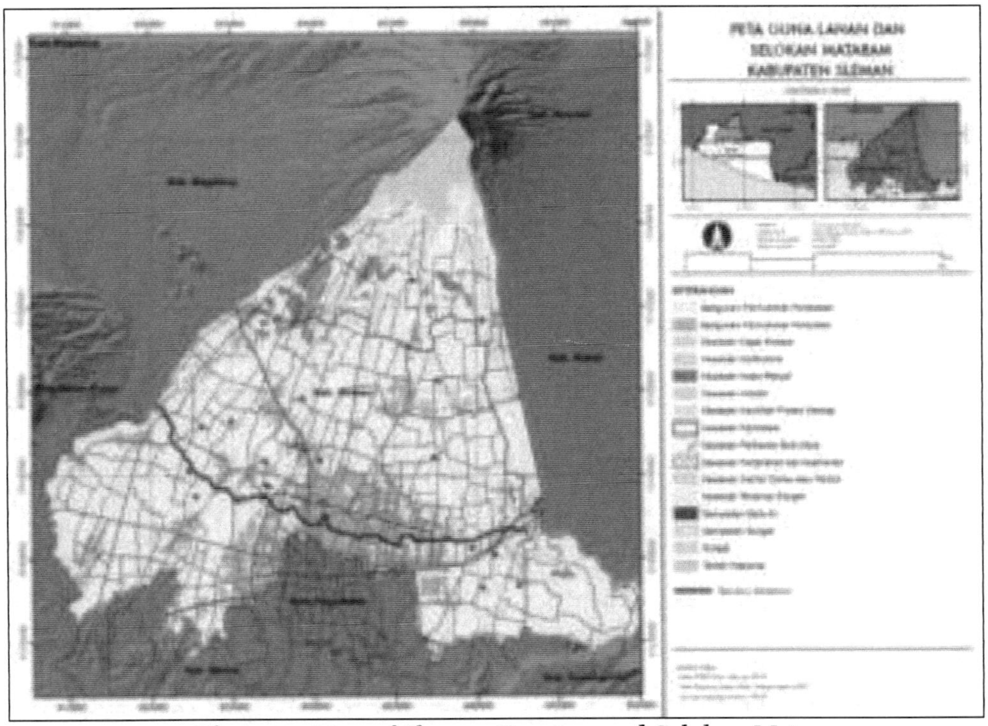

Figure 1 Map of Sleman Regency and Selokan Mataram

Table 2 Land use change along Selokan Mataram

No	Klasifikasi	1942-1950an	1960an	1980an	2000an	2020
1	Natural/Rural	90-100%	80-90%	60-80%	40-60%	30%
2	Sub-urban	10%	10-20%	20-30%	30-40%	30%
3	Urban	0%	10%	10-20%	20-30%	40%

Source: Analysis, 2020

3.4. Selokan Mataram as a cultural landscape

More than just its physical characteristics, the Mataram sewer area is a unity of cultural landscape, namely dynamic interaction between natural settings and the social and economic system of its inhabitants. According to Rahmi[14], as a cultural landscape, the Selokan Mataram sewer can be seen from its six forms described in Table 3 below.

Tabel 3 Selokan Mataram as a cultural landscape: Forms

No	Forms	Characteristics in Selokan Mataram	Photo
1	Natural forms	Two large rivers and a stretch of the plain that is part of Mount Merapi	

No	Forms	Characteristics in Selokan Mataram	Photo
2	Land utilization and land use	The dominance of agricultural/rural activities and began to interact with urban activities.	
3	Settlement Architectures	The dominance of agricultural/rural activities and began to interact with urban.	
4	Living tradition/ practices	Rural/traditional living ordinances that are beginning to affect the ordinances of urban life	
5	Artefacts	Typical canal structures and technical irrigation buildings are built with western/modern technology.	

Source: Field observation and analysis, *2020*

Furthermore, as a cultural landscape, Selokan Mataram can also be classified as a Cultural Reserve Area (*Kawasan Cagar Budaya*) based on the Cultural Heritage Act (UUCB) No.11 of 2010. As in Table 4 below, based on UUCB, there are six criteria to be defined as *Cagar Budaya*, which are: 1) Consists of two or more heritage sites; 2) A result of the cultural landscape with minimum 50 years old; 3) Has a clear spatial pattern with a minimum of 50 years old; 4) Shows previous human influences in the form of land use in a large scale; 5) Showing evidence of cultural landscape, and; 6) Having a soil structure which contains evidence of human activities.

Table 4 Heritage values of Selokan Mataram (Based on UUCB)

No	Criteria	Description	Notes
1	Consists of two or more heritage sites	Dam and 31.2 Km of canal	Although many pressures and threats, in general, the whole landscape of the Selokan Mataram meeting the criterias of authenticity, integrity, and relatively protection/conservation. Some areas are changing into urban landscapes

No	Criteria	Description	Notes
2	A result of the cultural landscape with minimum of 50 years old	Built in 1942-1944, so already more than 50 years old	could be regarded as dynamic cultural landscape, as well as the dominant function is still agricultural area. However, this trend should be carefully managed.
3	Has a clear spatial pattern with minimum of 50 years old	Traditional rural pattern along the canal	
4	Shows previous human influences in the form of land use on a large scale	Human interventions in the forms of agricultural landscapes and traditional villages	
5	Showing evidence of the cultural landscape; and	Dynamic interaction between people and the natural environment in the form of traditional rural-agricultural activities	
6	Having soil structure which contains evidence of human activities or fossil.	There exist some temples in the eastern parts of the Selokan Mataram.	

Source: Analysis, 2020

In terms of its rating, Selokan Mataram can meet the provincial *Cagar Budaya* rating following UUCB No. 11 of 2010. As shown in Table 5 below, the province's ranking means Mataram Sewer has a cross-region value, namely the entire region and community of the Special Region of Yogyakarta. This fact will certainly further support the region's specific values currently continuously developed after the Implementation of Law No. 13 of 2002, on the Special Region of Yogyakarta.

Table 5 Ranking of *Cagar Budaya* Selokan Mataram based on UUCB No. 11 Th. 2020

No	Criteria	Assessments	Notes
1	Represent heritage conservation cross-border regions	Yes, Selokan Mataram is located in two administrative boundaries; Sleman and Yogyakarta	Essential to support 'keistimewaaan' or the special status of the region
2	Represent creative product of the province	Yes, the reactive ideas of the Sultan HB I	Parts of the special status of the province
3	Rare and unique in the province	Yes, only one in the province	Part of ht history of the region
4	Evidence of living human civilization in the province	Yes, until now is still utilize and maintenance as it is part of the people living tradition (farmers)	Under changes and pressures
5	Has associated with the existing tradition	Yes, Selokan Mataram is still used by many people in the province, as a source of water	Under pressures.

Source: Analysis, 2020

3.5. Conservation issues and government interventions

As outlined above, Selokan Mataram as a cultural landscape, needs to be conserved. While in various urban developments, increasingly the existence of this historic channel. Conservation efforts must be carried out so that the values of Selokan Mataram as a heritage cultural landscape can be maintained. As stated in the UUCB, these conservation efforts include: 1) protection, 2) development, and 3) utilization, and Table 6 shows conservation efforts for Selokan Mataram that have been done by the Government of Sleman Regency. In terms of protection, the main thing to note is how to ensure the design of Selokan Mataram as an irrigation channel that guarantee the conservation of agricultural land, especially existing rice fields. Data obtained from the Provincial Statistical Bureau office in 2017 stated that Sleman Regency has the most extensive agricultural land of 21.84 thousand hectares or about 39.5% of the total agricultural land in the Special Region of Yogyakarta (Table 7). However, from 2010 to 2016, the land conversion rate in Sleman was the highest than other regencies, around 945 hectares (78.82%), or on average 160 hectares. Study by Kusumawati et al[15] documents that the water quality in this Selokan Mataram is deteriorating.

Table 6 Conservation issues and government efforts

No	Conservation aspects	Issues	Government efforts/ Present status	Notes
1	Protection	Landuse change; Loss of *sawah*; Farner tradition; Food security Visual landscape	LP2B (Perda..) RTRW; Incentives for farmers; Subsidy for farmers; Dana Desa *Bumdes*	Should be more effective
2	Development	Urban growth Tourism Agro-industry Services Infrastructures	Infrastructure development along Selokan Mataram; land use changes	Should be better controlled
3	Utilization	Tourism Education Leisures Local economy Cultural festivals	Developing and supporting tourism villages; supporting the local economy	Should be more innovations in utilizing Selokan Mataram for local economic development

Source: Field observation and analysis, 2020

Table 7 Agricultural land by regency/city in the Special Region of Yogyakarta 2010-2015 (ha)

Regency/City	Rice field		Dry field	
	2012	2016	2012	2016
Kulon Progo	10,304	10,366	35,027	34,933
Bantul	15,453	15,150	13,42	12,9923
Gunung Kidul	7,865	7,875	117,834	117,332
Sleman	22,786	21,841	16,624	20,617
Yogyakarta	83	60	187	16
S.R.of Yogyakarta	56,491	55,292	183,114	185.821

Source: Statistical information analysis on the development at the Special Region of Yogyakarta Province, 2017 [16]

4. Conclusion

This paper ended with four points. First, it argues that Selokan Mataram is a 'cultural landscape ' that is important to Yogyakarta. It has and fulfills essential values to be registered a cultural heritage. Based on Culturel Heritage Law (UUCB) Number 11. 2010, Selokan Mataram meeting six criteria as a cultural landscape as follow: 1) Consists of two or more heritage sites; 2) A result of the cultural landscape with minimum 50 years old; 3) Has a clear spatial pattern with a minimum of 50 years old; 4) Shows previous human influences in the form of land use in a large scale; 5) Showing evidence of cultural landscape, and; 6) Having a soil structure which contains evidence of human activities. As a heritage cultural landscape, Selokan Mataram could be regarded as Provincial Cultural Landscape because it accordance to five criteria: 1) Represent heritage conservation cross-border regions; 2) Represent creative product of the province; 3) Rare and unique in the province; 4) Evidence of living human civilization in the province, and; 5) Has associated with the existing tradition.

Second, as one crucial cultural landscape, Selokan Mataram and the surrounding area are threatened by the rapid development of the city. This is mainly in forms of urban development. Urban development in the area has manisfested in several threats or pressures to the SM. Such pressures including: 1) land use change; 2) Loss of *sawah*; 3) Farmer tradition; 4) Food security; and 5) Visual landscape. Third, the local government is not fully able to manage and control it. Several effort that already innitiated has able to maintain and protect SM from destruction. However, there is a clear tendency that some fundamental values of Selokan Mataram has deteriorated, particularly the rural/farming practice which is becoming so important even in this changing global world.

Lastly, it is recommended that more effective efforts should be done to conserve Selokan Mataram as an important heritage cultural landscape in the region. This includes: 1) *sawah* conservation through many schemes; 2) rural living which can become potentials experienced economic; 3) rural visual landsca[pe; 4) environmental protection, particularly water catchment and reservation; and 5) open and green spaces as important source of CO2 for the growing city.

References

[1] Plachter H and Rossler M 1995 Cultural Landscapes: Reconnecting Culture and Nature, in von Droste B, Plachter H and Rossler M *Cultural Landscapes of Universal Value*, Gustav Fisher Verlag, New York
[2] *Piagam Pelestarian Pusaka Indonesia* 2003 Jaringan Pelestarian Pusaka Indonesia
[3] UNESCO World Heritage 2003 *Cultural Landscapes: The Challenges of Conservation* World Heritage Papers 7
[4] UNESCO World Heritage, 2009, *World Heritage Cultural Landscapes: A Handbook for Conservation and Management.* World Heritage Papers 26
[5] Brown J 2015 Stewardship of Protected Landscapes by Communities: Diverse Landscapes, Diverse Governance Models in Taylor et al (ed) *Conserving Cultural Landscapes. Challenges and New Directions,* Routledge: New York
[6] Taylor K 2012 Landscape and Meaning: context for a global discourse on cultural landscape values, in Taylor K and Lennon 2012 *Managing Cultural Landscapes*, Routledge: New York
[7] Amin J J A 2012 Cultural Landscape of Java, in Taylor et al. (ed). *Conserving Cultural Landscapes. Challenges and New Directions.* Routledge: New York
[8] Taylor K 2015 Introduction: Cultural Landscape. The First Century Coservation Opportunities and Challenges, in Taylor et al. (ed) *Conserving Cultural Landscapes. Challenges and New Directions,* Routledge: New York
[9] Longstreeth R 2008 Introduction: The Challenges of Cultural Landscape for Preservation, in Longstreeth R *Balancing Nature and Heritage in Preservation Practice. Cultural Landscapes* 2008 London: University of Minnesota Press
[10] Irawati H and Ragil H 2015 Perubahan Fungsi Lahan Koridor Jalan Selokan Mataram Kabupaten Sleman, in *Jurnal Teknik PWK* Volume 4 Number 2, 2015
[11] Ratih G S 2011 *Penataan Ruang Jalan Pada Jalan Inspeksi Selokan Mataram Ruas Seturan-Jalan Wahid Hasyim,* Final Project Undergraduate Program in Planning, Department of Architecture and Planning
[12] Hadianti A 2010 *Pola Penggunaan Ruang pada Kawasan Sempadan Selokan Mataram Yogyakarta,* Final Project Undergraduate Program in Planning. Department of Architecture and Planning
[13] Muzaki A 2010 *Perubahan Guna Lahan dan Fungsi Bangunan setelah Peningkatan Ststus Jalan Indpeksi Selokan Mataram,* Final Project Undergraduate Program in Planning, Department of Architecture and Planning
[14] Rahmi D H 2015 The Cultural Landscape of Borobudur: Borobudur Villages – continuity and changes, in Kanki K, Adishakti L, Fatimah T (editors) *Borobudur as Cultural Landscape: Local Community Initiatives for the Evolutive Conservation of Pusaka Saujana Borobudur*, Kyoto University Press, Kyoto
[15] Kusumawati P, Achmad A R and Eko S 2019 Potensi Selokan Mataram: Ulasan Keadaan Fisik dan Kualitas Airnya, in *Jurnal Pendidikan Geografi* Tahun 24 Nomor 2 2019
[16] Statistical information analysis on the development at the Special Region of Yogyakarta Province 2017

The political economy of soil erosion in Cirasea sub basin, upstream Citarum basin

F Nurysyifa[1], Kaswanto[1] and H Kartodihardjo[2]

[1] Department of Landscape Architecture, Faculty of Agriculture, IPB University, Indonesia
[2] Department of Forestry Management, Faculty of Forestry, IPB University, Indonesia

E-mail: [1] nurysyifarhana@gmail.com; [2] kaswanto@apps.ipb.ac.id

Abstract. The current problems of the Upstream Citarum basin, particularly in the Cirasea Sub-Basin, are near related to economic factors and the low political position of the community. One of the reasons is the biophysical aspect, which influenced the *erfpacht* rights for Dutch and British plantation companies in the *Agrarische Wet* policy at the end of the 17th century. When Indonesia became independent, the Government had worked on environmental rehabilitation as well as dealing with land conflicts, but rehabilitation activities often failed to meet the primary needs of the community. Therefore, rehabilitation efforts in various programs often fail. The objective of this this study is to figure out the preferences of interests and motives from the government and other access authorities which ultimately affect the lives of other communities through the formulation of an environmental rehabilitation program. The approach is Bernstein Political Economy Analysis. The results show that political economy is be able to reveals the problem of erosion, which is always imposed on groups with a low political position compared to other groups who are more vital in reaching access. Even though the community has limitations in implementing environmentally-friendly agricultural practices. Moreover, the political economy can reveal the government's interests behind the land rehabilitation program which often sided with big investors and even tended to repeat the pattern of conflicts in the colonial era. Therefore, erosion can be an important element in describing the conditions of poverty that occur in rural areas.
Keywords: colonial, cultivation, erosion, military, political-economic

1. Introduction

River basin is one of the units of analysis of natural resource management which has a role not only in ensuring quality and distribution of water, but also in maintaining the preservation of ecosystem life, providing a source of food and energy availability, etc. But the river basin condition is getting worse from year to year, including the upstream area of the Citarum Basin, so that rehabilitation efforts are needed. Protected areas in the upstream river basin, when referring to Regional Spatial Plan (RTRW) Bandung Regency of local regulation number 27/2016, generally involves conservation forest areas, protected forest areas, water infiltration, springs, and river boundaries. Meanwhile, natural disaster areas, adjusting to the conditions of Cirasea Sub Basin, include (a) rehabilitation of land and water; (b) forest and critical land rehabilitation; and (c) controlling erosion.

One of the environmental problems that have been highlighted in the upstream area of the Citarum Basin is the problem of erosion. Technically, erosion is eroding of the topsoil layer by rainwater, which is influenced by rain erosivity, vegetation types and management, soil erodibility, and slope. Erosion causes the depletion of the topsoil layer, which reduces soil fertility and infiltration, land degradation and even produces sedimentation which will reduce reservoir function and threaten energy supply [1].

The Upstream Citarum Basin, which plays an essential role in maintaining the downstream environment sustainability, is still shackled by erosion problems. Tthe upstream Citarum Basin had an average erosion rate of 32,073,995.34 tons/year [1]. The erosion rate of the Cirasea Sub-Basin and the analysis showed a value of 24,930,983 tons/year with a very terrible erosion index status in 2008 [2]. Of course, this is inseparable from the growing area of critical land. According to data from the NGO of Raksa Wahana Citarum, the damage of the Mount Wayang area in Tarumajaya Village reached 412.05 ha from total land area of 645.00 ha that spread over the area of Petak 73 (Situ Cisanti Area) covering an area of 69.0 ha, Petak 71 covering an area of 55.0 ha, Petak 69 as large as 58.0 ha, Petak 18 covering an area of 125.0 ha, and Petak 19 reaching 105.5 ha. Data from the West Java Forestry Service stated the erosion that occurred in the Upper Citarum Basin impacted on the increase of sedimentation phenomenon in the Saguling Dam, which reached 8,467,926 tons. In accumulation, from 1987 to 2007, the total sediment was 84,644,878 m^3.

Not only physical aspects, but also social aspects can be used to explain the problem of soil erosion. Erosion often occurs due to inequality in gaining access between the majority of rural communities affected by erosion and other groups who are more vital in gaining access. Besides, soil erosion indicates a symptom of underdevelopment due to the economic limitations of marginalized groups in applying the agricultural sector by conservation principles. This social aspect makes the phenomenon of massive erosion occur in developing countries, particularly in Citarum Basin.

Therefore, soil erosion can be an essential element in measuring poverty in rural areas. Besides, the role of political, social, and economic science can be used to prevent the communities are affected by erosion from migrating to new areas to leave the agricultural sector. Of course, the problem is not only analysing "what is there" explicitly but also "who ordered what" in a phenomenon [3].

The management of the Citarum River Basin often faces failure to solve environmental damage even though it has involved many donor agencies with many programs. The government programs in rehabilitation efforts are often not in line with the community's interests as natural resource users. Also, marginalized communities are often imposed greater responsibility for damage the natural resources than those with excellent access. Therefore, the objective of this this study is to figure out the preferences of interests and motives from the government and other access authorities which ultimately affect the lives of other communities through the formulation of an environmental rehabilitation program. Furthermore, this study also essentially to understand the social symptoms of marginalized communities who tend to ignore environmental sustainability.

2. Methodology

The research has conducted in Cirasea Sub-basin, Upper Citarum Basin. Geographically, the Cirasea Sub-Basin in between 107° 37' 49,1747" E – 107° 48' 30,8923" E dan 6° 59' 32,9636" S – 7° 14' 35,2305" S. However, the research focuses on administrative areas that have major role in land rehabilitation, particularly forest areas, namely Kertasari, Pacet, Ibun, and Paseh Sub-districts with the following boundaries:

- North: Ciparay Sub-district and Majalaya Sub-district
- East: Leles Sub-district, Garut Regency
- South: Gunung Wayang dan Gunung Papandayan
- West: Pangalengan Sub-district

The orientation map of research location is presented in Figure 1. The study was conducted from July to August 2020. The study revealed the practice of erosion land rehabilitation based on a political economy perspective and presented through a descriptive approach. The data used in this study include secondary data collected through several data collection methods, namely interviews with several informants, literature study, and spatial data from several agencies and NGOs. The research uses Bernstein Political Economy Analysis [4] which includes four key questions that concern the social relations of production and reproduction: (1) who owns what; (2) who does what; (3) who gets what; and (4) what do they do with it?

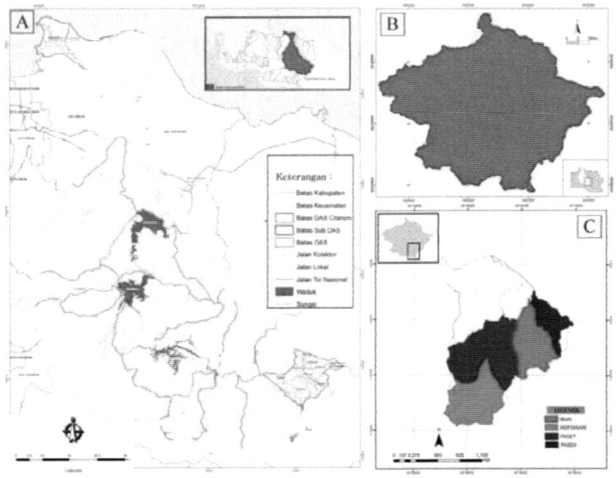

Figure 1 Research Location

(a) Citarum Basin; (b) Upstream Citarum Basin; and
(c) Cirasea Sub Basin

3. Results

3.1. The landscape of Bandung Basin formation process

The formation of the landscape Citarum flow, especially the Bandung highlands, is inseparable from a geological change process that occurs dynamically and lasts a very long time. This process can simultaneously explain the disaster-related phenomena currently occurring in the Bandung basin, especially in the Cirasea Sub-Basin area.

In the Tertiary era, approximately 25-30 million years ago, the north coast of Ancient Java was located roughly around Pangalengan, the north of Pangalengan was inundated by the sea. In the pre-quaternary era, around 2-14 million years ago, the northern coast of Ancient Java was increasingly shifting towards the north so that mountains began to appear above sea level and volcanoes began to be active [5]. Mount Jayagiri (pre-Mount Sunda) is the first giant volcano in the Bandung area which erupted 560,000-500,000 years ago and on one of its sides appeared Mount Sunda. Furthermore, the massive eruption of Mount Sunda between 210,000-105,000 years ago formed Mount Tangkubanparahu and Ancient Bandung Lake.

Mount Tangkubanparahu experienced two categories of eruptions and the material widened to the south. When the Ancient Lake Bandung was inundated, as shown in Figure 2, the lake water began to come into contact with a series of elongated hills. The area affected by the eruption of Mount Tangkubanparahu in the form of faults and collapsed areas supported the erosion, which was relatively high until it finally broke into the West Ancient Lake Bandung in the Cukangrahong area. After the West Ancient Lake Bandung began to shrink, the water of the Ancient Lake Bandung began to erode the rocks in Pematang Tengah. Finally, the East Ancient Lake Bandung receded with an outlet point in the area at Curug Jompong, Nanjung, and West Bandung Regency 16,000 years ago. The receding of the ancient lake resulted in a basin that became the forerunner to the location of Bandung, as shown in Figure 3.

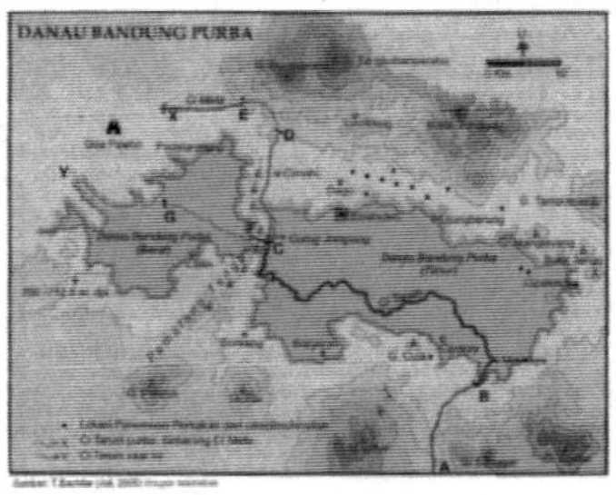

Figure 2 Ancient lake Bandung

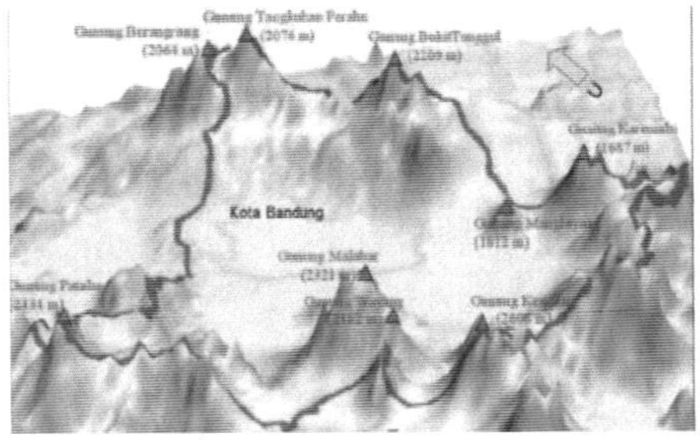

Figure 3 Bandung basin

The Bandung Basin, which is surrounded by mountains, condenses easily at the top of the mountain, thereby increasing rainfall. However, the lower vegetation density in the upstream area, as shown in Figure 4, causes the falling rainwater not to be infiltrated by the soil. The material in the basin area is dominated by volcanic material whose soil is relatively loose. So, the flowing water quickly erodes the upstream basin, and the material is trapped in the downstream basin, as shown in Figure 5.

The process of forming a landscape basin explains that erosion events are commonplace, even having a significant role in reducing the Ancient Lake Bandung. There is an argues that conservation issues, particularly erosion, will only be discussed if erosion seriously disrupts the accumulation interest of the most powerful industrial capitalists. That argues following the conditions of the upstream Citarum, which are currently being held responsible for erosion materials deposited in the downstream area. The consequence of erosion is the silting at the three large dams in Citarum, which will threaten the electricity availability for Java and Bali.

Figure 4 Land cover in Mount Wayang

Figure 5 Impact of soil erosion

3.2. Land use conditions in land rehabilitation efforts

The ratification of the Agrarische Wet 1870 triggered the Dutch to grant *Erfpacht* Rights condition for the opening of large private plantations in Sumatra and Java, including in the Cirasea Sub-basin area. This is the initial momentum for the entry of capital in the region. In 1915, the Cirasea Area, by adjusting the research location, entered the *Residentie Preanger Regentschappen, Afdeling Bandoeng, District Tjiparaij, Patjet, and Paseh.* Based on data from *Department van Binnenlandsch Bestuur* (1915), for the Cirasea Sub-basin area that obtained *Erfpacht* rights, only found in the *Tjiparaj* District with examples of plantation companies that obtained permits are presented in Table 1 [6].

Table 1 Example of *Erfpacht* Right in *District Tjiparaj*

Plot	*Erfpachtsrecht* Date	Company	Commodity
Tjikembang 1	June 20 1882	*Rotterdamsche Kinamaatschappij Tjikembang*	Quinine
Santosa 1	June 19 1904	*N. V. Maatschij. Santosa*	Quinine and tea
Kertasarie III	June5 1911	*N. V. The Kertasarie Tea company Ltd*	Quinine
Lodaja I	1882	*Cultuurmaatsch. Lodaja*	Quinine
Toetoep Noord	June 21 1904	*Cultuurmaatschappij Sedep*	Tea
Kertasari V	Jaanuary 5 1911	*Kertasari Tea Company Limited*	Tea
Taloen	July 4 1902	*N. V. Assamthee onderneming*	Tea and Quinine
-	-	Gunung wayang	Coffee and Rubber

Through the *erfpacht* policy, the liberalists hoped that the existence of economic freedom would result from the transformation of the village into a modern economy. Farmers can imitate the administrative management of large plantations in a modern way. However, things that happened was the opposite because many farmers became laborers and it increased inequality [7]. During the Old Order era, foreign plantations experienced nationalization. Almost all *Afdeling* belong to PT Perkebunan Nusantara VIII, except for the Kertasari plantation, which was once owned by the British, whose Commercial Use Rights/Hak Guna Usaha (HGU) was taken by a private company, namely PT PP London Sumatra.

The central government has determined the status of the Mount Wayang area as a Production Forest. Starting from 1960 until 2003, forest products were managed by Perum Perhutani as a state-owned company/Badan Usaha Milik Negara (BUMN). Perhutani's activities include timber production and logging with the concept of selective logging. To maintain the existence of the forest, Perhutani involves the community in conducting intercropping activities. People who participate in this program are allowed to plant vegetables every three years to wait for the Perhutani tree to harvest.

Perhutani's collaboration with the community began in 1982. This activity was initially quite useful in Mount Wayang at 18, 19, 20, 69, and 71 plots because the community could still be invited to cooperate. In 1989, the locations that had been planted with timber by Perhutani began to be disturbed because the community prioritized planting vegetables that were considered more profitable. Besides, the sanctions that were supposed to be imposed by Perhutani did not work, so that the community began to feel the state land ownership.

In 1992, the Department of Agriculture play along with the Japan International Cooperation Agency (JICA) built the UPT Potato Seed Development Center/Balai Pengembangan Benih Kentang (BPBK), which aims to improve the community's economy. Besides, through the program of JICA-JABAR (SJJB) Potato Seed, the government and JICA have provided training on potato seed production to ten farmers four times in the Kertasari sub-district [8].

In 1996, the planting of vegetables on forest land became more massive, so that the intercropping program was not sufficient. Finally, the New Order government closed all activities on the Mount Wayang land which impacted on people's livelihoods. After the HGU obtained by PT Perkebunan Nusantara (PTPN) VIII expired on December 31, 1997, and the extension process experienced a bottleneck. The community used this momentum to take up plantation land on a large scale on Mount Wayang. Even the practice of planting vegetables is carried out on steep land.

The provincial government started to be firm against the community by issuing a Governor's Decree Number: 522 / BINPROD / 2003 concerning the Prohibition of Intercropping Vegetables in the Mount Wayang Area. However, efforts to control community behavior only lasted three years due to the economic demands of the community increase. In 2009, the government, through the Minister of Forestry, issued a decision related to changes the status of the Gunung Wayang area, which was previously a Production Forest, become a Protection Forest.

The change in the forest area's status has triggered a change in the authority of the institution, which was previously held by Perhutani, now being taken over by the Basin and Protected Forest Management Agency (BPDASHL). One of the Forest and Land Rehabilitation (RHL) programs carried out by BPDASHL is continuing the Perhutani program through Agroforestry activities that began in 2015. One of the major activities includes 'ngajejer disisi' planting (planting trees along the edge of the potato cultivation). Also, unwritten rules apply to prohibit the use of the roundup and gromoxone herbicides as they will exacerbate erosion rates. Other Use Area (APL) are under the authority of the West Java Provincial Government and these areas are the main focus of the Citarum Harum Program.

The land conflicts that occurred in Kertasari District are certainly not without cause. The domination of HGU and the intercropping program of Perhutani with low enforcement of sanctions started many problems because conservation action could not guarantee the economic needs for the community. Also, the Human Development Index (HDI) in Kertasari District in 2012 was in the 30th position out of 31 Districts in Bandung Regency. Even Kertasari had to be complacent about accepting an investment of only 1.4 bilion. Armed with farming, the regional revenue (PAD) of Kertasari District is in the lowest position of the 31 Districts in Bandung Regency, which is only IDR 62 million.

The condition of Kertasari sub-district different from other subdistricts. Each sub-district does have a forest area and agricultural activities are still found, but spatially, Pacet, Ibun, and Paseh sub-districts are adjacent to Majalaya District which is known as a sizeable regional source of PAD for Bandung Regency so that people have various alternatives profession. On the other hand, the Kertasari Sub-district location, which is relatively far from the industrial center, makes it difficult for the community to access other jobs, considering that the community is only skilled in farming.

Protected forest areas that are not found in the Social Forestry Program area and the dominance of HGU make it more difficult for the Kertasari community to utilize the area. According to Safitri *et al.* [9], Other Use Areas outside of HGU which only 3,111.26 ha, from the total sub-district area of 15,121.36 ha, are contested by 77,157 residents. The rest is controlled by large plantation companies and protected forest areas. Figure 6 presents the distribution of land tenure that can be accessed by the community in a white area, even though this does not include the Perhutani's HGU area.

3.3. *Practices of deploying military personnel*

One way to project the interests behind the land rehabilitation program in the basin area is through the DAS rehabilitation program which is currently ongoing. For the Citarum case, the Commander of Kodam III / Siliwangi, Doni Monardo, initiated basin restoration through the Citarum Harum Program and President Jokowi responded quickly to the Presidential Decree 15/2018. In the program, the Coordinating Ministry for Maritime Affairs, Luhut Binsar Panjaitan, became the Chief Steering Officer in implementing the Task Force led by the Governor of West Java. In carrying out technical duties, the governor is assisted by the Indonesian National Armed Forces (TNI). A total of 1,700 TNI personnel divided into 22 sectors, which were determined based on the availability of 22 colonels who did not work so that they were ready to assist the basin rehabilitation task, were deployed to fix the Citarum problem. The government has budgeted IDR 605 billion for the Citarum Harum Program and IDR 300 billion to finance the TNI. The government has received funding from the World Bank of IDR 1.4 trillion to deal with the garbage problem.

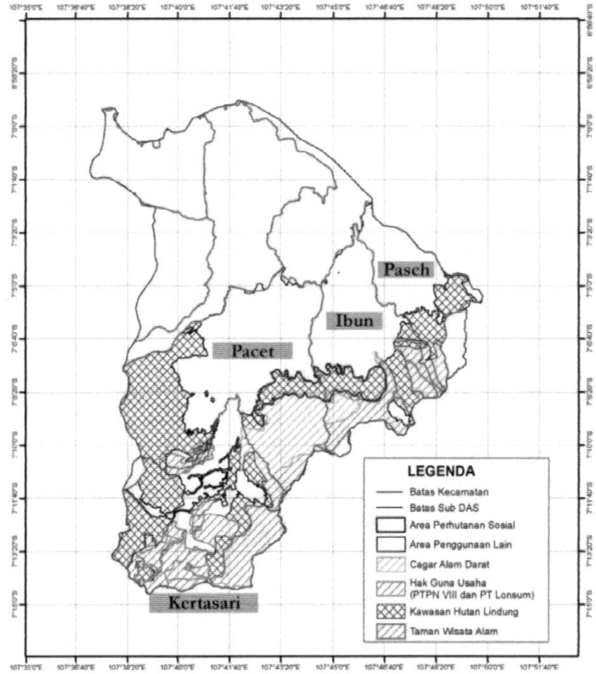

Figure 6 Distribution of land tenure

Governments in the third world countries [10] that were once colonized have mostly used guerrilla warfare strategies with a territorial command system (Koter). Therefore, after the country became independent and began to do a non-military role, the state often adopted a territorial function due to the comprehensive definition of Koter which did not yet have specific functions. For example, as happened in the mid-1980s, The Indonesian Military (ABRI) functioned as one of the tools of power in the Soeharto government by establishing ABRI's position in various government bureaucracies until the regional governments. To support the task, ABRI built a shadow government by forming territorial institutions consisting of the Kodam (Province), Korem (Regency / City), Kodim (Distric), Koramil (Sub-district) to the Babinsa (Village) level. The formation of 22 sectors of Citarum Harum is also adopted this method.

Military performance has succeeded in creating political and security stability and accelerating the modernization process during the New Order Era. However, the paradigm of implementing guerrilla warfare, which still dominates in non-military tasks, often runs inefficiently because it involves many personnel. Also, military's deployment in large numbers supports the re-strengthening of political positions within the civilian sphere. Therefore, many parties consider that the involvement of the TNI in the civilian development process, in non-military missions, can obstruct the democratization process [10].

Besides, the entrenched guerrilla warfare tactics made the Indonesian military very open to enemy attacks. This condition at risk when this culture is carried over when the TNI performs non-military duties because the TNI's work orientation will refer to the definition of 'enemy.' For example, in the case of land rehabilitation in the Cirasea Sub-basin area, especially in the Kertasari District area, the main problem highlighted is the wrong planting practice in the upstream area. So, the definition of 'main enemy' refers to the peasants. As a result, farmers will be considered incompetent in managing nature and many cultivated lands will be confiscated because the cultivated land belongs to the state. That accusations generally without considering the economic conditions of the community, including those the farmer's access and economic capacity in implementing environmentally friendly cultivation practices.

After the reformation, the focus of the TNI changed from the Praetorian Military, namely a military involvement in politics/government, to a professional military. TNI is faced with the challenge of submitting to civilian supremacy [11]. However, the process of changing the corps of the military from a conquering group to a professional group is certainly not an easy thing. Military professionalism has faced significant challenges in the Jokowi era.

Presidents with civilian backgrounds generally do not have much bargaining power to sustain their power. The visible indication is that many of Jokowi's policies have greatly benefited the interests of the TNI. The most prominent thing is related to the problem of the excess number of personnel. President responded with the realization of the Citarum Harum Program which was dominated by the military.

Also, the practice of planting coffee by TNI does not really follow conservation principles. For example, in planting coffee, TNI does not use shade plants and the planting principle uses a monoculture system. Coffee [12] is a plant that requires less light, so the suitable planting method is carried out with an agroforestry system. The legume tree is used as a shade for coffee trees.

In addition to causing erosion and deforestation, coffee cultivation [13] using a monoculture system creates new problems in the water footprint aspect. The water footprint blue refers to the volume of groundwater evaporated by plants [14]. The green water footprint refers to the consumption of rainwater. Also, the grey water footprint refers to the volume of water required to assimilate pollutant loads based on ambient water quality standards. In the case of Kertasari, both tea and coffee consume a lot of water. If stakeholders do not consider water consumption by plants, the area will find it increasingly challenging to obtain groundwater. Table 2 provides data to help imagine the water consumption by commodities available in the Cirasea Sub-Basin area. Planting coffee on state-owned land [15] will create new problems because the longer the planting period for a commodity, the more prone to claims of land tenure by the society. Especially when the citarum harum program will has been completed.

Table 2 The *average water footprint of* agricultural commodities that dominates in the Cirasea sub-basin area

Commodity	Global Average *Water Footprint* (m^3/ton)			
	Green	Blue	Grey	Total
Tea	7,232	898	726	8,856
Green Coffee	15,249	116	532	15,897
Roasted Coffee	18,153	139	633	18,925
Potato	191	33	63	287

Ma'arif's [10] recommendations for realizing ideal military professionals. That is transfer of TNI businesses to the government accompanied by clear compensation for the welfare of soldiers. But, for the case in Cirasea Sub-Basin, the TNI is still indicated to have a business element from the dominance of the TNI in Situ Cisanti. Moreover, TNI has started to build networks by involving the role of non-profit foundations. For example, the foundations that dominate the nursery in the Kertasari area are the Budiasi Foundation, owned by General Doni Monardo, and Artha Graha Peduli (CSR), where the TNI, through the Kartika Eka Paksi Foundation, has big role in the establishment of the business since the New Order Era.

3.4. *Land rehabilitation with classical methods strengthens shackles of land conflict*

Blaikie [3] describes that the ideological structure of land conservation in the colonial period not able to define erosion problems comprehensively. However, specifically for the Citarum Harum Program empirically, the Government continues this paradigm. A comparison of criteria and empirical conditions is presented in Table 3.

Table 3 Criteria and empirical conditions of classical soil conservation model

Criteria of Classical Soil Conservation Model	Empirical Conditions
The problem of soil erosion has been identified as an 'environment one' rather than a complex 'socio-environmental' problem. The 'coescion' method was applied by the colonial rulers that ignored the social problems that were the leading causes of soil erosion.	Although the main program of Citarum Harum seeks social engineering, the 'coescion' exacerbate the conflict that occurred in Kertasari District. Even violence be a way to facilitate access to land only for military purposes. This criterion happened when the community cultivation land (former PTPN VIII area, called Bongkor area) has grabbed by the military for the nursery.
Land users are often accused of being most responsible for erosion problems. Even most land users as marginal groups are identified as lack of education.	The criteria happened in Kertasari Sub-district because the TNI generally underestimates the capabilities of society and only involved the society in cleaning activities.
Overpopulation as a cause of soil erosion	The government had planned a transmigration program in the village of Tarumajaya, Kertasari in 2003, but the society returned to Tarumajaya.
involve the society, particularly farmer, more strongly in the market economy.	The main program of Citarum Harum, particularly in the upstream area, is increasing coffee productivity in order to meet global market demands

If the task force in any land rehabilitation program does not understand the social aspects and tends to use force in resolving conflicts, the government only repeat the colonial government's agenda behind the *Agrarische Wet* 1870 policy. The interests of the government and the military unilaterally have revived the rules of the *Domein Verklaring* in *Agrarisch Besluit,* section 1, which states that "all land which is not proven to have absolute property rights (*eigendom*) is the *domein* (absolutely owned) of the state."

4. Conclusions

Even though the Citarum Basin rehabilitation practice has been going on for a long time, there are still many task force parties who still have a classic soil conservation paradigm. This paradigm certainly has obstacles because social aspects are still neglected in the effort to solve problems. The program based on the military's need for jobs due to the consequences of applying military professionalism and the government efforts to give back to the military in maintaining political stability. Although the principle of "mental revolution" is used in the Citarum Harum program to ensure basin management sustainability, the program tends to focus only on increasing coffee productivity to increase economic growth and overcome poverty problem. Ironically, the economic improvement in the profit-sharing mechanism in the Citarum Harum program is still unclear and people access to the natural resources in their regions is getting more limited which resulating the basin rehabilitation program. Finally, the program only adds to the conflicts that occur in the society.

References

[1] Chaidar A N, Soekarno I, Wiyono A, and Nugroho J 2017 Spatial analysis of erosion and land criticality of the upstream citarum basin *Intern. J. of GEOMATE* **37** *133-140*Milson R, Coley A, Pravda V and Pravdova A 2004 Alignment and algebraically special
[2] Risdiyanto I 2020 “Kondisi Erosi dan Sedimentasi DTA Waduk Saguling,” ResearchGate, June 2020 <https://www.researchgate.net/publication/305520762_Kondisi_Erosi_dan_Sedimentasi_DTA_Waduk_Saguling> (3 September 2020)
[3] Blaikie P M 1985 *The political economy of soil erosion in developing countries* (London and New York: Longman Development Studies)
[4] Bernstein H 2010 *Class dynamic of agrarian change* (Nova Scotia: Fernwood Publishing)
[5] Mulyadi A 2003 Sumberdaya geowisata bandung purba *J. Geo. 'GEA'* **5** 1-9
[6] Departement van Binnenlandsch Bestuur 1937 *Cultuuradresboek voor Indië.* (Batavia: Landsdrukkerij)
[7] Wiradi G 2009 *Reforma agraria: perjalanan yang belum berakhir* (Jakarta, Yogyakarta: Insist Press, KPA dan Pustaka Pelajar)
[8] Hendra A and Wulandari E 2020 sumber pembiayaan dan produksi benih kentang di kecamatan Kertasari kabupaten bandung *J. Pemikiran Masy. Ilm. Berwawasan Agrib* Januari **6** 113-119
[9] Safitri H, Saeful Z, Prawiranegara I and Bachriadi D 2020 Land contestation and military intervention in Indonesia's Citarum River ecological project research gate, March 2020 <https://www.researchgate.net/publication/340183831_Land_Contestation_and_Military_intervention_in_Indonesia's_Citarum_River_Ecological_Project_1> (16 July 2020)
[10] Nugraha A and Hasanuddin 2014 Pergeseran militer politik ke militer profesional; studi tentang keberadaan komando teritorial era reformasi. *Jurnal Online Mahasiswa Fakultas Ilmu Sosial dan Ilmu Politik Universitas Riau.* **1** 1-21
[11] Ma'arif S 2014 Prajurit profesional-patriot: menuju tni profesional pada era reformasi *J. Sos. MASY.* **2**(19): 257-286
[12] Sobari I, Sakiro, and Purwanto E H 2012 Pengaruh jenis tanaman penaung terhadap pertumbuhan dan persentase tanaman berbuah pada kopi arabika varietas kartika I *Buletin RISTRI 3* **3** 217-222
[13] Chapagain A K and Hoekstra A Y 2007 The water footprint of coffee and tea consumption in the Netherlands *Ecolog. Econom.* **64** 109 – 118
[14] Mekonnen M M and Hoekstra A Y 2011 The green, blue and grey water footprint of crops and derived crop products *Hydrol. Earth Syst. Sci.* **15** 1577–1600
[15] Hall D, Hirsch P and Li T M 2020 *Kuasa Eksklusi: Dilema Pertanahan di Asia Tenggara* (Yogyakarta: INSISTPress)

IOP Conf. Series: Earth and Environmental Science **879** (2021) 012014 doi:10.1088/1755-1315/879/1/012014

Attaining food self-sufficiency through the Edible Landscaping (EL) Technology: Assessing the Promotional Activities of EL in the Philippines, amidst the COVID-19 Pandemic

F C Sanchez, Jr., M C Ilang-Ilang, M C E Balladares, B V Apacionado, R R P Tayobong, N G Medina, R O Torres
Institute of Crop Science, College of Agriculture and Food Science, University of the Philippines Los Baños, College, Laguna 4031, Philippines

Email: fcsanchez@up.edu.ph

Abstract. The sustainability of the food supply seems to be a never-ending concern of many households in the Philippines. This concern is most especially felt in this time of pandemic wherein many Filipinos are struggling on where to get their food to feed their families. The adoption of edible landscaping has become more in demand because of the current situation caused by the pandemic. As such, having a sustainable edible garden is very much recommended because it can provide the household a "garden to table" food which can help in making the household food self-sufficient. There is a current need to promote edible landscaping technology primarily through trainings and seminars and it should be intensified to inform and educate the people on food self-sufficiency amidst the COVID-19 pandemic. Edible landscaping (EL) as an approach that merges the science of crop production and the art of landscaping, is continuously being promoted using different strategies and one of which is the conduct of trainings and seminars all over the Philippines. This study assessed the different ways by which the Edible Landscaping Team of the University of the Philippines Los Banos delivered promotional activities such as webinars and e-training amidst the context of the COVID-19 pandemic. This research was conducted using the qualitative approach, specific case studies, and the data gathered were analyzed using thematic analysis. More than 39,000 individuals in the Philippines have already been reached and educated on the technology through the conduct of e-trainings and webinars. The presentations and lectures on edible landscaping were modified based on the needs of the requesting institution and were discussed using a combination of English and Filipino languages. The duration of the eight webinars ranged from a minimum of 30 minutes to a maximum of almost 5 hours (in discussion is written 'almost 4 hours). Based on the comments, most of the online viewers gave affirmative responses.
Keywords: edible landscaping, food sufficiency, promotion

1. Introduction

With the growing number of coronavirus disease (COVID-19) cases all over the world, the World Health Organization declared on 11 March 2020 the COVID-19 outbreak as a pandemic [1]. On 3 September 2020, the Philippines' Department of Health reported an additional 1,987 confirmed cases making the total cases 228,403. Out of the 228,403 total cases, 65,240 are active cases on that day[2].

The pandemic caused by the coronavirus disease (COVID-19) did not only cost lives but also the livelihood of people. In the Philippines, when the government first declared an enhanced community quarantine or lockdown in Luzon, the movement of the people was restricted, only allowing a limited percentage of the population to go outside. With the numerous times of community quarantine extension which were not anticipated by many, source of food has become a problem to many. In addition to this, many individuals have lost their jobs. According to the Philippine Statistics Authority, the unemployment rate rose to 17.6 percent in April 2020 with 7.2 million unemployed individuals [3]. The Department of Labor and Employment even estimated that 10 million workers may lose their jobs due to the pandemic [4]. Since not everyone is privileged enough to still earn a living despite the lockdown, workers who are no-work-no- pay struggle where they would get their everyday living. Thus, the dilemma on the source of food. As the government's response to the dilemma on the source of food, the Philippines' Department of Agriculture implemented the Plant, Plant, Plant program which is a "centerpiece initiative to ensure sustained food production and availability, food accessibility and affordability amidst the threats of COVID-19" [5]. With this, edible landscaping can be considered as a way to ensure food self-sufficiency.

Edible landscaping (EL) is an approach that merges the science of crop growing and the art of landscaping as it follows the basic tenets of designing as its guiding principles. It has some differences with conventional landscaping and backyard vegetable gardening. The difference between EL with conventional landscaping lies in the planting materials used. In conventional landscaping, ornamental plants are used as major planting materials. In EL, the major planting materials are vegetables, fruits, herbs, and medicinal plants. On the other hand, the difference between EL with backyard vegetable gardening is based on the area it occupies. In backyard vegetable gardening, vegetable plots occupy the back spaces of home gardens while EL occupies the premium spaces like the front spaces of the house and public areas of residential lots. EL has four goals namely, aesthetics, functionality, health and wellness, and food-sufficiency. According to de Guzman et al. [6] with aesthetics, people can be encouraged to engage in gardening through the beautiful and attractive environment that can be created; with functionality, edible gardens can be established based on the needs of the end-users; with health and wellness, end-users will be provided with safe and nutritious food; and with self-sufficiency, households will be provided with garden to table food. EL potentially provides a way of addressing food self-sufficiency at the household level. The adoption of edible landscaping has become in demand in this time of pandemic wherein food source has become one of the major concerns of many. Considering the current situation, having a sustainable edible garden is very much recommended because it can provide the household readily available food right in??their yards.

To attain self-sufficiency, the following promotional strategies and activities are being done: (a) exhibits; (b) conventions/ conferences/ publications; (c) media; (d) promotional products; (e) trainings and seminars and; (f) establishment of demonstration gardens. Since social gatherings are prohibited, because of the pandemic, only a few promotional strategies can be done at the moment. As such, the focus is given on the conduct of online trainings or webinars. This paper documented and explored the different ways by which the EL Team of the University of the Philippines Los Banos (UPLB) promotes the EL technology amidst the context of the COVID-19 pandemic. In general, this paper argues that the onset of pandemic poses an opportunity rather than a challenge in the conduct of training and webinars on edible landscaping technology. The pandemic specifically has called for the conduct of more extensive and intensive, as well as sectorally responsive promotional activities apart from the training and seminars to contribute to the attainment of the goal of food self-sufficiency.

2. Methodology

To achieve its main objectives, this paper utilized a qualitative approach, specifically case studies of the eight (8) webinars or e-seminars and e-trainings conducted from June to August 2020. Also included in the study are some descriptive statistics that provide observations on reactions, views, comments, and suggestions relating to the webinars, and other quantifiable aspects of the activities/webinars conducted. The information and set of data obtained from the webinars are analyzed and described narratively, while also providing broad overall themes when apparent. Quoted individuals in the discussion section were also given pseudonyms to protect their identity. The observations made were then used as the basis for?? recommendation, mainly about the ways by which the content and delivery of the webinars can be improved.

3. Results

This section specifically discusses the nature and other main characteristics of the webinars and e-trainings conducted for the promotion of edible landscaping technology in the Philippines, at the onset of the pandemic. Narrated in the case format are the webinars and e-training conducted by, or in collaboration with the members of the EL project, being the main and leading experts of the topic in the Philippines. The webinars are arranged based on the chronological order or date of their conduct.

Case 1: A Webinar for the General Filipino Public

Figure 1 ATI@Home: Live Seminar on Edible Landscaping (Tanim na Pagkain, Gawing Maka-Sining [Artistic Way of Crop Production]), last 11 June 2020 [7]

The Agricultural Technologies thru Internet at Home, or ATI@Home, featured live seminar series on urban agriculture as a support to the DA's Plant, Plant, Plant Program. The seminar, entitled, "ATI@Home: Live Seminar on Edible Landscaping (*Tanim na Pagkain, Gawing Maka-Sining* [Artistic Way of Crop Production])" was held on the 11th of June, 2020, via Facebook. The resource speaker from the UPLB EL Team used a mix of English and Filipino languages in discussing the brief history of edible landscaping, its goals, components, phases, and some ideas in doing an edible landscape garden. There was no hands-on activity for the webinar. However, between discussions, there was an interactive part which was a review portion wherein the online participants answer the prepared activity questions through the comment section. The live seminar, which lasted 1 hour, 40 minutes, and 47 seconds, has gathered more than 1,600 views, 1,100 comments, and 566 reactions from the online participants/ viewers. There was also a question-and-answer portion wherein the resource speaker answered the questions in the comment section. Most of the other comments in the comment section commended the resource speaker for a very informative webinar.

Cases 2 & 3: Recorded Pre-COVID-19 EL Trainings

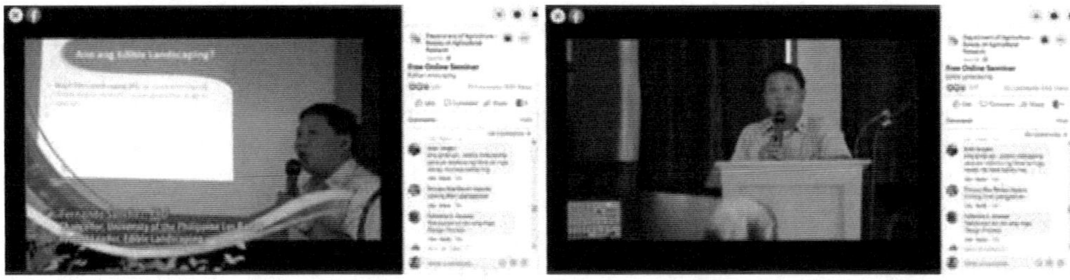

Figure 2 Online Seminar on Edible Landscaping Part I, last 23 June 2020 [8]

The Department of Agriculture-Bureau of Agricultural Research has streamed seminars on urban agriculture in June 2020, in celebration of the Farmers and Fisherfolk month. Edible landscaping was one of the three topics that were streamed in two parts. The first part of the seminar tackled the basics of edible landscaping specifically its definition, goals, and phases. It lasted 32 minutes and 43 seconds, with more than 8,900 views on Facebook, 93 comments, and 324 reactions from online viewers. Meanwhile, the second part of the seminar focused on edible landscaping ideas. The recorded video lasted 56 minutes and 25 seconds, which obtained more than 5,600 views on Facebook, 106 comments, and 298 reactions from online viewers. The English and Filipino languages were used in both seminars. Since it was just streaming of recorded EL trainings, the UPLB EL Team answered questions through the comment section. No hands-on activity took place.

Figure 3 Online Seminar on Edible Landscaping Part II, last 30 June 2020 [9]

Case 4: A Webinar on the Potential of EL Technology on Sustainability amid Population Growth

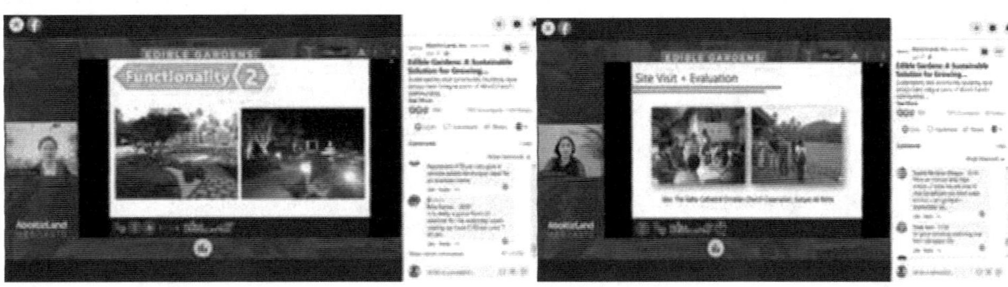

Figure 4 Edible Gardens: A Sustainable Solution for Growing Communities, last 17 July 2020 [10]

Edible landscaping has also been considered to be one of the featured episodes of the AboitizLand Webcasts of AboitizLand, Inc. To uphold its core of sustainability and community-building, the UPLB EL Team was invited to discuss the setting up of edible

IOP Conf. Series: Earth and Environmental Science **879** (2021) 012014 doi:10.1088/1755-1315/879/1/012014

gardens in a community and the importance of a sustainable edible garden amid a pandemic. The webcast was live-streamed via Facebook on 17th of July, 2020. The webcast lasted around 1 hour, and 39 minutes, with more than 14,300 views on Facebook, 395 comments, and 384 reactions from online viewers. The viewers' comments range from their excitement to questions on how to start an edible garden. The resource speakers used a combination of English and Filipino languages for the whole duration of the webcast.

Case 5: EL Technology as Plant-based Adaptation for Climate Change

Figure 5 Good Nutrition and Edible Gardens in the New Normal: A Webinar on Plant-based Solutions for Climate Change, last 22 July 2020 [11]

Edible landscaping has also been the subject of a particular webinar that addressed climate change through plant-based solutions or adaptations. Plant-based solutions were elucidated as involving the aspects of consumption and food production, and during the particular event, diets leaning on plant foods, as well as the production of food plants were the primary modes of adaptation practices that are being promoted to address the threat of climate change, especially in the context of vulnerable countries like the Philippines. This event was held via the video conferencing platform, Zoom in coordination with the Philippine Department on Environment and Natural Resources (Region IV-A), as well as with the Food and Nutrition Institute of the Department of Science and Technology on 22nd of July, 2020. The webinar lasted one (1) hour, seventeen (17) minutes, and fifty-one (51) seconds, and gathered around more than 1,500 viewers across the archipelago through the social media networking site, Facebook. During its live stream, the event also received fourteen (14) comments and thirty-four (34) reactions from the online audience. A combination of the official languages of the country, English and Filipino, was utilized as the main medium of instruction and discussion during the webinar. The event did not include any hands-on activities. In the comment section, viewers mainly inquired on what type of food plants or crops can be planted in urban areas with limited spaces, as well as the costing estimates in constructing an edible garden. Others commented or sought to inquire on when there will be a training session for edible landscaping in their province or near their locale.

Case 6: Practicing the Edible Landscaping Technology in Philippine Schools

Figure 6 Lecture on Edible Landscaping and School Garden Planning for School + Home Gardening cum Biodiversity Enhancement e-training series, last 13 August 2020 [12]

School-based vegetable gardening is embedded in the fabric of the educational system in the Philippines and is mainly institutionalized through the "*Gulayan sa Paaralan Program*" (School Vegetable Garden) or GSPP. Implemented in primary and secondary schools, this specific program hopes to attain the objective of establishing school gardens and consequently, ensuring the continuous supply of vegetables for the school-based feeding programs. While edible landscaping has been promoted and widely popularized as ideal in households and community spaces, its practice in school settings has still a lot of room for expansion and growth. It is in this regard that the South East Asian Regional Center for Graduate Study and Research in Agriculture coordinated with the Edible Landscaping project, and the regional office of the Department of Education in organizing and conducting edible landscaping and school garden planning session as part of the series of e-training on 13th of August, 2020. The training was planned and delivered for the 200 teachers of the seven schools (primary and secondary) from the island of Busuanga, province of Palawan. While the events' main participants were gathered via Zoom, the training session was live live-streamed on Youtube and has garnered one-hundred and fifty-seven (157) views, as well as five (5) comments. The e-training lasted for around four (4) hours, fifty-nine (59) minutes, and twenty (20) seconds. Just like with the previous webinar engagements, the resource speakers from the edible landscaping team utilized English and Filipino as mediums of instruction during the e-training. During the event, concepts and principles in edible landscaping were explained and a discussion on the utilization of indigenous plants as softscape materials in school edible gardens were included. The event also featured hands-on training focusing on the creation of a school-based edible landscape garden, as well as a garden plan to ensure sustainability. Designs were assessed by the resource speakers and they gave suggestions and recommendations. The e-training received favorable reactions from the audience and has inspired other attendees in organizing similar events on the same topic.

Case 7: EL in the Context of Alternative Food and Livelihood Sources

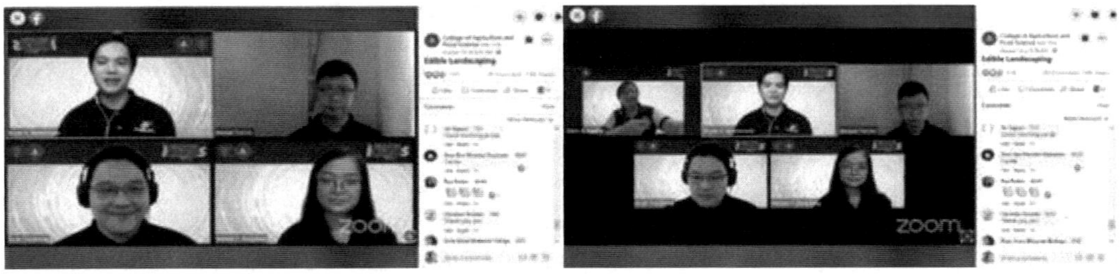

Figure 7 Aggie Ps Talk: Pagkain at Pangkabuhayan sa Panahon ng Pandemya (Food and Livelihood in the midst of Pandemic), last 14 August 2020 [13]

The College of Agriculture and Food Science (CAFS) at the University of the Philippines Los Banos reaffirmed its commitment to the Filipino people and orientation towards public service by organizing and conducting a webinar series entitled, "Aggie Ps Talk". The series primarily aimed towards the dissemination of knowledge on agricultural techniques and technologies developed in the college to inform the general public of the possible measures by which they can manage the socioeconomic impacts of lockdowns and restrictions on mobility due to the COVID19 pandemic. The edible landscaping technology is among these techniques and practices deemed as useful and adaptive measures as the availability and supply of food is threatened. The EL technology was featured in the mentioned webinar series on 14th of August, 2020. The said webinar lasted for around one (1) hour, twelve (12) minutes, and thirty (30) seconds, and gathered more than three thousand (3,000) viewers during its live stream on the social networking site, Facebook. Additionally, the video live stream garnered one-hundred and nineteen (119) Facebook reactions, and fifty-four (54) comments from the online audience. "Taglish" or the mix of the English and Filipino language served as the medium of communication during thisevent. The lecture presentation included a discussion on engaging with edible landscaping at the community-level. Towards the end of the webinar, one of the online participants asked a question related to community-based edible landscape gardens. Discussing the benefits of engaging in edible landscaping at both the household and community level has drawn interest among cause-oriented individuals and groups who attended the webinar. The webinar also managed to gather positive feedback and reactions from its audiences based on the responses that can be observed in its comment section.

Case 8: A Student-led and Organized Webinar on EL

Figure 8 Agventure with Hortsoc: A webinar about edible landscaping, last 29 August 2020 [14]

Providing an example of how the Filipino youth can demonstrate initiative and involvement in agriculture-related topics, the members of the university-based organization, UP Horticultural Society, organized a webinar series that featured topics in the discipline of horticulture, aptly titled, "Agventure with Hortsoc: An Agricultural Webinar". The series of webinars also served as the contribution of the student organization in helping to address the problems posed by the COVID19 pandemic and its consequent lockdowns in the country. On the 29th of August, 2020, the webinar series featured the edible landscaping technology and tapped a resource speaker from the Edible Landscaping team of UPLB. The student-led webinar on edible landscaping lasted two (2) hours, four (4) minutes, and thirty-nine (39) seconds, and was live-streamed through Facebook. It gathered around more than 3600 viewers and an audience of varied backgrounds, from all over the country. The webinar also managed to draw positive reactions based on the 196 reactions, and a total of two-hundred sixty-seven (267) comments was received. Some of the webinar participants hoped that their local governments would adopt the EL technology, while others were inspired to transform their existing ornamental gardens into one that is edible.

Discussion

This section provides a summary of the main patterns observed based on thematic analysis. At the time of the COVID-19 pandemic, the webinars and e-training conducted by the Edible Landscaping project team demonstrated variation in terms of content and themes, number of viewers, and length of duration. The delivery of the presentations varied based on the type of audience as exemplified by the first (general public) and fourth (schoolteachers) cases. Online audience and viewers totaled around 39,000 but varied across the webinars and e-training included in the study. The webinar and e-training sessions also varied in terms of duration, ranging from as short as thirty-minute audio-visual pre-recorded presentations to full-scale training sessions that lasted for almost four (4) hours. The e-training (Case 4) included in the study was the only one with a hands-on activity, while the webinar with an interactive feature was Case 1. Incorporating hands-on and interactive activities in webinars and e-training proved to be challenging as the internet connectivity of some audiences is unstable, and due to the general viewers attention limit span. The background of the collaborating and organizing teams also ranged from national and local government institutions, private companies, to university-based student organizations.

On the other hand, there were also some similarities among the eight (8) webinars and e-training analyzed. The lectures and discussions were delivered in both English and Filipino languages as these are the two official languages of the Philippines. The webinars and e-trainings were live-streamed mainly on the social networking site, Facebook, while others also posted video recordings via YouTube. In general, however, most of the online webinars involved the use of the popular video conferencing software, Zoom. These promotional activities also received generally positive reactions and feedback from the audiences online. The comments sections of the live-streamed videos have also shown that the delivery of the presentations was effective in evoking interest and inspiration among the audience. Inquiries about the technology ranged from technical concerns to ones that involved costing and wide-scale adoption. While the conduct of technology promotion activities such as webinars and e-training has been more challenging at the onset of the pandemic. The heightened awareness and interest of the public towards edible landscaping technology have provided an opportunity to showcase the importance and relevance of EL to the present needs of the country.

4. Conclusion

Ever since 2009 when the DA-BAR funded EL, the UPLB EL Team has been working closely with different agencies, schools, and institutions to promote the technology and address food self-sufficiency at the household level, especially in this time of the pandemic. Its use and importance were more highlighted at present due to the pandemic wherein food sufficiency is a concern. The overwhelming requests for webinars show that the potential of EL in sustaining food sufficiency has already been realized by many. The affirmative responses during the webinars and e-trainings have also demonstrated that the promotional activities conducted by the edible landscaping team, were in some way effective. The conducted webinars and e-training have also demonstrated the expansive utility and versatility of the edible landscaping technology in addressing social problems currently being experienced by the country as a whole. Through the promotional activities of EL that can help in addressing food sufficiency, it is expected that local communities and even those at the household levels will be encouraged to replicate and practice EL in the confines of their homes for them to have a readily available garden to table food. Hence, the UPLB EL Team's tagline, "no Filipino should be hungry".

References

[1] Coronavirus disease (COVID-19) pandemic [Internet]. WHO Regional Office for Europe. Available from: https://www.euro.who.int/en/health-topics/health-emergencies/coronavirus-covid-19/novel-coronavirus-2019-ncov

[2] COVID-19 Tracker [Internet]. Department of Health. Available from: https://www.doh.gov.ph/covid-19/case-tracker

[3] Employment Situation in April 2020 [Internet]. Philippine Statistics Authority. Available from: https://psa.gov.ph/statistics/survey/labor-and-employment/labor-force-survey/title/Employment%20Situation%20in%20April%202020#:~:text=e)%20Unemployment%20rate%20was%20estimated,million%20underemployed%20in%20April%202020.

[4] Yumol D. DOLE estimates 10 million workers will lose jobs this year due to COVID-19 pandemic [Internet]. CNN Philippines. Available from: https://www.cnn.ph/news/2020/5/21/dole-estimates-ten-million-workers-will-lose-jobs-covid-pandemic.html

[5] DA-AFID. DA assures enough food supply thru "Plant, Plant, Plant program" [Internet]. Department of Agriculture. Available from: https://www.da.gov.ph/da-assures-enough-food-supply-thru-plant-plant-plant-program/

[6] De Guzman R, Sanchez F C Jr, Balladares M C E, Mora J A M M, Tayobong R R P, Medina, N M, Aquino E C 2018 *Edible Landscaping Technology Dissemination in the Philippines: An Evaluation.* Journal of the International Society for Southeast Asian Agricultural Sciences (ISSAAS). Vol. **24**, No. 2: 150-158

[7] Apacionado B V 2020 *Edible Landscaping.* [Video]. Available from: https://www.facebook.com/bryan.apacionado/videos/10220610055434360 [Accessed 21 January 2021].

[8] Department of Agriculture – Bureau of Agricultural Research 2020 *Free Online Seminar. Edible Landscaping.* [Video] Available from: https://www.facebook.com/watch/?v=661006891117004 [Accessed 21 January 2021].

[9] Department of Agriculture – Bureau of Agricultural Research 2020 *Free Online Seminar. Edible Landscaping with Maria Charito Balladares.* [Video] Available from: https://www.facebook.com/1601722270068337/videos/2671188129806232 [Accessed 21 January 2021].

[10] AboitizLand, Inc 2020 *Edible Gardens: A Sustainable Solution for Growing Communities.* [Video] Available from: https://www.facebook.com/134024489998663/videos/735263483966238/ [Accessed 21 January 2021].

[11] DENR CALABARZON 2020*Webinar on Good Nutrition and Edible Garden in the New Normal.* [Video] Available from: https://www.facebook.com/DENR4AOfficial/videos/651990495401929 [Accessed 21 January 2021].

[12] SEAMEO SEARCA 2020 *Session 5: E-training on School and Home Gardens cum Biodiversity Enhancement for Busuanga Island.* [Video] Available from: https://www.youtube.com/watch?v=jlMojH_Pt5M [Accessed 21 January 2021].

[13] College of Agriculture and Food Science 2020 *Edible Landscaping.* [Video] Available from: https://www.facebook.com/cafsuplbph/videos/705606890297232 [Accessed 21 January 2021].

[14] UP Horticultural Society 2020 *Agventure with Hortsoc: Edible Landscaping.* [Video] Available from: https://www.facebook.com/watch/live/?v=791853924974321&ref=search [Accessed 21 January 2021].

IOP Conf. Series: Earth and Environmental Science **879** (2021) 012015 doi:10.1088/1755-1315/879/1/012015

Floodplans: landscape plan for a flood resilient municipality of Marilao, Bulacan, Philippines

G L Aquino[1] and H Palarca[2]

[1]Author, College of Architecture, University of the Philippines Diliman, Quezon City, Philippines
[2]Co-Author, Environmental Landscape Studio Laboratory College of Architecture, University of the Philippines Diliman, Quezon City, Philippines

E-mail: gizelle.aquino120@gmail.com

Abstract. According to the Philippine Atmospheric, Geophysical, and Astronomical Services Administration (PAGASA), an average of 20 typhoons a year visits the Philippines; and one of the significant repercussions of typhoons is flooding. Due to worsening and intensifying typhoons, the flooding problem is also getting worse, and two of the main reasons for this are land subsidence and sea-level rise. Marilao, Bulacan, is one of the Philippines' municipalities experiencing worsening flood levels. This study tackles an environmental approach to making the municipality adaptive and resilient to flooding by finding the most appropriate use of the land areas of Marilao – to benefit both the environment and its users. The study identified and analyzed the causes of flooding, using relevant maps such as hazard maps, existing land-use plans, and projected land-use plans to draft an effective plan for the municipality. After which, generating projections of the land conditions in the next century was done; to ensure that the proposed plans will accommodate future scenarios within the landscape. These data, along with four possible scenarios on the problems of the site, were analyzed. With this, the study took upon one of the approaches that led to the 2060 Environmental Protection Plan, which primarily focuses on addressing landscape requirements.

Keywords: flooding, landscape projections, land subsidence, sea-level rise

1. Introduction

In the Philippines, flooding has been an increasingly evident and severe problem [1]. One of the leading causes of flooding is the occurrence of typhoons in the country. According to the Philippine Atmospheric, Geophysical, and Astronomical Services Administration (PAGASA), an average of 20 tropical storms enters the Philippine Area of Responsibility per year. Moreover, among those 20 tropical storms, an average of 10 make landfall. The data shows that the Philippines is the most storm-exposed country on earth [2], making the country vulnerable to flooding effects and consequences. With Luzon and Visayas as the most susceptible to typhoon paths traverse the country each year [3].

Marilao, located at the south end of the province of Bulacan, Philippines, is a municipality that is part of the province's 4th district (Figure 1). It stretches in a northeast to southwest direction. Its southwest end is about 10km from Manila Bay's edge, and its northeast

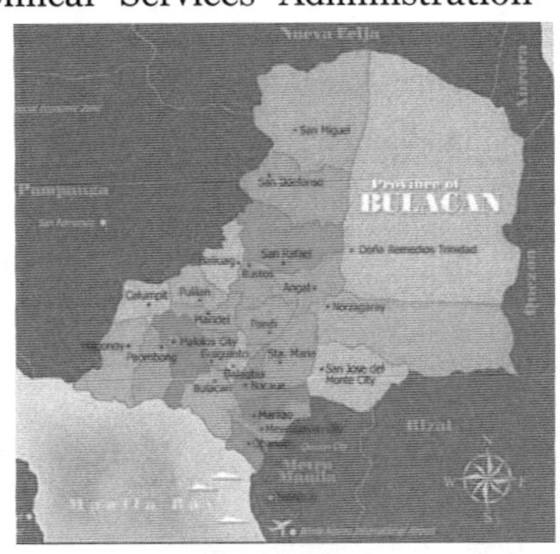

Figure 1 Province of Bulacan Municipal boundaries
Source: Peoples Development Institute Org

IOP Conf. Series: Earth and Environmental Science **879** (2021) 012015 doi:10.1088/1755-1315/879/1/012015

end is part of a pronounced alluvial plain beside the Sierra Madre Mountains. Higher elevations enclose the municipality on its landlocked northeastern side, while its southwestern parts are mostly flatlands of much lower elevations [4]. The water's general flow throughout the municipality is from the northeast end towards the southwest end, in which waters essentially flow toward Manila Bay; this is also how the water flows on the municipality's prominent river: the Marilao River, which traverses the municipality's entirety [5].

The municipality's geophysical characteristics make flooding an inevitable occurrence, mostly when large amounts of rain occur in the area because the municipality serves as a pathway of water coming down from higher elevations to the lower elevations, in which water should ultimately end up in Manila Bay.

As typhoons that hit the country become more robust due to global warming [6], the expectation is that flood levels will worsen, and its damaging effects are then also magnified. With this, specific solutions to flooding that are neither sustainable nor are adaptive to the changing dynamics of the environment may only provide temporary solutions. Reevaluating planning strategies for land-use concerning the current landscape and how it evolves through time ensures that proposed solutions do not resist nor restrain the force of nature but instead adapt to the environment so that communities may flourish despite the changing landscapes.

With this in mind, the study's main guiding question is: how can the municipality of Marilao, Bulacan, be transformed into a flood adapted municipality with regards to the 100-year flood level?

2. Methodology

The methodology follows a landscape planning framework that consists of two parts. Figure 2 [7] shows that the left side focuses on figuring out community considerations that support the study's landscape plan; this is done by first gathering relevant data from the community of Marilao, after which is the application of the two theories related to the community from the theoretical framework. This results in assessing what the community needs regarding flood resiliency and collating it into considerations based on the community.

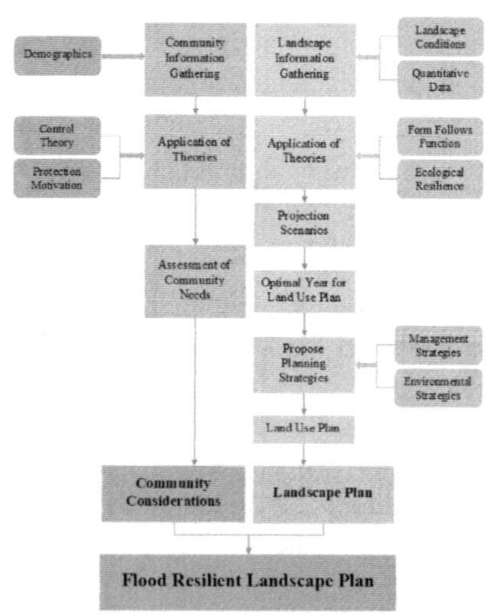

Figure 2 Research methodology

While the right side of the framework focuses on producing a land-use plan reflected into a landscape plan for the municipality of Marilao, the first step to achieving the landscape plan is extensive data gathering about the current conditions of the landscape of the municipality. After collecting data regarding the landscape, the next step is applying theories regarding the landscape to projection scenarios that may happen in the future. Projection scenarios reveal what would happen with the landscape conditions (relating to flood levels and intensity) if the landscape is left alone. Determining the optimal year for designing a land-use plan and a landscape plan results from knowing the projection scenarios. Finally, in drafting the landscape plan, the chronological execution of the proposed planning strategies regarding management strategies and environmental strategies is essential. The integration of the left and the right sides: the community considerations with the landscape plan shall result in the study's goal to design a flood resilient landscape plan.

3. Findings

3.1 Existing Landscape Conditions
Figure 3 [8] is a closer look at the municipality's boundary and how the Marilao River passes through the municipality's entirety. Also highlighted in figure 4 are the major roads that cross the municipality.

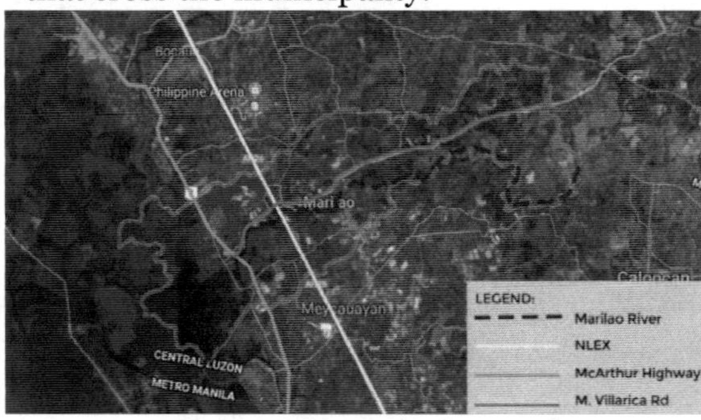

Figure 3 The boundary of Marilao, Bulacan
Source: Google Maps (2020)

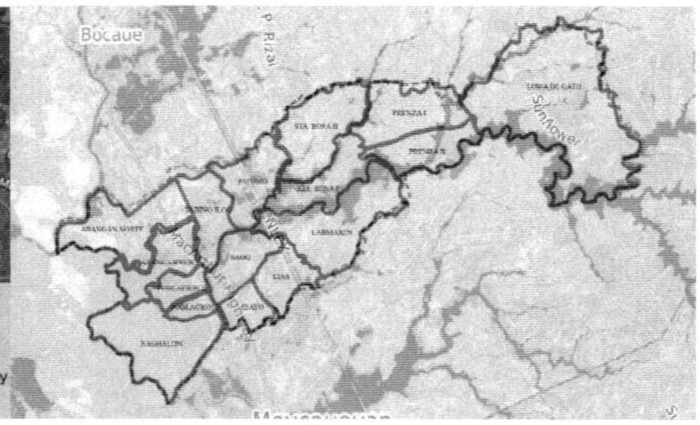

Figure 4 Barangay boundaries overlay in 100-year flood hazard map
Source: Google Maps, DOST-UP DREAM and Phil-LiDAR 1 Program

Overlaying the municipality's 100-year flood hazard map on the barangay boundaries defines which barangays are most affected by flooding. Findings showed that the majority of barangays in Marilao are severely affected by flooding. From the overlay map [9], the bulk of flood-prone barangays is at the southwestern end of the municipality, which is the lowest elevations of mostly at only 3 meters above sea-level, and the areas nearest Manila Bay [9]. As per areas on the northeastern part, although these areas are of significantly higher elevation peaking at 64 meters above sea-level, flooding still occurs – but only at specific sections that are directly adjacent to the river and its arteries.

3.2 Flooding Causes
Two situations result in flooding in Marilao, Bulacan: first is when the Marilao River overflows due to excess rainwater, and the second is when high tide surges. Typically, during typhoons or when the monsoons bring about large amounts of precipitation, water in the Marilao River's corridor swells until it inundates surrounding areas and then rushes to the lower southwest side of the municipality. Ideally, water drains to Manila Bay. Still, what happens in Marilao, the water gets trapped in the lower-lying areas because of poor drainage, siltation in bodies of water, and overflowing of Manila Bay. Flooding caused by tidal surges happens during high tide, wherein the lowest points of the municipality get flooded even without precipitation.

To effectively plan for flooding adaptability and resiliency of Marilao, it is not enough to look at the landscape's current conditions. Changes in the landscape is a factor impacting flood conditions and levels. Thus, to ensure that proposed strategies will still be relevant in the future, the landscape conditions and how it changes in the next century shall be investigated by this study.

3.3 Land Subsidence and Sea-Level Rise
Currently, Marilao has one of the highest rates of land subsidence in the country [10]. Based on a 2011 study by Kevin Rodolfo and Fernando Siringan, Marilao's land subsidence rate reaches about 55-60mm/year. Compared to the country's natural land subsidence rate

of around 0.7-2mm/year [10], which means that in less than 17 years, land in Marilao would have subsided by 1 meter already. This data is crucial because solutions applied to problems like flooding usually involve constructed engineering solutions such as building floodwalls, which would be ineffective in the long run because of land subsidence.

As per sea-level rise (SLR), According to PAGASA, sea-level rise in the Philippines is about 5-7mm/year [11], which is on the higher end of SLR globally. RCP 8.5, which is the high-end scenario projection of the sea-level rise in the Philippines made by PAGASA [11], factors in different causes of sea-level rise and is the projection used in this study.

3.4 Landscape Projections

Land subsidence throughout the next century is indicated below in Table 1. For sea-level rise, data from RCP 8.5 is shown below in Table 2.

Table 1 Projected land subsidence in Marilao

Year	Land Subsidence
2020	+0.6 m
2040	+1.8 m
2060	+3 m
2080	+4.2 m
2100	+5.4 m

Table 2 Sea-Level rise from 2020 to 2100

Year	Sea level Rise
2020	+0.1 m
2040	+0.2 m
2060	+0.4 m
2080	+0.8 m
2100	+1.1 m

3.5 Landscape Projections Cross-Sections

Changes in the landscape concerning the base water level from 2020 up to 2100 are graphically shown in the study's cross-sections [12]. The water level shown in the cross-sections below does not include the flood water level of the 100-year flood scenario but only shows the base tide level.

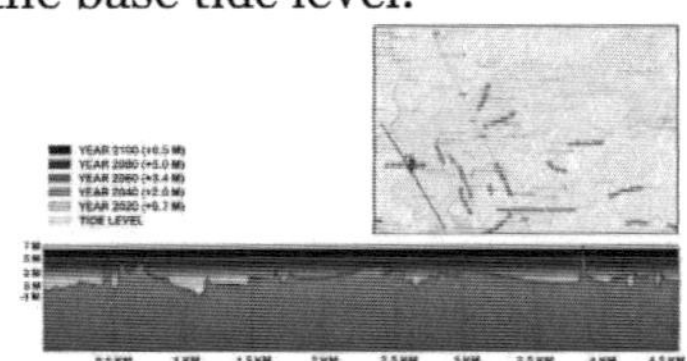

Figure 5 Cross section of southwest edge of Marilao (Fishpond Areas)

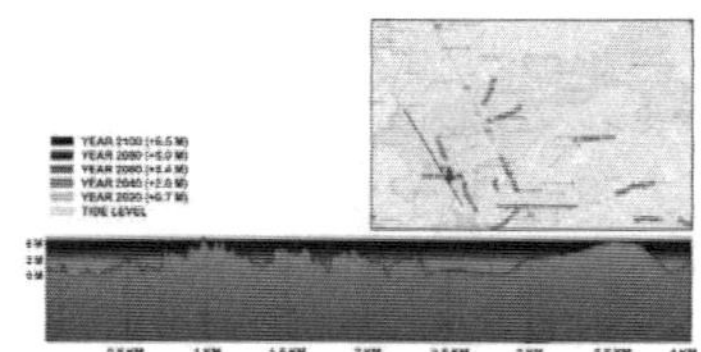

Figure 6 Cross section of residential areas at the lower southwest side

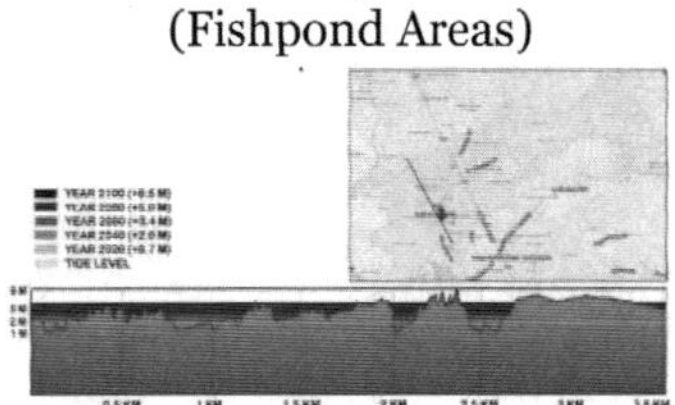

Figure 7 Cross section by macarthur highway

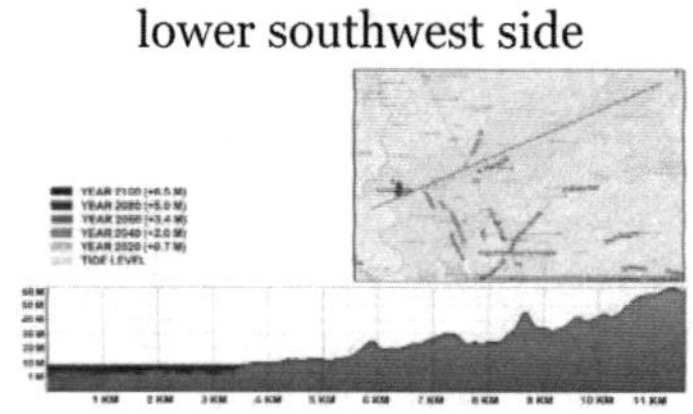

Figure 8 Longitudinal section

3.6 Population Projection

The residents of Marilao, Bulacan will be the ones primarily affected by the changes of the landscape in the next century. As the landscape changes, the population changes with it, and as such, the projection for the growth of the population in the next century is also calculated. The population projection is computed with the formula below, and the result is shown in figure 9 [13].

Exponential Change	**Exponential Growth Rate**		
Pt = Total Population Last Year * (e ^ r*t) where t = 1	r = (1/t)*ln(Total Population This Year / Total Population Last Year)	t = ln(Total Population This Year / Total Population Last Year) / r	1/t
245047	4.7493091%	1	1

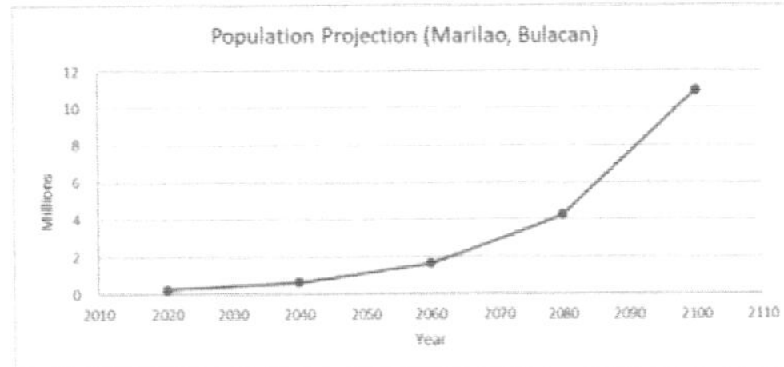

Figure 9 Line graph of population increase from 2020 to 2100

4. Results and Discussions

4.1 Discussion on Planning Strategy and Approach

In dealing with changing dynamics affecting the landscape, such as land subsidence and sea-level rise, choosing a planning strategy and approach is essential because it needs to handle high uncertainty events. Therefore, the planning strategy should be adaptable enough to prepare for more than a single outcome in the future. In designing the appropriate response to uncertainty, the approach should depend on the "degree of uncertainty and the degree to which a system can be controlled"[14].

According to the planning strategy matrix in Figure 10, Scenario Planning is the appropriate response "when control is difficult, and uncertainty is high" [14]. This is the case for this study; factors concerning flooding and its effects on the municipality of Marilao in the next century are highly uncertain and uncontrollable.

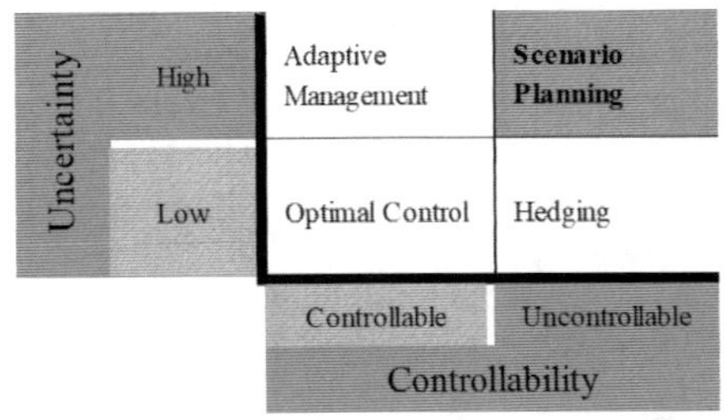

Figure 10 Planning strategy matrix from Peterson et al. (2003)

4.2 Formulation of Planning Strategy Framework

Adapting Peterson et al.'s Planning Strategy Matrix (Figure 10), Scenario Planning is used in formulating a planning strategy framework for Marilao, Bulacan. Scenario Planning deals with a variety of plausible futures but is not meant to predict the most probable future; instead, it has the objective to generate and examine decisions for each of the plausible futures [14]. With this method, scenario planning essentially aims to prepare for multiple futures, which equips significant actors with the "ability to recognize, adapt to, and take advantage of changes over time" [15]. In creating a variety of plausible futures, the study shall adapt the five-step process indicated in the study by Weeks (Figure 11) [15].

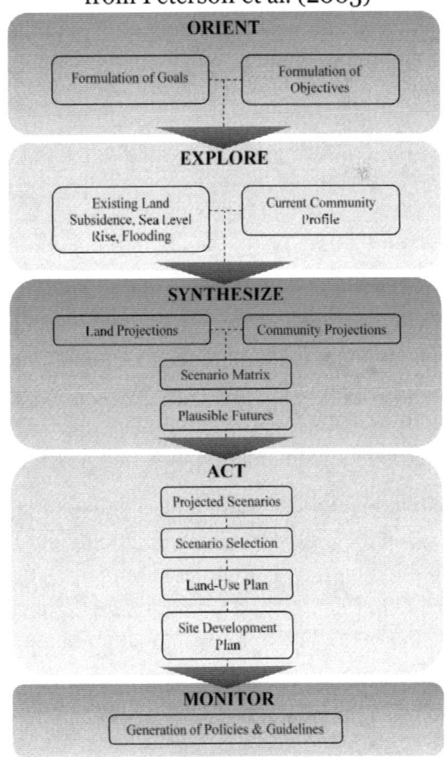

Figure 11 Planning strategy framework derived from Weeks et al. (2011)

4.2 Generation of Planning Schemes

Following the Planning Strategy Framework, generating the planning schemes begins with step three: Synthesize – wherein a scenario matrix is drafted by using a 2x2 matrix approach with the identified essential variables. The two variables used for the matrix are specifically: population growth for the *uncertain variable* regarding the community and land subsidence and sea-level rise for the *uncontrollable variable* regarding the landscape

[16]. The matrix generates four plausible futures A, B, C, and D – of which inferences are made, as shown in Table 3 below.

Table 3 Plausible futures inferences

Four Plausible Futures' Inferences:			
A	**B**	**C**	**D**
High Population Growth	*Low Population Growth*	*Low Population Growth*	*High Population Growth*
Low Land Subsidence, Sea-Level Rise	*Low Land Subsidence, Sea-Level Rise*	*High Land Subsidence, Sea-Level Rise*	*High Land Subsidence, Sea-Level Rise*
More usable land area	More usable land area	Less usable land area	Less usable land area
More people to be affected by flooding.	The landscape will be able to support its current intended use.	More on engineering solutions to control flooding	The landscape will not be able to support its assigned function.
Flood levels will not worsen exponentially.	The problem of flooding will persist if no solutions are applied.	Population affected by flooding could potentially relocate within the municipality.	The population will be exposed to more hazards or threats.
The adaptive strategy will be focused on equipping the community to cope with flood conditions.	Flooding solutions will be an immediate-reaction-response type.	Flood adaptive strategy for the community.	Solutions for both the landscape and the community issues are needed.
The development will sprawl to accommodate a high population.		The focus will be to control the flooding.	Flood levels will worsen.
The carrying capacity of landscape will be an issue.			High risk of overcrowding.
Municipality to become a city.			

In forming the four plausible futures' inferences, the guiding question is: What will happen to the municipality, concerning flooding, in the next century given the variables? Plausible futures A, B, C, and D are all given a specific scenario solution name and are assessed for risk and vulnerability, as shown in Table 4 below.

Table 4 Scenario solution name with risk & vulnerability assessment

Plausible Future	**Scenario Solution Name**	**Risk & Vulnerability Assessment**
A	Metropolis Shift Plan	Medium Risk & Vulnarebility
B	No Alteration Plan	High Risk & Vulnerability
C	Hard Engineering Solutions Plan	Medium Risk & Vulnarebility
D	Environmental Protection Plan	Low Risk & Vulnarebility

4.3 Selection of Planning Schemes

The explanation for the four different scenarios and the SWOT (Strengths, Weaknesses, Opportunities, and Threats) analysis per scenario is below (shown in Figures 12 to 15) [17].

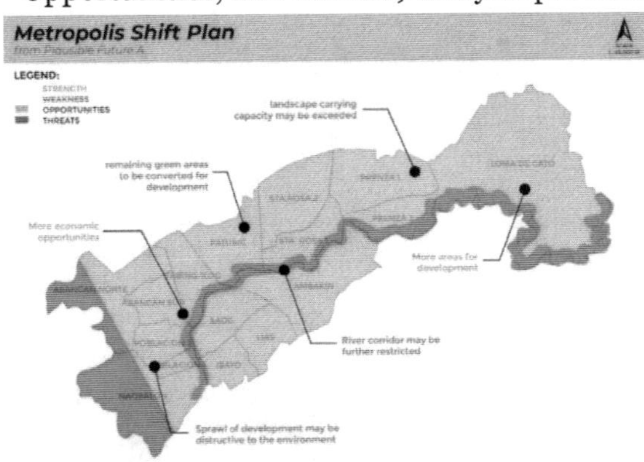

Figure 12 Metropolis shift plan

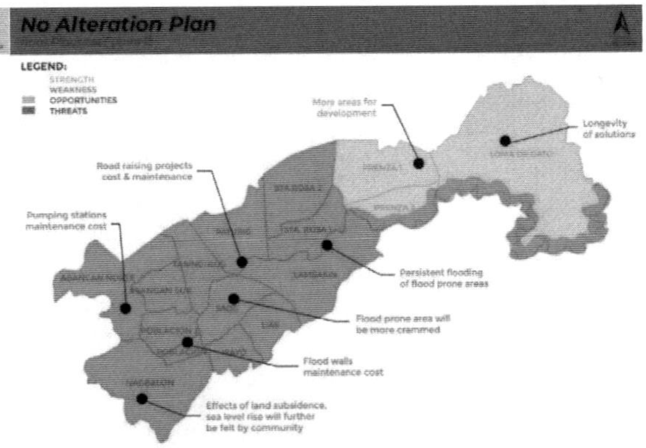

Figure 13 No alteration plan

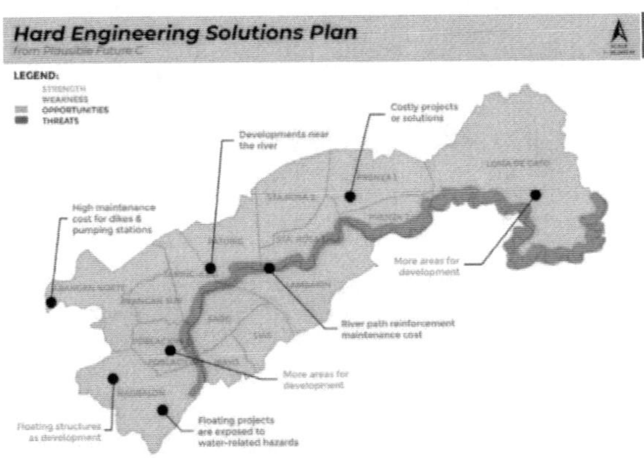

Figure 14 Hard engineering solutions plan

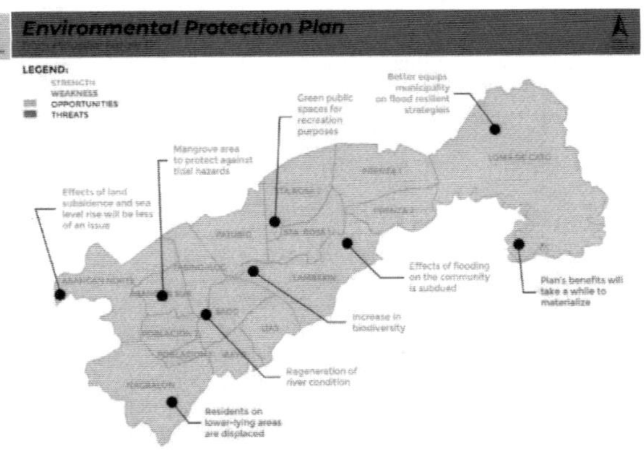

Figure 15 Environmental protection plan

From the evaluation of the four scenarios, an environmental approach is a priority for the solutions of the provisions related to the municipality's landscape conditions. Seeing that *Plausible Future D:* ***Environmental Protection Plan*** [17] has more strengths and opportunities with low risks and vulnerability and dealing with the extreme values of the variables considered, this strategy is applied in the study.

4.4 Environmental Analysis

As per the findings on land projections, each municipality's barangay was analyzed if it would be submerged or not in the next century. Table 5 below reflects data from land projections and cross-sections of the municipality. The (X) mark indicates that the barangay is already submerged by the respective year, and the (O) mark indicates that the area is still available for use.

Table 5 Land projection assessment from 2020 to 2100

Barangay	2020	2040	2060	2080	2100
Abangan Norte	O	O	X	X	X
Abangan Sur	O	O	X	X	X
Ibayo	O	O	O	O	O

Barangay	2020	2040	2060	2080	2100
Lambakin	O	O	O	O	O
Lias	O	O	O	O	O
Lias	O	O	O	O	O
Loma De Gato	O	O	O	O	O
Nagbalon	O	O	X	X	X
Patubig	O	O	O	X	X
Poblacion I	O	O	X	X	X
Poblacion II	O	O	X	X	X
Prenza I	O	O	O	O	O
Prenza II	O	O	O	O	O
Saog	O	O	O	O	O
Sta. Rosa I	O	O	O	O	O
Sta. Rosa II	O	O	O	O	O
Tabing Ilog	O	O	O	X	X

From the analysis of the landscape projections concerning land subsidence and sea-level rise, and with environmental conditions analysis of the site, the year 2060 is the ideal year to propose a landscape plan for a flood resilient municipality of Marilao, Bulacan. The year 2060 is chosen because it is the year wherein the municipality will experience drastic changes as five barangays are projected to submerge by 2060. Another basis for choosing this year is that there would be time to plan and shift the municipality's landscape from 2020-2060.

4.5 Land-Use Transition

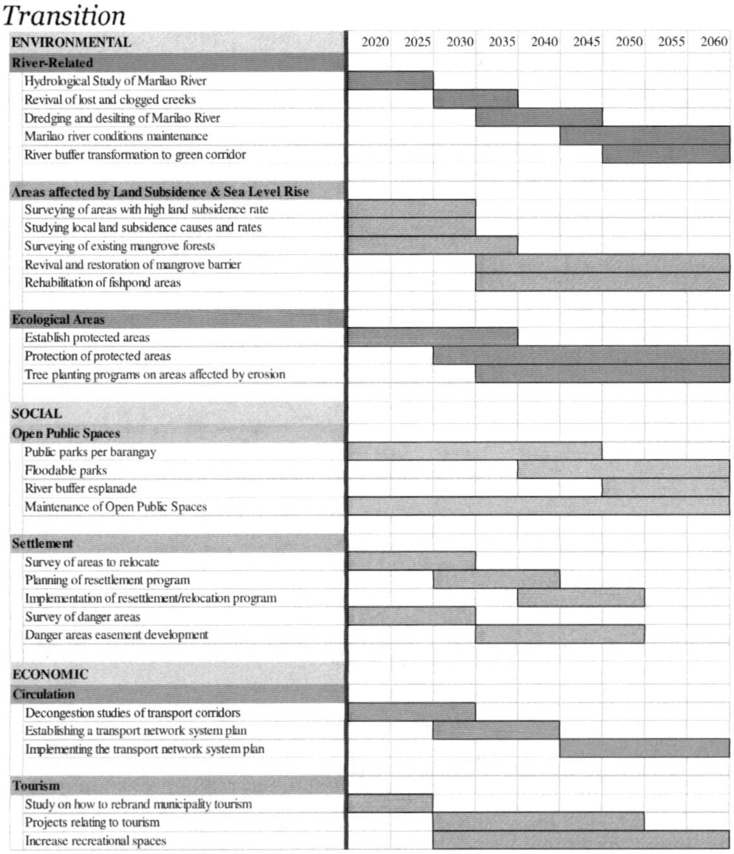

Figure 16 2020 to 2060 Gantt Chart

Figure 16 above shows a proposed Gantt chart on how the municipality of Marilao will shift from 2020 to 2060 as per the 2060 Environmental Protection Plan [17].

4.7 2060 Site Development Plan

The proposed master plan for Marilao, Bulacan, reflects the 2060 land-use plan [18] and the approach and proposals of the 2060 Environmental Protection Plan (Figure 17). The master plan recommends areas to be developed into green spaces to aid in the municipality's flood resiliency regarding the 100-year flood hazard map and the landscape projections in the next century.

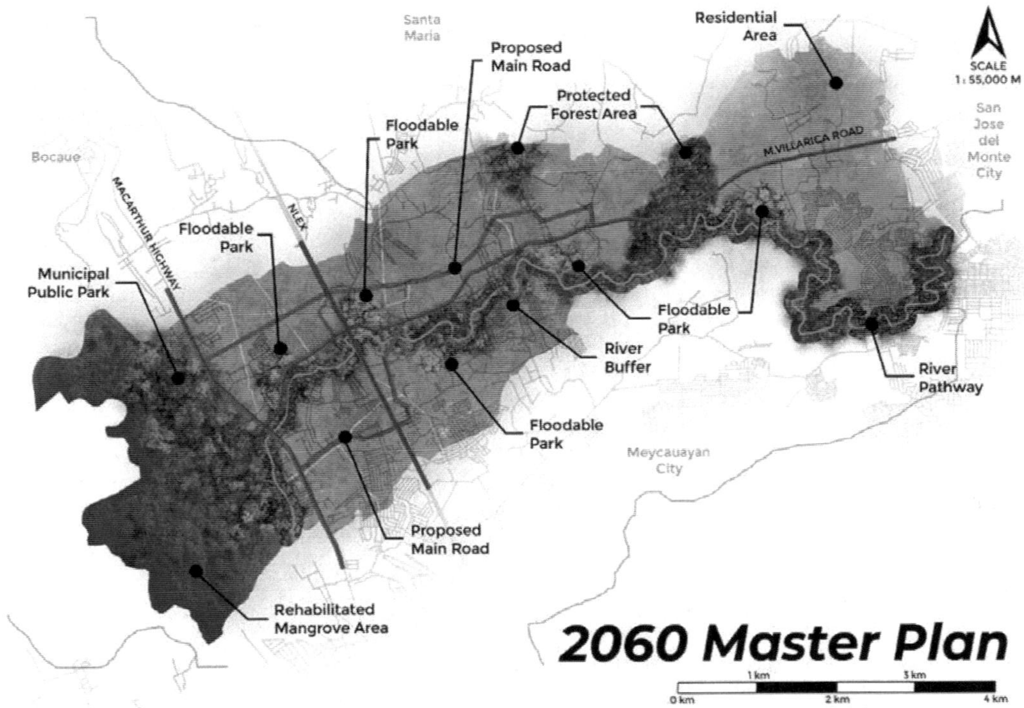

Figure 17 2060 Master Plan

4. Conclusion

The study takes on an environment-centered approach to adaptation and resiliency to flooding. The study is envisioned to serve as a preliminary body of information that approaches the drafting of solutions for flood-related problems by the projection and analysis of future scenarios and developing appropriate planning strategies.

The study's main problem, "How can the municipality of Marilao, Bulacan be transformed into a flood adapted municipality regarding the 100-year flood level?" was answered by creating the 2060 Environmental Protection Plan. The plan drafts actions that the municipality can implement and enforce to improve the municipality's flood resiliency and adaptivity by factoring how the landscape will transform in the next century.

References

[1] Bankoff G 2003 Constructing vulnerability: the historical, natural and social generation of flooding in Metropolitan Manila disasters [Internet] **27**(3):224–38. Available from: http://doi.wiley.com/10.1111/1467-7717.00230

[2] Brown S 2013 The Philippines is the most storm-exposed country on earth [Internet] TIME USA Available from: https://world.time.com/2013/11/11/the-philippines-is-the-most-storm-exposed-country-on-earth/

[3] Warren J Typhoons in the Philippines: a historical overview [Internet] ArcGIS Available from: https://www.arcgis.com/apps/MapJournal/index.html?appid=586f9150ae87491a8c7f1b86db7952a9#detail
[4] Marilao Planning Office 2020 Marilao, Bulacan elevation map and slope map Zenodo doi:10.5281/zenodo.4014926
[5] Google Images 2020 Marilao river flow direction doi:10.5281/zenodo.4014930
[6] Mei W and Xie S P 2016 Intensification of landfalling typhoons over the northwest Pacific since the late 1970s *Nat Geosci* **9**(10):753–7
[7] Aquino G L 2020 Floodplans research methodology chart doi:10.5281/zenodo.4014940
[8] Google Images 2020 Marilao Bulacan municipal boundary map doi:10.5281/zenodo.4014944
[9] Google Maps, DOST-UP DREAM and Phil-LiDAR 1 Program 2020 100-Year flood hazard map with municipal boundaries overlay Zenodo doi:10.5281/zenodo.4014946
[10] Rodolfo K and Siringan F 2011 Global sea-level rise is recognised, but flooding from anthropogenic land subsidence is ignored around northern Manila Bay, Philippines
[11] Cinco T, Villafuerte M I, Ares E, Manalo J, Agustin W, Aquino K et al 2018 Observed climate trends and projected climate change in the Philippines [Internet] Philippine Atmospheric, Geophysical and Astronomical Services Administration 2–11 p. Available from: https://icsc.ngo/wp-content/uploads/2019/07/PAGASA_Observed_Climate_Trends_Projected_Climate_Change_PH_2018.pdf
[12] Aquino G 2020 Floodplans cross-sections of Marilao, Bulacan doi:10.5281/zenodo.4014955
[13] Aquino G 2020 Marilao, Bulacan population projection from 2020 to 2100 doi:10.5281/zenodo.4014965
[14] Peterson G D, Cumming G S and Carpenter S R 2003 Scenario planning: A tool for conservation in an uncertain world *Conserv Biol.* **17**(2):358–66
[15] Weeks D, Malone P and Welling L 2011 Climate change scenario planning: A tool for managing parks into uncertain futures Park Sci [Internet] **28**(1):26–33 Available from: papers3://publication/uuid/8AE42E00-84A9-4E96-A746-F0BDDF0DD9ED
[16] Aquino G 2020 Floodplans generating scenario matrix doi:10.5281/zenodo.4014975
[17] Aquino G 2020 Floodplans planning schemes doi:10.5281/zenodo.4014977
[18] Aquino G 2020 2060 Proposed land-use map doi:10.5281/zenodo.4014981
[19] Buijs J, Boelens L, Bormann H, Restemeyer B, Terpstra T and Der T Van 2018 Adaptive planning for flood resilient areas: dealing with complexity in decision-making about multilayered flood risk management Proc 16th Meet Adapt Plan Spat Transform Groningen, Netherlands 23–5
[20] Chermack T 2011 Scenario Planning in Organizations: How to Create, Use, and Assess Scenarios Berrett-Koehler Publishers
[21] Favato G and Vecchiato R 2017 Embedding real options in scenario planning: A new methodological approach *Technol. Forecast Soc. Change* **124**:135–49
[22] Hutter G 2007 Strategic planning for long-term flood risk management : Some suggestions for learning how to make strategy at regional and local level *Int. Plan Stud.* **Volume 3(**No. 3):273–89
[23] Gunderson L H 2000 Ecological resilience — in theory and application

Analysis of *Telajakan* characteristic and the existence of ritual plants in Canggu Village and Penglipuran Village

I M P D Natawiguna[1*], H S Arifin[2], R L Kaswanto[2],
[1]Graduate Student of Landscape Architecture Department, IPB University, Bogor, Indonesia
[2]Landscape Architecture Department, Faculty of Agriculture, IPB University, Bogor, Indonesia

Email: mademonde13@gmail.com

Abstract: The life of Balinese people can never be separated from ritual activities or religious ceremonies. The existence of *telajakan* is essential to preserve. Aside from being a traditional Balinese identity, it also provides space for ritual plants and biodiversity. So that effort is needed to manage the existence of *telajakan*. This research is expected to analyze the characteristics of telajakan and the existence of ritual plants found in *telajakan*. Field observation was undertaken to inventory distribution point of remains *telajakan* in Canggu village. The sampling method uses purposive sampling by tracing along the main road and marking the visible point of *telajakan*. Changes function of *telajakan* was found in the Canggu area, and in Penglipuran villages, there is still preserved. Following this study, 49 samples of remains *telajakan* were collected from Canggu village and 69 samples from Penglipuran village. The characteristics of *telajakan* can be observed with the structure and size. The result shows, the structure of *telajakan* can be any combination of an open ditch, covered ditch, pedestrians track, open ground, grass, and fence wall. Size of *telajakan* can be grouped into small size (width <50 cm), medium size (width 50-99 cm), large size (width 100-149 cm), and extra-large size (width > 150 cm). Based on survey research and observation, various plants are found in each sample. Seventy-seven species of plants were found in 49 samples of *telajakan* in Canggu village. *Plumeria accuminata, Codiaeum Sp.*, and *Bougainvillea spectabilis* is the most common species. Meanwhile, in Penglipuran village, 117 species of plants were found in 69 samples of *telajakan*. Furthermore, the most common species found were *Plumeria accuminata, Alternanthera ficoidea, Cordyline terminalis, and Cananga odorata*. Result show both Canggu village and Penglipuran villages have high species diversity index (H'>3). Canggu village species diversity index amount 3.88 and Penglipuran village amount 4.12. The utilization of *telajakan* as space for ritual plants can preserve Balinese tradition and cultures' local identity.
Keywords: Canggu, green space, Penglipuran, ritual plants, *telajakan*

1. Introduction

Indonesia, as a developing country, has rapid growth. Economic growth from 2010-2019 around 5-6%, and population growth of 1.25%. Based on the 2020 population census, Indonesia's current population is 270 million [10]. Fast economic and population growth increased the need for land for economic development and residence. Land-use changes inevitable, especially changes of green space into built area. Some major cities in Indonesia, such as DKI Jakarta, Bandung, and Yogyakarta, have decreased the green space in 1972-2013

[9]. Bali, as one international tourist destination, also experience land-use changes. The existence of traditional green space in Bali has begun to changes. It is urgently for keep Balinese traditional identity. Rapid land-use change from green area to build area in Bali can reduce green space. Canggu, as a favorite tourist destination in Bali, runs into significant land-use change. During 2012-2017, the increased number of tourists cause land conversion from an agricultural area into buildings and tourism accommodation in Canggu [1]. Land conversion directly decreased agricultural production and undoubtedly impacted the environmental aspect [2]. Greenspace has an essential role in maintaining environmental sustainability. Green spaces can provide growing space for various plant species. Land conversion into build spaces should be balanced with the provision of green space.

Bali has local wisdom to managing human living space in harmony with nature. The philosophy of *Tri Hita Karana* (THK) and *Tri Mandala* (TM) form the spatial configuration in Bali's spatial layout. Balinese traditional housing has *telajakan* that have been become an identity in traditional housing pattern in Bali. The existence of *telajakan* as a green space with a strong linear pattern shape green corridor in Bali's traditional villages [3]. Efforts to increase the presence of green space in tourism areas can be made by restoring the existence of *telajakan*. THK philosophy guides the Balinese to living harmoniously with nature. Balinese respect nature by utilizing the existence of plants as a ritual facility for the religious ceremony. The plants that are used for a ritual facility called ritual plants. Various plant species are used as a raw material for the ritual facility. Approximately there are 300 kinds of plants used in Bali Hindu rituals [4]. Utilizing the ritual plants directly maintains the diversity of plants and a sustainable environment.

In Indonesia, Constitution No. 32 of 2009 [5] explains environmental management can be based on local wisdom. The development of *telajakan* as a green space in the tourism area can strengthen Bali's local identity. As an example, the tourism area in Penglipuran villages maintains *telajakan* as a part of the attraction. Canggu villages and Penglipuran villages show the different characteristics of *telajakan*. Canggu village with high community dynamics, and Penglipuran village with low community dynamics. This study will explain the structural configuration of the *telajakan* and the existence of the ritual plants in both villages.

2. Material and Methods

2.1 Study site and time

The study was conducted in Canggu villages, Badung Regency and Penglipuran villages, Bangli Regency, (Bali Figure 1). Sample of remains *telajakan* in Canggu was collecting alongside *Jalan Raya Batu Bolong,* and Penglipuran was collecting on the village's main road (*Rurung Gede*). The reason for choosing these two villages because Canggu villages and Penglipuran villages are famous tourism areas with different community dynamics and land-use changes. This study starts from January 2019 until May 2020.

2.2 Tools and Materials

The material used in this study is spatial data such as Google satellite images year 2019 obtains from Google Earth and literature data from the Encyclopedia of Balinese Ritual Plants. At the same time, the tools used are the Global Position System (GARMIN GPS map 62s), laptops (Intel core i7 6700 HQ), and smartphones (OPPO F7). The software that was used is ArcGIS (ArcMap 10.3), Microsoft Office 2016, AutoCAD 2018, Sketchup 2019, and Adobe Photoshop CS6.

2.3 Study Methods

This study's object is *telajakan* in Canggu village (Figure 1) and Penglipuran villages (Figure 2). This study used direct survey, observation, and measurement. A literature study was conducted to determine the variety of ritual plants used by Balinese. Data collection in-site study consists of general condition, sample point of *telajakan*, and existing plant species. Furthermore, the sampling method uses purposive sampling by tracing along the main road and marking the visible point of *telajakan*. The stages of this study include preparation, inventory, analysis, and output.

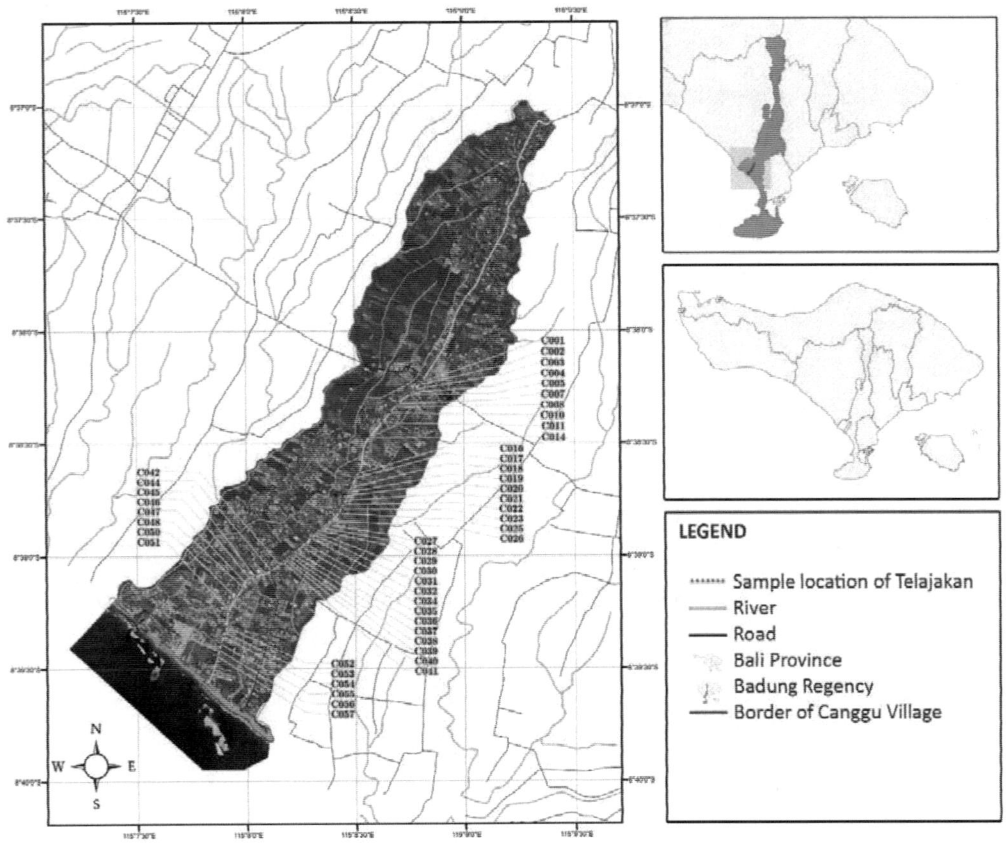

Figure 1 Study Site and Sample Location in Canggu Village

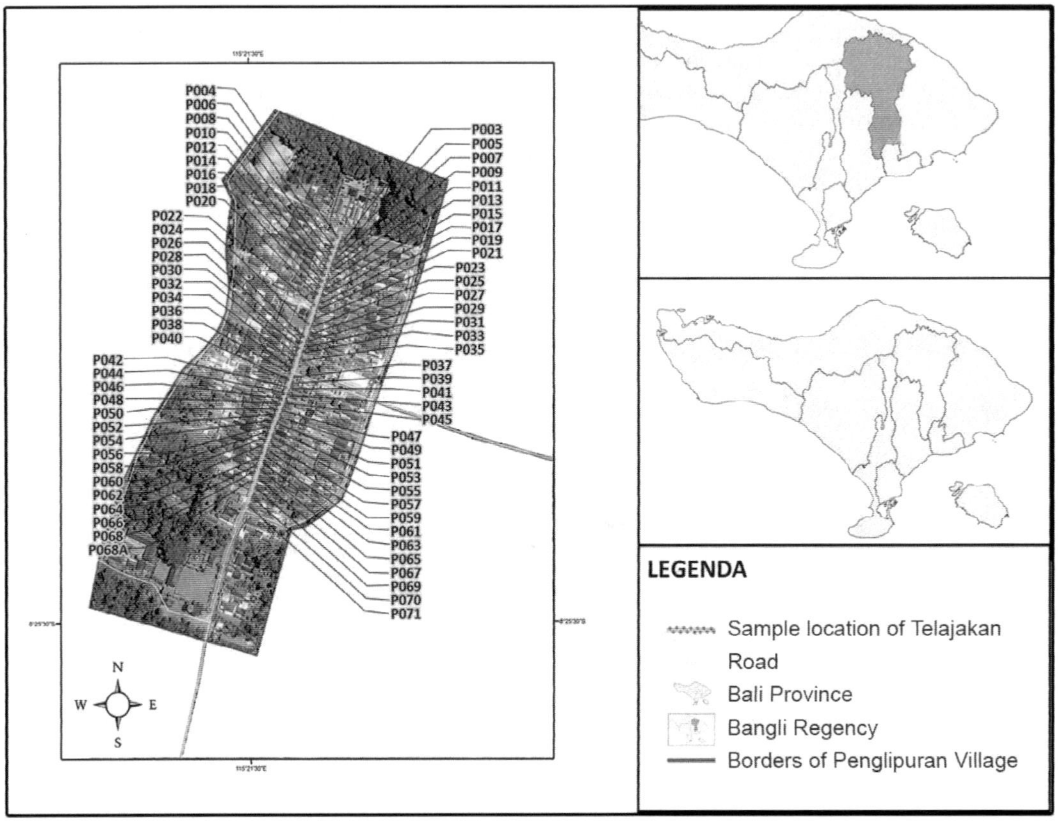

Figure 2 Study Site and Sample Location in Penglipuran Village

2.3.1 Preparation

Preparation starts from administrative and technical preparation. Pre-survey was conducted before selecting the location of the sample. Selection of potential study site by quick observation of the remains *telajakan* and pedestrians' crowd.

2.3.2 Inventory

In the inventory stages, the sample point of the *telajakan* collecting by GPS. Recording the plant species and documentation of each sample point is also carried out at this stage. For the *telajakan* characteristics, 49 samples of remains *telajakan* were collected from Canggu village and 69 samples from Penglipuran village.

2.3.3 Analysis

Data analysis on collected samples were compared based on the shape typology, and each sample was grouped according to the same shape. Descriptive qualitative analysis approach used to know the structural configuration of the *telajakan*. Structure configuration was conducted by observing the ditch, open ground, grass, pedestrians track, and fence wall. Size observation was also carried out at this stage. For the ritual plants' existence, the descriptive

IOP Conf. Series: Earth and Environmental Science **879** (2021) 012016 doi:10.1088/1755-1315/879/1/012016

quantitative analysis was used to find plant species diversity. This analysis generated the value of the species diversity index and list of ritual plants. Species diversity index can be calculated with equation Shannon and Weinner [7]

$$H' = \sum_{i=1}^{s} \text{Pi} \ln \text{Pi}$$

Information:
H' = Species diversity index
ni = Number of individual species
N = Total number of individuls
Pi = ni/N

The value calculation of the diversity index (H') show high species diversity ($H'>3$), medium species diversity ($1 < H' < 3$), or diversity low species ($H' < 1$). High species diversity has high community stability, while low species diversity has low community stability in an ecosystem.

2.3.4 Output
This study's output is the general condition of Canggu village and Penglipuran villages, characteristics of *telajakan*, the value of plants species diversity index, and a list of *telajakan* ritual plants in both villages.

3. Result and Discussion

3.1 General Condition
Canggu village is located in southwest Bali Island, specifically in North Kuta district, Badung regency, Bali province, Indonesia. The name Canggu famous as a top destination for international surfer. Canggu Beach and Batu Bolong beach becomes the most visited tourist spots. Topography condition mostly flat with a slope less than 3%. The total area of this village is 5.23 Km^2 and 1.265 People/ Km^2 for the population density. Average temperature ranges around 25-29 ^oC and annual average rainfall amount to 82.75 mm/year. The landscape structure dominates rice fields in the northwest area and the coastal tourism landscape in the southwest area. *Jalan Raya Batu Bolong* as main access road in the Canggu area (Figure 3) direct access to the beach. Alongside this road can be rice fields, houses, hotel villas, art shops, café, and restaurants.

Figure 3 *Jalan Raya Batu Bolong*

Penglipuran village is located in Bali Island, specifically in Bangli district, Bangli regency, Bali province, Indonesia. This village provides examples of spatial management following the T*ri Mandala* (TM) philosophy. The linear pattern in the Penglipuran villages following the mountain's axis in the north and sea in the south. TM philosophy explained mountain represents the pure space, and the sea represents the impure space [6]. Topography condition relatively flat with a slope of less than 5%. The site location is 500-625 meters above sea levels. Average temperature ranges around 18-27 °C and annual average rainfall amount 2000 mm/year. The landscape structure dominates with a bamboo forest in the northside and mix garden around the village. As a tourist village, the main road (*Rurung Gede*) is only accessible by walking. The parking area provides in the center and the north of the village (Figure 4).

Figure 4 Main Road and Parking Area

3.2 Characteristics of Telajakan

Following this study, 49 samples of remains *telajakan* were collected from Canggu village and 69 samples from Penglipuran village. The characteristics of *telajakan* can be observed with the structure and size. The result shows, the structure of *telajakan* can be any combination of an open ditch, covered ditch, pedestrians track, open ground, grass, and fence wall. Size of *telajakan* can be grouped into small size (width <50 cm), medium size (width 50-99 cm), large size (width 100-149 cm), and extra-large size (width > 150 cm).

The *telajakan* structure configuration in Penglipuran villages has an original form and applied in all found. This proves that the *telajakan* structure in these villages is still preserved. The format is an open ditch, open ground, grass, and fence wall. For the size of the *telajakan*, these villages have a large size (width 100-149) on average. While in Canggu village, 11 structure configuration was found (Table 1). This structure configuration can be used for the development of *telajakan* models in areas with high community dynamics. STc code means structure configuration, and the following number one until eleven is a total configuration found during the observation. STc-2 consists of ten samples that were the highest number found. Structure configuration of Canggu *telajakan* mostly in the form of STc-2. This happens because, in Canggu, the pedestrian track is more often used by foreign tourists. So, the structure of the pedestrian track, visible in most samples. Construction of pedestrian's way closed the ditch alongside *telajakan,* thereby further strengthening the presence of STc-2.

Table 1. *Telajakan* Structure Configuration

Structure of *Telajakan*						configuration	
	covered ditch	pds-track		grass	fence wall	STc-1	**Canggu village**
	covered ditch	pds-track	open ground		fence wall	STc-2	
	covered ditch		open ground		fence wall	STc-3	
	covered ditch	pds-track	open ground			STc-4	
	covered ditch	pds-track	open ground	grass		STc-5	
	covered ditch	pds-track	open ground	grass	fence wall	STc-6	
	covered ditch		open ground	grass		STc-7	
	covered ditch		open ground			STc-8	
	covered ditch				fence wall	STc-9	
open ditch				grass	fence wall	STc-10	
open ditch			open ground	grass	fence wall	STc-11	

Structure of *Telajakan*						configuration	
open ditch			open ground	grass	fence wall	STc-11	**Penglipuran village**

Note: pds-track means pedestrians track

In Canggu villages, the average size of *telajakan* in range medium size (width 50-99 cm). Pedestrians track occupied the original width of *telajakan*. The high density of pedestrians reduces green space on *telajakan*. The covered ditch also combines with the open ground to provide commercial space. An interesting finding was seen in sample C056 with Stc-5. This sample have extra-large size (width > 150 cm) *telajakan* and 16 species plants. This is the largest size and the most varied plant species found in Canggu villages. This sample's existing condition is a hotel that concern with green space—modification of *telajakan* in Canggu village highly influenced by the rapid development of tourism. Economic benefits are obtained easily by utilizing the space as a tourism support area. The remaining *telajakan* is community creativity for preserving *telajakan*.

3.3 Plants Diversity in Telajakan

Based on survey research and observation, various plants are found in each sample. Seventy-seven species of plants were found in 49 samples of *telajakan* in Canggu village. *Plumeria accuminata, Codiaeum Sp.*, and *Bougainvillea spectabilis* is the most common species (Table 2). Meanwhile, in Penglipuran village, 117 species of plants were found in 69 samples of *telajakan*. Furthermore, the most common species founded were *Plumeria accuminata, Alternanthera ficoidea, Cordyline terminalis, and Cananga odorata* (Table 3). Result show both Canggu village and Penglipuran villages have high species diversity index ($H'>3$). Canggu village species diversity index amount 3.88 and Penglipuran village amount 4.12. This value showed the variety of plants in Penglipuran village higher than Canggu village. The community in Penglipuran has increased awareness of the environment affected by local wisdom and local law (*awig-awig*) that is still respected [11]. Thereby the existence and the diversity of plants in Penglipuran *telajakan* close to the community.

Likewise, in Canggu, there is a high value of species diversity index obtained with an awareness of the community. Although the development of tourist facilities in Canggu is relatively rapid, the community is still trying to maintain the existence of the ritual zone [12]. So, the existence of *telajakan* and ritual plants can be pursued.

Table 2. Frequent plants species found in *telajakan* of Canggu Village

No	Scientific Name	Local name	Frequent species found in *telajakan*
1	*Plumeria accuminata*	Jepun bali/kamboja bali	28
2	*Bougainvillea spectabilis*	bougenvil	12
3	*Codiaeum sp.*	Puring	11
4	*Cocos nucifera L.*	Kelapa	9
5	*Dypsis lutescen*	Palem kuning	9
6	*Syzygium oleina*	Pucuk merah	9
7	*Cordyline terminalis*	Hanjuang	8
8	*Dracaena draco*	Pandan bali	8
9	*Hymenocalis speciosa*	Bakung mancur	8
10	*Veitchia merillii*	Palem putri	8

Table 3. Frequent plants species found in *telajakan* of Penglipuran Village

No	Scientific Name	Local name	Frequent species found in *telajakan*
1	*Alternanthera ficoidea*	krokot merah kerdil	40
2	*Plumeria accuminata*	Jepun	40
3	*Cordyline terminalis*	Hanjuang	29
4	*Cananga odorata*	sandat/kenanga	27
5	*Chlorophytum comosum*	lily paris	21
6	*Zinnia elegans*	Kembang kertas	18
7	*Impatiens balsamina*	pacah	17
8	*Hibiscus rosa sinensis*	Bunga kembang sepatu	16
9	*Hydrangea macrophylla*	Bunga pecah siu	16
10	*Codiaeum sp*	puring	13

3.4 Existence of Ritual Plants in Canggu village and Penglipuran village

Literature study referring to the Encyclopedia of Balinese Ritual Plants [8]. The plant species data collected in-site research was compared to the literature and selected plant with a ritual function. Result show 56.4 % of plants species that found in Penglipuran village have functioned as ritual plants. *Plumeria accuminata, Cananga odorata, Impatiens balsamina, Hibiscus rosa sinensis, and Hydrangea macrophylla are common* in this area. This composition of flowers usually used in *canang* and *kwangen* ritual facilities (Figure 5). *Canang* is a daily ritual facility that is easily found in the Bali environment. *Cordyline terminalis* and *Codiaeum sp* are the typical leaves found in Penglipuran villages. The community uses these leaves to complete *penjor* ritual facilities. The use of *penjor* symbolizes the mountain and community usually made for Galungan and Kuningan's special day.

Figure 5 *canang, kwangen, and penjor* ritual facilities

The percentage of ritual plants found in Canggu village was 45.4 %. The most common plant species found, namely *Plumeria accuminata, Bougainvillea spectabilis, Codiaeum sp,* and *Cocos nucifera L. Jepun* flowers used, were the same as those found in Penglipuran villages. Bougainvillea flowers have various colors that are also usually used in many ritual facilities. Utilizing the *cocos* leaves often seen on primary material of ritual facilities. Supplementary material that is traditionally used is banana leaves (*Musa paradisiaca*), but in this study, this plant is not in *telajakan.*

4. Conclusion

Following this study, 49 samples of remains *telajakan* were collected from Canggu village and 69 samples from Penglipuran village. The characteristics of *telajakan* can be observed with the structure and size. The result shows, the structure of *telajakan* can be any combination of an open ditch, covered ditch, pedestrians track, open ground, grass, and fence wall. Size of *telajakan* can be grouped into small size (width <50 cm), medium size (width 50-99 cm), large size (width 100-149 cm), and extra-large size (width > 150 cm). Based on survey research and observation, various plants are found in each sample. Seventy-seven species of plants were found in 49 samples of *telajakan* in Canggu village. *Plumeria accuminata, Codiaeum sp.*, and *Bougainvillea spectabilis* is the most common species. Meanwhile, in Penglipuran village, 117 species of plants were found in 69 samples of *telajakan*. The most common species founded were *Plumeria accuminata, Alternanthera ficoidea, Cordyline terminalis, and Cananga odorata*. Result show both Canggu village and Penglipuran villages have high species diversity index (H'>3). Canggu village species diversity index amount 3.88 and Penglipuran village amount 4.12. Result show 56.4 % of plants species that found in Penglipuran village have functioned as ritual plants. *Plumeria accuminata, Cananga odorata, Impatiens balsamina, Hibiscus rosa sinensis, and Hydrangea macrophylla are common* in this area. The percentage of ritual plants found in Canggu village was 45.4 %. The most common plant species founded were *Plumeria accuminata, Bougainvillea spectabilis, Codiaeum sp*, and *Cocos nucifera L.* The utilization of *telajakan* as a space for ritual plants can preserve Balinese tradition and cultures' local identity.

Reference

[1] Lanya I, I N Dibia, I W Diara, and D G Suarjaya. *Analysis of Subak Landuse Change Due to Tourism Accommodation Development in North Kuta Sub-district, Badung Regency, Indonesia*. IOP Conference Series: Earth and Environmental Science. 2017; 98:012024.

[2] Irene, Pinta Lizti. Farmland Conversion in Karawang, Indonesia: Discourse Analysis. Dissertation Faculty of Bioscience Engineering, Universitiet Gent. 2015

[3] Yudantini, N M Natah and Telajakan: *The Role and Identity in Indigenous Villages*. Pages 179-187 *in* Proceedings of International Seminar on Place Making and Identity. 2012

[4] Sardiana I K, and K K Dinata. Studi pemanfaatan tanaman pada kegiatan ritual (upakara) oleh umat Hindu di Bali. Jurnal Bumi Lestari. 2010; 10:123-127

[5] [PRI] Pemerintah Republik Indonesia. Undang-Undang Republik Indonesia Nomor 32 Tahun 2009 tentang Perlindungan dan Pengelolaan Lingkungan Hidup. 2009

[6] Arifin H S, and N Nakagoshi. *Landscape ecology and urban biodiversity in tropical Indonesian cities*. Journal of Landscape ecological engineering. 2009; 7:33-43.

[7] Magurran A E. *Ecological Diversity and Its Measurement*. New York (US): Chapman & Hall. 1991

[8] Sardiana I K, I W Windia, I K Sundra, I K K Dinata, I S M Sarwadana, and D I W Sukersa. Taman Gumi Banten Ensiklopedi Tanaman Upakara. Swasta Nulus. 2014

[9] Budiman A, Sulistyantara B, Zain A F. Deteksi perubahan ruang terbuka hijau pada 5 kota besar di Pulau Jawa (Studi kasus: DKI Jakarta, Kota Bandung, Kota Semarang, Kota Jogjakarta, dan Kota Surabaya). Jurnal Lanskap Indonesia. 2014 May 23;6(1):7-15.

[10] BPS-Statistics Indonesia. Statistical Yearbook of Indonesia 2020. BPS-Statistics Indonesia. 2020

[11] Qolby MT, Alhaq MT. Kajian kepedulian masyarakat berbasis kearifan lokal dalam upaya pelestarian lingkungan di desa penglipuran bali. Jurnal Ilmiah Pendidikan Lingkungan dan Pembangunan. 2019 Sep 28;20(02):1-2.

[12] Nurjani N P, Dwijendra N K. *How Traditional Balinese Houses Can Adjust and Cater for International Tourists in the Canggu Area, Bali Indonesia*. International Journal of Psychosocial Rehabilitation. 2020;24(03).

Planning for alternative road development connecting East Balikpapan - North Balikpapan

I Setiabudi[1] ,and I M A Hermawan[2]
[1]Polytechnic of Indonesian Land Transport, Ministry of Transportation, Jl. Raya setu No.89 Cibitung Bekasi 17520, Indonesia,
[2]Polytechnic of Indonesian Land Transport, Ministry of Transportation, Jl. Raya setu No.89 Cibitung Bekasi 17520, Indonesia,

E-mail : iqrasetia@gmail.com; arkahermawan@gmail.com

Abstract Balikpapan City is one of the largest cities in the East Kalimantan region and a city that is developing quite rapidly. These developments include a high population number accompanied by a large number of new activity growths which create a considerable traffic pull around the city center. According to the 2012-2032 City Spatial Plan for Balikpapan City, the priority plan for road development is on Outer Ring Road III and other alternative roads which will become new activity centers in Karang Joang and Manggar sub-districts in Balikpapan City.
The road network currently has the highest volume of 4288 pcu / hour with an average speed of 42 km / h and reaches a VCR of 0.8 which has 35 existing roads and 17 intersections. This study intends to improve traffic performance and aims to plan an alternative road network connecting East Balikpapan with North Balikpapan.
Simulation analysis uses Visum software to produce the best conditions for handling traffic in Balikpapan City arteries, collectors and local roads where the current condition of the traveler has to go through the East, South, Central and North Balikpapan District network. The method used is the calculation of existing traffic performance and simulation to determine network volume, network speed, total distance and total travel time. The simulation results with the Visum model are the best conditions of traffic performance with an average volume of 762 pcu / hour, an average network speed of 42 km / hour, a total distance of 148487 km, and a total travel time of 3369 hours.
Keywords: distance and travel time, network, speed, volume

1. Preliminary

The development of a certain area and / or location has an influence on the traffic around it. The transportation planning system is used to predict whether the transportation infrastructure is affected by this development and whether it can serve the existing traffic, plus the traffic that the development generates or attracts. If the existing infrastructure cannot support the traffic, a study on the handling of the infrastructure or the regulation of the traffic must be carried out, particularly those that are thought to have a significant impact on the road network system in the area. [1]

Balikpapan City is one of the big cities that continues to carry out development according to its position in the National Spatial Plan (RTRWN) as a National Activity Center (PKN) which functions or has the potential to be a National Marine Transportation Node, National Air Transportation Node, Mainstay Area, and Development Area. Integrated Economy (KAPET) which is equipped with the development of a highway network from Balikpapan - Samarinda (RTRW Kota Balikpapan, 2012). Meanwhile, the position of the City of Balikpapan in the Provincial Spatial Planning (RTRWP) of East Kalimantan is as the Primary Service Center, better known as the gateway to the Province of East Kalimantan.

Based on the analysis, the Balikpapan City Field Work Practice Team can find out from the results of traffic loading on the primary arterial roads around the road network development plan that has a V / C ratio> 0.58. so that there will be a surge in travel demand which of course will generate traffic and increase the traffic load on these roads which are alternative routes.

2. Literature Review

2.1 Transportation System

Transportation planning as a process can also be said to be the existence of processing activities (processing) of one or more inputs / inputs to obtain one or several outputs. As in other planning sectors, in transportation sector planning we use a general planning approach, where we analyze all the factors related to the existing problems, the general approach in the transportation sector can be said to be a 'systems' approach to transportation. [6]

Activity systems, network systems, and movement systems in their interactions will be interconnected. Changes in the activity system will affect the network system through changes in service levels in the movement system. Likewise, changes to the network system will be able to affect the activity system through increased mobility and accessibility of the movement system, so that these three micro systems interact with each other in the macro transportation system. [6]

2.2 PTV Visum Application

PTV Visum (Visum) is a leading software program used for [4]:

a. Analysis and forecasts for traffic and transportation systems.
b. Model private transport and public transportation in one integrated model.
c. Allows management of GIS data for private and public transport so that it can be managed consistently with a network editor.
d. Accomodate vehicle restrictions to optimize vehicle use and to analyze costs and revenues.
e. Support planning to develop policy measures and determine the impact of those policy measures.
f. Equipped with a microscopic traffic simulation system from PTV VISSIM (VISSIM).
g. Planning of public transport networks and services that are demand-based and service-oriented.
h. Easy-to-interpret graphical display covering strategic and operational processes throughout public transportation planning.
i. Model transportation demand and load onto the network. The principle of PTV VISUM modeling is a 4-step model (Four Step Modeling), namely: TG: Trip Generation (production & attraction); MS: Split Capital; TD: Trip Distribution .; TA: Trip Assignment.

2.3 Performance assessment

a. Road Section Performance Indicators [2]
   Road performance indicators consist of road capacity, volume, v / c ratio (traffic volume / capacity), and speed.
b. Road network performance indicators [2]
   The Road Section Performance Indicator consists of the average network volume, the speed of the road network, the average total mileage and the total average travel time.

3. General Description

According to the 2012-2032 Regional Spatial Plan (RTRW), the corridor planned for the third Balikpapan outer ring road will connect the East Balikpapan District, namely Manggar towards the north, namely Soekarno Hatta Street, and then further connect the City of Balikpapan with Kutai Kartanegara Regency. Stretching from north to south, the total length of the Balikpapan-Samarinda outer ring road (III) is approximately 12 km from the Manggar TPA (Final Disposal Site) junction to the end of the new Balikpapan-Samarinda road (Higway) (Km 12.9) . [5]

4. Research Methodology

This research was carried out based on the research methodology from the initial stages of problem identification, problem formulation, primary data collection and secondary data, data processing and analysis, traffic modeling during the morning and evening rush hour using PTV Visum Student Version 2020 software, model validation using the method (Chisquare) and (Mape), predicting the size of the upcoming trip in 2019-2024 and 2019-2029, until the final stage of problem solving alternatives with 8 traffic handling proposals or recommendations to anticipate the continuation of traffic performance.

5. Analysis and Problem Solving

5.1 Research Coverage Area

In general, the Balikpapan City Alternative Road Network area is a sub-center for service activities, which consists of 35 roads and 17 interconnected intersections (Figure 1).

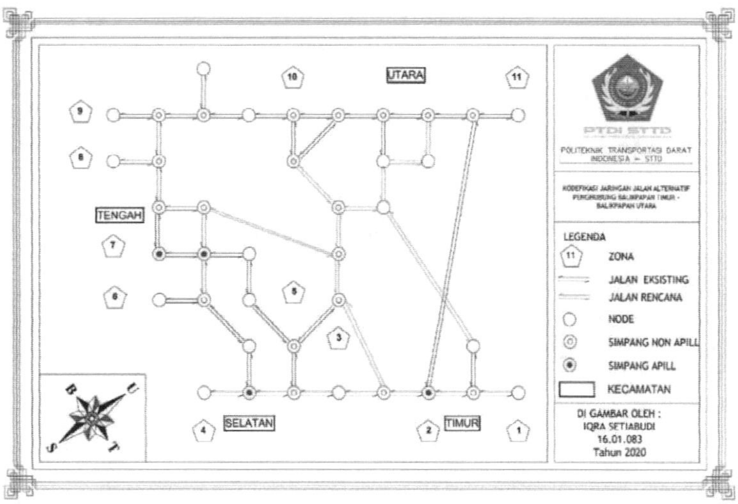

Figure 1 Codification of the existing alternative Road Network area in 2019
Source: analysis results, 2020

5.2 Zoning of Existing Travel

Before conducting the research, what was done was dividing the study area into several traffic zones. This zone is created based on the flow of traffic in and out ofBalikpapan City Alternative Road Network areawhich is used to identify the number of trips from one zone to another to get attraction and trip generation. The following table is the zone grouping contained inBalikpapan City Alternative Road Network area (Table 1).

Table 1 Balikpapan City Alternative Road Network area zone

Zone	Access	ADMINISTRATION
1	JL MULAWARMAN 3	Manggar Baru
2	JL TOLL BALIKPAPAN – SAMARINDA	Manggar
3	JL SEPINGGAN BARU 2	Sepinggan
4	JL MARSMA R. ISWAHYUDI 3	Sepinggan Raya
5	JL RUHUI RAHAYU 2	Sepinggan Baru
6	JL LETKOL POL. HM ASNAWI ARBAIN 1	Gunung Bahagia
7	JL RUHUI RAHAYU	Damai Baru
	JL MT HARYONO 3	
8	JL INDRAKILA 1	Gunung Samarinda Baru
9	JL SOEKARNO HATTA 5	Batu Ampar
10	JL KM 5,5 KARIANGAU 1	Graha Indah
11	JL SOEKARNO HATTA 11	Karang Joang

Source: analysis results, 2020

5.3 Trip Generation

Trip generation is the number of trips generated by the study area (Table 2). Ofyar Z. Tamin states that to estimate the number of generator trips that will come, a model is needed [6]. Based on volume of morning rush hour and evening rush hour.

Table. 2 Trip Generate – Trip Attraction Traffic Volume 2019

Zone	Access	Generate (Morning)	Attraction (Morning)	Generate (Afternoon)	Attraction (Afternoon)
1	JL MULAWARMAN 3	1818	3575	1907	2496
2	JL TOLL BALIKPAPAN – SAMARINDA	1887	736	470	381
3	JL SEPINGGAN BARU 2	1102	996	975	1343
4	JL MARSMA R. ISWAHYUDI 3	3126	2810	2103	2520
5	JL RUHUI RAHAYU 2	1469	1565	1525	1485
6	JL LETKOL POL. HM ASNAWI ARBAIN 1	1016	1019	2058	1241
7	JL RUHUI RAHAYU	1464	1965	1799	2004
	JL MT HARYONO 3	1705	1325	1689	2037
8	JL INDRAKILA 1	1378	996	1500	1584
9	JL SOEKARNO HATTA 5	2198	2515	2089	2223
10	JL KM 5,5 KARIANGAU 1	2284	2173	1582	1714
11	JL SOEKARNO HATTA 11	1503	1166	963	1141

Source: analysis results, 2020

5.4 Trip Distribution

The distribution of travel transport is the stage where trips between zones are calculated based on the study of the origin and destination of trips from each zone in a study area. At this stage, consider determining the interaction relationship between the number of zones based on the size of the trip generation and the trip attractions that have been carried out in the previous stage.

The trip distribution is carried out with the aim of predicting the number of trips that will be distributed in the planned year, namely (2019-2024) and (2019-2029). This loading estimate is carried out using the model used in transportation planning.In analyzing the spread of this trip, information is needed about the origin and destination matrix of the current state (Figure 2).

(morning)

O/D	1	2	3	4	5	6	7	8	9	10	11	Pi
1	0	434	171	267	96	109	143	74	106	113	177	1690
2	784	0	198	305	110	126	165	91	123	131	205	2239
3	352	225	0	411	150	190	232	131	182	168	199	2239
4	589	372	441	0	279	285	391	235	283	281	310	3465
5	219	139	166	287	0	125	160	98	121	112	115	1541
6	219	140	186	259	110	0	210	141	174	167	175	1781
7	439	281	347	546	217	323	0	260	355	335	352	3455
8	183	125	158	265	107	174	209	0	269	227	216	1932
9	257	164	214	311	129	210	278	262	0	406	356	2587
10	238	152	172	268	103	175	228	193	353	0	389	2271
11	225	144	123	178	64	111	145	111	186	235	0	1520
Aj	3505	2175	2175	3098	1365	1828	2160	1596	2153	2175	2495	24725

(afternoon)

O/D	1	2	3	4	5	6	7	8	9	10	11	Ai
1	0	618	296	141	245	111	221	356	159	117	202	2467
2	332	0	166	78	136	62	124	214	89	66	113	1379
3	176	183	0	124	218	110	205	361	155	99	129	1761
4	79	82	117	0	109	45	93	175	65	45	54	864
5	226	234	339	180	0	150	293	562	214	137	155	2490
6	141	147	236	101	207	0	240	499	191	127	147	2033
7	188	196	295	142	271	161	0	615	260	170	197	2494
8	284	317	486	249	487	315	577	0	716	418	439	4288
9	120	125	197	88	175	114	230	675	0	224	216	2163
10	105	109	149	71	133	90	178	468	266	0	223	1792
11	146	152	157	70	121	84	167	396	207	180	0	1680
Aj	1797	2163	2439	1243	2103	1241	2329	4321	2323	1582	1876	23417

Figure 2 Matrix of Origin of Existing Travel Destinations in 2019 (vehicle / hour)
Source: analysis results, 2020

5.5 Modal Split
the process of changing units of vehicle / hour into units of pcu / hour. In this study, the process begins with a trip generation / attraction through the existing traffic volume taken during the morning and evening rush hours so that the travel distribution matrix using the vehicle / hour is obtained, then to get the proportion of the mode used to change the origin destination matrix in times the percentage of use. Each mode in order to obtain a matrix of origin and destination units of vehicles / hour for each road transport mode. Furthermore, the traffic performance is indicated by the volume of traffic in passenger car units (pcu / hour), then the matrix of origin of destination for each mode is multiplied by units of emp for each road transport mode (Figure 3).

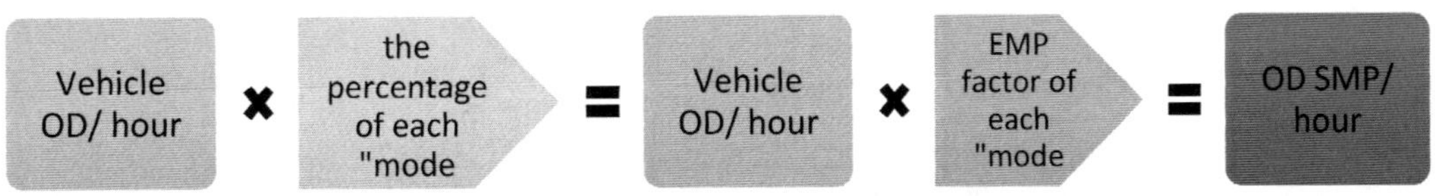

Figure 3 Change of vehicle / hour to pcu / hour
Source: analysis results, 2020

5.6 Trip assignment
The loads carried out in this analysis are as follows:
a. The present traffic assignment on the existing road network system.
b. Traffic assignment for the future (forecase) with a fixed road network system (without any changes to the road network).
c. Traffic assignment for the future (forecase) on the proposed road network system.

The loading model used in this study is the "Equilibrium Assignment" method. To make it easier to analyze this traffic assignment process, the author uses the Vissum 2020 software, which is a computer application program for transportation planning that has the ability to forecast travel demand modeling (Demand Transport Forecasting Model).

The following is a visualization of traffic assignments at 07.00 -08.00 WITA and 16.00-17.00 can be seen in the following loading image (Figure 4).

(Morning) (Afternoon)

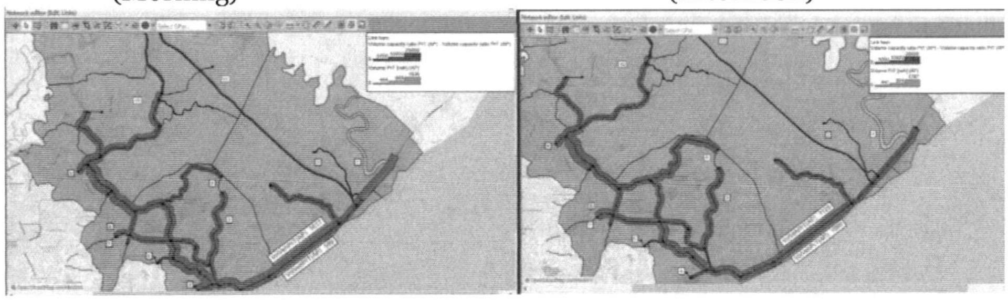

Figure 4 Visualization of 2019 traffic assignments
Source: analysis results, 2020

Table 3 Existing Road Network Performance 2019

No.	Parameter	Morning performance	Evening performance
1	Volume (pcu / hour)	762	661
2	Network Speed (km / h)	42	42
3	Total Travel Distance (km)	1063	991
4	Total Travel Time (hours)	23.3	22.4

Source: analysis results, 2020

5.7 Road Network Model Validation

The stages of traffic assignment are carried out using the PTV Vissum 2020 software, where the results obtained are road performance and road network and are equipped with visualizations. Traffic assignment is obtained from the results in the field (survey) which will later be compared with the model.

Model validation is carried out to prove whether the model that will be used to forecast trips in the planning year can be used or not. The validation used is by using the Chi square and MAPE test (Figure 5).

(morning)

I. HYPOTHESIS
H0 : Models with Survey Aligned
H1 : The Survey Model is out of sync

II. Trust Level Value α = 9 95% 0,05

III. Degrees of Freedom (v) = (k-1) = 34

IV. the Chi Square Value of the table (χ2 tabel) 49,80185

V. Calculate χ2 count = 45,47

VI. Decision Rules : H0 is accepted if χ2 is counted < 49,80185
H1 is accepted if χ2 count > 49,80185

VII. Decision :
Ho Accepted

k 35

(afternoon)

I. HYPOTHESIS
H0 : Models with Survey Aligned
H1 : The Survey Model is out of sync

II. Trust Level Value α = 9 95% 0,05

III. Degrees of Freedom (v) = (k-1) = 34

IV. the Chi Square Value of the table (χ2 tabel) : 49,80185

V. Calculate χ2 count = 18,27

VI. Decision Rules : H0 is accepted if χ2 is counted < 49,80185
H1 is accepted if χ2 count > 49,80185

VII. Decision :
Ho Accepted

k 35

Figure 5 Hypothesis table The validation test uses the Chi Square method
Source: analysis results, 2020

From the table above it can be seen that Ho: is accepted because X ^ 2 count <49.80, which has a value of 45.47 and 18.27, so the data from the model is accepted and can be used further for data analysis.

The smaller the MAPE value, the smaller the error of the estimation results, on the contrary the greater the MAPE value, the greater the error of the estimation results. The results of a prediction method have very good forecasting ability if the MAPE value is <10% and have good predictive ability if the MAPE value is between 10% and 20%

Based on the validation test using the MAPE method in the morning, it can be seen that the total is 2.29 and 1.71. this shows that the MAPE validation test requirements fordata from the model is received and can be used further for data analysis.

5.8 Transportation Forecasting

Performance forecasting in the planning year is needed to analyze traffic engineering management that will be or has been implemented in order to find out whether an engineering management that has been made can still be applied in the planning year. In forecasting the number of vehicles in the planning year, secondary data is used on the number of vehicles in the last 5 years to determine the growth rate. The formula used to predict the variables affecting trips in the coming year is as follows:

$$Pt = Po \; x \; (1+i)n \quad (1)$$

Information:
Pt = Variable number of plan year (YAD volume)
Po = Number of base year variables (existing volume)
i = Traffic growth rate
n = Number of years

In forecasting the volume of planned vehicles, data on the volume of the current year and the rate of traffic growth are needed. To find out the volume of vehicles used the formula for the traffic growth rate.

The following table is the road network data for the Balikpapan City Alternative Road Network area in the planned year, namely in 2024 and 2029 which is located in the alternative road network area of Balikpapan City at 07.00 - 08.00 WITA and 16.00 - 17.00 WITA (Table 4).

Table 4 Road Network Performance Planning year

No.	Parameter	Morning performance 2024	2024 afternoon performance	Morning performance 2029	2029 afternoon performance
1	Volume (pcu / hour)	944	820	1199	1044
2	Network Speed (km / h)	39	41	37	39
3	Total Travel Distance (km)	1448	1261	1850	1614
4	Total Travel Time (hours)	32	28.5	41	36.6

Source: analysis results, 2020

Based on the comparative data above, there is a significant difference in the loading of traffic carried out during the peak hours of the morning so that there is an increase in the volume in the area, travel distance and travel time compared to the loading that is carried out during the peak hours of the afternoon.

5.9 Traffic Handling Proposal Recommendations

Recommendations are needed in solving a transportation problem in an urban area, one alternative problem that can be done is by optimizing the existing facilities and infrastructure. Some of the problems that exist in the area of the Balikpapan City Alternative Road Network need to be improved in order to provide a service to the community in the form of convenience which is also to improve the performance of sections and road networks. several scenarios that can be done, among others.

The best scenario selection is based on the road network performance comparison in each scenario. Comparison of the performance of each scenario in 2019-2024 and 2019-2029 the development of the Balikpapan City Activity Service Sub Center at 07.00 -08.00 WITA and 16.00 -17.00 WITA can be seen in the Figure 6 and Figure 7.

(morning)

NO	Parameter performance	Volume (pcu / hour)	Network Speed (km / h)	Total Travel Distance (km)	Total Travel Time (hours)
1	Eksisting	944	39	1448	32
2	Skenario 1	942	37	1851	41
3	Skenario 2	948	40	1446	31
4	Skenario 3	858	38	1288	29
5	Skenario 4	788	42	1194	28
6	Skenario 5	852	39	1285	29
7	Skenario 6	860	39	1292	29
8	Skenario 7	789	42	1182	27
9	Skenario 8	780	44	1180	26

(afternoon)

NO	Parameter performance	Volume (pcu / hour)	Network Speed (km / h)	Total Travel Distance (km)	Total Travel Time (hours)
1	Eksisting	820	41	1261	29
2	Skenario 1	818	41	1259	29
3	Skenario 2	824	42	1264	28
4	Skenario 3	748	40	1135	26
5	Skenario 4	676	42	1017	23
6	Skenario 5	746	40	1132	26
7	Skenario 6	751	40	1138	26
8	Skenario 7	664	42	1015	23
9	Skenario 8	660	43	1012	22

Figure 6 Comparison of Road Network Performance Between Scenarios in 2024
Source: analysis results, 2020

(morning)

NO	Parameter performance	Volume (pcu / hour)	Network Speed (km / h)	Total Travel Distance (km)	Total Travel Time (hours)
1	Eksisting	1199	37	1850	41
2	Skenario 1	1212	37	1846	42
3	Skenario 2	1200	38	1850	41
4	Skenario 3	1085	36	1637	37
5	Skenario 4	1015	39	1538	34
6	Skenario 5	1081	37	1636	37
7	Skenario 6	1086	37	1642	38
8	Skenario 7	996	40	1530	35
9	Skenario 8	989	42	1533	33

(afternoon)

NO	Parameter performance	Volume (pcu / hour)	Network Speed (km / h)	Total Travel Distance (km)	Total Travel Time (hours)
1	Eksisting	1044	39	1614	37
2	Skenario 1	1056	39	1611	37
3	Skenario 2	1049	40	1614	36
4	Skenario 3	949	38	1445	34
5	Skenario 4	852	41	1294	30
6	Skenario 5	947	38	1444	33
7	Skenario 6	1013	38	1532	35
8	Skenario 7	852	41	1294	30
9	Skenario 8	847	42	1290	29

Figure 7 Comparison of Road Network Performance Between Scenarios in 2029
Source: analysis results, 2020

From several comparison tables above, a ranking of several existing conditions will be carried out and the selection of the best scenario in 2019-2024 for the development of the Balikpapan City Activity Service Sub Center.

After analyzing the eight scenarios above, the best scenario can be selected, namely the application of scenario 8, namely increasing road capacity in the form of road widening on problematic sections and the construction of alternative roads (manggar-sepinggan - ksatrian) in Balikpapan City, which is marked by a drastic reduction in problematic roads. with an av / c ratio above 0.7 with an average speed of 38.64 km / hr. This is in accordance with the plan of the Balikpapan City Government to make it easier for people to move, especially those that have not been reached by the road network and equitable development in the City of Balikpapan.

6. Conclusion

Based on the analysis and problem solving described in the previous section, it can be concluded that:

1. In 2019 the traffic conditions in the alternative road network area in East Balikpapan sub-district and North Balikpapan sub-district had 3 sections that could potentially be a traffic problem with a VC ratio above 0.7. Jalan Mulawarman V / C the ratio is 0.76, Jalan Syarifudin Yoes V / C ratio is 0.83, Jl Soekarno Hatta V / C is 0.82.

2. The design of transportation modeling for the development of the sub-center area of Balikpapan City service activities uses modeling using the PTV Vissum Student Version 2020 software. The results of volume-based traffic assignment at peak hours of the morning and peak hours of the evening. Modeling results obtained Traffic Network Performance as follows:
   Morning 07.00 - 08.00 2019
   a. Volume : 762 pcu / hour
   b. Network Speed : 42 Km / hour
   c. Total Travel distance : 1063 km
   d. Total Travel Time : 23 hours
   Afternoon 16.00 - 17.00 2019
   a. Volume : 661 pcu / hour
   b. Network Speed : 42 Km / hour
   c. Total Travel distance : 991 km
   d. Total Travel Time : 22 hours

3. Forecasting the demand for transportation every year has a trend of increasing traffic volume and the impact on traffic is the highest in the morning on Jalan Syarifudin Yoes V / C ratio of 0.94 and the highest in the afternoon on Jalan Mulawarman V / C with a ratio of 0.87 with the condition 5 years (2019-2024) when the development of the sub-center for Balikpapan City activities services. The highest is in the morning on Jalan Mulawarman V / C with the ratio of 0.98 and the highest in the afternoon is on Jalan Syarifudin Yoes V / C, the ratio is 0.93 to 10 years (2019-2029) after the development of the Balikpapan City activity service sub-center area.

4. traffic network performance in Balikpapan City in the planned yearWith the optimization of infrastructure and road network development in the morning and evening simulations, the best assessment is in scenario 8, namely increasing the capacity of problematic roads and adding alternative road networks (manggar-sepinggan - ksatrian). The modeling results obtained from the Traffic Network Performance as follows:
   Morning 07.00 - 08.00 in 2024
   a. Volume : 660 pcu / hour
   b. Network Speed : 43 Km / hour
   c. Total Travel distance : 1012 km
   d. Total Travel Time : 22 hours
   Afternoon 16.00 - 17.00 2024
   a. Volume : 780 pcu / hour
   b. Network Speed : 44 Km / hour
   c. Total Travel distance : 1180 km
   d. Total Travel Time : 26 hours

7. Suggestion
In accordance with the results of the analysis that has been done, the authors can propose the following suggestions:
1. anticipating traffic problems that occur in the planning year, when the development of the Balikpapan City sub-activity center area is implemented (2019-2024), and after the

development of the Balikpapan City sub-center activity area (2019-2029) by building alternative roads (manggar-sepinggan) -ksatrian) and additional capacity for problematic roads (JL Mulawarman - JL Marsma R iswahyudi - JL syarifudin yoes - JL MT Haryono - JL Soekarno Hatta (scenario 8)

2. The application of these eight scenarios needs to be considered from the legal aspect regarding land acquisition for the widening of several roads and the construction of a ring road in the sub-district of Balikpapan Timur - Balikpapan Utara.
3. It is necessary to carry out further analysis regarding the feasibility of the project, financial analysis, as well as from an economic point of view, it can provide positive results and impacts for the 10-year plan.

References

[1] Adisasmita, sakti adji, 2008. *Pengembangan Wilayah Konsep dan Teori*. Yogyakarta: Graha Ilmu.
[2] Adisasmita, sakti adji, 2012. *Jaringan transportasi teori dan analisis*. Yogyakarta : Graha ilmu.
[3] Badan Perencanaan Pembangunan Daerah. 2012. *Rencana Tata Ruang Wilayah Kota Balikpapan 2012-2032*. Balikpapan: Badan Perencanaan Pembangunan Daerah Kota Balikpapan.
[4] Group, PTV. 2020. *PTV Visum 20 Manual*. PTV AG: Karlsruhe.
[5] Pt.Nuansa Citramandiri , 2007.*Proposal Pembangunan Jalan Outer Ring Road III Kota Balikpapan*. Balikpapan
[6] Tamin, Ofyar. Z, 2008. *Perencanaan, Pemodelan, dan Rekayasa Transportasi*. ITB. Bandung: Penerbit ITB.

Landscape design of "Tanjung Budaya Pemuteran" coast area based on landscape engineering

I Abdullah[1] and P Nuryanti[2]

[1] Student, Department of Landscape Architecture, Faculty of Agriculture, Bogor Agricultural University, Indonesia
[2] Lecturer, Department of Landscape Architecture, Faculty of Agriculture, Bogor Agricultural University, Indonesia

E-mail: [1]ismail.jra@gmail.com; [2]pingkannuryanti@apps.ipb.ac.id

Abstract. "Tanjung Budaya Pemuteran" is a place of coastal tourism and alternative tourism based on community culture in Pemuteran Village, Bali. The landscape characteristic and inherent local wisdom make this bay is a unique place. *Gebug ende* is the native rain harvesting ritual of Pemuteran society, which can be suggestive evidence. Currently, Tanjung Budaya Pemuteran landscapes not suitable to support the site function. This research aims to identify the existing landscape to make Tanjung Budaya Pemuteran use optimally without disregarding inherent local wisdom. Landscape analysis spatial with the quantitative and qualitative approach used to know the potential and constraint. The landscape engineering approach is used to measure the volume of rainwater to optimize the result of *Gebug ende* rain harvesting. The basic concept of this landscape design is named "Identities of Pemuteran". Design concept generated through the transformation of *Gebug ende*. The block plan becomes a schematic design product that consists of a welcome area, a culture tourism area, a recreation area, and a service area. Each area is connected by primer and secondary circulation. Plants on-site are regulated based on vegetation groups of *Barrington*, *Pes caprae*, Mangroves, and existing. Visuals on the site are also arranged according to the design plan. The landscape engineering method produced the construction of a rainwater harvester with a volume of 10 m3. Site plan interpretations are explained in more detail through the site section, design details, and some illustrations.
Keywords: local wisdom, spatial design, sustainable design

1. Introduction

Tanjung Budaya Pemuteran is a public space located in Pemuteran Village, Bali. Tanjung Budaya has functioned as a place of coastal tourism and alternative tourism based on community culture. Besides, this place is also often used to implement traditional rituals, which is developed into an annual festival named *Pemuteran Bay Festival*. Besides the potential it has, the functions and possibilities of Tanjung Budaya Pemuteran are not supported by existing conditions. It can be seen from the irregular arrangement of landscape elements and unfocused management [1]. The place is also included in one of the areas with the longest rainless days in Indonesia, 249 days. *Gebug ende* becomes one of many iconic rituals held in this place. The villagers believed this ritual could summon rain when the drought-hit Pemuteran village.

The research aims to identify the potential and constraint Tanjung Budaya Pemuteran that will be used to make sustainable landscape design as consideration for solving existing problems. This landscape design is also designed to support local village customary rituals, especially in utilizing the potential for calling the rain named *Gebug ende*. Landscape engineering methods are also used to predict the optimal volume of rainwater that can be accommodated.

The research objective is to identify existing conditions, analyze and synthesize the potential and constrain of the site, and make a concept of sustainable landscape design of Tanjung Budaya Pemuteran based on landscape engineering. Hopefully, the research result can be used in developing and maintaining the landscape of Tanjung Budaya Pemuteran.

2. Method

2.1 *Research Site and Period*
The research was carried out in Tanjung Budaya Pemuteran, Pemuteran Traditional Village, Bali, Indonesia, from December 2019 to September 2020 (Figure 1).

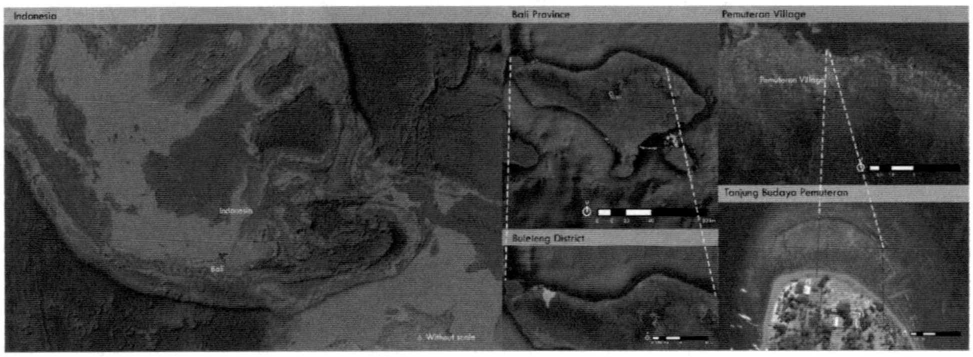

Figure 1 Research Site

2.2 *Research Method*
The method used in this design research activities is as follows: 1. Data Inventory include: (a) survey; (b) distribution of questionnaires to the 30 respondents; (c) interview (snowball sampling); and (d) observation; 2. Analysis and synthesis include: (a) physical and biophysical aspect; (b) levels of comfort; (c) tourism aspects; (d) social and cultural aspect (e) oceanography; (f) carrying capacity; (g) and landscape engineering. The measure of comfort level used the following formula:

$$\text{Thermal Humidity Index (THI)} = (0{,}8\ T + RH\ T)/500$$

T: air temperature (^{o}C)
RH:relative humidity of air (%)

2.3 *Research stage*
The method used in the study combines the six stages of planning and recreational landscape design by Gold and the six stages of the planning and design process by Simond [2]. The modifications resulted in the following four stages of landscape design as follows: (1) Preparation; (2) Data inventory; (3) Analysis and synthesis; and (4) Design.

3. Result and Discussion

3.1 *General Condition*
Tanjung Budaya Pemuteran designated as coastal recreation area. After Pemuteran Bay Festival was held, the government of Pemuteran Village starts to develop this place by built some Balinese-themed facilities. Patung Baruna becomes the newest facility that was built in 2019. However, existing facilities are not quite adequate and well maintained.

3.2 Analysis and synthesis

3.2.1 Physical and Biophysical Aspect

Location and Site Boundaries. Tanjung Budaya Pemuteran is located in Pemuteran Village, Buleleng District, Bali, Indonesia. This research area is 2.3 ha. On the north and east sides, the research site is directly bordered by the Sea. Glass House Hotel borders the site on the western part. The garbage treatment plant borders the site on the southern side.

Climate. The tropical comfort criteria are in the temperature range of 21-27°C [3]. The Tanjung Budaya Pemuteran has an average temperature at the value of 26.30°C with a humidity level of 84%. Based on those data, the calculation shows the importance of THI (Thermal Humidity Index) on site is 25.46. These results prove Tanjung Budaya Pemuteran is convenient area.

Topography, Slope, and Soil. Topography is influential in placing the object to be built. The slope of Tanjung Budaya Pemuteran is at a value of 0 - 20% with an altitude of -3 - 6 mas. Land with a 0-8% slope is the soil classified as flat land [4]. The land is ideal for active open spaces and massive buildings. The open land area on the site gets into this classification. Besides, flat land has a variety of monotonous views [5]. It can be overcome by adding vertical elements to the site. The diversity of opinions has been represented by the shape of the coastal landscape and Pantung Baruna.

Hydrology. In general, Tanjung Budaya Pemuteran land is covered by sandy soil. It causes the infiltration process on the site to run well. Based on observation results, the runoff flows from the south of the site to the north. There are two puddles caused by damaged asphalt and soil erosion. These puddles can be resolved by fixing the asphalt and filling the soil.

Accessibility and Circulation. Tanjung Budaya Pemuteran can be accessed from various directions, either via the Pura Dalem Pemuteran street or from the west and east sides by following the Pemuteran coastline. The place can be visited using a wheeled vehicle with a distance of 32.9 km from Gilimanuk Harbor and 134 km from Ngurah Rai International Airport. The circulation on the site consists of vehicle circulation and pedestrian circulation. There is damage to the vehicle circulation that needs to be repaired.

Visual and Acoustic. Objects with high visuals in Tanjung Budaya Pemuteran are views of the hills, a mangrove tree, a swing, and the Baruna Statue. Those objects are often used as photo spots. Low visuals are owned by the garbage treatment plant and the entire toilet on the site. Tanjung Budaya Pemuteran has a source of noise that comes from waste processing activities. The noise that occurs does not interfere with user activity. It is due to the open character of the site and the sound of the waves, which tend to be calm.

Vegetation and Animals. The selection of vegetation on the site has an essential effect on on-site development. The selected vegetation type will later be adjusted based on the functions of the vegetation itself. The existing vegetation selection was adapted from the three plant formations on the site, namely the *Barringtonia*, *Pes caprae*, and *Mangrove*. Birds and fish dominate the animals found on the site. Also, there are dogs (*Canis familiaris*) which are the pets of visitors or the community. The vegetation arrangement is maintained according to the existing conditions to preserve the habitat of the animals.

3.2.2 Tourism

Facilities and Utilities. Existing facilities and utilities on the site are not yet fully managed. It can be seen in the toilet facilities, food stores, and signboards on the site. The light utility and wifi hotspot on the site is not operating correctly. Periodic maintenance is necessary to maintain the condition and function of existing facilities and utilities.

Object and Tourist Attraction. The research site has a significant draw on cultural attractions for the people. The community's cultural attractions are displayed at each Pemuteran Bay Festival as the main agenda. The activity was initiated based on the commitment of the Pemuteran Village community to environmental conservation and the development of alternative tourism based on community and cultural empowerment. According to Jero Ketut Wirdika as a traditional leader, cultural attractions displayed at Tanjung Budaya Pemuteran are the main activities for alternative tourism in Pemuteran Village. Cultural attractions that can be seen on the site are the *Gebug ende*, the *Dewa Ayu* dance, and the *Ngaben*. *Gebug ende* and *Dewa Ayu* dance are cultural attractions that are often featured at the Pemuteran Bay Festival.

3.2.3 Social and Culture
User. The results of the questionnaires that have been distributed show several perspectives of visitors to the site. As many as 50% of respondents stated that the site conditions were excellent. Site cleanliness needs to be improved. It can be seen from 51% of respondents who said their site was still dirty. The additional facilities that the respondents wanted were signage and photo spots with a percentage of 30% and 25%.

The design of a research site requires consideration of the preferences of the end. Respondents most like the calm atmosphere with a percentage of 45%. There are 67% of respondents stated that the Baruna Statue object was a distinctive element on the site. The majority of respondents wanted a natural landscape design with a questionnaire percentage of 70%. It is also evident from the ratio of 73% for selecting natural materials and 46% for leafy plants to put into the landscape design. Besides, respondents wanted the use of paving blocks as land cover with a percentage of 53%.

Activities. Based on the questionnaire results, relaxing activity is the main goal of respondents visiting the site with a large percentage of 53%. The majority of tourists who fill out the questionnaire visit Tanjung Budaya Pemuteran in the afternoon, reaching 77%. There are 63% of respondents see the beach on weekends and 37% on weekdays.

Management. Tanjung Budaya Pemuteran is managed directly by the village government and Pemuteran Village customary stakeholders. The current management focuses on the cleanliness handled by the employee garbage treatment plant and site safety aspects. Employee garbage treatment plant is assigned to keep the area clean. The Security aspect itself is guarded directly by *Pecalang segara*. All management parties Tanjung Budaya Pemuteran, using *Awig-awig* in carrying out its duties.

Culture. The interview results depict Pemuteran Village implementing the *Tri Hita Karana* and *Tri Angga* philosophy on spatial planning in Pemuteran Village. In a macrocosmic order, Tanjung Budaya is included in *Palemahan*, where this area describes the relationship between humans and nature. Apart from that, The management has been regulated in the norms of community life of Pemuteran Village (*Awig-awig*).

3.2.4 Oceanography
This aspect is carried out through the matrix suitability of beach tourism for recreation from Yulianda [6]. As a result, the site has a total index value of 67 [Table 1]. The value shows that Tanjung Budaya Pemuteran Beach is a suitable beach area for recreational activities. The availability of raw water on the site needs to be considered for landscape engineering.

Table 1 Matrix of the Suitability of Beach Tourism for Recreation

No.	Parameter	Condition	Point	Class	Score	Iv
1	Depth	1 – 2 m	5	S1	3	15
2	Ocean base material	Fine black sand	3	S1	2	6
3	Beach type	Beach with coral	5	S2	1	5
4	Beach width	22-25 m	5	S2	2	10
5	Ocean current	0 – 0.17 m/s	3	S1	3	9
6	Beach slope	11.30°	3	S2	2	6
7	Sea brightness	8 – 9 m	1	S2	2	2
8	Land cover	Tall shrubs	1	S2	2	2
9	Dangerous biota	-	3	S1	3	9
10	Availability of raw water	<500 m	1	S1	3	3
Total index value						67

Very suitable (S1)= 71 – 90 Not Suitable (S3)= <50 Suitable (S2)= 51 – 70

3.2.5 Site Carrying Capacity

The carrying capacity is based on a space plan that suits the needs of the user's activity. Space is divided into four spaces, the welcome area, cultural tourism area, recreation area, and service area [7]. The site carrying capacity refers to international standards and can be seen in Table 2.

Table 2 Site Carrying Capacity

Area	Facilities	Σ	Spacious (m^2)	Standard	Capacity
Welcome area	*Signage*	1	11.4	n/a	n/a
	Car park	1	121	25.28 m^2/unit	4
	Motorcycle park	1	121	5 m^2/unit	24
	Entrance	1	20	1.5 m^2/person	13
	Bale	1	40	2 m^2/person	20
	Gazebo	1	6.25	2 m^2/person	3
	Bench	5	0.675	0.27 m^2/person	12
Service area	Toilet	8	4	2 m^2/person	16
	Food store	1	105	1.5 m^2/person	70
Cultural tourism area	Outdoor theater	1	2,356.80	6.2 m^2/person	380
	Curve bench	1	58.86	0.27 m^2/person	218
	Baruna statue	1	395.15	6.2 m^2/person	63
	Shelter	1	8	2 m^2/person	4
	Sand box	1	112	1.86 m^2/person	60
	Bale	1	40	2 m^2/person	20
	Gazebo	1	6.25	2 m^2/person	3
Recreation area	Swing	1	0.32	0.27 m^2/person	1
	Beach	1	13,628	12,96 m^2/person	1,051

3.2.6 Landscape Engineering

The landscape engineering analysis of this research focuses on rainwater harvesting. This is based on the potential of *Gebug ende* traditional rituals and site constraints included in one of the areas with the longest rainless day in Indonesia.

Water Needs. According to the standard from the Ministry of Public Works and Public Housing of Indonesia, which rea, it is assumed that the site requires 0.15 L/sec/ ha to fulfill the site's water needs [8]. The land area of 0.75 ha is used in calculating this water requirement. Based on the plan, the site's operational time is 06.00-21.00 WITA. Therefore the total water requirement obtained is 6,075 L/day.

Frequency of Rainfall. Frequency analysis uses rainfall data for the last ten years of the study area. The distribution test was carried out using the Chi-square and Smirnov-Kolmogorov test approach. The Result Chi-squared test shows that lognormal distribution is acceptable because of x <x, with value x= 4 and critical x= 5,991. Testing with the Smirnov-Kolmogorov test with results obtained dmax<dcritical with dmax= 0.18 and dcritical= 0.41. The rainfall used is probability P(x)= 80% or equal to 1.25 with rainfall of 44 mm.

Reservoir Volume Analysis. The site has a food storage facility with a roof area of 179 m^2. The potential volume of rainwater that the roof can accommodate after being influenced by the roof runoff coefficient of 80% is 6.301 liters. If every day the water stored is used to meet the operational needs of Tanjung Budaya Pemuteran, there is an excess of 226 L/day. The excess water will be held for five days (weekdays) as a reserve for holidays (weekends). The total volume of planned water collected after being harvested for five days is 7,431 liters.

3.3 Concept

3.3.1 Basic Concept
The basic concept of the Tanjung Budaya Pemuteran landscape design is "Identities of Pemuteran" (figure 2). This concept elevates Pemuteran Village's identity through the cultural aspects and natural fascination of Pemuteran Village. Cultural elements are chosen to introduce culture and customs in Pemuteran Village. The development of cultural factors is carried out through spatial planning that supports cultural activities. Nature will be preserved in the aspect of natural fascination.

Figure 2 Basic Concept

3.3.2 Design Concept
Gebug ende was chosen as a design concept in the Tanjung Budaya Pemuteran landscape design (Figure 3). The attribute used in the *Gebug ende* ritual is implemented into the design element. Rattan (*Penyalin*) becomes a weapon, which is transformed into a straight line. The shield (*Ende*) used to defend against attacks is transformed into a circle and arch. Apart from that, the knight uniform (*Saput Poleng*) is transformed into the square.

Figure 3 Design Concept

3.3.3 Development Concept

The concept of developing the Tanjung Budaya Pemuteran landscape design consists of the concept of space, the concept of circulation, the concept of vegetation, and the concept of view. The concept of space is divided into four main areas. The design of this study is reception areas, cultural tourism areas, recreation areas, and service areas.

The circulation concept used in the site design is divided into two categories, primary and secondary circulation. The primary circulation is designed using asphalt material on the main road and green pavement for parking. Secondary circulation is designed semi-modern. The material used in the secondary circulation is concrete, sand, and *conwood.*

The concept of on-site vegetation applies to the existing vegetation formation arrangement. Three formations are the vegetation formation of *Barrington*, *Pes caprae*, and Mangrove. Also, this concept is also designed to maintain vegetation that has been planted.

Potential views on the site are developed in the reception area, cultural tourism, and recreation. The view in the reception area is focused on the Tanjung Budaya Pemuteran new signage. View area cultural tourism is directed reciprocally from the direction of the Baruna Statue and in the opposite. The recreation area has a natural landscape on the north, east, side west sites. The view of the recreation area is also focused on an iconic mangrove tree that grows on the reef.

3.4 Schematic Design

The output of the schematic design is a block plan. Development concepts are generated by combining and overlying development concepts. The development concept combined includes the concept of space, the concept of circulation, the concept of vegetation, and the concept of view.

3.5 Design

The planning stage for the Tanjung Budaya Pemuteran tourism spatial plan begins with creating three alternative designs. The design alternatives referred to the development concepts, block plan and adjusted to the results. All three designs have the same design pattern. The difference between the three design alternatives lies in the material and landscape elements used. The first alternative design was chosen as the Tanjung Budaya Pemuteran site plan. This site plan has also been evaluated by a team of experts from coastal research and development, the Ministry of Public Works and Public Housing of Indonesia, and the Head of Pemuteran Village.

There is a cut and fill method in the land arrangement of the site. The cut and fill method is used to plain the for developing Cultural Plaza. This is done to simplify the process of building existing facilities. Also, it can support the implementation of more comfortable attractions for visitors. The cut and fill method also affects the shape of the land in the coastal area. The Block plan and site plan is provided in Figure 4.

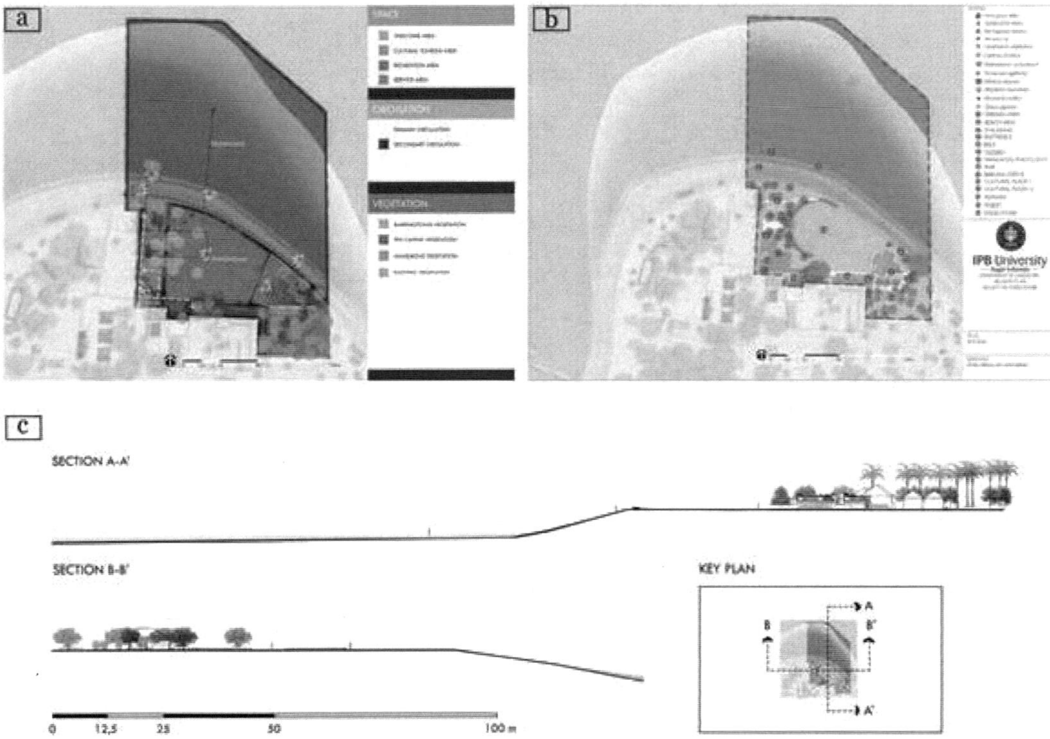

Figure 4 Image of (a) Block Plan; (b) Site Plan; and (c) Site Section

Tanjung Budaya Pemuteran landscape design focuses on beach recreation activities and community-based alternative recreation. For the recreation area, changes in landform are carried out using cut and fill to suit development needs. The facilities added to the welcome area are a parking lot for two-wheeled and four-wheeled vehicles, a small gazebo, an entrance, and signage. Rainwater Harvesting Building (RHB) is located in the service area, to the east of the food store. The harvested water will be connected to a toilet on the southeast side of the food store. Cultural Plaza I and II were built to support ritual activities and alternative tourism on the site. Cultural Plaza I made with partially cut circular shape with bench facilities that follow the pattern. This curved bench is designed as a seat that can be used to visually focus on the Baruna statue as the main stage of the show. Plaza Budaya II is made smaller on the southeast side of the site. The plaza functions as a playground, a place for cultural performances, a place for event preparation. The design illustration can be seen further in Figure 5.

Figure 5 Illustration of Landscape Design of Tanjung Budaya Pemuteran

Rainwater harvesting on-site is carried out in an integrated system (Figure 6). The food store roof functions as a rainwater catchment area. The rainwater is flown to the Rainwater Harvesting Building (RHB) on the east side of the food store through the gutter pipe's inlet. Once clean, the water is stored in an RHB. The stored water can be drained to the food store on the west side and toilets on the south side through a connecting pipe if it is to be used. If there is excess capacity, the water will be flowed out into the infiltration hole.

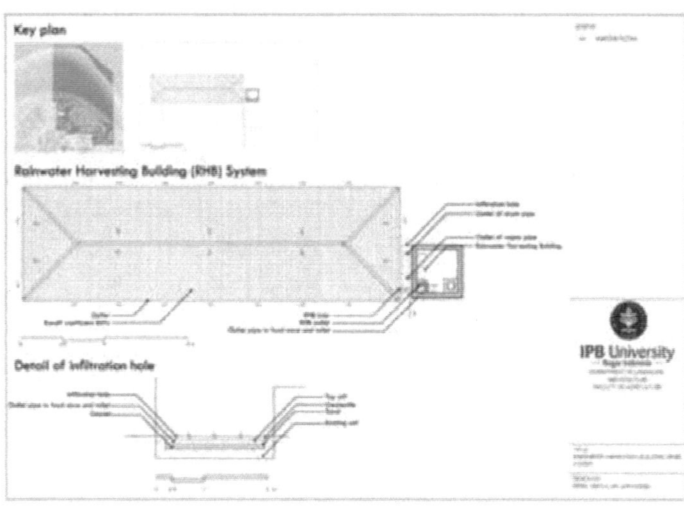

Figure 6 Rainwater Harvesting Building (RHB) System

4. Conclusion and Sugestion

4.1 Conclusion

Tanjung Budaya Pemuteran has potential that needs to be developed. Such potential is the character of the natural landscape and the local wisdom. The development aims to produce recommendations for landscape arrangements that can support user activities and site functions. This research landscape design combines physical and biophysical aspects, hydro-oceanography, tourism, culture, and landscape engineering to produce ideal and sustainable arrangements. It resulted in recommendations for Tanjung Budaya Pemuteran's landscape design with a Rainwater Harvesting Building with a capacity of 10,000 L as a place of coastal recreation and alternative tourism based on community culture.

4.2 Suggestion

This research is expected to be material for thought and reference for the managers of Tanjung Budaya Pemuteran. Besides, the local wisdom of an area needs to be developed through a scientific approach. It will produce new insights and outcomes useful for the community, especially in the scientific development of landscape architecture.

References

[1] Naufal A D 2019 Cuaca dan Iklim sepanjang 2019, Catat Soal Kemarau yang lebih Panjang [internet] Kompas.com
[2] Nafar S 2016 Desain ekologis taman koleksi pakis berbasis sistem klasisfikasi *bioregion di Ecopark, Cibinong Science Center-Botanic Garden* [skripsi] Arsitektur Lanskap IPB University Bogor (ID)
[3] Laurie M 1986 Pengantar kepada Arsitektur Pertamanan (Terjemahan) Intermatra Bandung (ID)
[4] Keppres No 32/1990 Indonesia
[5] Booth N K 1983 Basic Elements of Landscape Architectural Design Waveland Press Inc: Illinois
[6] Yulianda F 2007 Ekowisata Bahari sebagai Alternatif Pemanfaatan Sumberdaya Pesisir Berbasis Konservasi. *Seminar Sains pada Departemen MSP* IPB
[7] Harris C W and Dines N T 1998 Time-Saver Standards for Landscape Architecture: Design and Construction Data McGraw-Hill Publishing Comany New York(US)
[8] BPSDM Kementerian PUPR 2018 Modul Proyeksi Kebutuhan Air dan Indentifikasi Pola Fluktuasi Pemakaian Air [internet] bpsdm.pu.go.id

Lovina Beach landscape design in Buleleng Regency Bali based on landscape engineering

M F Ramadhan[1], P Nuryanti[2]

[1] Department of Landscape Architecture, Faculty of Agriculture, IPB University, Indonesia

[2] Department of Landscape Architecture, Faculty of Agriculture, IPB University, Indonesia

E-mail: faisal_rmdhn32@apps.ipb.ac.id

Abstract. Lovina Beach is a famous tourist attraction on the northern island of Bali. This beach is somewhat unique because it has an exciting dolphin viewing attraction. However, some lack elements on-site reduce the value of beach tourism, such as unattractive design, many idle facilities, empty spaces, and puddle. This study aims to analyze and redesign the beach based on landscape engineering to overcome sustainability. The design stage consists of project acceptance, research and analysis, design, and construction drawings combined with the scoring system of the hydro-oceanographic analysis. The analysis aimed to identify the value of coastal vulnerability indexes. The results of this study are site plan concepts, perspective drawings, cut images, planting plans, and detailed engineering designs, which is formed on the results of analysis and synthesis, and preferences score. The design of this coastal waterscape will be used as a recommendation to the beach management.

Keywords: Coastal vulnerability indexes, Dolphin, Scoring, Sustainable, Tourist attraction

1. Introduction

Bali, which is rich in biodiversity, natural ecosystems, and cultural values, provides a potential for developing the world of natural tourism. The character of the economic value of nature tourism objects in one place can be different from natural tourism objects in other places, which is due to differences in the landscape. This tourism is classified into North Bali and South Bali (Sulaiman 2010).

One of the choices that become a destination for local and international tourists for traveling is North Bali. Although its prestige is still inferior to South Bali, North Bali has many attractive tourist destinations. In carrying out the development of a good tourist area, there will be a demand for accommodation for visitors, the provision of equipment for visitor convenience, and the availability of adequate facilities. One of the tourist destinations that provide attractive attractions for visitors is Lovina Beach, Kalibukbuk Village, Singaraja District.

Lovina Beach is located about 10 km west of Singaraja city. Many foreign and local tourists visit to see unspoiled beaches and see dolphins abundant on this beach. One way to optimize the existing potential is by designing a coastal landscape based on engineering and scoring the coastal vulnerability index referring to Ozy et al. (2008) as a reference.

This design focus examines the needs of the coast more deeply based on the questionnaire and the consideration of supporting factors for the coastal density index. Developing and improving this landscape design based on landscape engineering reduces road-noise, protects soil, and creates runoff from structures. This research activity is expected to provide benefits as the right solution for the site, input to parties related to the Lovina Beach area and assist in determining the best use of the site, and become the basis for answers to problems on the site.

IOP Conf. Series: Earth and Environmental Science **879** (2021) 012019 doi:10.1088/1755-1315/879/1/012019

2. Method

The method used in this research is the descriptive method with the stages of field surveys and literature study by direct observation, interviews, and questionnaires. There are four stages in data collection: the preparation stage, data collection, data analysis, and design.

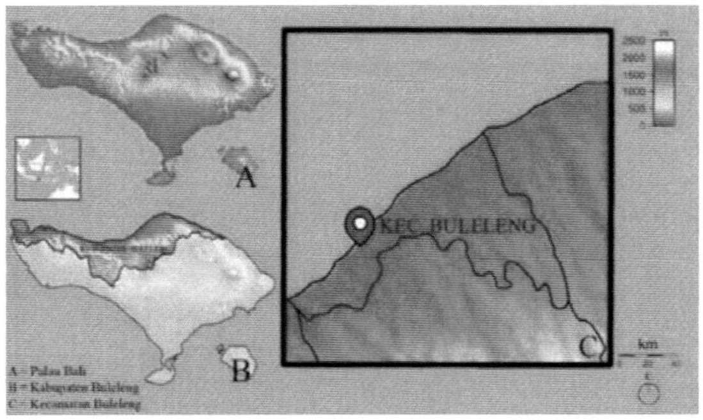

Figure 1 Research sites
Source: Topographical Map of Indonesia

3. Preparation Phase

This stage includes coordination and collaboration with the PUPR Office of Buleleng Regency, Bali, the spatial and territorial division for determining research locations, setting objectives, making research proposals, requesting permits to conduct research, and tools and materials for field surveys as references and literature studies.

4. Data Collection Stage

This stage uses data obtained from survey results in the field, interviews with related institutions, traditional leaders and local culture, and literature related to the research topic. The data used in this study were obtained from measurements and direct observations in the field.

5. Data Analysis Stages

The analysis stage is data processing analysis of physical and biophysical aspects and social and cultural factors. The research was carried out by identifying potentials and obstacles, and continue with a solution to the potentials and constraints, and formulating synthesis. Synthesis is developed into coastal concepts and designs.

6. Physical and biophysical aspects

The landscape analysis was implemented to identify potential and constraints related to topography, climate, vegetation, animals, facilities, and utilities. The research results are needed as a landscape design proposal for Lovina coastal waters based on landscape engineering.

7. Cultural and Social Aspect

This aspect analysis is needed to determine the elements that are expected to be on the site, referring to the wants and needs of the site users and the activities therein. The approach to this aspect is in the form of interviews and questionnaires.

The snowball sampling interview method and in-depth interviews, namely the sampling technique with the help of key informants, and from the key informants, the results are explored more deeply. Researchers define criteria as requirements to be sampled (Subagyo 2006). This technique selects traditional leaders or historians who will be critical informants and other informants competent to provide additional data. The in-depth interviews were conducted to prepare a list of questions that had been prepared (Moleong

2005). This stage will be a reference in the proposed Lovina coastal landscape design based on landscape engineering.

The questionnaire method used is non-probability quota sampling. A total of 30 respondents were randomly selected, including visitors and the surrounding community. The output of this questionnaire will be considered in the final design that is proposed. The form can be seen in the attachment.

8. Hydro-Oceanography Aspects

The vulnerability model that includes all the factors that influence coastal vulnerability consisting of physical, socio-economic, and environmental variables is seen as a complete vulnerability model at this time. The cultural parameters of this assessment model are referred to as integrated vulnerability and become the coastal vulnerability assessment concept that will be applied in this study. There are several references for assessing the vulnerability index used, namely countries that use the Coastal Vulnerability Index (CVI).

9. Design Stage

The design stage is the final stage and produces output in the form of a proposed design. The design process refers to Booth (1983), in which the design process stages consist of; 1. Project acceptance. 2. Research and analysis and engineering aspects with coastal vulnerability index scoring. 3. Design (Design). 4. Construction drawings. 5. Implementation (Implementation). 6. Post-construction evaluation. 7. Maintenance (Maintenance). However, the limitation of this research lasts until stage 4, while stages 5 to 7 will be submitted to Lovina Beach. Then the basic concept and planned design will continue to address the on-site problems until they are identified and produce a block plan. The illustration image is provided in a 3D model of CAD, SketchUp, and Lumion.

10. Results and Discussion

Location and Site Boundary

Lovina Beach is located in Kalibukbuk Village, Buleleng Regency, North Bali, Bali. Geographically it is located at the position 8003'40 "- 8023'00" LS and 114025'55 "- 115027'28" East Longitude. This research site has the following site boundaries (Figure 2).

- North boundary: Bali Sea
- Eastern boundary: Anturan Village
- Southern boundary: Kayuputih Village
- West boundary: Kaliasem Village

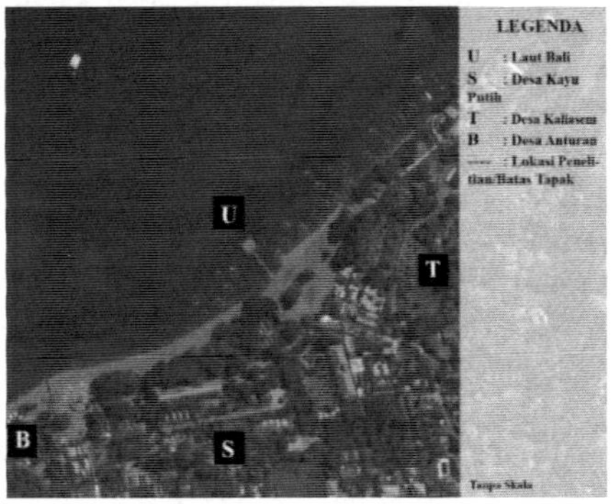

Figure 2 Orientation and site boundaries
(Source: maps.google.com 2019)

Analysis

Climate

Lovina Beach has an average temperature of 28° C, and average relative humidity of 82%.

The determination of the comfort index based on the THI method connects the weather and comparable humidity conditions in the developed study area (Nieuwolt 1977) in the form of the following equation:

$$THI = (0.8 \times T) + \{RH \times T / 500\}$$

T = Average THI temperature = 21-27(Comfortable) RH=Humidity THI> 27 (Uncomfortable)

The result of the THI calculation on Lovina Beach is 26.99. It shows that the THI of the site is in the comfort range. Hence efforts are needed to increase the associated comfort, stabilize the temperature and microclimate in the Lovina Beach area.

Topography and Slope

The district has recorded that it only has a slope of 0-25% (Table 2).

Table 1 The slope of Lovina Beach and the area of the slope area

District	**Slope (0-25%)**	**Aslant (25-40%)**	**Total**
Buleleng	4.694,00		4.694,00

(Source: Center for Meteorology, Climatology and Geophysics Region III Denpasar)

Hydrology and Soil

Puddles were found in the parking lot near the entrance. It shows that there are deficiencies in the site design so that the runoff problem still occurs. Then one of the related problems is the absence of adequate parking space. Vehicles that come are irregular, making the placement of drainage channels less possible.

Making drainage channels can be the answer to this problem if it is harmonized with parking arrangements. Additional drainage channels can be made and placed on the sides near the parking line or sidewalk if a parking area is completed.

The sand on Lovina Beach is blackish gray. It makes the beach tends to be hotter than other beaches in Bali. The addition of vegetation shelter for widening the garden is needed to reduce the heat on the beach.

Accessibility and Circulation

Lovina Beach is a pretty strategic tourist destination with its facilities that can be reached with various accesses. Visitors who use the following alternatives:

Accessibility

Access can be reached in two ways, namely air and land routes.

1. Air Line

   The plane to Denpasar and then travel about ±120 Km to the North.

Three alternative path choices, namely through;

a. Gilimanuk or the West line,

b. Bedugul through hills or access perpendicular to the North, and

c. East of the island of Bali via the city of Singaraja.

The time taken is approximately 3-4 hours from I Gusti Ngurah Rai Airport.

2. Land route

   Bus to West Bali Island, Gilimanuk, access from Java Island via the Bali Strait.

Circulation

There were three types of existing circulation include:

a. The Primary circulation for all motorized vehicles and pedestrians along every part of the coast.

b. The secondary circulation. It is inner-shore circulations, which are accessible to pedestrians.
c. The Tertiary circulation. It is the circulation that connects the primary circulation and the secondary circulation.

Facilities and Utilities

Activities on the beach can indeed vary. This is evidenced by the use of the beach, which can be used according to visitors. There is one activity of beach users, namely making the area on the beach's sand a multi-functional place, as shown in Figure 3.

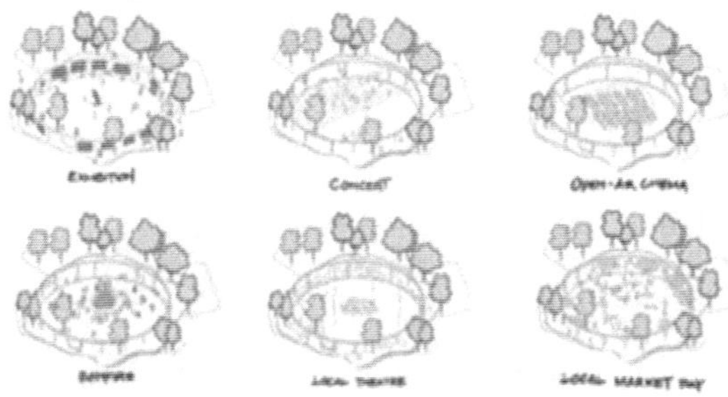

Figure 3 Sketch of Expression Space Usage Plan

Vegetation and Animals

The vegetation that already exists on the coast is quite diverse from the shade vegetation, directors, to displays that function for aesthetics. Besides the diversity of vegetation that is already on the coast, another aspect that is no less important is the optimization of the functions of the plant's existence. It takes some vegetation with various strata, for example, the function of shadow cover as a shade as shown below (Figure 4). Then in other aspects, the role of plants or vegetation can be a filter for vehicle noise

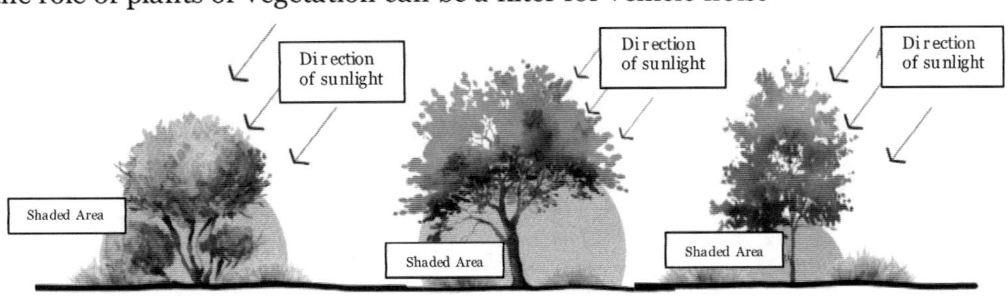

Figure 4 Analysis of high, medium, and low vegetation shade cover

Visual

The visual role is quite important for user comfort in order to see the site optimally (Figure 5).

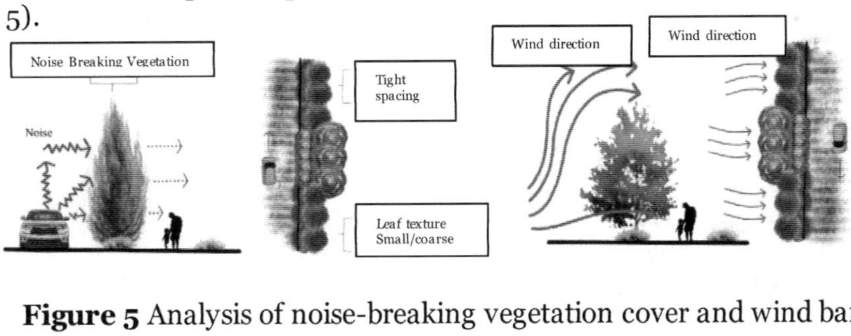

Figure 5 Analysis of noise-breaking vegetation cover and wind barriers

Lovina Beach has pretty good visuals. It was evidenced by the beach, which is in a calm choppy sea. Then the view is centered on the Lovina Beach entrance area, which has signage and a dolphin statue tower. Harris (1998), in the book Standard Time-saver for Landscape Architecture, suggests the optimal degree of view of the human eye, which is 280, so that a certain distance is needed to be able to see the entire scenery around it (Figure 6). This beach still has a lot of land for additional material that can attract visitors and beach users. The view on this beach can also be a good view because visitors can watch the sunset. Lovina Beach also still needs some supporting plants to provide optimal and aesthetic functions on the beach. Overall, this beach can be said to be a good view. However, there are some points which are bad views. Such as guardrails that are less neat and many are destroyed, trash cans that are disturbing near the sea, and rivers that look dirty. It makes the aesthetics of the beach diminish and must be rearranged into a good view.

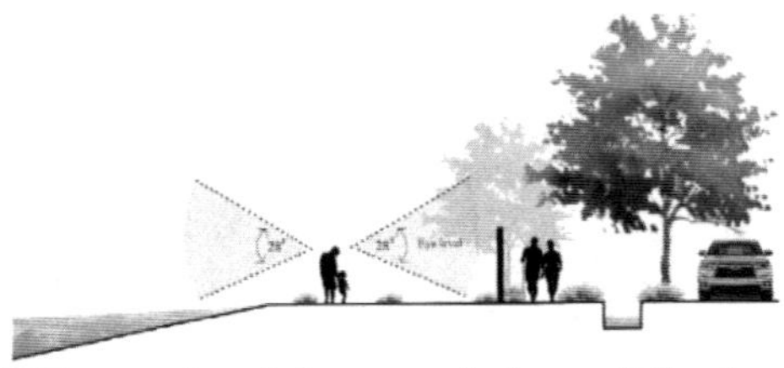

Figure 6 The optimal degree of view of the human eye

User activity

User activity uses a database as a result of interviews. The results of questionnaire interviews with respondents related to potentials, constraints, and user expectations of Lovina Beach were obtained, which were considered until the design stage. The number of respondents was 30 people, consisting of 20 men and ten women. The respondents under 15 years old are four people between 15-25 years old, eight people, 25-50 years 12 people, and over 50 years, six people. The activities carried out by users are dominated by users who are walking / sightseeing, then resting, and sunbathing. For tourist attractions that the user needs is natural scenery, while cultural tourism attractions, local traditions, cultural arts, and festivals. The following is the visitor identity data presented in the form of a diagram (Figure 7).

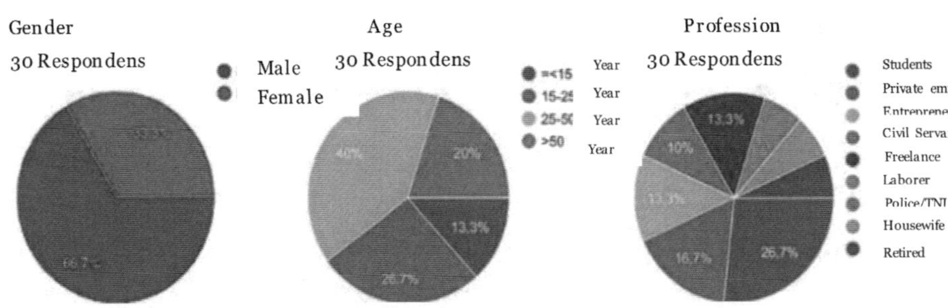

Figure 7 Graph of Lovina Beach Visitor Identity.

The most visitors found walking/sightseeing is important activities were 17 people. The desired activity to support activities on the beach is walking/sightseeing, equivalent to resting 12 people. The assessment of the beach conditions from the respondents showed that the beach conditions were classified as "Good," with a total of 46.7% of respondents. The most facilities needed by visitors are Gazebo/Shelter, mentioned by 12 people. The desired natural tourist attractions are natural scenic attractions, mentioned by 27 people.

Cultural tourism attractions desired by 23 visitors. For the type of tourism activity desired on Lovina Beach, it is more directed to family tours, mentioned by 24 people. Meanwhile, according to beach visitors, the condition of the beach that must be improved is the rubbish on Lovina Beach which is relatively small because only six people complained.

Vulnerability Index

The analysis for each parameter of the vulnerability index can show problem-solving and provide solutions for the coast. However, this is limited by human limitations in landscape engineering. The following is an analysis and synthesis of the variables that determine the value of physical and human parameters; The physical parameters of Lovina Beach get the following values:

1. Geomorphology (4), namely beaches of sand, coral, mud, deltas, mangroves;

These geomorphological parameters consist of steep cliffs, medium cliffs, sand dunes, crust rock beaches, and sand beaches or mud beaches. Geomorphological parameters reflect natural coastal landforms and their resistance to erosion.

2. The rate of change of the coastline (2) is -1 - 1;

This coastline rate can be episodic, seasonal, per decade, or per century. The rate of coastal erosion-rooting can be determined based on comparisons of old maps and aerial photographs.

3. The slope of the beach (4) is <0.3;

The description and evaluation of potential inundation and possible erosion is implemented in this research. Relatively sloping coastal areas are more rapidly flooded by tidal floods and eroded by erosion than steep beaches.

4. The rate of increase in sea level (2.5) is 2.51-3;

It is determined based on the rise and fall of the annual mean sea level over a certain period at the tide monitoring station on the coast. Global sea-level height ranges from 5-10 mm per year based on IPCC calculations (2007).

5. Mount Mount (3,5) which is 1-2;

The high tide mount provides low vulnerability. Conversely, low tide levels are highly susceptible. This view is based on the potential chance of extreme water levels due to storms. In micro-tidal coastal areas, beaches with tides of less than 2m, the chance of inundation are slight because the extreme water level during the storm is lower than the highest water level.

6. Wave Height(1) which is under 1.1.

Physical variables are crucial in determining the level of vulnerability of the coast. The crashing of the waves will cause erosion on the beach; the higher the waves, the greater the erosive power. The energy of the waves hitting the shore is directly proportional to the square of the wave height.

IKP F (Physical) (2,56) shows that the vulnerability value of the physical parameters of Lovina beach is moderate.

The Human Influence parameter gets a value, based on Solomon's 2010 data:

1. Coastal Protection (2), namely 20-30%, is good enough because it was found that there is at least a beach protection building in the form of a jetty. The existing condition is quite good, but it needs to be developed for maximum potential.
2. Sediment Tappers (1.5), namely 20-40%, not very good; it was found that there was a small river outlet but without the surrounding protection.
3. Coastal Vegetation (1,5), namely 30-50%, is dominated by shade plants. However, there are still some areas that are not shaded.
4. Land Allocation (4), namely the tourism industry, is quite good, and the maximum is seen from the many tourists offers on the spot.
5. Groundwater Consumption (1.5), namely 20-30%, Poor drainage, there are puddles in several places.

6. Total population (2.5), namely 2.01 to 2.25,
7. Culture (2), namely 2-3. Explain that there are at least three customary activities found.

IKP PM (Human Influence) (2.01) shows that the vulnerability value to human influence on Lovina Beach is moderate. The recommendation that can be addressed to reduce Lovina Beach's vulnerability in the future lies in the variable part of the human influence of coastal protection. This can be done by modifying the existing coast protection. Lovina Beach has a jetty-type beach protector. The condition of the jetty, which is currently on Lovina beach, is not too good, and its function can still be optimized. Considering the designation of this coastal land as a tourism industry, it will be more if the existing coast guard or jetty does not only function as a coast protector but has few attractions for Lovina beach.

Concept

The concept adopted for Lovina Beach is based on the dolphins that are characteristic of Lovina Beach, namely "Dolphin head Anatomy," which is the anatomy or composition of the inner dolphin head. This concept is a philosophy of how the dolphin head has intelligence and translated into the division of space and its functions. Then also means that all types of activities on the beach are ordered, directed, and neatly arranged based on this. Parts of the anatomy are the brain, skull, blowholes, melons, maxilla/mandible, and acoustics. The translation from basic concepts to design concepts used in the division of space (Figure 8).

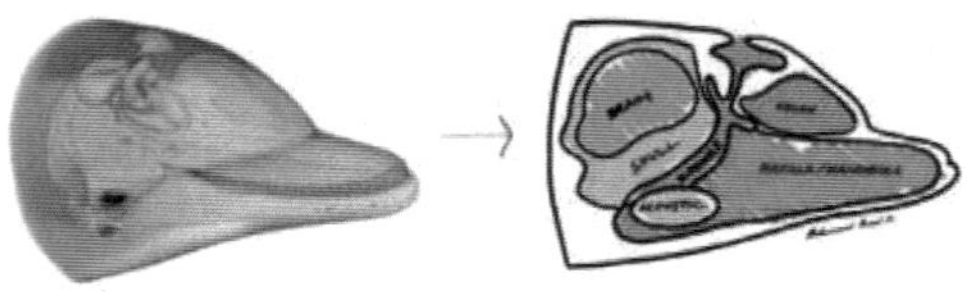

Figure 8 Schematic Design Concept

Development Concept

The development concept is an advanced stage of the design concept. It discusses space, circulation, vegetation, visuals, facilities, and lighting in more detail. The following is an explanation of the six development concepts.

Zoning

The concept of space zoning includes the Brain Room, a place for users to add insight into all things related to Lovina Beach and its history and information on supporting beaches. The skull space is translated into space to strengthen the body, namely sports. Then, the Blowhole Space is a respiration hole that is connected to every part of the dolphin head anatomy and translated as a circulation path. Then the melon is streamed, which is an essential part of the dolphin's head. It is translated into a place of worship. The maxilla/mandible or the upper and lower jaw of a dolphin means a dining area or restaurant and café. Finally, the acoustics are translated into theater spaces or expression spaces for various activities on Lovina Beach.

The design area is 0.6 ha, with a beach length of 175 m and a beach width of approximately 50 m. The carrying capacity of the facility's construction is also calculated by referring to the calculation of the area divided by the existing space standard or individual space standards. Nurisjah et al. 2003, argued that carrying capacity is calculated based on the average in m2/person. The goal is to obtain the optimal degree that can be accommodated on-site with various planned activities. The calculation of carrying capacity (DD) for the user is as follows:

$$\mathrm{DD} = \frac{\mathrm{A}}{\mathrm{S}}$$

Information: A = Area used (m2), S = Standard needs per person (m2/person).

Circulation concept

The concept of circulation is designed to optimize the interaction between the user and the beach. The concept is done by utilizing a groove made in such a way as to surround all coastal location points. The seaside location points are connected by connecting each primary, secondary, and tertiary route.

Visual concept

Several locations on Lovina Beach have been identified into several views, namely Good View, Bad View, and Spot View. The locations that have a bad view are view of trash, a bridge over the river estuary, and vacant land in the middle between a food stall and the beach border. Whereas the spot view locations are the Lovina Beach dolphin statue monument, Lovina Beach signage, views towards the sea, and views of the sea from the top of the jetty.

Vegetation concept

The concept of vegetation used is based on the functions of the vegetation itself, include 1. Shade from the sun; 2. Environmental engineering for hydrological systems and animal habitats; 3. The architectural function of vegetation is to divide and direct the site; 4. Aesthetic needs can be seen from the physical vegetation.

Block Plan

The block plan consists of a concept map of space zoning, circulation, visuals, vegetation, facilities and utilities, and lighting (Figure 9).

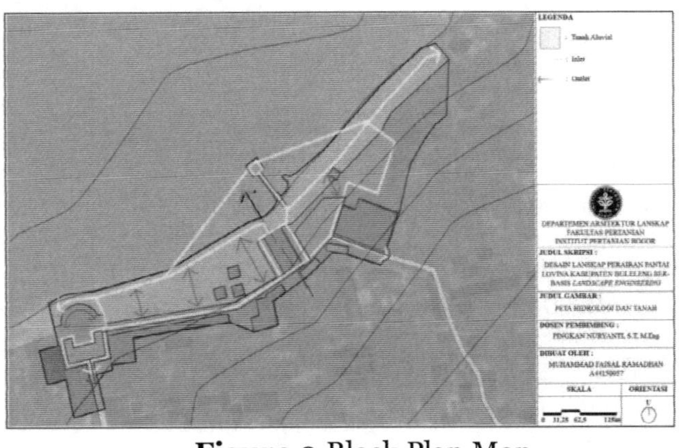

Figure 9 Block Plan Map

Design

Site Plan Drawing

This site plan focuses on the overall design that balances functionality and aesthetics (Figure 10). The use of several existing spots becomes a more useful place, seen in a more regular parking lot with dividing lines for each vehicle.

The design is eliminating boundaries between users who have activities when shopping or eating with the beach by adjusting the walkway and seating area. The rearrangement of the existing sports area became multi-functional and added a place to play for children around the area according to the results of the questionnaire interview. It can live up to the beach atmosphere with an expression area for various activities. The overall perspective drawing is a projection of the overall design of the beach made in three dimensions (Figure 11). The sectional view image is a longitudinal and transverse coastline design made with montage (Figure 12), and DED Jetty (Figure 13).

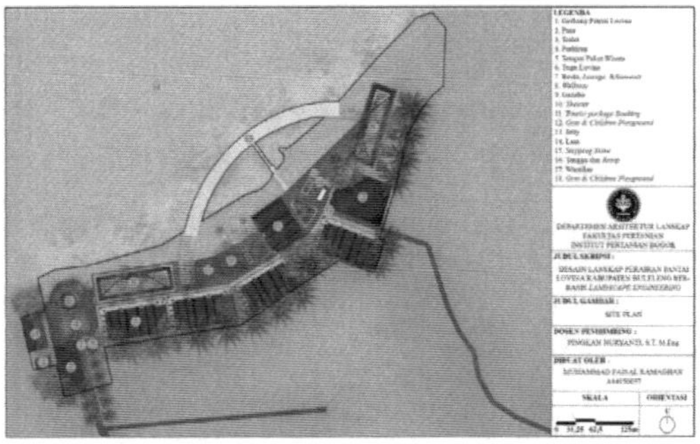

Figure 10 Map of the Site Plan

Figure 11 Overall Map

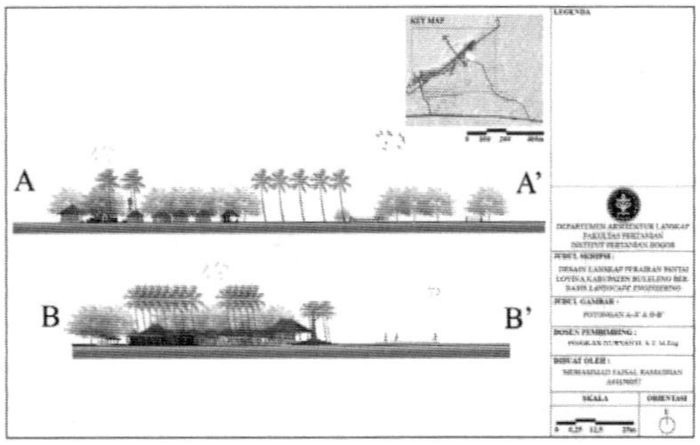

Figure 12 A-A '& B-B' pieces

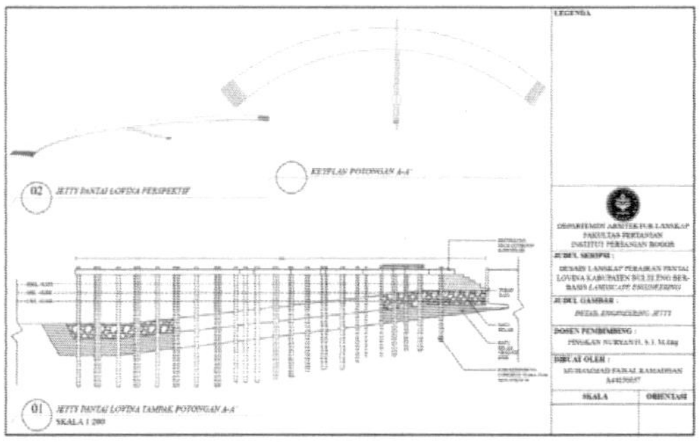

Figure 13 Detail Engineering Design Jetty

11. Conclusion

Lovina Beach has an area of 0.6 ha with a beach length of 175 meters and a beach width of approximately 50 meters. The condition of the existing beach sand, which is blackish gray, tends to make the beach hotter than other Bali beaches. Additional shelter and widening vegetation are needed to reduce the heat of the beach. Activities usually carried out by users based on observations are sports, walking/looking at the beach, sitting around, and sunbathing. Lovina Beach has quite good tourism potential in terms of natural attractions, as evidenced by field observations which show that the number of tourist visits is quite busy. However, this potential must be balanced because this beach has several obstacles. Based on the Integrated Coastal Vulnerability Index (IKP) results, the score is 1.51, which shows that this beach is classified as having a low level of vulnerability. The concept raised is "Dolphin head anatomy." The concept is implemented to the division of zoning on the coast. There are additional elements on the beach, such as entrance gates, signage, information boards, and the formation of restaurant café and lounge areas. The results of the coastal landscape design based on landscape engineering can be a recommendation for related parties and coastal managers for consideration.

References

[1] Booth NK, 1983. Basic Elements Of Landscape Architectural Design. New York (US): Waveland Press.
[2] Harris CW dan Nicholas TD, 1998. Time-saver Standar fo Landscape Architecture (second edition). USA: McGraw Hill Inc.
[3] Moleong, 2005. Qualitative Research Methodology. Bandung (ID): Youth. Rosdakarya.
[4] Nieuwolt S, 1977. Tropical Climatology, an introduction to the climate low latitude. John Willey & Sons, New York (US).
[5] Ozyurt G, Ergin A, & Esen M, 2008. Indicator Based Vulnerability Assessment Model to Sea Level Rise. Dubai (UEA)
[6] Subagyo, PJ, 2006. Research Methods in Theory and Practice. Jakarta (ID): Rineka.

The coastal design of the Sumur Tiga beach as a coastal tourism destination

M Kamal[1] and D R Anwar[2]

[1] Student, Department of Landscape Architecture, Faculty of Agriculture, IPB University, Indonesia

[2] Lecturer, Department of Landscape Architecture, Faculty of Agriculture, IPB University, Indonesia

E-mail: [1] mustofa_kamal@apps.ipb.ac.id; [2] dewian@apps.ipb.ac.id

Abstract. The East Coast of Weh Island, Aceh has good ecological conditions with high coral fish biomass. The coastal tourism activities spread over the east coast of Weh Island, especially at Sumur Tiga Beach. The tourism activities in Weh Island especially Sumur Tiga Beach can certainly contribute to the development and economic sectors and also have an impact on the surrounding environment. Thus, an alternative landscape design is needed for structuring the site to create environmental preservation, and sustainable tourism. This study aims to identify the elements and characteristics of the site, analyze-synthesize the potential and constraints of its coastal landscape, and create the concept and design of the landscape. The study was located in Sumur Tiga Beach, Sabang City, Ie Maulee District, Sukajaya Village, Weh Island, Aceh. The method includes several aspects, such as physical and biophysical, tourism, social, and hydro oceanography which is using both qualitative and quantitative descriptive and spatial analysis. The basic concept is "*Adat ngon hukom lagee dzat ngon sifeuet*" which is implemented into three concepts, such as *hablumminallah, hablumminannas, and hablumminalalam*. This concept is used to determine the design concept. then the development concept and block plan are determined then used as a reference in developing alternative designs for the site. And from these alternatives, the final design for the site is concluded. The design concept is a metaphor of "*Ragam Hias*" refer to the Acehnese motifs, such as *pucuk reubong, awan si on, awan meucanek, putra taloe dua,* and *sulur*. Such as gazebo, main gate, main hall, main signage by a form that comes from a *pucuk reubong* motif with the additional pattern of acehnese culture, also plantation and circulation patterns using a combination of *pucuk reubong, awan si on, awan meucanek, putra taloe dua,* and *sulur* motifs. There are also seawalls on the shoreline to avoid abrasion due to high sea waves and wind speeds that blow quite fast during the east season.

Keywords: coastal landscape, coastal tourism, sumur tiga beach

1. Introduction

Indonesia was an archipelago that consists of 17,508 islands and a shoreline of 81,000 km with potential for enormous coastal and marine resources. Indonesia's natural marine and coastal resources such as fishery resources, biological resources (biodiversity), and mineral resources. One of the potentials that can be utilized from these resources is coastal tourism. Coastal tourism activities are [1] recreational activities around the coastline while enjoying natural surroundings. Thus, it can be an option for utilizing coastal landscape resources in Indonesia. One of these coastlines was located on Weh Island, Aceh.

The beauty of the coastline founded in Weh Island has become the point of interest for tourism destinations for both local and foreign tourists [2] which was proved by the increasing number of foreign tourist visits to Weh Island in 2016-2018 (0.6%). One of these coastline zones in Weh Island was the east coast. The east coast has a coral reef ecosystem with 54.33% for the percentage of living in the conservation area and classified as "good" category, as well as around of conservation area with a percentage of 33.05% [3]. Meanwhile,

IOP Conf. Series: Earth and Environmental Science **879** (2021) 012020 doi:10.1088/1755-1315/879/1/012020

the abundance of reef fish reached 214 species in conservation areas and 159 species outside the conservation areas and classified as good conditions with high reef fish biomass. One of the beaches included on the east coast was Sumur Tiga Beach.

Sumur Tiga Beach was well known for its beautiful scenery on the east coast of Weh Island. It was a flat and white sandy beach with ocean waves and overgrown Barrington trees along with the site. Some of the tourist activities that can be done at Sumur Tiga Beach are swimming, playing, snorkeling, diving, relaxing, photography, walking along the beach, enjoying the beauty of nature, and more.

The high number of tourism visits can certainly benefit the surrounding beach. One of them was improving the welfare of the surrounding community, especially in the tourism sector. Also, increasing tourism activities can advance infrastructure and facilities that support these tourism activities. However, the increasing number of visitors harms the environment, such as shifting the existence of natural ecosystems on the site. The increasing development activities and the population on the mainland will lead to depletion of land resources, thus coastal areas and surrounding islands will become the foundation of hope for future economic development [4]. If there is no proper arrangement, especially on Sumur Tiga Beach, it will disturb the natural ecosystem on the site and decreasing the quality of the landscape. Thus, to overcome this problem, we must set the plan for designing Sumur Tiga Beach as a sustainable coastal tourism destination. One of them is through research on the landscape design of Sumur Tiga Beach on Weh Island.

2. Methods

2.1 Research location and term

The study was located at Sumur Tiga Beach, which is located on Weh Island, Ie Maulee Village, Sukajaya District, Sabang City, Aceh Province. The coordinates was 5 ° 53'35.54 "U & 95 ° 20'18.58" T (Figure 1). The research was held for ten months (November 2019 -August 2020).

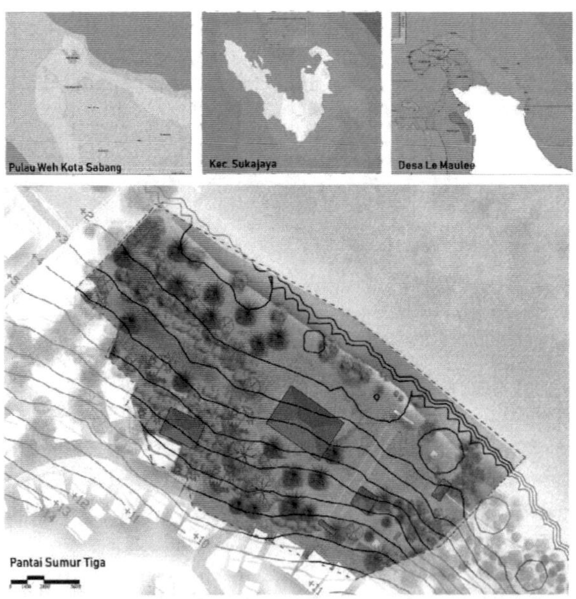

Figure 1 Research site

2.2 Research methods

There are four approaches used for the research, physic & bio-physic, social, tourism, and

IOP Conf. Series: Earth and Environmental Science **879** (2021) 012020 doi:10.1088/1755-1315/879/1/012020

hydro-oceanography. The research stages [5] are project acceptance, research/analysis, concept, construction drawing, implementation and post-construction, and evaluation maintenance. However, the steps are limited to construction drawing in this study. The concept stage started with basic concepts, then used to determine the design concept. Furthermore, the development concept is selected and the block plan is used as a reference in developing 3 alternative designs for the site then the final design for the site is determined from these alternatives.

3. Result and discussion

3.1 General condition

Sumur Tiga Beach is usually visited by tourists and local people on Weh Island. It is located in Ie Maulee Village, Sukajaya District, Sabang City, Aceh. Based on interviews from several stakeholders, it has a reasonably complex management system because of its position in the center of Sabang City and the presence of several facilities that have the potential to be developed into tourism interest. However, some potential aspects, both in terms of natural resources, tourist attractions, availability of facilities, and management which involving many parties, have not been able to make the site become good coastal tourism because of the management system is not running properly. This problem must be explored to find a solution to this problem and become a reference in structuring the landscape of Sumur Tiga Beach.

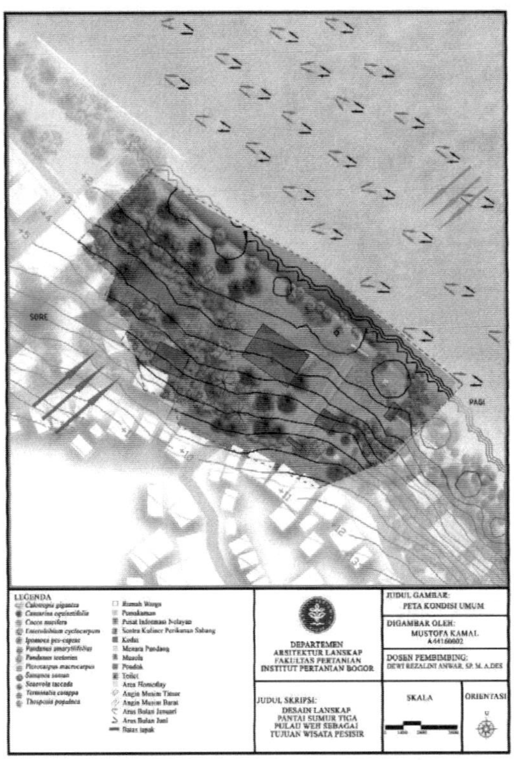

Figure 2 General condition

4. Analysis

4.1 Physical and biophysical aspect

4.1.1 Location and boundaries

Based on the survey, the site is close to residential areas and the city which eases the accessibility and ensures the availability of facilities and utilities. The existing element that has the potential to become a barrier is *Pterocarpus macropilus* which is planted in rows. The location of the site is included in the red zone for tsunami and abrasion disasters and also the red zone for ground movement disasters [9]. The level of disaster vulnerability is one of the limiting factors in the identification of conservation and tourism zones. The location of the beach in the disaster zone can be an inhibiting factor for the development of coastal tourism and needs to be considered carefully. The strategic location of the city can be an advantage in the landscape design of Sumur Tiga Beach by improving the visual quality of the entrance and utilizing natural boundary vegetation. However, disaster zone aspects need to be attended to in landscape development by controlling elements of the site that affect the disaster, such as slopes, vegetation, hydrology, and so on.

4.1.2 Climates

The criteria for comfortable temperature and humidity can be found through the Thermal Human Index (THI) equation. Comfort criteria have an index value of 21-27oC and uncomfortable> 27oC. Based on the data obtained, the average temperature on the site is 27.4oC with a humidity of 84.3%, and the comfort index value is 26.53 which is included in the comfortable category. The average wind speed reaches 7.02 knots (13.11 km/hour) and is included in the very weak wind category [10] on the Beaufort scale with the effect of the wind being felt on the face. The average rainfall is included in the moderate category, which is 142.9 mm3.

The higher the albedo value, the higher the temperature and reduces the level of comfort [11]. Based on the survey results, the site climate conditions which are dominated by vegetation cover have a low albedo value and have a good level of comfort. Some of the factors that reduce comfort on the site are sun intensity, wind speed and rainfall. High sun intensity with moderate rain can reduce comfort on the site, especially in arid land cover conditions, so it is necessary to provide shade in the form of a gazebo or shade vegetation. A practical canopy shape is used to prevent temperature increases, such as around and open canopy shape that can reduce the average temperature by about 2oC and increase RH by 5% [12] while a gazebo can block direct sunlight on the site. Wind speed affects wave speed and tourist activities on land, so it is necessary to add barrier vegetation and wave retaining walls along the shoreline to prevent abrasion.

4.1.3 Topography and slope

Topographic and slope analysis refers to slopes [13], topographic conditions, and slopes at site divided into 2 types, 2-5% slope, which is included in the flat, and sloping 15-40% which is included in the steep category. 0-5% slope is an ideal area with various spatial functions and does not need to face problems in terms of grading or land engineering [5]. However, it can give the impression of being monotonous because there is no variation in the surface of the ground, there is no room for protection from disturbances so that access to one's privacy is blocked. Also, sites that have a slope of 0-5% have a chance of problems with drainage because of the flat ground surface. The 10-15% slope of the land can provide a broad direction of view and is a potential for site development, but is considered too steep for various land uses.

According to the survey results, the site has a sloping and steep topography. Sloping topography is located around the welcome area and sandy area that has the potential to become the center of inland coastal tourism activities, while the steep topography is located in the middle of the site which is limited in its development. The synthesis of the analysis results needs to pay attention to the site slope conditions. Flat topographical sections can be utilized to support general footprint factors such as central buildings, parking areas, yards, externally making facilities difficulties. At a slope of 1% the footprint can be used for lawn use, while at a slope of 2% it can be used for making pavements and sidewalks. Steep topography requires site engineering through cut and fill or creating retaining walls to minimize and avoid erosion problems.

4.1.4 Hydrology and soil

Analysis of soil refers to the classification of soil types [13]. Types of soil composition at Sumur Tiga Beach consist of brown podsolic soil, red yellow podsolic, and grumosol which are sensitive to erosion and soil class 4 so that they can be an inhibiting factor in the landscape design of Sumur Tiga Beach. The synthesis of the analysis is the arrangement of other landscape elements such as the use of slopes, vegetation and different existing potentials to compensate for these weaknesses.

The source of water for residents at Sumur Tiga Beach which is located in Ie Maulee Village is obtained through springs and groundwater extraction / dug wells with a distance of 0.01-0.1 km from the coastline [14]. Based on survey, the drainage at site used open drainage system with inlets originating from settlements and outlets leading to the sea. There are two open drainage streams found on the site with size 20-50 cm in cement cover covered with leaf litter. This condition causes the possibility of blockage of the drainage outlets on the site. Also, there is a runoff flow on sloping and steep soil conditions. This runoff flow can cause inundation on sloping areas, and erosion on steep areas.

The synthesis of hydrological analysis is to maximize the absorption of rainwater to maintain the volume of groundwater around the coast. By preserving the volume of groundwater, it can prevent the erosion of fresh groundwater by salt water. The absorption of rainwater can be realized by maintaining vegetation in a steep area to increase the absorption of runoff flow into the soil and prevent erosion, constructing infiltration wells, and adding drainage networks.

4.1.5 Accessibility and circulation

Analysis of circulation and accessibility refers the level of site accessibility based on the total distance traveled [15]. The distance from Sabang City to the site is 2.1-2.3 km and Balohan Port to Sabang City is 10-11 km, so the accessibility from Sabang City to the site is in the high category, while the accessibility from Balohan Port to the Sabang City is in the very low category. The synthesis of the accessibility analysis is the addition of public transportation modes such as buses and public transportation to make it easier for tourists for accessing the site. Also, the provision of bicycle rental facilities by the Sabang City government can be a transportation solution to a site that is close to the Sabang City.

Based on the survey, the condition of circulation path on the site is divided into two types, pedestrian and vehicular. The pedestrian path is still natural in the form of land and grass cover without any apparent pattern. The vehicular line is only in the welcome area to the main parking and signage with truepave cover. The lack of circulation paths with a clear pattern causes user activity to focus only on the welcome area, as well as the sandy area along the beach. Synthesis in circulation analysis is defining the vehicular road type for the Pantai Sumur Tiga design must have a road width of 6.5 meters [13]. Based on the map of the Weh Island ring road development plan, there is a planned road area to pass through Sumur Tiga Beach, so it needs to be planned in the development of the landscape design.

4.1.6 Facilities and utilities

Based on the survey, it is necessary to utilize the Sabang fisheries culinary center building as a restaurant, marine gallery, and art market. The survey results also recommend to utilize a fishermen information center building as a management office and club house and a lawn area as a play and recreation area. It also recommends constructing several new facilities such as a wooden deck, welcome area, panoramic spots, gazebo, benches, shelters, gates, and wayfinding. The survey also recommends replacing prominent signage for supporting coastal tourism activities. Analysis of utilities explain the provision of electricity networks sourced from PLN or generators, providing lighting on the site, procurement of waste disposal systems to city landfills and addition of landfills on-site, and planning additional drainage such as absorption wells and water channels [16].

4.1.7 Visual and acoustic

Analysis of visual and acoustic is based on on-site observation. Good views become potential aspects that must be maintained and developed, bad views become an obstacle in the site. Good views on the site are available such as a white sandy beach with rows of coconut trees along the shoreline. This goodview needs to be maintained by preserving the vegetation on the beach and adding a panoramic spot as a place to watch the view. There is also a Bad view of garbage and household waste such as plastic, wood, paper, and others piled up on the beach and around the homestay. The bad view can be solved by adding trash can facilities at several points to prevent people from littering. Also, several facilities such as gazebos and viewing towers have been damaged so it is necessary to maintain the facilities by involving residents. The source of the acoustics on the site is the ocean waves sound from the site. These sources are potential that gives a natural impression on the site, thus the synthesis is maintaining the acoustic source and add other acoustic sources such as friction sounds such as coconut, sea cypress and others with the plantation around the coast.

4.1.8 Vegetations and animals

Analysis of vegetation refers to vegetation function [5] which consists of fundamental values, environmental values and visual values. In terms of the structural function of vegetation such as coconut (*Cocos nucifera*), beach naupaka (*Scaevola taccada*) planted on steep topography giving the barringtonia landscape character to site. Based on environmental functions, sea cypress (*Casuarina equisetifolia*), banyan (*Ficus citrifolia*), and ketapang (*Terminalia catappa*) on the site play a role in climate amelioration and windbreak. Barringtonia tree formations also function as a deterrent to erosion on steep topography and increase the absorption of rainwater. Vegetation rows of *Pteropilus macrocarpus* at several points on the site act as natural boundaries. In terms of visual functions, the stretch of biduri (*Calotropis gigantea*) around the fishermen's information center building provides a good visual. The expanse of *Zoysnia matrella* around the main signage gives an expansive visual feel.

The synthesis of vegetation analysis is maintaining and increasing the presence of existing vegetation and increasing vegetation with certain functions on the site, such as vegetation with visual functions in welcome areas and recreation, environmental functions in recreational areas as windbreaks, and structural functions in the form of barriers on the boundary of the footprint, as well as on the circulation route. Analysis of animal refers to site survey. Existing animals such as cows, goats and chickens are livestock that are kept by residents and spread in the site, causing some dirts which scattered on the site and reducing the beauty. The synthesis is the addition of a grassy area for livestock and the addition of vegetation boundaries in that area.

4.2 Tourism aspect

In this study, the tourism aspect [7] divides the tourism aspects into tourist attractions, tourism services, transportation, information, and promotions. Based on on-site observations, halal tourism services have met the requirements on the site. Various written rules [17] governing halal tourism in Sabang City and accordance with the values of local wisdom of the community. Transportation to Weh Island only through fast boats and ferries. Based on on-site observations, there are no public transportation due to the large number of vehicle rental services in Sabang City. Synthesis of this problem by making tourist rules that equalize vehicle rental prices, and tour packages coordinated by pokdarwis of each village in promoting tourism in their area. Information related to the advantages and natural resources of Sumur Tiga Beach is still limited regarding the sloping white sand, even though there are many other advantages can be known by users when visiting the site. The promotion of tourism activities at Sumur Tiga Beach can be increased through good cooperation between parties in providing tourism services, both government and non-government actors. One of them is through the Sabang Calendar of Events which features events at Sumur Tiga Beach. Promotional activities can also be carried out by the community using social media through the use of commonly used hashtags, such as *#visitsabang, #wonderfulsabang, #sumurtigabeach*, and others. The tourism activities proposed as tourist attractions are coast walk, camping site, culinary center, panoramic spot, and coastal garden.

4.3 Social aspect

Analysis of social aspect refers to the user's perception on the site which is based on the characteristics, hopes, desires, complaints and activities that they carry out on the site. Based on the results of questionnaire, the most familiar coastal tourism activity on the site is relaxing in the beach area (83.3%). Then playing activities around the beach and along the beach sand (75%) and (71.1%), enjoying culinary around the beach (61.7%), looking for shells and unique rocks (50%), swimming around the beach (45%), snorkeling (38.3%), staying at hotels / homestays (35%), diving (28.3%), and fishing (15%). Also, there are other activities such as camping and drinking coconut water. Most complaints when at Sumur Tiga Beach are a lot of rubbish (25%). Also, there are respondents' perceptions about expectations for landscape design of Sumur Tiga Beach. The questionnaire's results show that the most desirable type of land cover is grass, leafy vegetation, natural material, non-paved circulation, organic pattern, analogue colors, tropical garden style, and terraces land type.

4.4 Hydro-oceanography aspect

The hydro-oceanographic aspect is a physical parameter used in determining the suitability of resources for coastal tourism purposes. Study from propriety of resources for coastal tourism destinations Sumur Tiga Beach [1] divides the assessment into two categories, namely coastal tourism activities and marine tourism activities. The results of the analysis of the Tourism Suitability Index (IKW) value for the Sumur Tiga Beach tourism in tables 2,3 and 4 obtained the IKW value of 81.89% for the suitability of coastal tourism and 86% for the suitability of marine tourism. Based on the results of the analysis, the values of 81.89% and 86% are values that fall into the the appropriate category to be used as a coastal tourism area.

5. Concept

5.1 Primary concept

The concept in the development of Sabang City tourism [17] was stated as a development of marine tourism products with history and geology refer on the principle of halal values and local wisdom values. The connection between halal values and local wisdom in Aceh law cannot be separated from the slogan "*Adat ngon hukom lagee substance ngon sifeuet*" which is a reflection that local wisdom and Islamic Sharia law, become one such as substances and nature, and cannot be separated each other [18]. It because Aceh's culture was based on Islamic law which comes from the *Al Quran* and *Hadith*. This motto become the primary concept in developing the landscape design for Sumur Tiga Beach. From these concepts, there are three Islamic concepts in human behavior that can be adopted in the development of landscape designs on Sumur Tiga Beach. These are *hablumminallah* (human interaction with their God), *hablumminannas* (human interaction with each other), and *hablumminalalam* (human interaction with nature and surrounding creatures). These three interactions are applied to the three concepts of developing the landscape design for Sumur Tiga Beach.

5.2 Design concept

Based on the primary concept derived from "*Adat ngon hukom lagee substances ngon sifeuet*", it provides an overview of the design concepts that will be applied in the landscape design of Sumur Tiga Beach. The design concept must provide design and cultural linkages with cultural centers such as *Banda Aceh, Aceh Besar* and others. Thus the design concept adopted is a metaphor of *Ragam Hias* which originates from the richness of Acehnese motifs, such as *pucuk reubong, awan si on, awan meucanek, putra taloe dua,* and *sulur.* For example, Putra Taloe Dua as the roof formation in the gazebo, shelter,tents, main gate and hall roof, *sulur* on the main signage, and the combination of these patterns in planting patterns and circulation patterns.

5.3 Concept development

Concept development is compiled after determining the primary concept and design concept. In this research, several development concepts that will be described specifically consisting of the concept of space, the concept of circulation, the concept of sequences, the concept of vegetation, the visual concept, and the concept of hydrology. The concept of spatial development divides the landscape into zones according to their activities and functions on the site, which are reception zone, recreation zone, service zone, and conservation zone.

Circulation planned at the site is divided into primary, secondary, and tertiary circulation. The primary circulation is 5 meters wide to accommodate vehicular and pedestrian circulation with asphalt and Classico tile cover. This circulation is planned with a maximum slope of 4% and grading at several points to facilitate people with disabilities. Secondary circulation accommodates pedestrian circulation and becomes a link between one room and another so that it is planned that it is mainly located on the site. To support user activity, this circulation must be equipped with a ladder, ramp and handrailing. Tertiary circulation is a type of circulation that foot travelers can only enter, so it is only equipped with stairs and hand railings. The Circulation path utilizes materials such as pavement and wooden decks.

The concept of sequence describes the array of motions carried out by users in enjoying tours on the site. The planned sequence concept is 20-40 minutes in the reception zone, 2-3 hours in the recreation zone, 1 hour in the conservation zone, and 60-90 minutes in the service zone. With the concept of sequence, it is hoped that the user can understand the meaning and pattern of space in a sequential site arrangement in the tourist area.

The concept of vegetation divides the types of vegetation based on the functions that are on the site. There are five types of functions planned in the development of Sumur Tiga Beach design, display function, barrier function, shelter function, conservation function, and windbreak function. The visual concept developed refers to the synthesis of visual analysis on the site, by maintaining and increasing the presence of good views and eliminating bad views by adding landscape elements such as trash cans, vegetation screens, and others.

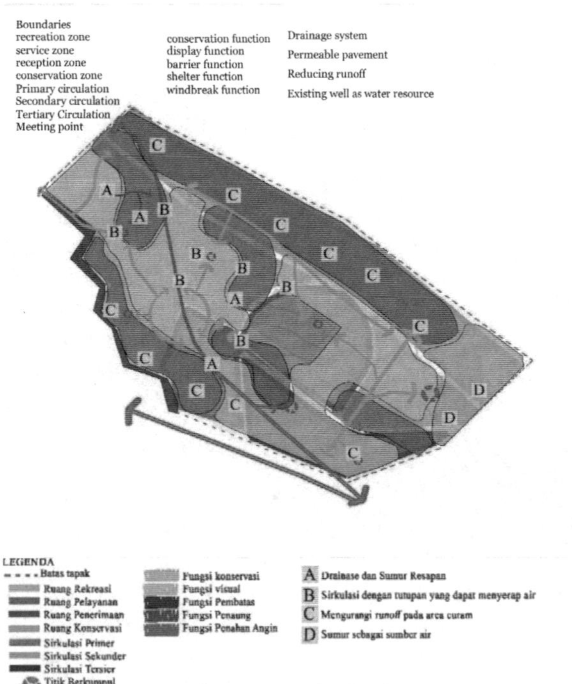

Figure 3 Block plan

The hydrological concept aims to preserve the availability and quality of water in the three old wells at Sumur Tiga Beach by increasing groundwater absorption when it rains. The vegetation was planted on steep topography to conserve groundwater. The effort to conserve groundwater are also obtained by constructing retaining walls, terraces, implementing permeable pavement material, infiltration wells, retention ponds, and open drainage networks that can drain water to infiltration wells.

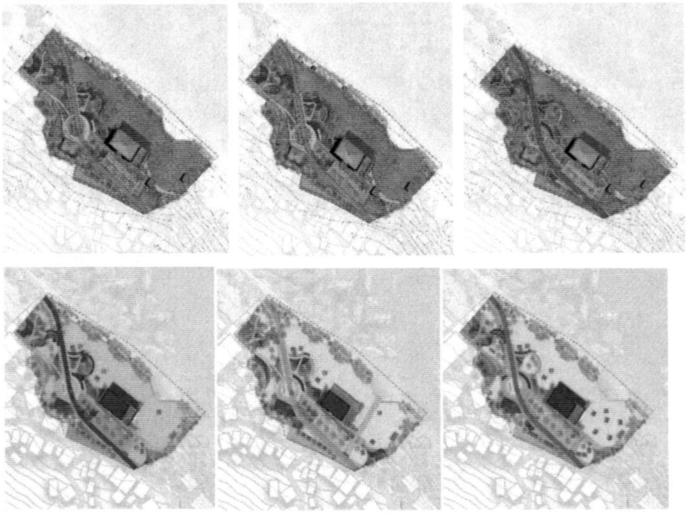

Figure 4 Alternative designs

6. Design

The final design siteplan begins with making three design alternatives by the results of the block plan and the concepts that have been made by considering the functions, activities, and supporting facilities. Thees three alternative designs are refer to design concept that is applied through decorative compositions in determining vegetation patterns, circulation, building forms, and landscape furniture. Furthermore, the three design alternatives were selected through the questionnaire results that had been distributed to 30 respondents.

Based upon those issues of poll respondents' preferences for those three alternative designs, the most chosen alternative design is alternative design 2 at 43.3%. The site plan made in this final design is based on the development of an alternative design 2. The landscape design of "Sumur Tiga Beach" was performed by a predetermined design concept, including the gazebo, main gate, main hall, prominent signage, the pattern of the plantation, and circulation networks. The form of the "*pucuk reubong"* motif with an additional pattern of Acehnese culture patterns was implemented in the main gate design. The form is a combination of *pucuk reubong, awan si on, awan meucanek, putra taloe dua,* and *sulur* motifs. There are also seawalls on the shoreline to avoid abrasion due to high sea waves and wind speeds that blow quite fast during the east season.

7. Conclusion and Suggestion

7.1 *Conclusion*

Sumur Tiga Beach is one of the white sandy beaches on Weh Island which has coastal tourism potential. There needs to be a landscape design to support tourism activities and maintain sustainability of the landscape. Sumur Tiga Beach has lying in a smooth plain and steep topography, therefore that is necessary to pay attention to the difference in height in the design process. The THI value received at the place, was 26.53, and classified as comfortable category, but the wave height at Sumur Tiga Beach is strong enough during the east season thuss seawall and coastal plantation are needed to prevent abrasion. The presence of abrasion and wave height at certain times means that any equipments must do planned with a more substantial construction to anticipate that. This landscape design of Sumur Tiga Beach has a basic concept of "*Adat ngon hukom lagee substance ngon sifeuet*" which means that local wisdom and Islamic Sharia law are one, such as substances and properties, which cannot be separated. The design concept used is a metaphor of *Ragam Hias*, such as *pucuk reubong, awan si on, awan meucanek, putra taloe dua,* and dan sulur to determine suitable patterns for landscape design, as well as for deciding vegetation patterns, circulation, and other elements. The development concept contains the concept of zones, the concept of sequence, the concept of circulation, the concept of vegetation, the concept of hydrology and the concept of visuals.

7.2 *Suggestion*

Sumur Tiga Beach has tourism potential. Sources observed in Sumur Tiga Beach remain a conservation area for the East Waters of Weh Island. So that in developing tourism, we need return awareness on the preservation of the landscape on Sumur Tiga Beach. One of the problems in Sumur Tiga Beach is garbage. Therefore, local government and the community to intervene to maintain the cleanliness of the beach and the development of tourism on Sumur Tiga Beach so that it can afford advantages from the economic side including the aspect of landscape preservation.

References

[1] Rakhmawati A 2002 Aplikasi penginderaan jauh dan sistem informasi geografis dalam penentuan kesesuian lahan ekowisata bahari di perairan laut taman nasional bali barat Skripsi Fakultas Perikanan Dan Ilmu Kelautan IPB
[2] [Disbudpar Aceh] Dinas Kebudayaan dan Pariwisata Aceh 2015 Pantai sumur tiga http://disbudpar.acehprov.go.id/pantai-sumur-tiga/# Diakses 6 Juni 2019
[3] Hastuty R, Yonvitner and Adrianto L 2014 Tutupan karang dan komposisi ikan karang di dalam dan luar kawasan konservasi pesisir timur Weh Island, Sabang Depik, Vol **3**(2):99-107 Institut Pertanian Bogor, Bogor
[4] Dahuri R, Rais J, Ginting S P and Sitepu M J 1996 Pengelolaan sumberdaya pesisir dan lautan secara terpadu PT Pradnya Paramita, Jakarta
[5] Booth N K 1983 Basic elements of landscape architectural design (New York (US): Waveland Press)
[6] Nurisyah S 1998 Rencana pengembangan fisik kawasan wisata bahari di wilayah pesisir Indonesia *Buletin Taman Dan Lanskap Indonesia Perencanaan, Perancangan dan Pengelolaan* Vol **3** 2 1997 (Bogor: 133 Studio Arsitektur Pertamanan Fakultas Pertanian Institut Pertanian Bogor)
[7] Gunn C A 1994 Tourism planning: Basics, concepts, cases (Washington DC (US): Taylor & Francis)
[8] Yulianda F 2007 Ekowisata bahari sebagai alternatif pemanfaatan sumberdaya pesisir berbasis konservasi Seminar Sains pada Departemen MSP, FPIK IPB 21 Februari 2007; Bogor, Indonesia Bogor (ID): Departemen MSP IPB
[9] [Walikota Sabang] 2018 Rencana tata ruang dan wilayah kota sabang Tahun 2017-2022
[10] Kartasapoetra A G 2004 Klimatologi pengaruh iklim terhadap tanah dan tanaman PT Bumi Aksara, Jakarta
[11] Nafar S 2016 Desain ekologis taman koleksi pakis berbasis sistem klarifikasi bioregion di ecopark, cibinong, science center-botanic gardens (Bogor(ID): Fakultas Pertanian Institut Pertanian Bogor) [Skripsi]
[12] Alfian R, Budiarti T and Nasrullah N 2016 Pengaruh bentuk hutan kota terhadap kenyamanan termal di sekitar hutan kota *J. Buana Sains* Vol **16** No 2:101-110
[13] [PUPR] 2007 Pedoman kriteria teknis kawasan budi daya permen pekerjaan umum No.41/PRT/M/2007
[14] Edyanto H 2000 Penelitian kualitas air danau aneuk laot di weh island provinsi nanggroe aceh darussalam *J. Teknik Lingkungan* Edisi Khusus Hal 115-124
[15] Suthanaya P 2009 Analisis aksesibilitas penumpang angkutan umum menuju pusat kota denpasar di provinsi bali *Ganec Swara* Vol **3**(3):88-93
[16] Umami I K 2013 Perancangan wisata bahari di pantai boom tuban: tema extending tradition (Doctoral dissertation, Universitas Islam Negeri Maulana Malik Ibrahim Malang)
[17] [Walikota Sabang] 2018 Qanun kota sabang No 2 Tahun 2018 tentang rencana pembangunan jangka menengah kota sabang Tahun 2017-2022
[18] Nurdin A R 2015 Beberapa catatan tentang kearifan lokal masyarakat aceh Diakses melalui https://maa.acehprov.go.id/news/detail/beberapa-catatan-tentang-kearifan-lokal-masyarakat-aceh?q=ukhuwah+islamiyah#0

Attachment
Siteplan

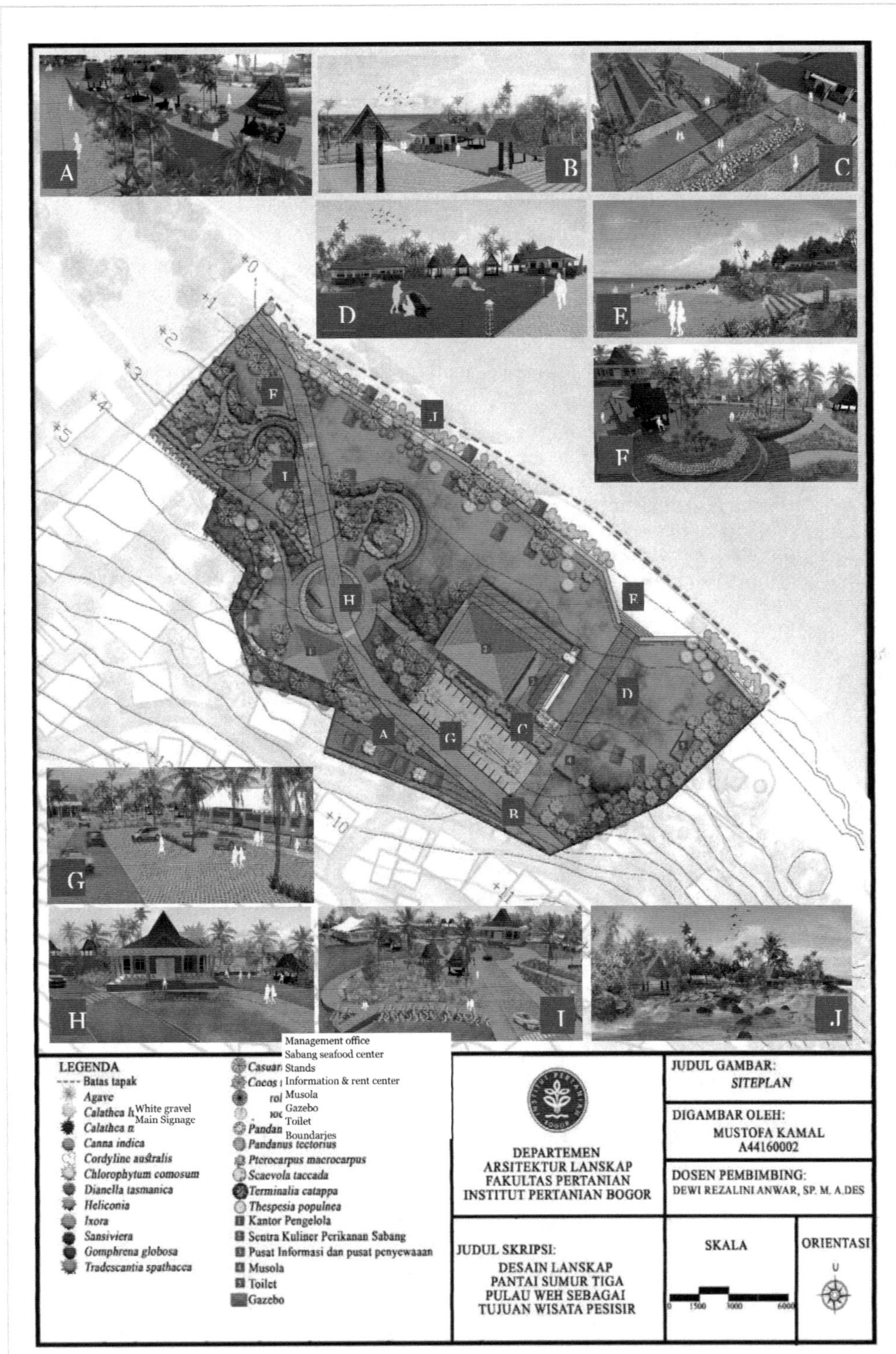

Environmental service assessment of trees at Yogyakarta International Airport open space using CITYGreen 5.2

R A Rahmadanti[1] and B Sulistyantara[2]

[1] Student, Department of Landscape Architecture, Faculty of Agriculture, IPB University (Bogor Agricultural University), Bogor, Indonesia

[2] Lecturer, Department of Landscape Architecture, Faculty of Agriculture, IPB University (Bogor Agricultural University), Bogor, Indonesia

E-mail: [1]riandini01@gmail.com@gmail.com; [2]bbsulistyantara@yahoo.co.id;

Abstract. Urban development has many negative impacts in the form of an increase in the concentration of CO2 in the air. One of them is due to the increase in the volume of vehicles. CO2 causes an increase in global temperature (global warming) through the greenhouse effect. Yogyakarta International Airport has a green open space planted with trees to support the existence of the airport itself. Trees have contributed to efforts to reduce CO2 in the air through CO2 absorption. Control of the amount of CO2 emissions in the air is done by adding carbon stocks on land so that the CO2 concentration does not continue to increase, therefore this research is necessary. The purpose of this study is to analyze the ability of trees to absorb pollutants and store carbon and estimate the value of environmental services that will be contributed by the tree green system at Yogyakarta International Airport based on existing planting plans. The research method used is modeling using CITYGreen 5.2 software to determine the value of tree services in the ability to absorb air pollution, carbon storage capacity, and cost savings that can be done by tree canopies. The method used in this study consisted of preparation, data inventory, and data analysis. The results of this study show that the value of environmental services provided by 125,72 acres of tree canopy cover in absorbing air pollution in the airport area is $ 9.045,54 / year or equivalent to Rp 133.356.587,10 per year, the total concentration of pollutants that can be absorbed is 3.813,30 lbs/year or equivalent to 1.729,68 kg/year, and the capacity of carbon storage is 1.326,46 tons with a sequestration rate of around 29,9 tons/year.

Keywords: air pollution, carbon storage, CITYgreen, environmental service, green open space, tree

1. Introduction

Various cities in Indonesia continue to experience development as well as infrastructure development, one of which is the airport. This activity also has a negative impact in the form of an increase in CO2 concentration in the air due to the increase in vehicle volume. Vehicle releases the chemical into the air becoming air pollution [1]. The quality and comfort of the urban environment are influenced by the availability and existence of tree canopies, both in the form of urban forests and urban green open spaces [2]. The existence of green open space has the potential to increase the availability of clean water and air [3]. Green open space in the airport landscape has an important role because it supports the existence of the airport itself [4]. Yogyakarta International Airport has green open spaces planted with trees to support the existence of the airport itself. The contribution of trees to improve the quality of health and environmental services is very large. Vegetation is one element of the ecosystem that can be manipulated in a planned and cost-effective way [5]. This research purpose to this study is to analyze the ability of trees to absorb pollutants and store carbon and estimate the value of environmental services that will be contributed by the green open space at Yogyakarta International Airport based on existing planting plans. The scope of the study is limited only to the trees. Sources of pollutants studied in this study are limited to vehicles, both on land and air. The value of tree environmental services in the area of Yogyakarta International Airport is known by using CITYGreen 5.2 extension which is only limited to studying aspects of the benefits of tree canopies in storing carbon and absorbing air pollution. The economic value of trees in this study refers to a cost indirect from social health and environmental damage. A cost that must be issued by the government in overcoming the problem of damage environment, in this case, is caused by O3, SO2, NO2, PM10 and, CO.

2. Methods

2.1 Description of the study area
The location of the study was conducted in Kulon Progo District, Special Region of Yogyakarta Province. This airport has a land area of 587,30 Ha with an undeveloped area of 251 Ha (open space). The area is divided into segments (lots). This is due to the large area of the airport and to facilitate calculations. The areas observed in this study are Lot A, Lot B, Lot C, and Lot E. Lot A and B are the main entry areas for the airport from the National Road. Lot C is the western supporting area, while Lot E is the eastern supporting area. Lot D is not included in the study area because it is a terminal building.

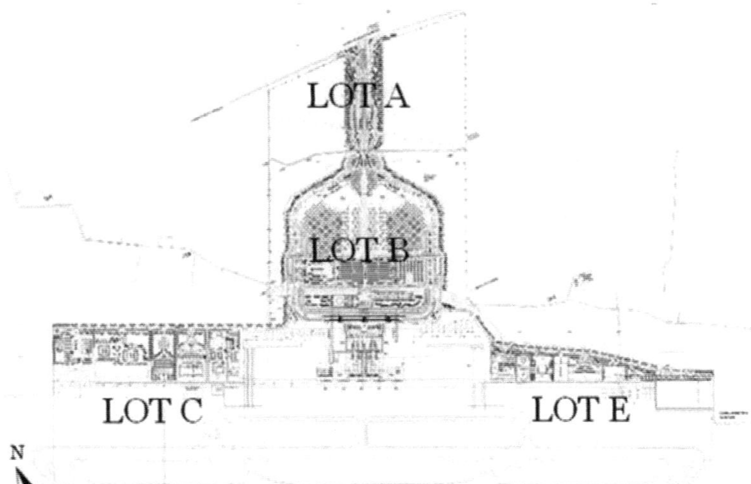

Figure 1 Study sites at Kulon Progo District
(7°54′27″S 110°03′16″E)
Source: PP KSO

2.2 Inventory
Inventory is carried out to obtain a complete picture of site conditions. At the inventory stage, satellite image preparation, site digitization, and attribute data collections for the canopy theme are carried out. These data are in the form of secondary data from the planting plan and masterplan obtained from PP KSO. PP KSO is a state-owned company that is a main contractor for the Yogyakarta International Airport project. In this study,

google earth satellite imagery was used. CITYGreen identifies land cover as canopy and non-canopy. Canopy is a theme that displays a tree canopy in the form of a view over the trees. Non-canopy is a theme that displays other than the tree canopy.

2.3 Analysis
The data that has been obtained will then be inputted. The research analysis method used is modeling using Arcview software version 3.3 with extension CITYGreen 5.2 to determine the value of tree services in the ability to absorb air pollution, carbon storage capacity, and cost savings that can be made by tree canopy. CITYGreen can estimate air quality which consists of pollutant concentrations O3, SO2, NO2, PM10 and, CO. The application of CITYGreen in the study of air quality and carbon storage has been widely used in research. [6] [7] [8] The analysis stage was limited to the vegetation aspect, namely trees, while shrubs and ground cover were not included in the analysis.

Table 1 Data Required for Analysis

	Required	**Values Acquired From Data Within CITYGreen and User Definable**
Air Quality	Tree canopy	Closest air quality city
Carbon Storage/ Sequestration	Tree canopy, trunk diameter (for individual trees)	

Source: User Manual CITYGreen 5.0

3. Result and Discussion

31. *Identify the type of vegetation*
Yogyakarta International Airport is a new airport, therefore many new trees are planted and all of them are in young age class. From table 2, it shows that the total number of trees based on the planting plan map in the airport area is 4.488 individual trees. It can be seen that the highest number of trees in the four areas is the Lot B area, which is 2.130 trees with a percentage of up to 48%. Meanwhile, Lot A has the least number of trees with a total of 586 trees with a percentage of 13%. There are 43 species of trees in the airport area. Those species are spreading in four lots. Even though Lot A only has the lowest percentage number of trees, the tree diversity in area A is the highest when compared to the other which is 26 tree species, following by Lot B which is 22 tree species, Lot E which is 13 trees, and Lot C which is 13 tree species. More complete tree species data is presented in Table 3.

Table 2 Recapitulation of The Number of Trees Found in All Areas

Area	Total	Percentage
Lot A	586	13%
Lot B	2.130	48%
Lot C	1.142	25%
Lot E	630	14%
Total	**4.488**	**100%**

Based on the data obtained from the planting plan map, the dominant tree species from the four research areas is Sapodilla trees (*Manilkara kauki* (L.) Dubard.) totaling 882 trees with a percentage of 18.32% and in second place is Spanish Cherry trees (*Mimusops elengi* L.) totaling 397 trees with a percentage if 8.85%. More complete data and the existence of each tree species in each area can be seen in table 3.

Table 3 The Existence of Each Tree Species

Plant species	Lot A	Lot B	Lot C	Lot E	Total	Percentage
Aegle marmelos Linn.	48	-	140	25	165	3.68%
Agathis dammara (Lamb.) Rich	-	74	-	99	221	4.92%
Alstonia scholaris L.	58	-	-	-	58	1.29%
Ardisia elliptica Thunb.	12	-	-	-	12	0.27%
Arenga pinnata Merr.	-	68	-	-	68	1.52%
Barringtonia asiatica (L.) Kutz	14	58	-	-	72	1.60%
Bauhinia purpurea L.	-	-	180	28	208	4.63%
Borassus flabellifer Linn.	23	-	-	-	23	0.51%
Cananga odorata (Lam.) Hook f & Thomson	-	36	-	-	36	0.80%
Cassia fistula L.	29	177	-	-	206	4.59%
Casuarina sumatrana (de Vriese) L.	-	-	98	129	227	5.06%
Chrysalidocarpus lutescens (H.Wendl.) Beentje & J.Dransf.	-	63	-	-	63	1.40%
Cocos nucifera L.	30	-	-	-	30	0.67%
Cocos nucifera var. Eburnea	24	38	-	-	62	1.38%
Cyrtostachys lakka Becc.	6	-	-	-	6	0.13%
Delonix regia (Hook.) Raf.	13	-	15	3	31	0.69%
Elaeocarpus ganitrus Roxb.	8	-	-	-	8	0.18%
Eucalyptus deglupta Blume.	19	38	189	19	265	5.90%
Eugenia uniflora L.	-	27	-	-	27	0.60%
Ficus benjamina L.	10	9	-	-	19	0.42%
Ficus elastica var. variegata	-	2	-	-	2	0.04%
Ficus religiosa (L.) Gasp.	10	10	-	-	20	0.45%
Fragraea fragrans Roxb.	28	-	-	-	28	0.62%
Garcinia mangostana L.	-	25	-	-	25	0.56%
Hibiscus tiliaceus L.	10	-	-	-	10	0.22%
Inocarpus fagiferus (Park.) Fosb.	21	-	-	-	21	0.47%
Jacaranda acutifolia Bonpl.	-	-	-	92	92	2.05%
Khaya senegalensis (Desr.) A. Juss.	-	-	80	43	123	2.74%

Plant species	Lot A	Lot B	Lot C	Lot E	Total	Percentage
Kleinhovia hospita L.	-	144	-	-	144	3.21%
Lagerstroemia indica L.	-	-	137	27	164	3.65%
Livistona rotundifolia (Lam.) Mart.	62	71	-	-	133	2.96%
Manilkara kauki (L.) Dubard.	10	643	128	41	822	18.32%
Melaleuca leucadendron L.	24	-	-	-	24	0.53%
Mimusops elengi L.	16	293	32	56	397	8.85%
Palaquium rostratum Burck.	-	227	-	-	227	5.06%
Phoenix dactylifera L.	-	13	-	-	13	0.29%
Plumeria acuminata Ait.	7	48	59	8	122	2.72%
Roystonea regia (Kunth.) O.F.Cook	41	-	-	-	41	0.91%
Samanea saman (Jacq.) Merr.	-	-	84	60	144	3.21%
Spathodea campanulata Beauv.	23	-	-	-	23	0.51%
Stelechocarpus burahol (Blume) Hook. & Thomson	21	35	-	-	56	1.25%
Tamarindus indica L.	19	-	-	-	19	0.42%
Wodyetia bifurcata	-	31	-	-	31	0.69%
Total	**586**	**2.130**	**1.142**	**630**	**4.488**	**100.00%**

3.2 Air pollution reduction benefits

According to research by Carpenter et al. [9], the plant can reduce air pollutants through the process of releasing oxygen into the atmosphere called oxygenation and the mixing of polluted air with clean air called dilution. Trees act as air filters [10]. The closer the distance between trees in the green open space, the better the ability of the green open space to absorb pollution. The use of CITYgreen to calculate the ecological benefits has been used in several studies abroad. Among them are an assessment of the carbon fixation function [11] and the benefits of increasing tree canopy through the planting and preservation of appropriate tree species [12]. The ecological benefits calculation for local trees in this site study was approached with the species name in the tree species database available at CITYgreen.

The results of the CITYGreeen analysis estimate the potential ability to absorb air pollutants as listed in Table 4. The total number of pollutants that can be absorbed by the existence of the trees in the four areas in the Yogyakarta International Airport area is 3.813,30 Lbs (1.729,68 kg) per year. The total amount of O3 that can be absorbed is 1.318,65 Lbs/year or equivalent to 598,13 kg (1lbs = 0,453592 kg), SO2 of 442,44 Lbs/year or equivalent to 200,69 kg, NO2 of 682,97 Lbs/year or equivalent to 309,79 kg, PM10 of 1.223,27 Lbs/year or equivalent to 554,87 kg, and CO at 145,97 Lbs/year or equivalent to 66,21 kg, where the heaviest pollutant that can be absorbed is O3 and the lowest pollutant concentration that can be absorbed is CO. Therefore, to further increase the absorbable pollutant mass, it is necessary to add vegetation following the criteria for absorbing pollutant particles.

Table 4 Air Pollution Reduction Benefits

Area	Air pollution removal (lbs/year)				
	Ozone (O3)	Sulfur dioxide (SO2)	Nitrogen dioxide (NO2)	Particulate (PM10)	Carbon monoxide (CO)
Lot A	197,42	66,24	102,25	183,14	21,85
Lot B	648,87	217,71	336,07	601,93	71,83
Lot C	303,5	101,83	157,19	281,55	33,6
Lot E	168,86	56,66	87,46	156,65	18,69
Total	**1.318,65**	**442,44**	**682,97**	**1.223,27**	**145,97**

3.3 Carbon storage capacity

The carbon storage capacity is estimated from the closure of the study area using CITYGreen 5.2 modeling. Based on the CITYGreen 5.2 recapitulation analysis, it can be seen that the total carbon storage capacity obtained by the Yogyakarta International Airport area is 1.326,46 tons with an annual increase of around 29,9 tons/year. This carbon storage capacity will be achieved by the existing planting plans. The planting program has not yet been fully completed and is expected to be like the planting plan in five to six years. The largest carbon storage capacity is in Lot B of 652,71 tons, while the lowest is Lot E of 169,86 tons/year. Lot B has the largest amount of carbon storage capacity due to a large amount of tree canopy cover in its area, so that the value of its carbon stock is large, while Lot E is the area with the lowest percentage of tree canopy cover, therefore the value of its carbon stock is the lowest. The main carbon stock stores are in tree biomass, floor vegetation, litter, and soil organic matter [13]. Stored carbon in the form of tree biomass constitutes the largest carbon pool.

Specifically for the aspects of carbon storage and absorption, there is no information available regarding the summary of economic benefits in CITYGreen. The economic benefits for this aspect are derived from the calculation of the amount of stored carbon (tons) multiplied by the carbon price based on international standards (ranging from $ 5.00- $ 15.00 per ton) [14]. In this study, it is assumed that the carbon price (ton) used is $ 5.00. With a carbon sequestration capacity of 1.326,46 tons/year, the economic value that can be received from this service is $ 6.632,30 / year, equivalent to Rp 97.778.672,44 / year (1 $ = IDR 14.742,80) or Rp 8.148.222,70 / month.

Table 5 Carbon storage benefits

Area	Benefits of Carbon Storage	
	Carbon storage capacity (tons)	Sequestration rate (tons/year)
Lot A	198,59	4,48
Lot B	652,71	14,71
Lot C	305,30	6,88
Lot E	169,86	3,83
Total	**1.326,46**	**29,9**

3.4 The economic value of air pollution reduction benefits

Based on the results of the analysis using CITYGreen 5.2 software, pollutants that can be absorbed by trees in the four areas in the Yogyakarta International Airport area are O3 of 1.318,65 Lbs/year or equivalent to 598,13 kg (1lbs = 0,453592 kg), SO2 442,44 Lbs/year or equivalent to 200,69 kg, NO2 of 682,97 Lbs/year or equivalent to 309,79 kg, PM10 of 1.223,27 Lbs/year or equivalent to 554,87 kg, and CO of 145,97 Lbs/year or equivalent to 66,21 kg, where the heaviest pollutant that can be absorbed is O3 and the lowest pollutant concentration that can be absorbed is CO. Therefore, to further increase the absorbable pollutant mass, it is necessary to add vegetation by the criteria for absorbing pollutant particles.

Table 6 The Economic Value of Air Pollution Reduction Benefits

Pollutant	Economic benefit value ($)
O3	4.047,65
SO2	331,82
NO2	2.096,62
PM10	2.506,96
CO	62,49
Total (USD/year)	9.045,54
Total (IDR/year)	133.356.587,10

The value of economic benefits is an external saving, namely, the value that must be spent by the government in overcoming the problem of environmental damage, in this case, is pollution by pollutants in the air. The value of economic benefits that can be saved in the Yogyakarta International Airport area is O3 of $ 4.047,65 or equivalent to Rp 59.673.694,40 (1 $ = IDR 14.742,80), SO2 of $ 331,82 or equivalent to Rp 4.891.955,90, NO2 for $ 2.096,62 or equivalent to Rp 30.910.049,30, PM10 for $ 2.506,96 or equivalent to Rp 36.959.609,90 and CO for $ 62,49 or equivalent to Rp 921.277,57. The greatest economic value that can be saved is O3 and the lowest cost savings is CO. The total economic value that can be saved in the Green Open Space at Yogyakarta International Airport is $ 9.045,54 / year or equivalent to Rp 133.356.587,10 / year or Rp 11.113.048,93 / month.

4. Conclusion

The results of the analysis of the ecological and economic benefits of this study illustrate how much the benefits of trees in a city green open space area. Trees affect the quality of the urban ecosystem in terms of 2 ecological benefits of trees, in terms of the potential for pollutant absorption and the potential for carbon storage. Based on CITYGreen 5.2 analysis, the potential ecological benefits of trees at Yogyakarta International Airport in reducing pollutants in the air are 3.813,3 Lbs or 1.729,68 kg a year. The value is worth the external cost savings that can be incurred community and government due to the pollutants of $ 9.045,54 or Rp 133.356.587,10. The carbon storage capacity obtained is 1.326,46 tons with an increase every year around 29,9 tons/year.

References

[1] Febrianti A M and Sulistyantara B 2020 Evaluation of physical function and air pollution tolerance of roadside tree in Bogor Botanical Garden's surrounding *IOP Conf. Ser.: Earth Environ. Sci.* **501** 012002

[2] Fatimah I S, Sinukaban N, Munandar A, Kholil 2011 Aplikasi Program Analisis CITYGreen 5.4 untuk Kajian Ruang Terbuka Hijau (RTH) dan Manfaat Layanan Terukur Ekosistem Kota Bogor *J. Ekologia* **11**(1): 1-11

[3] Joga N and Ismaun I 2011 *RTH 30%! Resolusi (Kota) Hijau* (Jakarta: PT Gramedia Pustaka Utama) chapter 6 p 36

[4] Yusaputri F 2017 Desain lanskap di area boarding room terminal 2D dan 2E Bandara Internasional

[5] Soekarno-Hatta [skripsi] Program Sarjana Institut Pertanian Bogor

[6] McPherson, E. G., D. J. Nowak, R. A. Rowntree 1994 Chicago's urban forest ecosystem: results of the Chicago

[7] Urban Forest Climate Project *Gen. Tech. Rep. NE-186* (Radnor, PA: U.S. Department of Agriculture, Forest Service, Northeastern Forest Experiment Station) chapter 1 p 1-2

[8] Lukmanniah P 2011 Manfaat kanopi pohon dalam upaya penyimpanan dan penyerapan karbon di kawasan perumahan Kota Bogor [skripsi] (Bogor : Institut Pertanian Bogor)

[9] Al-Hakim AH 2014 Evaluasi efektivitas tanaman dalam mereduksi polusi berdasarkan karakter fisik pohon pada jalur hijau Jalan Pajajaran Bogor [skripsi] (Bogor : Institut Pertanian Bogor)

[10] Soviyanti A N 2017 Penilaian jasa lingkungan pohon pada jalur hijau jalan sebagai upaya mereduksi polusi udara di Kota Bogor [skripsi] (Bogor : Institut Pertanian Bogor)

[11] Carpenter P L, Walker T D, Lanphear F O 1975 *Plants in the Landscape* (San Fransisco : W.H.Freeman and Company)
[12] Goodwin D 2017 *The Urban Tree* (Great Britain: Ashford Colour Press Ltd)
[13] Peng L, Chen S, Liu Y, Wang J 2007 Application of CITYgreen model in benefit assessment of Nanjing urban greenbelt in carbon fixation and runoff reduction *Chinese Journal of Applied Ecology **18(6)**: 1293-1298*
[14] Hilton B N 2007 Urban ecosystem analysis using GIS: The Claremont Colleges sustainability initiative *Claremont, CA, The Claremont Colleges Council of President*
[15] Ijazah M and Sancayaningsih R P 2018 Carbon Stocks in Pinus merkusii and Acacia auriculiformis Stands at Mangunan Protection Forest Dlingo Bantul Special Region of Yogyakarta *Conf. : Seminar Nasional XII dan Pendidikan Biologi FKIP UNS SP-017-10* p 830-837
[16] Fatimah I S, Sinukaban N, Munandar A, Kholil 2013 Valuation of Ecological Benefit of Greenery Open Space of Bogor City Using CITYGreen 5.4 Software *J. Nat. Resour. Environ. Manage.* **3**(1): 31-38

Application of Biophilia Concept on Greenery Layout Arrangement of Buperta Cibubur Tourism Landscape

S Z Alya[1], R Khrisrachmansyah[2]
1 Landscape Architecture Department, IPB University, Bogor, ID
2 Landscape Architecture Department, IPB University, Bogor, ID

E-mail: sarah_zt@apps.ipb.ac.id[1], rezky_kh@apps.ipb.ac.id[2]

Abstract. The concept of Biophilia shows how humans have a fundamental desire to connect and depend on nature, providing healing and positive impacts on emotional change. DKI Jakarta Province population number has increased significantly from year to year so that the need for green open space increases. Buperta City Forest Area has enormous potential to facilitate the needs of the green space in the City of East Jakarta, but the development of existing natural tourism is still not optimal. The concept of Biophilia can be a solution to increase the attraction, user experience, and increase relaxation benefits for Buperta Cibubur user in sustainable way. The objectives of this research are: (1) Identifying and studying the characteristics of the site and its users, (2) Developing criteria for the utilization of the Biophilia concept in green open spaces, and (3) Developing greenery layout arrangement recommendations. Furthermore, direct observation, interview, and literature study were used comprehensively to collect the data. Data processed using Biophilia approach, through descriptive analysis and spatial method of analyzing the potential and constraints from the site. Detail reccomendation developed on zones with strongest natural potential such as "Kempa", Recreation Area, and Pine Forest, with each different approach of using Biophilia concept according to the environment and user characteristics. The output of the research is in the form of planting plan and site plan.
Keywords: Biophilia, Buperta Cibubur, Tourism Landscape

1. Introduction

Based on BPS data in 2015-2045 Indonesian Population Projection Results of the 2015 Revised Edition of the Inter-Census 2015 Population Survey [1], the projection of the population of DKI Jakarta can reach up to 10,891,900 in 2025 and increase to 11,239,300 by 2045, so the need for green open space as ecological and recreational functions will continue to increase.

The largest urban forest in East Jakarta is the City Forest of Bumi Perkemahan dan Graha Wisata Cibubur (Buperta) with an area of 210 ha. Buperta Cibubur has huge potential in the field of natural tourism and nature conservation because it has a large area. Buperta Cibubur has been used as a campsite, sports, and nature tourism, but the development of natural tourism in Buperta Cibubur has not been optimal because the management funds are insufficient, which only comes from entrance fees [2].

According to [3] a tourist area is a commercial area devoted to recreational activities. Recreation provides relaxation to the mind, provides a new eye view (new view) so as to renew the mind's memory, forget about problems and relieve existing stress. Thus, with an approach, it can be said that the Buperta area to achieve the goals of recreation, needs to consider the arrangement that affects the mind of the user.

Over the past twenty years, studies have shown that there is a relationship between nature and increased productivity, increased healing value and even increased understanding of learning on a large scale [4]. The concept of Biophilia shows how humans have a basis to connect and depend on nature. These studies also state that the integration of the natural environment into the spaces where humans live can enhance healing and have a positive impact on emotional changes [5]. Empirical studies have reported that environmental experiences greater emotional recovery [6].

Consequently, the greenery layout arrangement and its optimization is crucial for the sustainability and benefit of Buperta Cibubur Landscape. The objectives of the study are recreating criiteria of biohilia in Buperta Cibubur, analyze which area that potential to be optimalized with biophilia concept, and make some recommendations in terms of concept in green layout arangement. Therefore, the researcher wants to compile criteria and greenery layout arrangement using the principles of Biophilia concept as a way to increase the attractiveness of the tourist area of Bumi Perkemahan and Graha Wisata Pramuka (Buperta) Cibubur through the experience of the relationship between human and nature.

2. Methods

2.1. Study Site and Time

The research was conducted in the tourist area of Buperta Cibubur that shows on Figure 1 for 9 months from December 2019 to August 2020.

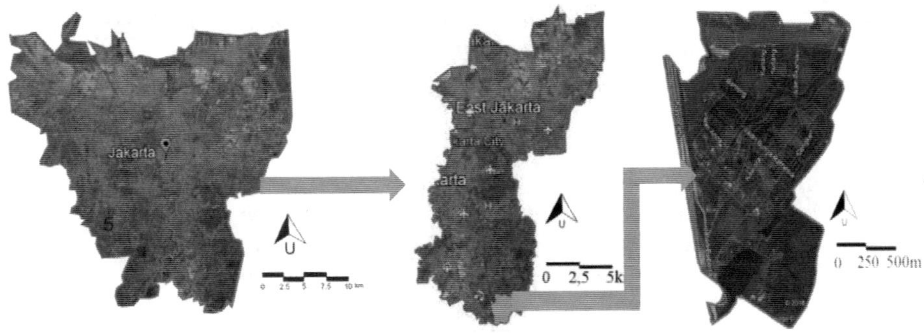

(a) (b) (c)

Figure 1 Study Sites: (a) DKI Jakarta Province (b) Jakarta Timur City (c) Buperta Cibubur
(Source: Google Earth Pro)

2.2. Study Methods

In this study, the overall process of methods are using Biophilia patterns by [6] as a consideration. In general, there are three platforms, comprising of nature in space, analogue of nature, and nature of space. Furthermore, there are seven patterns (P1-P7) in nature in space, three patterns (P8-P10) in analogue of nature, and four patterns (P11-P14) in nature of space. The detail of each pattern can be seen as follow:

Table 1 14 Patterns of Biophilia

14 Patterns of Biophilia		
Nature in Space	P1	Visual Connection with Nature
	P2	Non-Visual Connection with Nature
	P3	Non-Rythmic Sensory Stimuli
	P4	Thermal and Airflow Variablity
	P5	Presence of Water

	14 Patterns of Biophilia	
	P6	Dynamic and Diffuse Light
	P7	Connection with Natural System
Analogue of Nature	P8	Biomorphic Forms and Patterns
	P9	Material Connection with Nature
	P10	Complexity and Order
Nature of Space	P11	Prospect
	P12	Refuge
	P13	Mystery
	P14	Risk/Peril

Nature in space describing the patterns of the natural existing conditions and the connections among nature and human activities. In addition, analogue of nature is the pattern that transform all appearance of the elements into natural shape. Nature of space is describing the non-physical patterns of natural phenomena that can be plugged in the development areas.

2.2.1. Preparation
In the preparation stage, the researcher conduct a research about the location and research topic then prepares permits for the research location manager, prepares material tools and compiles technical details of the research.

2.2.2. Inventory
At the inventory stage, the researcher looks for the information needed for the analysis stage. The information obtained includes location, climate, topography, hydrology, visual, circulation, facilities, land use and activities, and vegetation. Information is obtained through visits to research sites, interviews with site managers, related stakeholder, and publications.

2.2.3. Analysis
The analysis stage in this study is divided into two approaches, spatial analysis and descriptive analysis. Descriptive analysis formulates potentials and constraints as reference materials for the application of the concept using general site condition data with Biophilia patterns as material for analysis. Then conducted a study with a spatial analysis approach to determine the zone with the strongest natural potential by vegetation and visual analysis.

Vegetation potential was assessed using the Normal Difference Vegetation Index (NDVI) method to see quality of the greenery on the overall site which uses the normalized difference between the near red and infrared reflectance values due to the reflection by [7]. The NDVI analysis technique measures the quality of vegetation using a Landsat 8 map with an accuracy of 30x30m, then categorized based on the vegetation quality:

$$NDVI = (NIR - RED) / (NIR + RED)$$

NIR = Band that has a near infrared length (band 5)
RED = Reflection in visible red band (band 4).

Visual potential is assessed using the Scenic Beauty Estimation (SBE) method, which is an assessment through a questionnaire on photos taken based on a grid and the presence of a biophilia pattern to get a comprehensive representation of the entire site and capture the biophilia potential on the site. Data were collected by researchers directly and assessed by landscape architecture students. This technique uses criteria based on [8] where landscapes including low visual quality have an SBE value of <-20, landscapes include moderate visual quality if they have an SBE value between -20 to 20, and landscapes including high visual quality have an SBE value> 20 with calculations:

$$SBEx = (ZLx - ZLs) \times 100$$

SBEx = SBE value of x-scene
ZLx = Average z value of the x-scene
ZLs = Average z value of standard scenes

SBE consists of three steps, namely survey and selection of landscape representations, taking questionnaires, and analyzing data. There are 30 landscapes presented using photos. Photos were presented using the online QuestionPro form with each photo given 8 seconds to grade on a scale of 1-8. Respondents who assess the visual quality of the landscape presented are students of the Department of Landscape Architecture IPB semester 8 who according to [8] are part of the community who are considered critical and care about their environment. The number of respondents who assessed was 30 people.

The results of the descriptive analysis are displayed in the form of a green arrangement recommendation table, while the results of the spatial analysis are displayed in the form of a map of potential zones classified in 5 classes, from low to high values according to the dominance of the analysis value in each zone.

2.2.4. Synthesis
The results of the NDVI and SBE analysis at this stage are overlaid, validated by assessment of potential biophilia patterns on each land use zone, then three zones with the most potential from the entire site are determined. The three selected zones then determined the sub-concept of the biophilia concept based on the table of recommendations for green arrangement that was made in the descriptive analysis stage.

2.2.5. Recommendation
The recommendation process is processing the sub-concepts that have been created at the synthesis stage and determining the alternative green layout arrangement designs for each site development. These recommendations are presented in the form of a concept and site plan.

3. Result and Discussion

3.1. General Site Condition
Bumi Perkemahan and Graha Wisata (Buperta) Cibubur has an area of 210 hectares, located in Pondok Ranggon Village, Cipayung District, East Jakarta City. After being classified based on the land use and activity, the Buperta area can be divided into 14 zones, i.e. main access, arboretum, sports area, support area, agricultural area, tourist area, lake, tourist house, pine forest, men's camp (Kempa), women's camp ("Kempi"), airport, main field, and swamp. The aspects that are inventoried are location, climate, topography, hydrology, visuals, circulation, facilities, zones and activities as well as green systems.

3.2. Descriptive Analysis

3.2.1. Location, Site Boundary and Accessibility
Buperta Cibubur has a strategic location. The site is the main access for the Cibubur T-junction which is right in front of the Jagorawi toll exit from the direction of Jakarta and the entrance door is too close to the Trans Jakarta and *Angkutan Kota* (Angkot) bus stops so

that the vehicle intensity that occurs is high during busy hours. Consequently, due to the existing conditions the pollutans in the study area are high and need to be considered to minimize the pollutants.

One of the Biophilia approaches that can be applied is Refuge (P12) [6] by separating the site from the surrounding area. Its implementation in green system arrangement can be in the form of utilization of vegetation which can reduce pollutants and noise.

3.2.2. Climate
The air temperature in DKI Jakarta province reaches a maximum temperature of up to 29.7°C with an average of 28.8°C, while the average air humidity is 74.4% [9]. According to [10], the relative comfort temperature in the tropics is 27-28°C and the ideal humidity is between 40% -75% In other words, Buperta Cibubur is not in the ideal human comfort and In the Buperta Lake area, air humidity is higher due to the presence of water bodies.

Biophilia patterns that can be applied is Thermal and Airflow Variability (P4) and Refuge (P12) [6] can be applied in several areas that require more comfort and high intensity of use. The implementation of this principle is in the arrangement of green systems, namely planting shade vegetation and ground cover to reduce surface temperature and adding a number of vegetation with wide crowns in order to reduce surface temperature without reducing the area of space or activity.

3.2.3. Topography
Using the modified criteria of Hardjowigeno and Widiatmaka (2007) in [11], for urban forest recreation, a land slope of 0-15% is a very suitable land characteristic, a slope of 15-25% is quite suitable, a slope of 25-45% is marginal, while a land slope greater than 45% is not suitable. Most of the land in Buperta has a slope of 0-15%, which mean flat and sloping so that the overall topography of the Buperta area is very suitable.

By using the Biophilia approach, this potential can be exploited by greenery layout arrangement with a combination of Prospect pattern (P11) and Mystery (P13) [6] in areas with a slope category of 0-25%, applications in green arrangement can be by highlighting the contour using vegetation.

3.2.4. Hydrology
The good condition of Lake Buperta and the surrounding environment which is environmentally preserved have the potential to become a source of clean water and a house with ideal conditions for animals, as well as being a model lake that can support the development and management of other water bodies in DKI Jakarta province. Most of land cover in Buperta Site is bare land and grass, so the water absorption rate at the site is high, and drainage is well distributed so that the potential for flooding is very small. Potential inundation can be found in areas with bare land cover, such as parking lots.

When associated with the Biophilia approach, we can increase the diversity and sanitation Lake Buperta Complexity and Order (P10) [6], mimicking the concept of natural water cleansing by vegetation. In addition, to overcome the inundation problem in parking lots, it would be better if provided pavement material is permeable or with a good slope.

3.2.5. Visual
A recreational urban forest must fulfill several facilities such as parking lots, security posts, hygiene facilities, toilets, water installations, electrical installations and drainage systems [12]. In general, the quantity and completeness of the facilities at the Buperta Cibubur site is quite complete. Some of the facilities with high usage intensity on the Buperta Cibubur site are well maintained, while some of the facilities with low intensity usage are poorly maintained.

With Biophilia approach, pattern that can be used is Mystery (P13) and Risk (P14) can be applied to provide a user experience to the site, with Prospect (P11) to guarantee its safety. Patterns of Non-Visual Connections with Nature (P2) can be added to interesting experiences and create connection patterns with natural systems [6]. In summarize, there are three patterns that can be used to optimize the visual context at the study area.

3.2.6. Land Use and Activity
The types of tourism activities in the Buperta area can be broadly divided into several main types of activities, namely camping, picnics, sports, education with small to large scale user groups. The effectiveness and visitor attractions have the potential to increase if the site arrangement is adjusted to the intended use and activities carried out on the site.

Using the Biophilia approach, the zone with camping activity can apply the pattern of Analogues of Nature, Complexity and Order (P10). Zone with picnic activities, can apply the Prospect (P12) which aims to create the feeling of comfort and safety needed in picnics while still maintaining the existing view. Zone with sports activity can apply Material Connection with Nature (P9) pattern on trails and activity areas to increase the fun aspect of recreational sports. Whereas in the education zone, Mystery (P13) can be applied because it can cause a sense of exploration or curiosity about what will happen or what will come afterwards [6]. Above all, there are four patterns that can be added in the study area to improving activities of recreation to nature.

3.2.7. Greenery Layout Arrangement
The diversity of vegetation types that exist in the Buperta area is a potential, however the vegetation is still arranged without a clear pattern, the large number of untreated weeds in some northern camping areas also make it less comfortable to be used for camping. As a result, the impression that created is that the site is not neatly arranged. Referring to these potentials and constraints found, patterns of complexity and regularity can be applied (P10) [6] by adding variation of complexity on greenery arrangement considering the purpose, also adding vegetation with a different shape from existing vegetation, and vegetation leading the view towards a good view found on the site.

3.3. Vegetation Potential Study
Buperta Cibubur site as one of the largest urban forests in East Jakarta has enormous potential in the aspect of vegetation. In developing the area it is necessary to take advantage of the potential of this vegetation, the Vegetation Potential Study functions to determine zones with the strongest vegetation potential for further development. This study uses the Normalized Difference Vegetation Index (NDVI) analysis technique.

In the results of this analysis, the lowest data obtained was -0.07 and the highest data was 0.52. The quality of vegetation in the Buperta area as a whole is dominated by high category with a percentage of 66.6% with an area of 139.86ha, followed by medium category at a percentage of 15% with an area of 31.5ha, high category at a percentage of 7.1% with an area of 14.91ha, and low category with a percentage of 6.6% with an area of 13.8ha, and very low category with a percentage of 3.9% with an area of 8.3ha.

3.4. Visual Potential Study
The Buperta Cibubur area as a tourist area, in order to increase the attractiveness of the area apart from utilizing its potential in the vegetation aspect, it is also necessary to consider its natural visual potential. Visual Potential Study serves to determine areas with the strongest visual potential for further development.

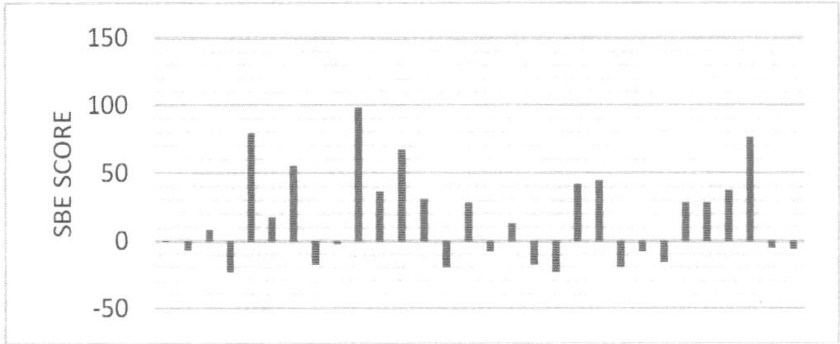

Figure 2 SBE Score Graph

In the results of this analysis, the lowest data obtained was -0.07 and the highest data was 0.52. The visual quality of nature in the Buperta area as a whole is dominated by very high category with a percentage of 37.7% with an area of 79.17ha, followed by medium category at a percentage of 37.3% with an area of 78.33ha, high category at a percentage of 21.3% with an area of 44.73ha, and low category with a percentage of 3.5% with an area of 7,34ha, and 0,2% very low category with an area of 0.42ha.

3.5. Biophilia Pattern Potential Scoring

The assessment of potential Biophilia patterns on existing sites can help complement research in assessing the strongest natural potential that can be used for Biophilia concept's site development. In addition to determining the strongest natural potential, Biophilia pattern potential scoring can also be used to determine patterns that can be prioritized in site development.

Result of this pattern potential scoring shows that arboretum zone categorized as a very strong potential on Biophilia pattern, followed by lake zone categorized as a strong potential, kempa zone, "Kempi" zone, tourism area zone with moderate potential, main field, sport area, and main access with low potential, and lastly sport area, agriculture area, and air field with very low potential.

3.6. Synthesis

Figure 3.a describes dark green means very high in natural condition, figure 3.b. shows that the dark pink means very high quality of visual, figure 3.c describes the dark turquoise means the area has more indicator of Biophilia patterns. In the overlay analysis results (Figure 3.d), it was found that most of the sites had a moderate category at a percentage of 50,2%v which is 105ha, then followed by high category at a percentage of 38.4% with an area of 80,64ha, very high visual and vegetation potential with a percentage of 6.5% with an area of 13,65ha and low category with a percentage of 4,8% with an area of 10,08ha, while the very low category is not found. Areas with the highest visual and vegetation potential categories are mostly found in the Arboretum, followed by Pine Forest, Tourism Area, Sport Area, "Kempa", and "Kempi" zones.

According to [6] Biophilia design patterns must be adapted to the surrounding environment and the predicted user population for the space. Each space will present different design challenges depending on programming, user types and dynamics, climate, culture, and various physical parameters, as well as existing or required infrastructure. To determine the 3 zones to be developed, the result of weighted overlay on figure 3.c is limited, considering land use zone, the type of environment and user characteristics, "Kempa" Zone, Tourism Area and Pine Forest are selected to be developed into the recommendation stage as seen of figure 3.d.

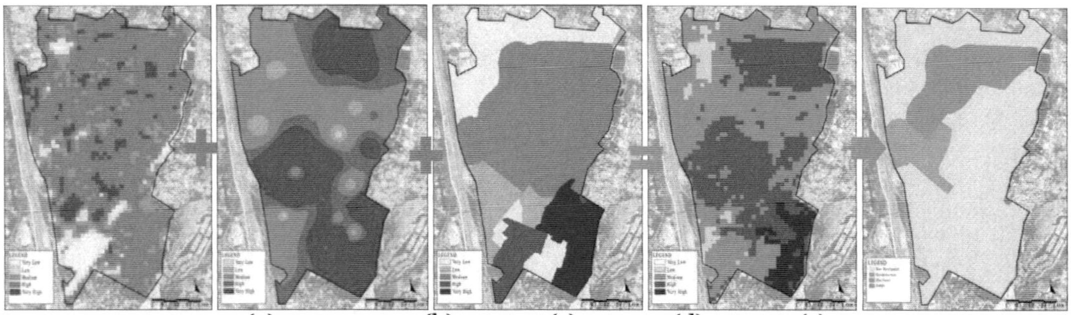

(a) (b) (c) (d) (e)

Figure 3 Natural Potential Study Process
(a) NDVI Analysis (b) SBE Analysis (c) Biophilia Scoring (d) Overlay (e) Development Plan

Based on the results of the descriptive analysis that has been conducted by considering the potentials and constraints on the site, the application of the Biophilia concept in greenery layout arrangement in each zone can be applied through the following sub-concepts as seen on Figure 4.

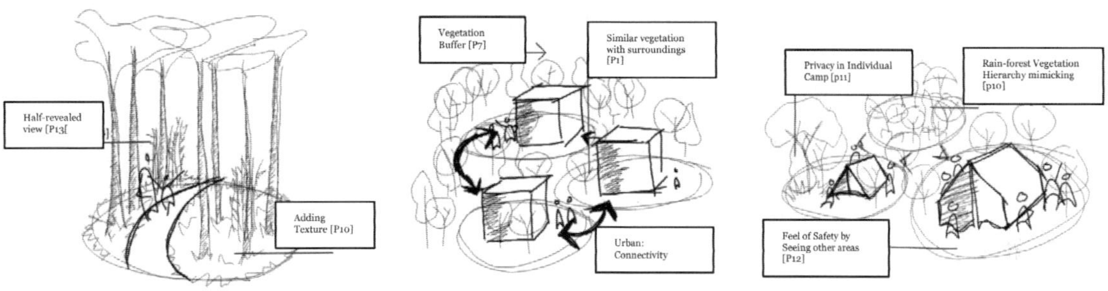

(a) (b) (c)

Figure 4 Concept Sketch (a) Pine Forest (b) Recreation Area (c) "Kempa"

3.6.1. Pine Forest: Wild Curiosity Trail
Wild in this concept describes the atmosphere to be achieved in the pine forest zone by this recommendation using a Non-Visual Connection with Nature (P2) pattern and Risk (P14), while Curiosity describes the impression to be achieved to support the educational function in the pine forest zone using Mystery (P13) pattern. The design of the activity area in the pine forest zone will be designed in the form of a trail, this path can be used by tourist visitors on foot or by bicycle and academics to be able to explore the pine forest associated with the arboretum, without destroying the existing vegetation. The design concept is applied in an organic form to make the experience more natural.

3.6.2. Recreation Area: Green Urban Miniature
Green in this concept is carried through the vegetation from outside the site (P1) which is obscured by the boundaries into the site by creating a vegetation buffer (P7), the vegetation that was previously dense outside the site when brought into the site is getting less and less. Meanwhile, the urban atmosphere is brought about in the connection between buildings and their respective areas to one another through the pedestrian circulation path which is built continuously.

3.6.3. "Kempa": Systemic Forest Camp Site
Systemic forest is a hierarchical and diverse system that exists in most forests in tropical areas, which has a random pattern and high diversity, consisting of various types of

vegetation with different shapes and heights (P10). This concept also emphasizes that this zone is a camp site where safety (P12), comfort (P11), space and the existence of facilities must be considered, so that combining the two concepts, Systemic Forest Camp Site will create a safe, comfortable and free camp area. without losing the forest-like atmosphere which is one of the main goals of camping.

3.7. Greenery Layout Arrangement Reccomendation
Green planning recommendations are made based on the results of the analysis and synthesis that have been carried out on the Bumi Perkemahan and Graha Wisata Cibubur areas. The vegetation types recommended on the site are partly vegetation types that are already in the Buperta area so that a connection with the natural system is formed (P7), and all of them can have a significant influence in the application of the formulated biophilia concept.

3.7.1. Pine Forest
In implementing the Wild Curiosity Track sub-concept, three types of circulation paths will be formed; pedestrian, bicycle and car lanes (existing). The bicycle path is formed around the entire site and has a dangerous character to support the wild atmosphere that you want to achieve in the form of obstacles, while the pedestrian path is made effective by considering the view that will be obtained.

In this recommendation, the recreation room is divided into sky (upper) and land (lower) recreation, taking into account the difference in the potential view to be achieved, the activities in the lower recreation room are in the form of picnics and camping, while in the upper recreation room in the tree house, namely sight seeing activities. The vehicle parking space is also planned to use a permeable pavement to avoid potential puddles that arise with land cover.

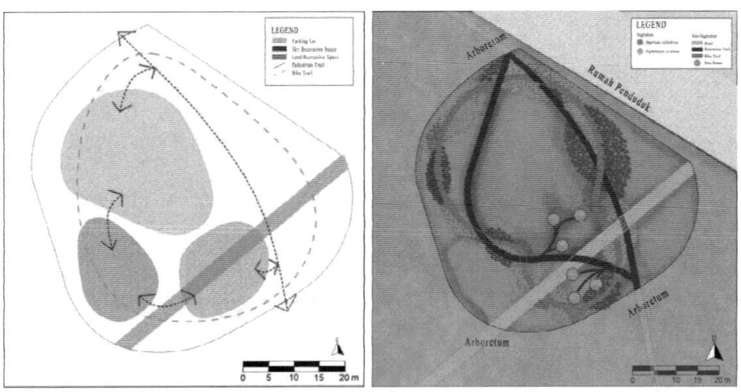

(a) (b)

Figure 5 Pine Forest Greenery Layout Arrangement Reccomendation
(a) Block Plan (b) Site Plan

The recommendation for green arrangement that is added to the existing green layout of the pine forest zone is the addition of shrubs with different textures on most sites (P10), such as *Imperata cylindrica* with a fine texture and *Nephrolepis exaltata* with a coarse texture. and a bold color as a texture enhancer and differentiator between activity and non-activity spaces, except in the lower recreation room and parking space where activities are carried out above it. In addition, around the parking space there is vegetation *Terminalia Mantaly* which plays a role in limiting the view (P13).

3.7.2. Recreation Area

The Green Urban Miniature concept emphasizes the connection between each part of the site, and also on the connectedness of the activity area with the surrounding natural environment. The tourism zone is divided into several zone, namely a prayer zone, a traffic garden zone, a guest house, and other tourist spaces (swimming pool, scouting education, and a hall). The connection system is implemented in the form of pedestrian circulation through plazas and pedestrians. At the center of activity, a shared parking space is formed that is associated with the entire site.

Most of the green planning recommendations that are added to the existing green layout of the tourist area have the aim of creating an integration of the surrounding environment into the activity area (P7). Most of the existing vegetation is of medium height so the added vegetation is of high and low elevation to create diversity (P10). Vegetations with high crowns were *Ceiba petranda*, *Swietenia mahogany*, *Samanea saman* and *Filicium decipiens*, and low canopy vegetation were *Carex morowii*, *Nephrolepis exaltata* and *Pennisetum rubrum*. *Terminalia Mantaly* was added to add an urban accent to the site.

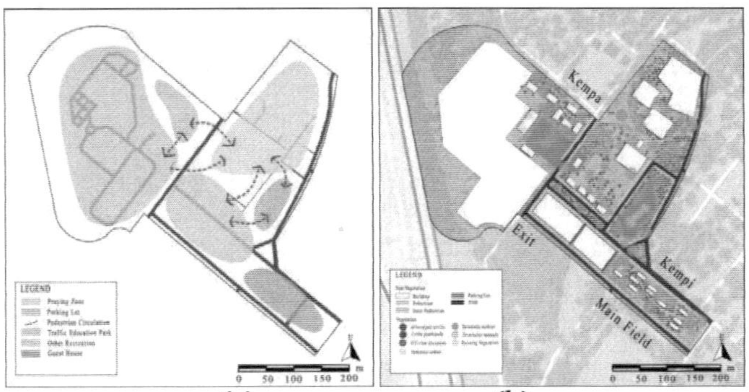

(a) (b)

Figure 6 Recreation Area Greenery Layout Arrangement Recommendation
(a) Block Plan, (b) Site Plan (b)

3.7.3. "Kempa"

The Systemic Forest Camp Site concept focuses on a camping experience that resembles a camping experience in a natural forest, but is guaranteed safety and security. Therefore the felt zone is divided into three types of space, namely closed and open individual campsites that can be used for families and individual, group campsites, and display rooms. However, open areas can be used during the day and closed areas with protection are more comfortable to use at night. The group campsite is for large groups, while the display space will be used and formed as a borrowing view of the surrounding camp spaces. While the green arrangement adapted the concept of camping arrangement by [13] as seen on figure 7.

(a) (b) (c) (d)

Figure 7 Camp Site Greenery and Tent Layout Arrangement
(a) Close Individual Camp 1 (b) Close Individual Camp 2 (c) Open Individual Camp (d) Group Camp
Source: [13] modification

Recommendations for the arrangement of green layout in group campsites focus on the formation of vocal points as a way of giving a little texture (P10) to the site using *Samanea saman* and *Erythia fusca* vegetation because they have a firm branch shape. Green planning recommendations for group campsites focus on the establishment of privacy (P12) and prospects (P11) using *Erythina fusca* vegetation because it has low branches but does not cover the view and *Necrolephis exaltata* because it includes vegetation that already exists on the site and can limit movement without close view. Meanwhile, the recommendation for green arrangement in the display room focuses on establishing a hierarchy of vegetation height (P10) using *Ceiba pentandra* and *Polyathia longifolia* vegetation because the canopy is high enough to be seen from a distance.

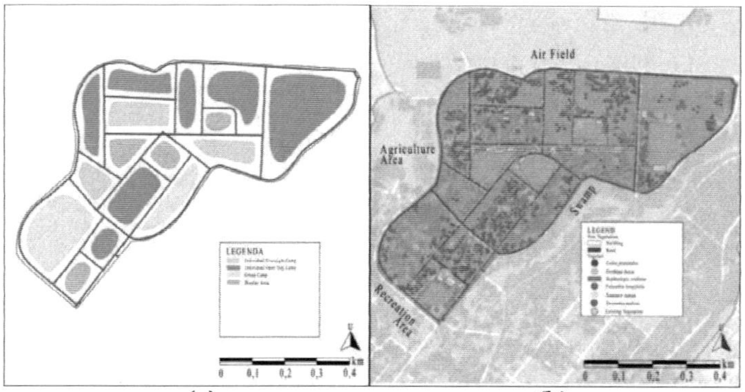

(a) (b)

Figure 8 "Kempa" Greenery Layout Arrangement Reccomendation
(a) Block Plan (b) Site Plan

4. Conclusion

In an effort to overcome obstacles and take advantage of existing potentials, this study uses the concept of Biophilia. In a site which already dominated by vegetation, the concept of biophilia or the formation of a relationship between humans and nature can be achieved not only by adding natural material to the site, but also can be achieved in the form of applying characters of nature to the site. Its application in green arrangement can take into account the shape of the canopy, height and texture of each vegetation and its influence on human feelings so that it can create natural environment characteristics. The concept of biophilia in its application in three zones with the highest visual and vegetation potential in the Buperta Cibubur area is broken down into three sub-concepts, Wild Curiosity Track in Pine Forest Zone, Green Urban Minature in Tourism Zone and Systemic Forest Camp Site in "Kempa" Zone. The three sub-concepts are determined based on the types of activities that are often found in the zone and aim to systematically maximize the potential that exists in the zone.

References

[1] [BPS] Central Bureau of Statistics Indonesia 2018 Proyeksi Penduduk Indonesia 2015-2045 Hasil SUPAS 2015 (Revision Edition). Jakarta (ID): BPS RI
[2] Hazrina K 2018 Estimasi Willingness to Pay Pengunjung dan Alternatif Strategi Pengelolaan Hutan Kota di Jakarta Timur [thesis]. Bogor (ID): Fakultas Ekonomi dan Manajemen, Institut Pertanian Bogor.
[3] Yates 1979 Sport Psychology [internet]. [downloaded March 2008]. Available at: https://e-psikologi.com
[4] Heerwagen J, Loftness V, and Painter S 2012 The Economics of Biophilia. New York (US): Terrapin Bright Green llc.
[5] Mustika A and Nuffida NA 2017 Aspek Alam sebagai Bagian Therapeutic Architecture pada Rumah. Jurnal Sains dan Seni ITS. 6(1):G-16.
[6] Browning WD, Ryan CO, Clancy JO, 2014 14 Patterns of Biophilic Design. New York (US): Terrapin Bright Green llc
[7] Burgan, Robert E, and Hartford RA 1993 Monitoring vegetation greeness with satellite data. Ogden (UT): Department of Agriculture, Forest Service

[8] Daniel T and Boster R, 1976 Measuring Landscape Esthetics: The Scenic Beauty Estimation Method. Colorado: USDA Forest Service
[9] Dinas Komunikasi Informatika, dan Statistik Provinsi DKI Jakarta 2019. Buku Statistik Provinsi Daerah Khusus DKI Jakarta. Jakarta (ID): DISKOMINFOTIK Provinsi DKI Jakarta
[10] Laurie M 1990 Pengantar kepada Arsitektur Pertamanan (terjemahan). Bandung (ID): Intermatra
[11] Faradilla E 2017 Analisis Kesesuaian Lahan untuk Ruang Terbuka Hijau dan Ruang Terbuka Biru di Sentul City, Bogor. Jurnal Lanskap Indonesia. 9(1): 101-109
[12] Hakim R and Utomo H 2004 Komponen Perancangan Arsitektur Lanskap Prinsip Unsur dan Aplikasi Desain. Jakarta (ID): Bumi Aksara
[13] Bell and Simon 1997 Design for Outdoor Recreation. Spon Press. London

IOP Conf. Series: Earth and Environmental Science **879** (2021) 012023 doi:10.1088/1755-1315/879/1/012023

Multicriteria potential mapping for Mangrove ecotourism in Batu Ampar, West Kalimantan

S Sisriany[1], **B Sulistyantara**[2], **and T Budiarti**[2]

[1] Graduate School of IPB University, Department of Landscape Architecture, IPB University, Bogor, Indonesia

[2] Department of Landscape Architecture, IPB University, Bogor, Indonesia

E-mail: bambang_sulistyantara@apps.ipb.ac.id

Abstract. Mangrove ecosystems provide a wide variety of vital ecological and economic ecosystem services to coastal communities, including ecotourism. Batu Ampar, where one of the most extensive mangrove forests lies, is encouraged to develop mangrove ecotourism. This study aims to determine the priority for ecotourism development in Batu Ampar districts by using the potential map based on the multiple aspects from tourism, environment, legal, and landscape value. This research's multicriteria use includes Natural Tourism Attraction, Mangrove Tourism Suitability, Biophysics Vulnerability, Landscape Values, and Spatial Policy, and weighted overlay based on Analytic Hierarchy Process (AHP) score. The results indicate that by using multicriteria for the mangrove ecotourism potential, 18% is the high potential area, 30% is moderate potential, 22.6%, and the rest are not potential. The high potential area should become the priority of ecotourism development in Batu Ampar, situated in five villages: Batu Ampar, Telok Nibong, Nipah Panjang, Tanjung Harapan, and Sungai Kerawang.
Keywords: AHP, Ecotourism, GIS, Mangrove, Potential Mapping

1. Introduction

Mangrove ecosystems provide ecological benefits, as nutrition supplier, a place for spawning and foraging for aquatic biotas, protecting shorelines from erosion, providing nurseries for many species of fish and crustaceans, preventing intrusion of seawater and strong winds, and blocking sea waves, as well as an economic benefit, among others, as a provider of various timber forest products and ecosystem services, and providing a place for mangrove nursery areas [1–3]. The mangrove ecosystem in the ecological landscape structure is in the ecotone area, an intermediate area between two different ecosystems. It has high biodiversity due to the combination of the two ecosystems and has a very high vulnerability.

Indonesia, as a tropical archipelago, has the largest area of mangrove forests in the world, with a total area of 4.25 x 10^4 km^2 (26% of world total), followed by Brazil (1.34 x 10^4 km^2), Nigeria (1.05 x 10^4 km^2) and Australia (1.00 x 10^4 km^2) [4]. One of the largest is the Batu Ampar Mangrove Biosphere, located in West Kalimantan, with a total area of 655.85 km^2. Batu Ampar mangrove forest has been claimed as one of the most extensive mangrove forests in Asia and has the most comprehensive mangrove species in the world.

Batu Ampar district is located in Kubu Raya Regency. This regency has claimed that mangrove is their identity, especially one particular species of mangrove name *Kandelia candel*. The local government concern with the ecotourism development in this area to support tourism. Unfortunately, these potentials have not affected the local community's social economy, and existing tourism has consistently not attracted tourists. Ecologically, the mangrove protected forest in Batu Ampar is good. Still, however, its sustainability is considered insufficient from the social and economic aspects [5]. An area is designated as a conservation area to provide three dimensions of benefits: ecological benefits, economic benefits, and social benefits.

Ecotourism is one of the activities in the Batu Ampar mangrove area. Ecotourism has been believed to encourage regional and local economic growth for the improvement of community welfare and preserve the natural resources and biodiversity as tourist objects and attractions. Quality ecotourism in the mangrove area of Batu Ampar has the potential to be sustainable tourism by maintaining the quality of objects and natural attractions. Also, its activities can impact community life values, such as providing income, welfare, and social life of local peoples. Spatial assessment is needed to study this large area properly, and integrated with multiple factors for sustainability, including ecological aspects, policy aspects, to social and economic aspects, otherwise a sustainable tourism will not be adequately actualized.

The mangrove ecosystem in Batu Ampar well distribute almost in every village in Batu Ampar, and every village is encouraged to develop ecotourism in their area. An assessment of mangroves' potential needs to be carried out, especially in the protected forest area of Batu Ampar, West Kalimantan, to determine the priority of ecotourism areas developed with a spatial approach. The primary purpose of this research is to model the potential map of mangrove ecotourism with an appropriate combination of multiple aspects from tourism, environment, legal, and landscape value. The appropriate combination is obtained by assign the weight calculated from Analytic Hierarchy Process (AHP) methods.

2. Methods

This research was conducted in the Mangrove Landscape of Batu Ampar District, Kubu Raya Regency, West Kalimantan Province (Figure 1), located at 0°43'06.1 S and 09 ° 42'34.0" E, in a total area of 2,002.00 km^2. This district consisted of 15 villages, as follows: Ambarawa, Batu Ampar, Muara Tiga, Nipah Panjang, Medan Mas, Padang Tikar I, Padang Tikar II, Sumber Agung, Sungai Besar, Sungai Jawi, Sungai Kerawang, Tanjung Beringin, Tanjung Harapan, Tasik Malaya dan Teluk Nibung.

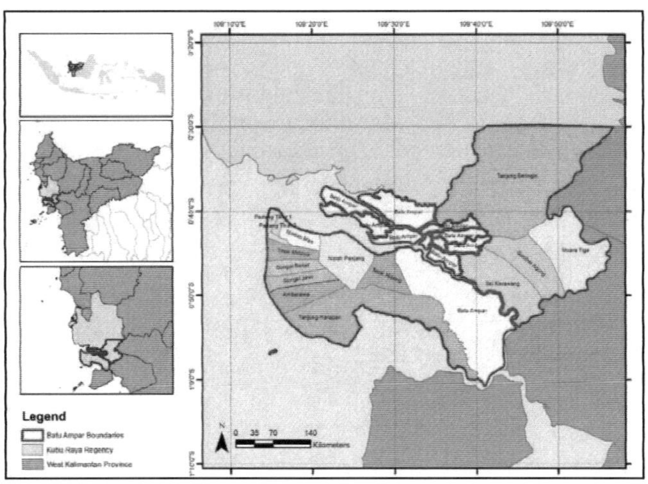

Figure 1 Case Study Location

The field survey was carried during August 2019 to collect primary and secondary data and carry out the questioner including the interview towards the local people and stakeholders such as local government and NGOs. The datasets used in this research include the SRTM-DEM for elevation and slope map, Indonesia's geospatial portal, and the multispectral satellite images from Landsat 8 OLI from March 20, 2017 (for low tide, z=-0.102m), and July 30, 2018 (for high tide, z=0.6m). The year 2017 and 2018 was purposedly chosen to match the collected secondary data related to this study such as mangrove forest from LAPAN (Indonesia's Aeronautics and Space Institute), and mangrove species, biota, village forest policy and document from SAMPAN (NGO) and LPHD Batu Ampar (Forest Management Group). The mapping and analysis processes are mostly conducted in ArcGIS Desktop 10.5, and the rest in other software such as QGIS 3.8 and Microsoft Excel.

Identification and mapping of mangrove ecotourism parameters in this study consist of multiple criteria from the four main aspects: mangrove ecotourism, biophysics vulnerability, legal aspects, and local people value, see Table 1. The criteria used in this study are as follows: 1) Natural Tourism Attraction [6] and Mangrove Tourism Suitability [7] for tourism aspects, 2) Mangrove Biophysical Vulnerability Index [8] for vulnerability aspects, 3) Landscape Value [9], for local people perceptions, and 4) Village Forest and Consensus Forest Use for the policy aspects.

Potential maps are generated by using weighted overlay analysis towards those multiple criteria. The weight score assigned to each map of the criteria are obtained by using the AHP methods [10] from the selected expert respondent, with Consistency Ratio (CR) = 2.3%, and principal eigenvalue = 4.062, see Table 1.

Table 1 Multicriteria parameters for mangrove ecotourism potential mapping

No	Aspects	Weight	Categories	Datasets
1	**Ecotourism**			
	a. Natural Tourism Attraction [6]	4	Feasible for Development - 3 Moderately Feasible - 2 Not Feasible - 1	Scenery, Natural Resource, Tourism Activity, Hygiene, Condition, Distance, Distance from City Centre, Road type, Accommodation and Room, Existing Facilities in the Area
	b. Mangrove Tourism Suitability [7]	17	High Suitability -3 Moderate Suitability -2 Low Suitability -1	Mangrove Density (NDVI), Mangrove Species, Biota Species
2	**Biophysical Vulnerability Index** [8] (Inundation & Erosion)	46	Low Vulnerability – 3 Moderate Vulnerability – 2 High Vulnerability - 1	Geology, Elevation and Slope (SRTM-DEM), Mangrove, Water Vegetation, Marsh
3	**Landscape Value** [9]	5	High Value - 3 Moderate Value -2 Low Value - 1	Land use base map, and Questionnaire for Sociocultural Value, Economy Value, Environment Value, and Aesthetic Value.
4	**Legal Aspects** – Spatial Policy	28	Village Forest – 3 Protected Forest, Other Usages -2 Production Forest - 1	Village Forest, and Consensus Forest Use Map

3. Result and Discussion

3.1 Ecotourism Aspects
Natural Tourism Attraction criteria are widely used for the ecotourism potential assessment, especially in Indonesia. Although these criteria have been handy and cover multiple ecosystems such as coastal areas and mountain areas, it still requires some adjustment and modification, particularly for this study on the mangrove ecosystem. Therefore, for ecotourism aspects is also used other criteria for Mangrove Tourism Suitability [7]. Contrary to the Natural Tourism Attraction criteria, this Mangrove Tourism index is specifically focused on the mangrove as the attraction. The combination of these two criteria expected to represent most of the tourism aspects.

3.1.1. Natural Tourism Attraction
For Natural Tourism Attractions assessments, the results show four villages are feasible to develop: Telok Nibong, Tanjung Harapan, Nipah Panjang, and Medan Mas. Meanwhile, two villages are not feasible to develop: Muara Tiga and Tanjung Beringin. And the rest are classified as moderately feasible (Figure 2d).

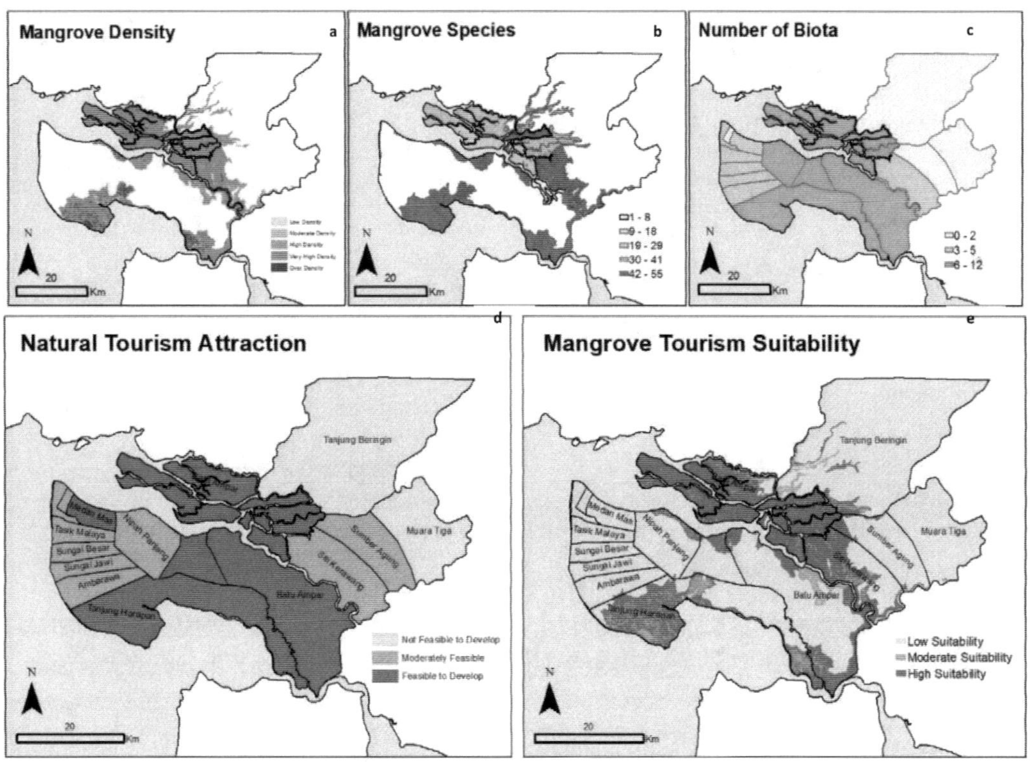

Figure 2 Ecotourism Aspects

3.1.2. Mangrove Tourism Suitability
For Mangrove Tourism Suitability, there are three parameters, mangrove density, number of mangrove species, and the number of biotas (Figure 2a, 2b, 2c). Based on these criteria, the area with mangrove covers has high suitability, although some of the mangrove covers moderate suitability (Figure 2d). Mangrove density maps were generated from NDVI, and only High, Moderate, and Low Density include in the calculation. Non-vegetation and Over High-Density mangrove areas are considered not suitable for ecotourism.

Table 2 List of Mangrove in Batu Ampar Districts

Classification[11]	**Conservation Status Category**[12]	**Mangrove Species**
True Mangrove	Least Concern	*Acanthus ebracateatus, Acanthus ilicifolius, Aegiceras corniculatum, Avicennia officinalis, Barringtonia racemosa Bruguiera cylindrica, Bruguiera gymnorrhiza, Bruguiera parviflora, Bruguiera gymnorrhiza, Ceriops tagal, Kandelia candel, Lumnitzera littorea, Nypa fruticans, Rhizophora mucronata, Rhizophora apiculata, Sonneratia alba*
	Near Threatened	*Brownlowia tersa, Ceriops decandra, Sonneratia ovata*
	Critically Endangered	*Bruguiera hainessii,*
Mangrove Associated	Least Concern	*Barringtonia asiatica, Cerbera manghas, Excoecaria agallocha, Dolichandrone spathacea, Cryptocoryne ciliata, Hibiscus tiliaceus, Pandanus tectorius, Pongamia pinnata, Terminalia catappa, Thespesia populnea*
	Unclassified	*Allophylus cobbe, Ardisia elliptica, Clerodendrum inerme, Dalbergia candenatensis, Derris trifoliata, Finlaysonia obovata, Ipomoea pes-caprae, Melastoma malabathricum, Merope angulata, Morinda citrifolia, Morinda citrifolia, Passiflora foetida, Pluchea indica, Planchonella obovata, Sarcolobus globosus, Scaevola taccada, Sesuvium portulacastrum*

Mangrove species in Batu Ampar are recorded around 46 species and distributed almost entirely in the mangrove forest. There is consist of true mangrove species and mangrove associates, see Table 2. Biota in Batu Ampar includes *Channa striata, Channa micropeltes, Trichopodus Trichopterus, Anabas testudineus, Helostoma temminckii, Macrobrachium rosenbergi, Scylla serrata, Apis Cerana, Trigona sp, and Orcaella brevirostris.*

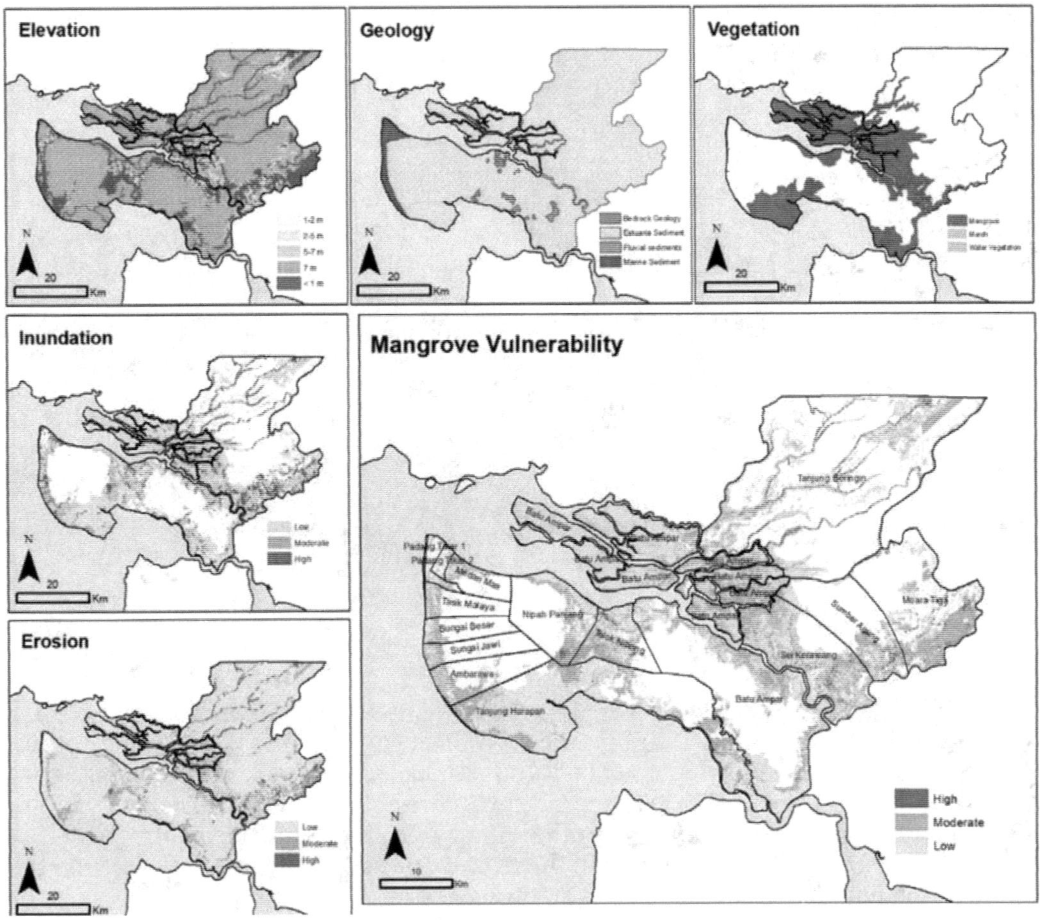

Figure 3 Biophysical Vulnerability Aspects

3.2 *Mangrove Biophysical Vulnerability Index*

Despite their high potential for ecotourism, mangroves have a very high risk of destruction and very vulnerable to changes, moreover to climate changes [13,14]. Therefore, it's essential to identify areas of mangrove that are prone to damage and vulnerable. The ecotourism activities will not disturb the vulnerable ecosystems and, alternatively, manage to conserve the high vulnerability areas.

Identification of the mangrove vulnerability in this research is divided into two effects, Inundation and Erosion (see Figure 3d & 3e), each consist of three parameters Elevation, Geology, and Vegetation [15] (see Figure 3a, 3b, and 3c). The map for these aspects is generated from multispectral satellite images (vegetation) using the decision trees classification for Marsh [16], Water Vegetation, and Mangrove [1] from various spectral indices, including NDVI (low tide), NDWI, NDMI (high tide), NDBI. Different from other aspects, high vulnerability indicates that not suitable for ecotourism and vice versa. Overall results (Figure 3f) shows that the mangrove area in Batu Ampar has a Low Vulnerability index.

3.3 *Landscape Value*

A total of 20 respondents from local people were asked to share their perceptions and preferences of the landscape in their environment. In this case, are divided into five classes: mangroves, settlement, agriculture, water bodies, and forest. This approach is modified from Ecosystem Services Value [9] since we exclude the photograph that represents the landscape considering the respondent are local people and understand well about these landscapes. Each data was calculated and imported into the land use map with the same five land use classes.

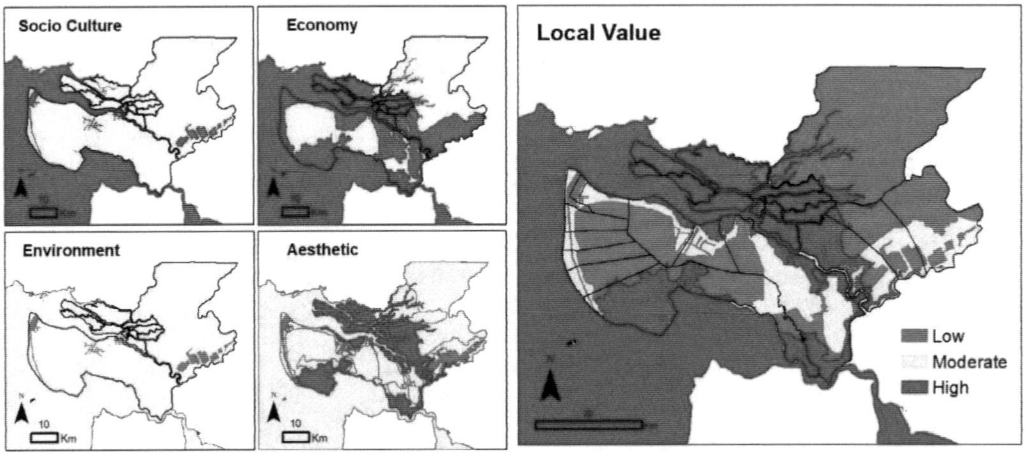

Figure 4 Landscape Value by Local People

The results show that local people highly value the water bodies as part of their sociocultural and economy. It suited well with the local people as the coastal community who depends their lives on the coastal natural resources, see Figure 4, in this case, not only the sea and the rivers, but also to the mangroves forest. Local people also highly value the mangroves for the aesthetics. On the other hand, it is important to encourage their value on the environmental aspects to avoid the overuse of the mangroves areas as one reason for the declining quality of the coastal environment [17].

3.4 Spatial Policy for Ecotourism

One of the best approaches to studying the potential and impact of ecotourism in the policy aspects is examining it according to the spatial scales at which ecotourism and policies are situated [18]. In this research, we focus on the study area's spatial policies that considerably influence most of the sites' activities. The policies related to the protected area in Batu Ampar use in this study is the Village Forest and Consensus Forest Use. Based on the Ministry of Environment and Forestry Decision in 2017, ten of fifteen villages in Batu Ampar received the right to manage the Village Forest: Batu Ampar [19], Teluk Nibung [20], Tanjung Harapan [21], Ambarawa [22], Sungai Besar [23], Tasik Malaya [24], Medan Mas [25], Padang Tikar Satu [26], Nipah Panjang [27], and Sungai Jawi [28].

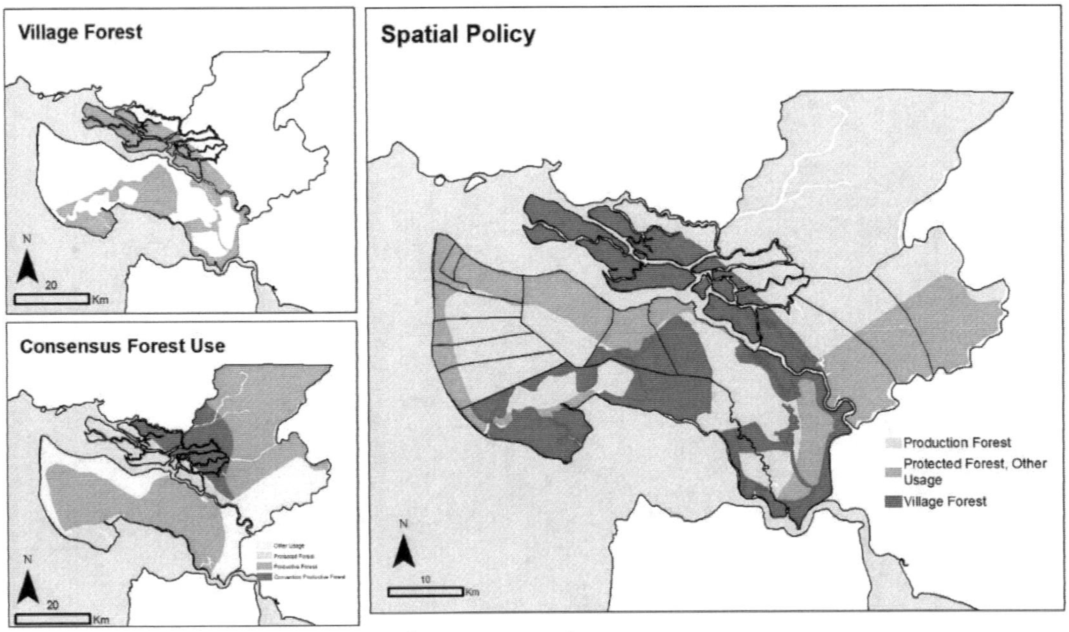

Figure 5 Legal Aspects

The Village Forest's management right includes the Forest Management Group's right to use the area to collect non-timber forest products and environmental services, such as ecotourism. Hence, the Village Forest Area score is the highest. Following that, two forest types from the Consensus Forest Use, the Protected Forest and Other Usage, are possible for ecotourism area but require plenty of permit and approval. On the other hand, the production forest is not supposed for ecotourism, but if the company decided to establish ecotourism, it also require a permit and approval (Figure 5).

3.5 Mangrove Ecotourism Potential Map

Overall, based on the weighted overlay from four aspects in this research, the results show that the study area is potential for mangrove ecotourism, where 414.67 km^2 (18%) with High Potential area, 697.28 km^2 (30%) Moderate Potential, 522.39 km^2 (22.6%), and the rest 672.33 km^2 (29%) are Not Potential area, from total area 2,306.57 km^2 of Batu Ampar District. High Potential areas for mangrove ecotourism are situated in Batu Ampar, Telok Nibong, Nipah Panjang, Tanjung Harapan, and Sungai Kerawang (Figure 6). Most of the village in Batu Ampar district has areas that moderately potential mangrove ecotourism.

Comparing the results for mapping the mangrove ecotourism using only Mangrove Ecotourism Sustainability (Figure 2) with the multicriteria results (Figure 6), it is clear that the results are less detailed if using only the mangrove ecotourism sustainability. The mangrove ecotourism sustainability underestimated the area with moderate potential for ecotourism, and this research method manages to delineate the moderately potential area. These results might be beneficial information for the area that has the moderate potential area of ecotourism.

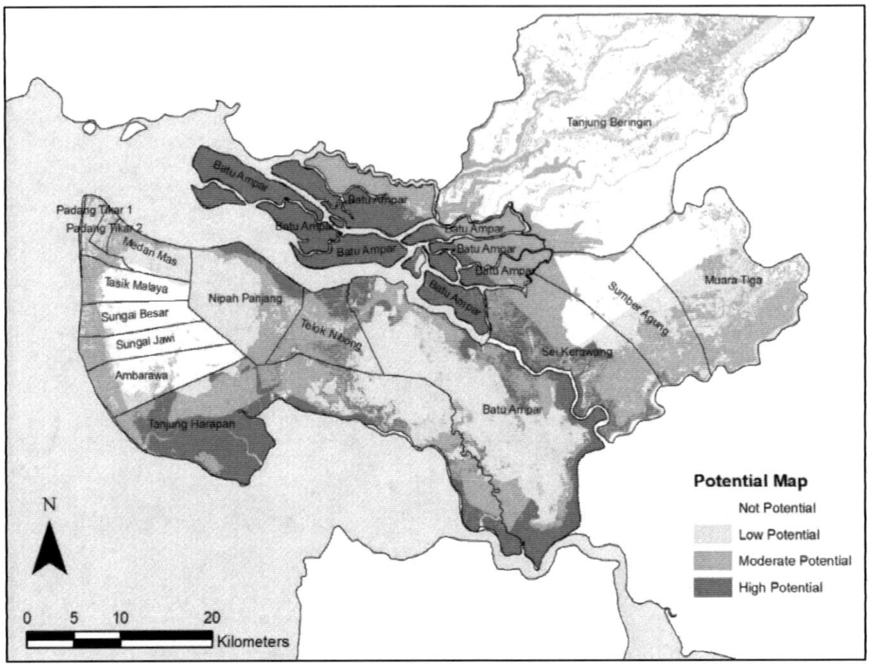

Figure 6 Mangrove Ecotourism Potential Map

4. Conclusion

This research demonstrated mangrove ecotourism potential mapping in Batu Ampar, Kubu Raya, West Borneo. The results indicated that the study area is potential for mangrove ecotourism. Based on this research, the priority for ecotourism development in Batu Ampar District should be on five villages with high potential areas: Batu Ampar, Telok Nibong, Nipah Panjang, Tanjung Harapan, and Sungai Kerawang. By using the multicriteria mapping, the potential area more specific and detail. Further research about ecotourism plan is needed to develop the ecotourism in Batu Ampar area, to bring benefits for the local community and the government in Batu Ampar, Kubu Raya, West Kalimantan.

Acknowledgement

This research was supported by the Ministry of Research Technology of Higher Education, Republic of Indonesia, through Master's Thesis Research Scholarship 2020.

References

[1] Zhang X, Treitz P M, Chen D, Quan C, Shi L and Li X 2017 Mapping mangrove forests using multi-tidal remotely-sensed data and a decision-tree-based procedure *Int. J. Appl. Earth Obs. Geoinf.* **62** 201–14
[2] Fennell D 2015 *Ecotourism* (Routledge)
[3] Spalding M, Blasco F and Field C 1997 *World Mangrove Atlas* ed M Spalding, F Blasco and C Field (Okinawa, Japan: The International Society for Mangrove Ecosytems)
[4] Valiela I, Bowen J L and York J K 2006 Mangrove Forests: One of the World's Threatened Major Tropical Environments *Bioscience* **51** 807
[5] Karlina E, Kusmana C and Bismark M 2016 Analysis of Sustainability of Mangrove Protection Forest Management in Batu Ampar, Kubu Raya Regency , West Kalimantan Province *J. Anal. Kebijak.* **13** 201–19
[6] Departemen Kehutanan R I 2003 *Pedoman Analisis Daerah Objek Oobjek dan Daya Tarik Wisata Alam (ADO-ODTWA)* (Bogor)
[7] Yulianda F 2007 Ekowisata bahari sebagai alternatif pemanfaatan sumberdaya pesisir berbasis konservasi *Makal. Semin. Sains*
[8] Rogers K, Mogensen L A, Davies P, Kelleway J, Saintilan N and Withycombe G 2019 Impacts and

adaptation options for estuarine vegetation in a large city *Landsc. Urban Plan.* **182** 1–11
[9] Zoderer B M, Tasser E, Erb K H, Lupo Stanghellini P S and Tappeiner U 2016 Identifying and mapping the tourists' perception of cultural ecosystem services: A case study from an Alpine region *Land use policy* **56** 251–61
[10] Saaty T L 2008 Decision making with the analytic hierarchy process *Int. J. Serv. Sci.* **1** 83
[11] Giesen W, Wulffraat S, Zieren M and Liesbeth S 2006 *Mangrove Guidebook for Southeast Asia* (Bangkok, Thailand: FAO and Wetlands International)
[12] IUCN 2020 The IUCN Red List of Threatened Species *Version 2020-3*
[13] Alongi D M, Murdiyarso D, Fourqurean J W, Kauffman J B, Hutahaean A, Crooks S, Lovelock C E, Howard J, Herr D, Fortes M, Pidgeon E and Wagey T 2016 Indonesia's blue carbon: a globally significant and vulnerable sink for seagrass and mangrove carbon *Wetl. Ecol. Manag.* **24** 3–13
[14] Mcleod E and Salm R V 2006 *Managing Mangroves for Resilience to Climate Change IUCN Global Marine Programme*
[15] Rogers K and Woodroffe C D 2016 Geomorphology as an indicator of the biophysical vulnerability of estuaries to coastal and flood hazards in a changing climate *J. Coast. Conserv.* **20** 127–44
[16] Mao D, Wang Z, Du B, Li L, Tian Y, Jia M, Zeng Y, Song K, Jiang M and Wang Y 2020 National wetland mapping in China: A new product resulting from object-based and hierarchical classification of Landsat 8 OLI images *ISPRS J. Photogramm. Remote Sens.* **164** 11–25
[17] Nickerson-Tietze D J 2000 Community-based management for sustainable fisheries resources in Phang-nga Bay, Thailand *Coast. Manag.* **28** 65–74
[18] Fennell D A and Downling R K 2003 *Ecotourism Policy and Planning* (London: CAB International)
[19] Kementerian Lingkungan Hidup dan Kehutanan R I 2017 *SK Nomor 515/MENLHK-PSKL/PKPS/PSL.0/2/2017 tentang Pemberian Hak Pengelolaan Hutan Desa Kepada Lembaga Pengelola Hutan Desa Batu Ampar*
[20] Kementerian Lingkungan Hidup dan Kehutanan R I 2017 *SK Nomor 516/MENLHK-PSKL/PKPS/PSL.0/2/2017 tentang Pemberian Hak Pengelolaan Hutan Desa Kepada Lembaga Pengelola Hutan Desa Teluk Nibung*
[21] Kementerian Lingkungan Hidup dan Kehutanan R I 2017 *SK Nomor 518/MENLHK-PSKL/PKPS/PSL.0/2/2017 tentang Pemberian Hak Pengelolaan Hutan Desa Kepada Lembaga Pengelola Hutan Desa Tanjung Harapan*
[22] Kementerian Lingkungan Hidup dan Kehutanan R I 2017 *SK Nomor 519/MENLHK-PSKL/PKPS/PSL.0/2/2017 tentang Pemberian Hak Pengelolaan Hutan Desa Kepada Lembaga Pengelola Hutan Desa Ambarawa*
[23] Kementerian Lingkungan Hidup dan Kehutanan R I 2017 *SK Nomor 520/MENLHK-PSKL/PKPS/PSL.0/2/2017 tentang Pemberian Hak Pengelolaan Hutan Desa Kepada Lembaga Pengelola Hutan Desa Sungai Besar*
[24] Kementerian Lingkungan Hidup dan Kehutanan R I 2017 *SK Nomor 521/MENLHK-PSKL/PKPS/PSL.0/2/2017 tentang Pemberian Hak Pengelolaan Hutan Desa Kepada Lembaga Pengelola Hutan Desa Tasik Malaya*
[25] Kementerian Lingkungan Hidup dan Kehutanan R I 2017 *SK Nomor 522/MENLHK-PSKL/PKPS/PSL.0/2/2017 tentang Pemberian Hak Pengelolaan Hutan Desa Kepada Lembaga Pengelola Hutan Desa Medan Mas*
[26] Kementerian Lingkungan Hidup dan Kehutanan R I 2017 *SK Nomor 523/MENLHK-PSKL/PKPS/PSL.0/2/2017 tentang Pemberian Hak Pengelolaan Hutan Desa Kepada Lembaga Pengelola Hutan Desa Padang Tikar Satu*
[27] Kementerian Lingkungan Hidup dan Kehutanan R I 2017 *SK Nomor 524/MENLHK-PSKL/PKPS/PSL.0/2/2017 tentang Pemberian Hak Pengelolaan Hutan Desa Kepada Lembaga Pengelola Hutan Desa Nipah Panjang*
[28] Kementerian Lingkungan Hidup dan Kehutanan R I 2017 *SK Nomor 525/MENLHK-PSKL/PKPS/PSL.0/2/2017 tentang Pemberian Hak Pengelolaan Hutan Desa Kepada Lembaga Pengelola Hutan Desa Sungai Jawi*

IOP Conf. Series: Earth and Environmental Science **879** (2021) 012024 doi:10.1088/1755-1315/879/1/012024

Determination of freight transport terminal location by using the AHP method for sustainable urban development

T A Putri [1] and N Sari[2]
[1]Student, Polytechnic of Indonesian Land Transport Academy, Ministry of Transportation, Bekasi 17001, Indonesia
[2]Lecturer, Polytechnic of Indonesian Land Transport Academy, Ministry of Transportation, Bekasi 17001, Indonesia

E-mail : tiara6917@gmail.com

Abstract. The rapid economic growth in Samarinda encourages higher demand for logistics that required vehicles and infrastructure for freight transport. The movement of goods in Samarinda is not supported yet by strategic freight transport. This cause many of freight transport parked and unload along the roadside. Writers did some research to determine the best location for freight transport terminal development using an AHP (Analytical Hierarchy Process) method. AHP (Analytical Hierarchy Process) method combines several considerations such as accessibility, traffic performance, regional pattern of transportation to reach multi criteria problem solving. The result of the analysis points out the score for Simpang Pasir Regency which is , the evaluation quality of accessibility 64%, traffic performance 23%, and regional pattern of transportation 12% . With that result Simpang Pasir Regency is chosen as the best alternative location for freight transport terminal development.
Keywords: Analytical Hierarchy Process, freight transport terminal, sustainable urban development

1. Introduction

1.1 *Background*

Samarinda is the capital city of East Kalimantan with increasing economic growth. The growth caused the demand for logistics also increased. The increase of logistic demand in Samarinda causes the development of tools and infrastructure of freight. The private sector and company compete to gain profit to fulfill the demand for freight transportation so that the in and outflow of logistic in Samarinda could be fill up. The high volume of goods movement with the domination of internal and external trips has not been supported by the existence of a strategic freight terminals. It caused a lot of on-street parking and unloading activity on the roadside and caused the decreasing of a road segment performance. The decreasing performance of road segment happened in Jalan Jakarta, Jalan Ir. Sutami, and many more. The problem can be solved with a few solutions; one of them is to build a freight terminal. Samarinda had already had a freight terminal that was built in 2010 and launched in 2011. Still, it didn't work well while operated because the location is not strategic enough with the road network that passed by freight, and it becomes unfit to operate anymore.

It needs to consider the aspect of location considering the freight terminal has a more elaborate scale and there are a lot of activities inside while planning the freight terminal. Freight terminal is expected to become a place that is representative and adequate to accommodate the land transport activity, especially goods distribution.

1.2 *Research Problem*
The high volume of goods movement has not been supported by the location of a strategic freight terminal. This caused many on-street parking and unloading goods on the roadside. It made the performance of the road segment decreased. The performance reduction of road segment can be found in road passed by freight such as Jl. Jakarta, Jl. IR. Sutami, and many more.

1.3 *Research Focus*
The determination of several locations that have the potential to become an alternative for the construction of a freight terminal in Samarinda based on the selection factor using a Likert scale and analyze the selection of the most appropriate alternative development location using the Analytic Hierarchy Process (AHP) method.

2. Literature Review

2.1. Terminal
Based on Regulation of the Minister of Transportation of the [1], terminal is base of a general motor vehicle that used to arrange the departure and arrival, load and unloading people and, or goods, and mode transfer.

2.2. Determination of Location
Based on [2] the existing literature, which describes and evaluates the aspects of quality and productivity of terminals, the necessary criteria to define how the intermodal terminal's network were selected. In selecting the criteria, several sources from actual studies and the literature are being analyzed, and a list of the criteria and their subcriteria were created. Based on the critical analysis and the method of surveying traffic experts (road, rail, water, and air traffic), the relevant criteria and subcriteria will then be selected. The recognized criteria include legislative, environmental; goods flows, spatial, technical-technological, and organizational. The legislative and environmental criteria are not evaluated in parallel with the other criteria, since they depend on individual states' legislation. Thus, it is first necessary to check whether specific locations meet the required legislative and environmental criteria to establish intermodal terminals in these areas.

2.3. Transportation and Freight
Based on on Technical Guidelines for the Implementation of Freight Transportation, Transportation is moving people and, or goods from one place to another by vehicle and Car for goods is every motor vehicle other than a motorcycle, passenger car, and bus.

2.4. Metode *Analytical Hierarchy Process* (AHP)
Based on [3} Analytical Hierarchy Process (AHP) method was developed by Thomas L. Saaty in 1970 in Warston School. AHP method is a method that used in decision making with several consideration such as perception factors, preference, experience, and intuition. AHP combines valuation and personal value in one logical way.

Analytic Hierarchy Process (AHP) can solve complex multi-criteria problems into a hierarchy. A complex problem is that it consists of many criteria (multi-criteria), unclear problem structure, indefiniteness opinion from a decision maker. According to Saaty, hierarchy is a representative from a complex problem in a multilevel structure in which the first level in purpose, followed by factor level, criteria, sub criteria, and until last level of the alternative. With hierarchy, a complex problem could be divided into a group that later will create a hierarchy so the problem will be more structured and systematic.

This method is a frame to make an effective decision to simplify decision making and quicken the decision making by solving problems into parts, arrange the part or variable into a hierarchy structure.

3. Research Methodology

3.1. Collecting data
Data collected in this research consists of two kinds of data, which are primary and secondary data. The collecting of primary data done with observation and interview. The collecting of secondary data aims to collect the data needed to support research. Data such as RTRW of Samarinda City, RDTR Samarinda. City, Road Network Map, Freight Network Map, Landscape Map, Administrative Map of Samarinda City, and Topography Map.

3.2. Method of Analysis
According to one of the research purposes that has been appointed is to analyze the determination of freight terminal construction location that suitable with criteria of freight terminal construction. The determination of factor that used to analyze the potential location of freight terminal is Likert scale consists of The availability of Spatial Land, Condition of Land Use, Topography Condition, Population and Road Network.

The location determination of freight terminal used the AHP method that combined various considerations to solve multi-criteria problems. The main purpose of the AHP method is to determine priority and judgment score to find out how important a location factor compared to other location.

The determination criteria are used to choose the construction location of the freight terminal with AHP method consists of Accessibility Criteria, Road Performance Criteria, and Transport Regional Pattern Criteria.

4. Research Analysis

4.1. Collecting Data
Selection of Potential Alternative Location for Freight Terminal. The result of selection was made to obtain five potential locations. Then the selection will used Likert scale to get the score result from each criteria and statement indicators to get the alternative location of the freight terminal.

Determination factors that are used in analyze the potential alternative location of freight are using Likert scale involve of The availability of Spatial Land, Condition of Land Use, Topography Condition, Population and Road Network.

After analysis of each selection factor determining locations of freight terminal construction, the result of assessment shows the top three alternative locations of freight terminal with the biggest score based on the total score of each selection factor of freight terminal construction in Samarinda which zona 10 (Kelurahan loa bahu), zona 20 (Kelurahan Bukuan), dan zona 22 (Kelurahan Simpang Pasir).

4.2. Location Determination of Freight Terminal Construction with Analytical Hierarchy Process Method
There is a lot of decision-making method in the location determination of freight terminal. Therefore to get the most suitable location, each location should have all the expected criteria. The difference of importance level of each location factor become a problem, because not all specified criteria has the same fulfillment with the sum number of fulfilled criteria.

The method that can be used to measure the priority or the quality of importance level of location factor against location determination is Analytical Hierarchy Process (AHP)

Method[4][5][6][7]. The main purpose of the AHP method is to determine the priority to find out how the comparison of each locations importance. It will be used for the further given relative assessment of each factor[8][9].

5. Result and Discussion

5.1. *Determination of Criteria Priority*

The analysis steps were made to determine the priority of criteria with the *Analytical Hierarchy Process* (AHP) method[10][11][12]. Based on the results of the questionnaire survey done in 10 respondents, a comparative value obtained then shows in table 1 below:

Table 1 Pairwise Comparison Matrix Criteria for Determining the Location of Freight Transport Terminals in the Samarinda

CRITERIA	Accessibility	Traffic Performance	Transport Regional Pattern
Accessibility	1	3	5
Traffic Performance	1/3	1	2
Transport Regional Pattern	1/5	1/2	1

The table above shows that the highest scale score of comparison of each criterion is criteria 1 (Accessibility) and criteria 3 (Transport Regional Pattern), which is equal to 3. Generally, the comparison scale score of criteria 1 (Accessibility) and other criteria show the dominant number. The result shows that most of the respondents assumed that the freight terminal should be located near the location of distribution goods such as the market, warehouse, and port of goods.

Table 2 is the beginning matrix, the initial matrix form of comparison between criteria, then the priority vector value is calculated by normalizing the initial matrix. So that the results of the priority vector calculation are as follows.

Table 2 Normalization Matrix and Priority Vector Criteria

CRITERIA	Accessibility	Traffic Performance	Transport Regional Pattern	Priority Vector
Accessibility	0.652	0.667	0.625	0.648
Traffic Performance	0.217	0.222	0.250	0.230
Transport Regional Pattern	0.130	0.111	0.125	0.122

Based on the calculation in the table above, the results show that the priority vector column shows that criteria 1 (Accessibility) is the most important criteria among others because it has the most significant grade, which is 0,648 (64,8%), followed by criteria 2 (Traffic Performance) which is 0,230 (23%) and criteria 3 (Transport Regional Pattern) which is 0,122 (12,2%). The result shows the respondents' perception stated that to determine the location of freight terminal, the most important criteria is accessibility from and to the industry location and distribution such as market, warehouse, the port of good.

The next step is to do a consistency test with firstly calculate the maximum eigenvalues (λ maximum) with formula IV.1 and index consistency (CI) using formula IV.2 then calculate consistency ratio (CR) using formula. The calculation result of CI and RI which tabulated in table 3 below.

Table 3 Calculation of Consistency Test (CR) Criteria

CRITERIA	**Accessibility**	**Traffic Performance**	**Transport Regional Pattern**	**Eigen Vector**	**E vector/ Priority Vector**
Accessibility	0.652	0.667	0.625	1.948	3.007
Traffic Performance	0.217	0.222	0.250	0.690	3.003
Transport Regional Pattern	0.130	0.111	0.125	0.366	3.001
λ maks = rata-rata(e-vector/priority vector) = 3,003					
$CI = \frac{\lambda maks - n}{n-1}$; $n = 3 \rightarrow CI = 0,0018$					
$CR = \frac{CI}{RI}$; $RI = 0,98$ (Tabel IV.3) → CR = 0,001					

Based on the calculation in the table above, the result of consistency ratio is 0,0018 which already fulfilled the requirements of consistency test which is smaller than 0,1. Consistency is needed for the results of weighting to be accepted. After determining the weight in each stage starting from performance weight, sub-criteria weight until each weight of alternative toward sub-criteria, then recapitulation of weighting result or priority vector was done which is the result of priority weight of each step.

The result contains two kinds of weight assessment which is criteria and sub-criteria weight assessment. The calculation was done to get the real weight of each sub-criteria. Accessibility criteria have 0,64 weight while sub-criteria with distance from the market has 0,17 weight. Furthermore, multiplication is carried out to determine the results in the table below, which is the final weight from each sub-criteria where for distance to market sub-criteria has a final weight of 0,11. Likewise, for the sub-criteria, then the same calculation method is carried out so that the final results of the weight assessment of each alternative location against the sub-criteria are as shown in the the table 4.

Table 4 Result Of Recapitulation Criteria Weight And Sub-Criteria Weight, And Alternative Based On Sub-Criteria

Criteria	**Accessibility**			**Traffic Performance**			**Regional Transport Patterns**		
Criteria weights	0.64			0.23			0.12		
Sub-criteria	Distance to market	Distance to port	Distance to warehouse	V/C Ratio	Speed	Density	Volume HV	Volume LV	Volume UM
Weight of Sub-criteria	0.17	0.35	0.48	0.35	0.15	0.50	0.63	0.26	0.11
Alternative 1	0.64	0.23	0.52	0.16	0.37	0.30	0.16	0.28	0.58
Alternative 2	0.23	0.65	0.30	0.30	0.45	0.16	0.30	0.22	0.11
Alternative 3	0.12	0.12	0.17	0.54	0.20	0.54	0.54	0.50	0.30

Tabel 5 Final Result Of Location Determination Of Freight Terminal Construction Weighting

Sub-criteria	Distance to market	Distance to port	Distance to Ware-house	V/C Ratio	Speed	Density	Volume HV	Volume LV	Volume UM
Final Weight	0.11	0.22	0.31	0.09	0.03	0.12	0.08	0.03	0.01
Alternative 1	0.65	0.23	0.52	0.16	0.37	0.30	0.16	0.28	0.58
Alternative 2	0.23	0.65	0.30	0.30	0.45	0.16	0.30	0.22	0.11
Alternative 3	0.12	0.12	0.17	0.54	0.20	0.54	0.54	0.50	0.30

After each alternative location weight against sub-criteria were obtained, then an assessment is carried out for each of the alternative locations to obtain priority ranking to become the chosen alternative location for the construction of a freight terminal. The calculation from each alternative location is obtained from the multiplication results between each sub-criteria weight with the results of the assessment of alternative locations for each sub-criteria. The following is a table of the results of calculating the total score of each alternative location.

Based on table 5 above, the first rank of the alternative location for the construction of a freight terminal is alternative one, which is located in Zone 22, Simpang Pasir Village in JL. Trikora section with total score of 0,372. The second rank is alternative two, which is located in Zone 20, Bukuan Village in JL. Ampera section with total score of 0,353. At the same time the third rank is alternative three, located in Zone 10, Loa Bahu Village in JL. Ring Road section. So, in the analysis of determining the freight terminal location using the Analytical Hierarchy Process method, the results of the selected location are obtained, namely Alternative 1.

6. Conclusion

Freight transport route planning affects the economic development of a region [13,14,15]. Based on the analysis and recommended recommendations described in the previous chapter, the following are the conclusions:

1) The weighting of the criteria for the construction of a freight terminal using the Analytical Hierarchy Process (AHP) method obtained the highest weight on the accessibility criteria of 64.8%, the traffic performance criteria were 23%, and the criteria for regional transport patterns were 12.2%. The results of the weighting of the sub-criteria for the construction of a freight transportation terminal using the Analytical Hierarchy Process (AHP) method obtained the highest weight on the distance to warehousing sub-criteria by 31%, while the lowest criteria for the sub-criteria for non-motorized vehicles (UM) was 1%;
2) There are three alternative locations selected for the construction of freight transport terminals, namely in zone 10 (Loa Bahu Village), zone 20 (Bukuan Village), and zone 22 (Simpang Pasir Village). The results of the priority ranking assessment obtained that the location in the first rank is alternative one, which is located in Zone 22 of Simpang Pasir Village on the JL. Trikora section with a total score of 0.372. The second rank is alternative two, which is located in Zone 20, Bukuan Village on the Jl. Ampera with a total score of 0.353. At the same time the third rank is alternative three, which is located in Zone 10, Loa Bahu Village on Jl.Ring Road with a total score of 0.273. Alternative one comes out as the best alternative for freight terminal location.
3) Analysis of the determination of the main facilities and supporting facilities at the freight terminal to be built related to serving the loading and unloading activities of goods and

intermodal transportation. Based on the results of the analysis, it is found that the primary and supporting facilities for freight transportation terminals refer to the Government Regulation of the Republic of Indonesia Number 79 of 2013 on Road Traffic and Transportation Network, concerning the need for goods transportation terminal facilities.

4) To improve safety, [16] roads that are used as routes for freight transport need to be equipped with signing and marking.

References

[1] Kementrian Perhubungan, *2018. PM 102 of 2018*
[2] Brnjac Nikolina 2015 Inland Intermodal Terminals Location Criteria Evaluation: The Case of Croatia Transportation Journal Vol. **54** p **4**
[3] *Kementrian Perhubungan, 2004. Surat Keputusan SK.727/AJ.307/DRJD/2004*
[4] Saaty, Thomas L, 1990. The Analytic Hierarchy Process: Planning, Priority, Setting, Resource Allocation. University of Pittsburgh
[5] Adisasmita, Rahardjo. 2011. *Transportasi dan Pengembangan Wilayah.* Penerbit Graha Ilmu: Yogyakarta
[6] Aditya, Faris Prima. 2012. *Analisis Penentuan Lokasi Pembangunan Terminal Angkutan Barang di Kota Pekanbaru.* STTD Bekasi
[7] Kurniawan, Fahri. 2012. *Analisa Penentuan Letak dan Fungsi Terminal Angkutan Barang Kota Cirebon.* STTD Bekasi
[8] Morlok, Edward K. 2005. *Pengantar Teknik dan Perencanaan Transportasi.* Erlangga: Jakarta
[9] Ortuzar, J and Willumsen LG. 1990. *Modelling Transport.* John Wiley & Sons: Toronto
[10] Richard. 2018. *Analisis Penentuan Lokasi Terminal Angkutan Barang Di Kabupaten Lombok Tengah,* STTD Bekasi
[11] Putri, Sherly Nandya. 2018. *Penentuan Lokasi Pembangunan Terminal Angkutan Barang di Kawasan Perkotaan Sampit,* STTD Bekasi
[12] Tamin, Ofyar Z. 2000. *Perencanaan dan pemodelan Transportasi.* Penerbit ITB : Bandung
[13] Behrends , S. 2011. *Urban Freight Transport Sustainability The Interaction of Urban Freight and Intermodal Transport Thesis Department Of technology Management and Economics Chalmers University of technology.* Sweden
[14] Doblanc, L. 2011. *City Distribution, a key Problem for the urban economy: guidelines for practitioners", C. Macharis and S. Melo (ed.), City Distribution and Urban Freight Transport,* Edward Elgar, Cheltenham, UK, pp.
[15] Engstrom R 2016 *The Roads' role in the Freight system Transportation Reaserch Procedia* 14 (2016) 1443-1452 doi: 10.10.16/j, trpro.2016.05.217 published by Elsevier B.V Sweden
[16] Sari, N. 2020. Transport Management of Sustainable Urban Deve lopment at Tambun Market Area IOP Publishing doi: 10.10.88/1755-1315/501/1/012034

Towards sustainable and regenerative territories: ecological restructuring of phosphorus retention and mobilization based on substance flow analysis (sfa) in Quirino Province

T K Tecson[1], N Navarra[2] and J D Villajuan[2]

[1] Undergraduate student, Environmental Landscapes Studio Laboratory, UP College of Architecture, University of the Philippines-Diliman
[2] Faculty, Environmental Landscapes Studio Laboratory, UP College of Architecture, University of the Philippines-Diliman

E-mail: tomkirbytecson@gmail.com

Abstract. Phosphorus (P) is an essential and finite nutrient element that cannot be substituted by any other substance. To overcome limited supply of mineral Phosphate rocks (PR) deposits and to contribute to its sustainable management, a better knowledge on P flows is needed. In this study, a Substance Flow Analysis (SFA) for P was conducted for the Quirino Province. Different subsystems were investigated such as forestry, agriculture, anthropogenic consumption, organic wastes, and wastewater management. The researcher found that within the system under study, there is an excess total P inflow into the three (3) major river watersheds system in Quirino Province. The aim of the study was to analyze the flow of Phosphorus that enter, transform, pass through, and leave the territorial subsystems in order to create a knowledge bank that will aid in the transition of Quirino Province from linear to circular metabolism.
Keywords: substance flow analysis, circular economy, phosphorus recycling

1. Introduction

The phosphorus fertilizers that our farmers use today in food production came from mining of phosphorite, or phosphate rock. Estimates show that Earth's crust contains a total of 65 Million Metric tons phosphate rocks, of which are concentrated to Morocco and Western Sahara. This figure, however, only covers the phosphate rocks that are "sufficiently rich in phosphate for conversion to useful products by present methods" (Britannica, 2019). For commercial purposes, phosphate rocks are acidized to form water-soluble phosphorus compounds which are essential additions to fertilizer.

The concern is that "phosphorus is wasted on its journey from mining to being eaten by humans, and the wasted phosphorus ends up in waterways where it can cause algal blooms." (Britannica, 2019). With increasing population, comes increasing demands for food. And increasing food production requires fertilizers for more crop yield. Increased phosphorus usage will deplete the nonrenewable supply of phosphate rock. In fact, researches [1][2][3][4] have shown that the global reserves of phosphate rocks will be depleted in the next 50-100 years. Phosphorus, being faced with lot of challenges in geographic constraints, phosphorus- induced water pollution, increasing usage as nonrenewable resource, calls for sustainable management [5].

According to Nest and Cordell, increasing the efficiency and management of phosphorus can minimize the adverse effects of eutrophication, chemical pollution, climate change and etc. Further, "'flow analyses' have proven to be a valuable tool" in assessing the efficiency of flow of materials and/or energy [6]. In article written by Weinstock entitled "The Metabolism of a City: The Mathematics of Networks and Urban Surfaces", cities "are not arrays of material structures, but are regarded as analogous to living beings, as they consume energy, food, water and other materials, excrete wastes and maintain themselves down through the generations." He also added that metabolism can be related to cities through networks and scaling phenomena. Meaning, mathematically, the cities can be networks of metabolic processes: Rates of resource consumption can be measured through mathematical parameters. Movement of resources within a city is like that of movement of nutrients in the circulatory system of a living thing. This field of study is under industrial ecology and it is called "Urban Metabolism." However, in an article written by Yan Zhang, the existing multi-scale studies about urban metabolism have been primarily theoretical. Zhang expounded that "application of this particular research to actual planning has been rare." [7].

What this research is trying to achieve is providing a framework and research model that can be replicated in a local setting to improve nutrient management at macro level. By proving interventions and recommendations in sustainable management of phosphorus, the theme if this research is adhering to the United Nations Sustainable Development Goals of Circular Economy. In connection, this research aims to fill in the gap of urban metabolism research and actual planning in rural territory by providing an extensive knowledge base about the phosphorus metabolism of a territory.

2. Methodology

The methodology came from the set of the literature reviewed: The general approach of this paper was adapted from [7]; the research tactics and more detailed versions of crude phosphorus metabolism model were derived from other literatures reviewed [8][9][10][11][12]. The general approach of the study was illustrated below (Figure 1).

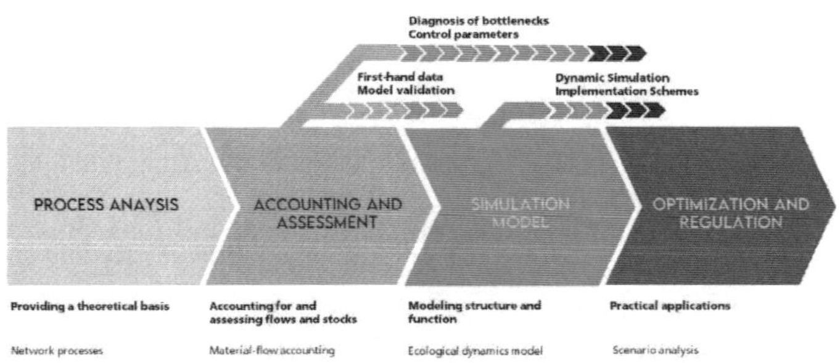

Figure 1 Research framework adapted from Zhang Y. urban metabolism

The first step, the process analysis, focused on the identifying the different actors within a metabolism. By identifying the possible linear input and output processes and uniting it to produce model of a network that revealed the system's inner details [7]. For second step, the accounting, Substance Flow Analysis (SFA), one of the research models of Material Flow Analysis (MFA) that focuses on quantifying the flow of a given substance across a region, has been employed in this phosphorus metabolism analysis. This tactic provided approximations of the following: (i) inflows, phosphorus that are entering the system/subsystems/metabolic actors through a particular material in a given time; (ii) inflows, phosphorus that are exiting the system/subsystems/metabolic actors through a particular material in a given time; and (iii) stock, the difference between the total inflows and total outflows in a particular system/subsystems/metabolic actor. However, some P flows were neglected, due to their statistic were not available, and/or not clearly traceable.

The third step was simulation model, which refers to the representation of the dynamics

IOP Conf. Series: Earth and Environmental Science **879** (2021) 012025 doi:10.1088/1755-1315/879/1/012025

and system's internal operating mechanisms that shows metabolic processes and factors that influence them [7]. In this study, ecological dynamics modelling was used through comprehensive Sankey diagram to show how the phosphorus was metabolized in different social, economic, and natural elements of the system.

For the last step, optimization and regulation encompassed the identification of factors that drives the (in)efficiency of metabolic processes and minimizes the environmental impacts. However, this study will only cover the first three steps as the main objective of this paper was to provide knowledge base about the current phosphorus metabolism of Quirino Province.

2.1 System boundary and temporal scale

The system boundary was defined as the administrative border of the Quirino Province (Figure 2). The metes and bounds used were from the Provincial Development and Physical Framework Plan (PDPFP) 2011-2030, which was complemented by map of National Mapping and Resource Information Authority (NAMRIA).

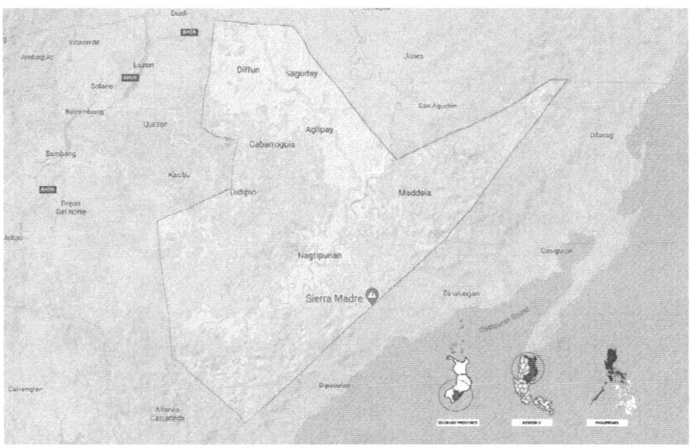

Figure 2 Administrative map Quirino Province

In addition to that, this study was carried out a static substance flow analysis. Meaning, all the datasets that have been considered and phosphorus flows have been computed were only good a specific year. For the temporal scale, the year 2018 was selected because it provided the researcher the latest and the greatest number of statistical data available for secondary data research. Moreover, a static SFA was made due to the limitation of time and logistics. In the events wherein there was no data available for 2018, the previous or following year/s was referenced.

2.2 Data collection

In this type of analysis, there are two types of data required: material mass flows and P concentration. In this study, the material mass flows datasets were called and under the umbrella term *parameters*. Material mass flows pertains to values and data which characterizes the amount of raw materials, goods, and products circulating within the territorial system. There were also indicators of the different societal factors of the territory, such as production values, land area, human and animal population, which were included in the *parameters*. Ergo, these are the values which were gathered by the government statistics offices and other census-concerning agencies.

In general, the parameters were multiplied with their phosphorus concentration values to obtain the elemental form of phosphorus. However, not all P flows were easily determined by just multiplying the mass flows with its corresponding P concentration value/s: Instead, supplemental equations were employed to obtain the P values (in kg). Shown below are the three (3) general equations, which were adapted from [11], used in the SFA.

The complete list of all parameters and coefficients used in this study is in the supplementary data, Table 2 and 3, respectively.

Fixed flow equations:

$F_i = P_a \times C_a$ (1)

F_i is a P flow from one subsystem to another, P_a is a parameter, and C_a is a phosphorus coefficient. Example of this Flow 12 (F_{12}) that shows the wind erosion from croplands to atmosphere.

Dependent flow equations:

$F_i = F_a \times P_a$ (2)

This type of equations shows how one P flow is dependent on another flow. The Flow 16 (F_{16}) shows how the P content of post-harvest residues is dependent on the crop production.

Balancing equations:

$\sum F_{input} = \sum F_{output}$ (3)

Balancing equations adheres to the Law of Mass Conservation. Some unknown P flows can be obtained by subtracting the unknown P flow to the overall inflow or export flow. The Flow 83 (F_{83}) shows how the estimation of remains of wild animals in the forest vis-à-vis to their foraging. The complete list of all parameters and coefficients used in this study is in the supplementary data, Table 2 and 3, respectively.

3. Results and discussion

To effectively visualize the findings of the phosphorus metabolism of the whole Province of Quirino, a Sankey diagram was formulated based on the information of Table 1. Shown below is the Figure 3, which contains the 82 simulated P flows that entered, exited, transformed through the total of 20 metabolic actors of five (5) subsystems of the Quirino Province system. The width of each arrow was directly proportional to the quantity of the phosphorus per flow: The thicker the arrow, the greater the value. Each flow (F_n) can be referenced back to the Table 1 of Supplementary Data.

3.1 Province of Quirino (main system)

The total P that circulated within the Quirino Province for 2018 was 75 669 018.49 kg-P. Based on Table 3, the largest P flows came from F_{24} and F_{27} which represented the Forest uptake, and Forest residues, respectively. These flows were heavily dependent on the published values of above-ground biomass densities of tropical forests, and average Carbon-Phosphorus uptake ratio in tropical ecosystems. The magnitude of these forest P flows relatively to the whole system went beyond the researcher's expectations because the hypothesis and goal setting were geared towards built-up areas and agriculture.[1]

[1] *(Note for the following illustrations of Sankey Diagrams: The width of the arrows does not represent the actual value of phosphorus (kg-P) but it represents the amount of P in each flow in relation to the whole system (% of total P) with a scale of 1%:0.15 cm:28pt (A1 size). Percentage values amounting to 2% and/or lower had a default width of 0.25pt for ease of visualization. High-resolution versions of the following diagrams are in the Supplementary Information)*

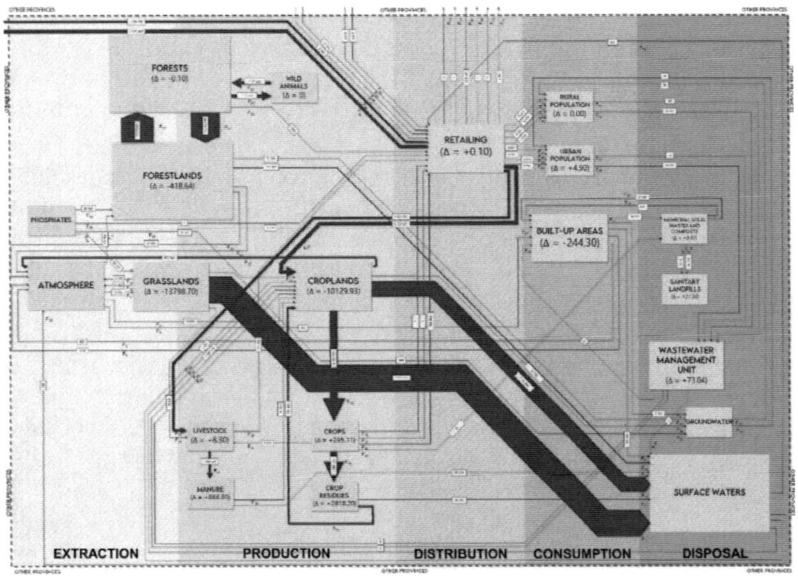

Figure 3 Comprehensive sankey diagram of phosphorus metabolism in Province of Quirino for 2018

3.2 Forestry

The present analysis revealed that in 2018, Forestry subsystem had elevated outflows (Δ= -14217.34 MT-P) associated with immense flow from F_{31}: Erosion and runoff from grassland, accounting 95.75% of total P outflows of the subsystem (14 403.32 MT-P). Following this is a mass flow from F_{24}: Forest trees uptake, and F_{27}: Forest residues (20.4259%) which was mainly due to nutrient cycling of the forest and forestlands.

Based on the data gathered, the main concern was on the subsystem of grasslands due to erosion and runoff. The grasslands possess high negative delta (-13798.70 MT-P). There is a huge loss of nutrients over time, mainly due to surface runoffs and erosion, and subsurface runoff and nutrient leaching. Grasslands is the leading landcover which is losing P through surface runoff and soil erosion (13 791.29 MT-P/year). This thesis referenced the works of local scientists--Francisco, H., & delos Angeles, M. According to their report, it is recorded that in the Philippines, grasslands / pasturelands records an estimated annual soil loss amounting to 267 tons per hectare (P_{234}).

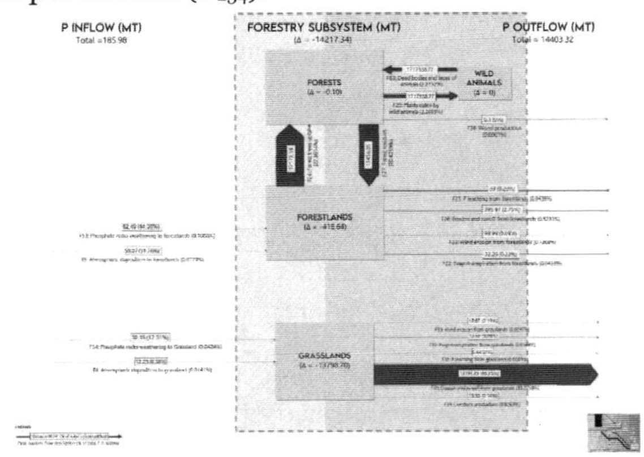

Figure 4 Sankey diagram of phosphorus metabolism in forestry subsystem of Province of Quirino for 2018

3.3 Agriculture

Of the eight (8) Phosphorus inflows of the agricultural subsystem, F_{64}: Mineral fertilizer consumption and F_{65}: Feed additive consumption were the two largest inflows, with the total of 97.47% of P inflow value of 2843.93 MT-P. F_{64} had a 61.08% share in inflows, while by F_{65} had a total inflow of 36.39%.

Moreover, four (4) out of five (5) metabolic actors in this subsystem indicated an apparent phosphorus accumulation: in the year 2018, crop residues accumulated a total of 2941.20 MT-P, making it the biggest "gainer" in the subsystem. Additionally, manure was comprised of 866.89 MT-P. The accumulation in these two metabolic actors were products of F_{16}: Post-harvest residues for the crop residues, and F_{37}: Livestock and poultry excreta for the manure. Along with these, Livestock (Δ = +8.30) and Crops (Δ = +295.11) were also considered as "gainers". Nonetheless, the overall change in balance remains negative (Δ = -6042.56) due to a phosphorus depletion of 10129.93 MT-P in the croplands.

The biggest contributor to the outflows in the agriculture subsystem was F_{15}: Erosion and runoff from croplands, consisted of 7775.89 MT-P or 87.5% of total outflows. It was then followed by F_{12}: Wind erosion from croplands.

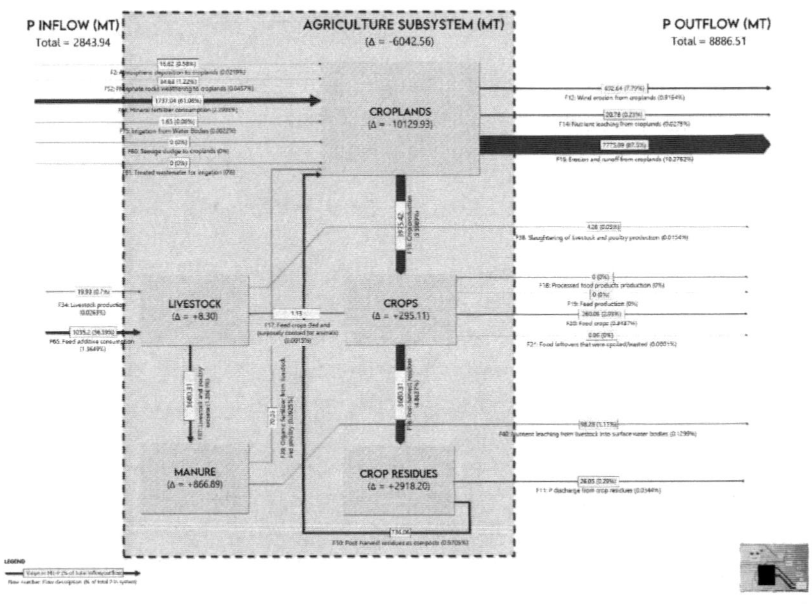

Figure 5 Sankey diagram of phosphorus metabolism in agriculture subsystem of Province of Quirino for 2018

3.4 Retailing

An assumption that there were no storage facilities available for products was made in this subsystem—however, to avoid generalization, each variable was examined. For instance, under the livestock and poultry group, eight (8) commodities were included. The production of each commodity—such as chicken eggs, was subtracted by the rural and urban population consumption. If the resulting value is negative, it indicates a negative supply—imports. Contrary, if the equation returns a positive value, then there is a surplus and thus exports. By utilizing the SUMIF function in Microsoft Excel, values were added according to the sign convention. This process was done to all variables to avoid confusion between 'net movement of Phosphorus per commodity group' and 'total imports and exports of particular commodity group.' Despite the assumption that retailing subsystems do not have storage capabilities, a delta value of 0.10 MT-P was recorded due to the lack of P flow that characterized the consumption of F_{28}: Wood production (0.10 MT-P).

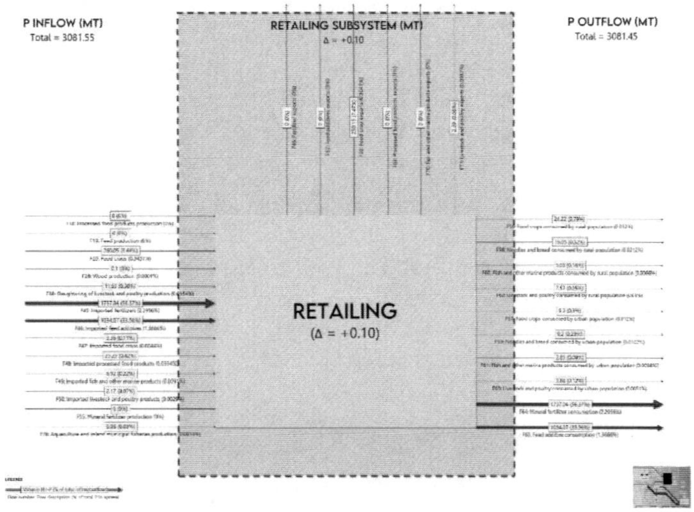

Figure 6 Sankey diagram of phosphorus metabolism in retailing subsystem of Province of Quirino for 2018

3.5 Anthropogenic consumption

Due to time and logistical constraints, the researcher was unable to secure data sets on the consumption of detergents and soap but this shortcoming was supplemented by securing a data regarding the average household wastewater generation in rural and urban households in the Philippines. Assuming that even if the P flows were incomplete, the yielding output of detergent usage will still be reflected on the wastewater generation. Based from these, were human excreta and wastewater generation— F72, F73, F77, and F78 were the main outputs of the population, accounting the 20.43% of total outflows. On the other hand, majority (44.65%) of the phosphorus inflows for this subsystem went to rural population despite this metabolic actor having zero-delta value. Overall, the inflow-outflow ratio of this subsystem was 0.331.

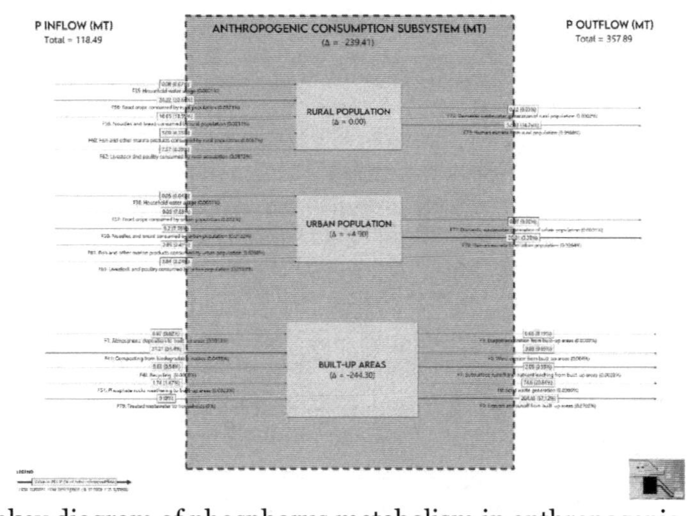

Figure 7 Sankey diagram of phosphorus metabolism in anthropogenic consumption subsystem of Province of Quirino for 2018

3.6. Waste and wastewater

Among the analyzed subsystems, waste and wastewater sector had a different trend. It was observed that this subsystem was an overall accumulator (Δ = +109.86). This data result was expected due to: *i*) inclusion of the sanitary landfills in this subsystem (SLF), and *ii*) according to an FGD conducted with officials from Quirino Provincial Natural Resources and Environmental Office (PNREO), the Province of Quirino have had no wastewater treatment plant (WWTP). 50.51% of the total inflows were produced by F_8: Solid waste generation, followed by F_{73}: Human excreta from rural population (35.77%) and F_{78}: Human excreta from urban population (13.55%). In addition, about 98.32% of the estimated outflow of this subsystem was credited to F_{41}: Composting from biodegradable wastes (37.21 MT-P).

The inflow-outflow ratio of wastes and wastewater subsystem was 3.903, the highest of them all.

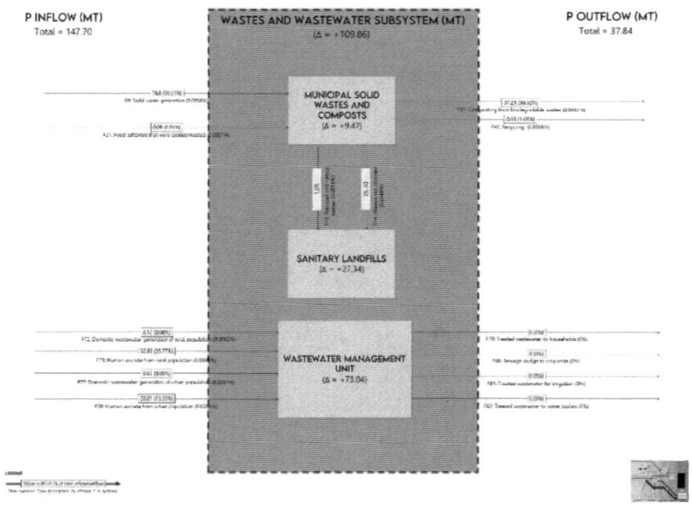

Figure 8 Sankey diagram of phosphorus metabolism in wastes and wastewater subsystem of Province of Quirino for 2018

4. Conclusion

This model can provide quantitative basis to support policymaking at the local government unit (LGU) level. This paper creates a proposal on methodical approach of analyzing the mass flows and phosphorus mobilization at a provincial level based on thorough and holistic secondary research. Thereupon, the forestry subsystem has a high imbalance of input and output with the estimated input/output ratio of 0.0129. This indicates a massive outflow vis-à-vis inflow. The main priority area for this subsystem was grasslands and other wooded lands. The biggest contributor to the outflows in the agriculture subsystem was F_{15}: Erosion and runoff from croplands, consisted of 7775.89 MT-P or 87.5% of total outflows. It was then followed by F_{12}: Wind erosion from croplands. Based on this ecological modelling, the forestry and agriculture subsystem produce high imbalance in terms of input and output that creates apparent impact to the water environment.

References

[1] Cordell D, Drangert J O and White S 2009 The story of phosphorus: global food security and food for thought *Global environmental change* **19**(2):292-305

[2] Vaccari D A 2009 Phosphorus: a looming crisis *Scientific American* **300**(6):54-9
[3] Cooper J, Lombardi R, Boardman D and Carliell-Marquet C 2011 The future distribution and production of global phosphate rock reserves *Resources, Conservation and Recycling* **57**:78-86
[4] Cordell D and White S 2014 Life's bottleneck: sustaining the world's phosphorus for a food secure future *Annual Review of Environment and Resources* **39**:161-88
[5] Carpenter S R 2008 Phosphorus control is critical to mitigating eutrophication *Proceedings of the National Academy of Sciences* **105**(32):11039-40
[6] Chowdhury R B, Moore G A, Weatherley A J and Arora M 2014 A review of recent substance flow analyses of phosphorus to identify priority management areas at different geographical scales *Resources, Conservation and Recycling* **83**:213-28
[7] Zhang Y Urban Metabolism
[8] Verger Y, Petit C, Barles S, Billen G, Garnier J, Esculier F and Maugis P 2018 AN, P, C, and water flows metabolism study in a peri-urban territory in France: The case-study of the Saclay plateau *Resources, Conservation and Recycling* **137**:200-13
[9] Chen M, Chen J and Sun F 2008 Agricultural phosphorus flow and its environmental impacts in China *Science of the Total Environment* **405**(1-3):140-52
[10] Chowdhury R B, Moore G A and Weatherley A J 2018 A multi-year phosphorus flow analysis of a key agricultural region in Australia to identify options for sustainable management *Agricultural systems* **161**:42-60
[11] Fan Y, Hu S, Chen D, Li Y and Shen J 2009 The evolution of phosphorus metabolism model in China *Journal of Cleaner Production* **17**(9):811-20
[12] Theobald T F, Schipper M and Kern J 2016 Phosphorus flows in Berlin-Brandenburg, a regional flow analysis *Resources, Conservation and Recycling* **112**:1-4

Multi-functional utilization of suburban green space : *text mining to analyze "Forest around the Mountains" Satoyama activities*

Q Y Xiang[1*], Y M Mao[1], and K Furuya[1]

[1] Graduate School of Horticulture, Chiba University, 648 Matsudo, Matsudo-shi Chiba 271-8510 Japan

E-mail: yumulisha@gmail.com

Abstract. Due to the acceleration of urbanization, many green spaces are facing the fate of abandonment, especially the urban and suburban green space. In Japan, these suburban green spaces can also be used as a Satoyama. Satoyama is a multi-purpose ecosystem, including some secondary forests, farmlands, lakes, marshes, and so on. They are affected by human beings and benefited each other. It is proof of harmonious coexistence that human beings have been groping for in nature since ancient times. This study collected the records of the Satoyama activity group in Matsudo City, Chiba Prefecture, Japan from 2008 to 2019, and used KHCoder3 to mine the characters. The purpose of this study was to identify that it is found that people use suburban green space in variety nowadays through the activities of volunteers in Satoyama. From the records, the green space has a series of changing stages after becoming Satoyama, in which 2008-2011 is the first stage, and 2016-2019 is the other stage, but 2012, 2014, and 2015 were grouped separately. In addition to the daily maintenance of the forest, there were also some special words that can be seen that people attach great importance to the publicity of Satoyama as a place for Children's environmental education. For the future, Satoyama activities can afford good references to urban green space multi-function, maintain ecological balance, and keeping sustainable development.
Keywords: multi-function, suburban, text mining, Satoyama, volunteer

1. Introduction

In this time of global urban densification, urbanization is developing more and more rapidly, and the concentration of labor force forms densification [1]. More and more abandoned green lands appear in farmland, mountainous and suburban areas [2]. However, many studies show that human beings who have been dependent on nature need green space since ancient times [4][5] both physically and psychologically due to the existence of biological affinity [3]. The famous theories are stress recovery theory (SRT) and attention recovery theory ART [6], shown that the experience in nature can bring a good effect on relieving stress and restoring attention. There is a contradiction between the human demand for green space and the ignoring of green space. It is an important part of the current project to make effective and reasonable use of green space in the city. In particular, the small suburban green space, which is close to the city and easy to be ignored, will be used because it is mistaken for private territory [7]. In Japan, small suburban green space can be used as sustainable development and utilization of resources as Satoyama [8].

"Satoyama" has the significance of agricultural land opposite the remote mountains, which can be used in agricultural production as fuel, compost, grass ash, and other supplies [9]. In the past, it was the existence of heat sources. However, with the development of urbanization, secondary forests were not fully utilized due to many miscellaneous forests are replaced by new energy [10]. Now people not only pass on the traditional functions of the forest to the next generation, but also can reuse Satoyama with new ways in order to build a sustainable development environment and make Satoyama proper management.

The main significance of Satoyama activities includes: (1) the sustainable management and utilization of resources;(2) the living and cultural development of symbiosis with nature;(3) [12] the sustainable use of space protecting the biodiversity of the secondary forest. [13][14]. In addition, environmental education, forest bathing, natural observation, and forest operation can be the parts of Satoyama.[15].

But most of these studies are paying attention to typical activities of Satoyama Mountain in one region or often far away from the city. Satoyama activities in the urban-suburban green space will be a better choice for the urban residents [16]. In some studies, it is common that researchers participate in Satoyama activities for a short period of time before reporting [13], but we cannot know how the volunteer group establishes Satoyama in a short period of time. On the other hand, for Satoyama activities, researchers will focus on it from the perspective of normal citizens [17]. That is not common to do a long-term follow-up from the perspective of volunteers.

Therefore, the purpose of this study is to find out what the specific activities in the multi-functional place like Satoyama and to know the how a volunteer group change and develop a suburban-urban green space of Satoyama through text mining of "Forest around the Mountains" reports. we want to show the diversified and the supplementary utilization of the suburban green space as Satoyama in this study from the special perspective of volunteers. We can also understand how the area is developed as a place to hold Satoyama activity through text mining. Finally, the diversity of activities in Satoyama in Japan described in this study can be used as a reference for other countries, and the detailed time course can help us have a deeper understanding of the development of suburban green space.

2. Methodology

2.1. Study Area

2.1.1. History of the Forest

This forest is named as "囲い山の森" in Japanese and has the same meaning as "Forest around the Mountains" in English, which is located in the area between Goko and Kanegasaku, Matsudo City, Chiba Prefecture. The total area is about 200 meters from north to south, 100 meters from east to west, about 2 hectares. In 1782 of Edo period, the squire of Kawagoe Domain began to immigrate here and develop. Until about 1960, the forest has been used as a place to obtain building materials, logs for firewood and charcoal, fallen leaves for composting, and dead branches for fuel. However, since the use of oil, plastics, and chemical fertilizers, the forest has largely lost its traditional use. A lot of small shrubs such as Aucuba japonica and broad-leaf bamboo became more and more without management for a long time. Because sunlight was seldom coming in, the trees also grew denser than before. (Figure 1)

IOP Conf. Series: Earth and Environmental Science **879** (2021) 012026 doi:10.1088/1755-1315/879/1/012026

Figure 1 The satellite map of Forest around the Mountains by using Google engine digital Earth Technology. (50 m)

2.1.2. History of the Group of the Forest around the Mountains

Volunteers come from the second "Introduction of Satoyama Volunteer Lecture", which was held by the Matsudo city government, which was focusing on graduates to set up new Satoyama activity groups. At that time, the new group was named "the group of the forest around the mountains". It has been looking for sample plots and carrying out activities since the spring of 2005. And it began to record the activities on October 10, 2007. Since then, they were starting the activity here on the first Saturday and the third Tuesday of every month. At first, the forest was dark that could not be illuminated by sunlight so many bamboos in front of the forest could not be seen at all. During this period, volunteers devoted themselves to open the activity venues centering on making the forest bright. In order to protect biodiversity here, less than half of the human activity area was built, and the rest was the living places of many animals and plants like orioles, Chinese bamboo partridge, Japanese Woodpecker, Japanese tit, and so on. (Figure 2)

IOP Conf. Series: Earth and Environmental Science **879** (2021) 012026 doi:10.1088/1755-1315/879/1/012026

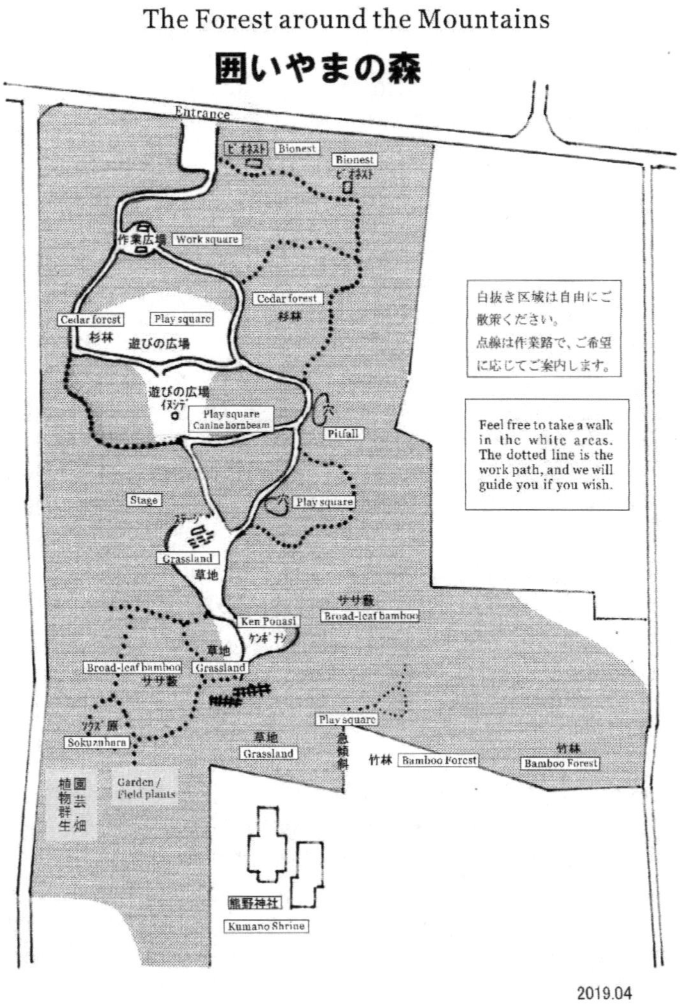

Figure 2 The study location map about the layout of the Forest around the Mountains

2.2. Methods and tools

The research procedure is divided into three steps: sorting out, analyzing, and interpreting. First of all, for "sorting out", go to the official website of the Forest around the Mountains. In the activity record, each year was divided into four parts by 3 months, like January to March will be a part. And one part contains 8 to 13 records. Each record contains activity time, date, weather, participants, nature observation, Satoyama activity, and next plan. Secondly, about analyzing, we need to extract the Satoyama activities and classify them by year. Thirdly, KH coder 3 [18] was used for text mining, and two types of maps of "co-occurrence" and "correspondence" were obtained. Finally, using the keyword in context (KWIC) concordance to select keyword searches to interpret the meaning of keywords in the figure.

3. Results and Discussion

3.1 High frequency of words for descriptive analysis

In the description of Satoyama activities, volunteers' work was extracted as the main body of analysis. Overall, KH coder3 was used to perform preprocessing, and 625 paragraphs and 1,777 sentences were confirmed after simple statistics. In addition, the total number of extracted words (the total count of all words contained in the object file) was 38,499, and the number of distinguished words that can be used was 16,148. In addition, auxiliary words, auxiliary verbs, and other general words in the article were excluded. As the words used in the analysis, 4,441 words were extracted (3,775 different words). (Table 1)

Table 1 Basic information after KHcoder3 analysis

Database stats	Tokens (in use)	38,499 (16148)	Documents Units	Sentences	1,777
	Types (in use)	4,441 (3775)		Paragraphs	625

Among them, the current frequency of the first 100 frequent words was shown in Table 2. Among the 100 frequently used words, the verb "Cut (172)" appeared most frequently. There were 20 words related to nature, and the most frequently used word was "Forest" 141 times. Most of the words for plants are "Forest (141)", "Tree (114)", "Branch (74)" and other general nouns. Among them, there are also words such as "Cedar (37)" which specifically indicated plant names. There was a noun for a natural disaster like "Hurricane (28)". What's more, there was a word for quantifiers like "Hon, a kind of Japanese quantifiers about trees (46)", words for action "Fall down" and adjectives like "High (32)" and "Withered (17)".

Secondly, there were 22 verbs that closely related to human activities, such as "Cut (172)", "Logging (138)", "Work (75)", etc.; there were 20 words indicated place, such as "Square (140)", "Passage (66)", There are 16 nouns of behavior, such as "Observation (57)", "Events (50)", "Treatment (49)"; and 5 nouns related to time like "Day (29)", there were 5 words associated with Children, such as "Swing (30)" and "Hammock (22)". There were 10 nouns in another category: the representative was "Rope (70)" as a tool; the "Music (17)" derived from the activity; the word "Danger (18)" indicated that the safety was important in the activity; "Litter (37)" had the most frequent occurrence in this category.

In addition to reflecting the biodiversity of the Satoyama environment, volunteers also need to maintain the environment after the disaster. Logging and pruning took up most of the time, for example, the word "Rope" was accompanied by logging because high trees need to be pulled down with ropes after using a saw. This can be seen in the description of the "Fall down" of trees. In addition, "Observation" of "Nature" was also very important for Children to play. This can be seen in the emergence of "Swing", "Hammock" and other simple game facilities. The most obvious thing that we need to know about a lot of nouns and verbs had a high frequency of people's actions. But we need to look at the contents shown in the figure of co-occurrences to understand the relationship between words and dig deeper into the meaning of Satoyama activities.

Table 2 Documents frequency of the top 100 words

Words	**DF (h5)**	**Words**	**DF (h5)**	**Words**	**DF (h5)**	**Words**	**DF (h5)**
Cut	172	West	35	Nature	25	*Cornus controversa*	19
Forest	141	Preparation	34	Before	25	Information	19
Square	140	Weeding	34	Surroundings	24	Enclosure	19
Logging	138	Fall down	34	Playground equipmt	24	Child	19
Tree	114	Confirmation	33	Schedule	24	Time	19
Work	75	Activities	33	Meeting	23	Cleaning	19

Words	DF (h5)	Words	DF (h5)	Words	DF (h5)	Words	DF (h5)
Branch	74	Collection	33	Think	23	Corresponding	19
Rope	70	Central	33	Mowing	23	Separation	19
Passage	66	Tamping	33	Shiitake mushroom	22	Broad-leaf bamboo	18
Observation	57	High	32	Hammock	22	Palm	18
Removal	56	Maintenance	32	Ball	22	Danger	18
Area	52	Southwest	31	Hands	22	Examination	18
Events	50	Northern side	31	Last time	22	Down trees	18
South	50	Aoki	30	Part	22	South side	18
Treatment	49	Swing	30	Pay	22	North	18
Warehouse	49	*Quercus myrsinifolia*	29	Stage	21	Name	18
Roads	49	Participation	29	Sharp corner	21	*Styrax japonicus*	17
Entrance	48	Day	29	To mow	21	Net	17
Hon	46	Hurricane	28	Implementation	21	Music	17
Nearby	42	Bamboo	28	Little	21	Below	17
Edge	39	Direction	28	Break	21	Flower	17
Arrangement	39	Month	27	To stretch	21	Hole	17
litter	37	Harvest	26	Cut down	21	Withered	17
Cedar	37	To surround	25	Climb	20	Grassland	17
Installation	37	Enjoy	25	Repair	20	Experience	17

3.2 The centrality for co-occurrences of words and analysis of Satoyama activities

The "co-occurrence network" was used when we were analyzing the data. To ensure the words appeared meaningful, the minimum number of words appearing in "term frequency" was set as 20, and no upper limit. Document frequency did not set the upper and lower limits. Only noun, SA variant noun (In Japanese grammars these words are classified as サ変 sa-hen, an abbreviation of サ行変格活用 sa-gyō henkaku katsuyō, sa-row irregular conjugation), adjective, verb, adjective, noun B, and noun C need to be checked in the selection column. The word "Top 60" was selected in "Filter edges", and its frequency was divided into 50, 100, 150, and 200, four levels. After these steps, we described a network diagram (Figure 3, Figure 4) connecting words with similar patterns (i.e., words with a strong co-occurrence degree). In the co-occurrence network diagram, the stronger the co-relationship was, the thicker the lines were. And the words with higher frequency would be described with larger circles. In the following, the words in the co-occurrence network will be explained with the original excerpt of the Satoyama volunteers' reports by using KWIC concordance to confirm how the words are used from the original text (keywords would be into bold). (Figure 3)

According to the number of connecting black lines, the vocabulary can be divided into three categories. First, there were 7 groups with weak connectivity that can express little meanings, because of the words connected with 2 or fewer black lines. Secondly, the groups connected with three to five black lines were medium connectivity groups. There were two groups in the research. Finally, only one group's words with five or more black lines connected in figure 3. At last, we chose meaningful groups with a strong middle connection to be explained.

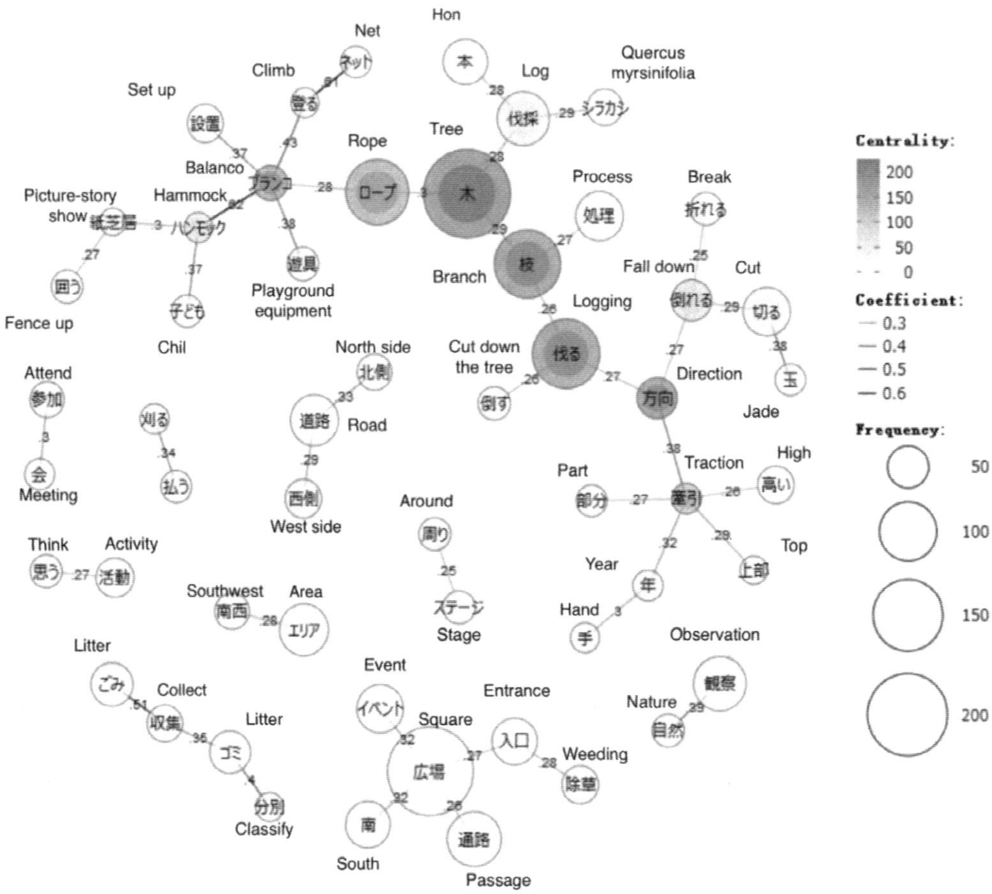

Figure 3 The central co-occurrence of reports in the Forest around the Mountains

In the group with strong connectivity, we can see that "Tree" is the center. Judging from the size of the circle, it was also the word with the most frequent occurrence, extending to the left and right sides. To the left, "Rope", "Swing" as the bridge, and were connected with the words like "Set up", "Climb", "Net", "Hammock", "Child", "Picture-story show", "Enclose", "Playground equipment". We searched "Child" as the keyword, you can see the sentences containing these words: "playing on the cable car in the central square: **Swing**, **hammock**, vertical **net**, **climbing net** on the tree, etc. adults and Children enjoy it very much."; "The **picture-story show** is very good! I was deeply impressed, by the way the Children played freely and happily." When extending to the right, it used "Branch", "Logging" and "Direction" as a bridge to connect with other words used in Satoyama to show preparation, for example, the words used to describe tree felling are: "Fall down", "Tow", "Cut", etc. When we input "Direction" in KWIC concordance, we can get some sentences like: "Together with the treatment of shrubs, we should investigate the **falling** of the tree by **towing**. The priority of operation is keeping safe." Combined with figure 4, we know that the meanings of these actions were about transforming Satoyama into a playground that can provide environmental education for Children. Volunteers tried their best to develop facilities in limited venues, including climbing, artificial making, picture-story show, etc. The establishment of multi-functions was related to the daily management of the Forest around Mountains.

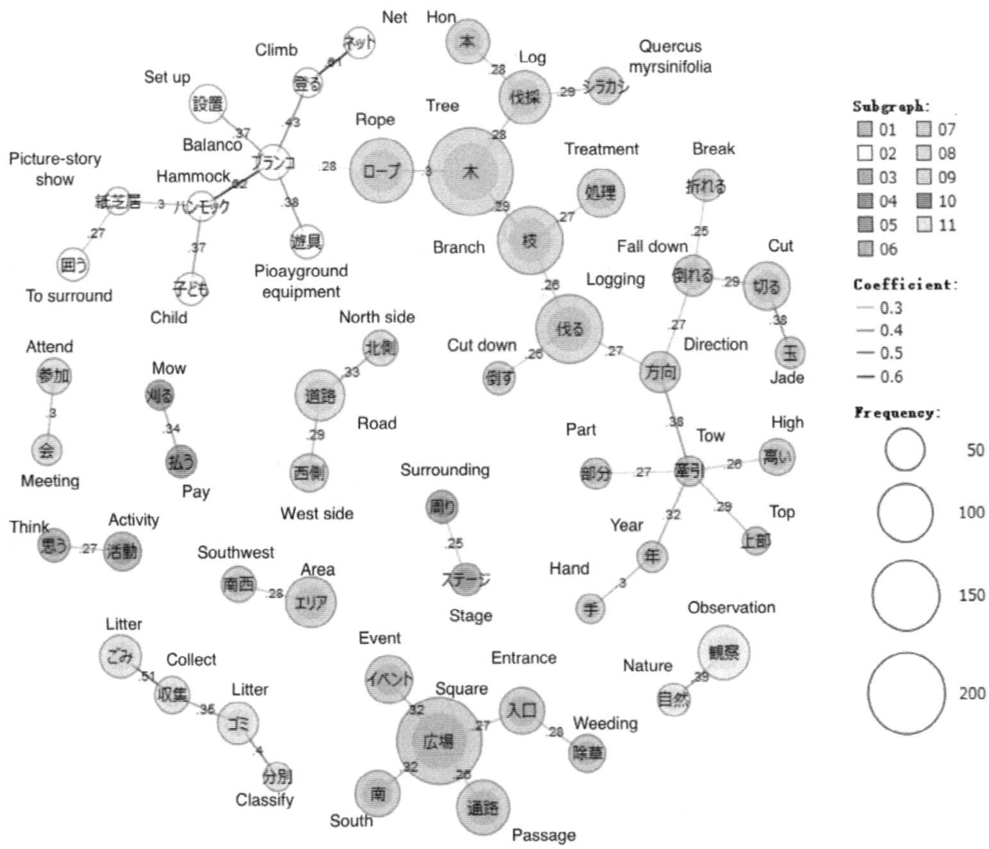

Figure 4 The co-occurrence modularity of reports in the Forest around the Mountains

Two groups with medium connectivity expressed their meanings clearly: One is "Collection", "Litter" and "Classify". With "Litter" as the keyword search, the sentence you can get is "The final **collection** and **classification** of the littler: in the whole activity area, we usually pay attention to **litter collection**." "Today, we will focus on **classification** and treatment, and we have been fighting with **litter** almost all day. The statistics of **litter collection** are shown on a separate sheet."Another was "Weeding" for the "Square" as the main body of the "Event", and a simple description of the square was given. The orientation of the squares was at the "South" side, "Entrance" or the "Passage". "Square" was regarded as the keywords to search, the relevant sentences were "along the north side of the road, **Entrance Square**, **South side Square** of the implementation of weeding." "Mowing the **North / South Square**." "The work of making **passage** in the grassland of Southwest." These can become a multi-functional place, cannot exist without the maintenance of volunteers. Nearby residents or passers-by will throw litters in. On the other hand, Satoyama is a part of nature where many weeds grow. That will bring weed cover and make Satoyma more invisible (there is no garbage can on the roadside in Japan, someone throw garbage into the weeds to cover it.) As a public place managed by people, if there is no volunteer maintenance and management, this situation will continue in a vicious circle, and the forest will gradually become a dump.

3.3 Correspondence analysis of words every year of Satoyama activities

In the KH coder, the "Correspondence analysis" instruction was used. To ensure that words appeared meaningful, the minimum frequency of words in term frequency was set to 20, and no upper limit. Document frequency does not set the upper and lower limits. Only noun, SA variant noun, adjective, verb, adjective, noun B, and noun C were checked in the selection column. For the correspondence analysis option, we chose the option with words and variables to classify words by date. The size of the circle is related to the frequency. When we followed the steps, the bubble plot with labels was expressed in figure 5. It can be seen that the green space was becoming a series of changes from the initial to the later stage after being Satoyama, in which the period from 2008 to 2013 was one part, and the period from 2016 to 2019 was the other part. Because the words of 2012, 2014, and 2015 were relatively special, which were separated into groups respectively.

The words from 2008 to 2011 were “Square”, “Fallen leaves (deciduous)”, “Log”, “Cut”, “*Quercus myrsinifolia”,* “Broad-leaf bamboo”, “High”, “Event”, “Withered”, “Bag”, “South”. These words were used as keywords to search for sentences, for example, “Because there is no leaf of **deciduous** trees, the crown is cedar, the middle layer is **Quercus**, at the bottom, you can see Aoki who started to grow again.”; “In the afternoon, a **log** (Mukunoki), which can be the pillar of the signboard.”; “Get acquainted while listening to the sounds of insects in the **southern** grassland (**Event Square**).”; “Quantity **bags** of garbage: bottles/17 **bags**, 5 cans, 5 **bags**, 40 plastics, 3 combustible rubbishes, 3 umbrellas, 15 large junk.” These sentences clearly expressed the preparation stage of the initial work, including: “cutting down trees”, “weeding”, “arranging activity square”, “clearing garbage”, etc. (Figure 5)

In contrast to the other group, the words presented from 2016 to 2019 were “Maintenance”, “Cleaning”, “Work”, “Nature”, “Harvest” and so on. We can see the sentences with these words like “**Maintenance** around and inspection of the stage and seats for the 'Music Concert’ in the forest.”; “**Cleaning** dead trees scattered by typhoons” “After that, in the spring sunshine, while observing the **nature** and carefully observing the newly opened working channels in the Southwest area, we enjoyed the luxuriant **natural** scenery of 'Forest around the Mountains'.”; “Taro potato **harvest**, garbage disposal.”

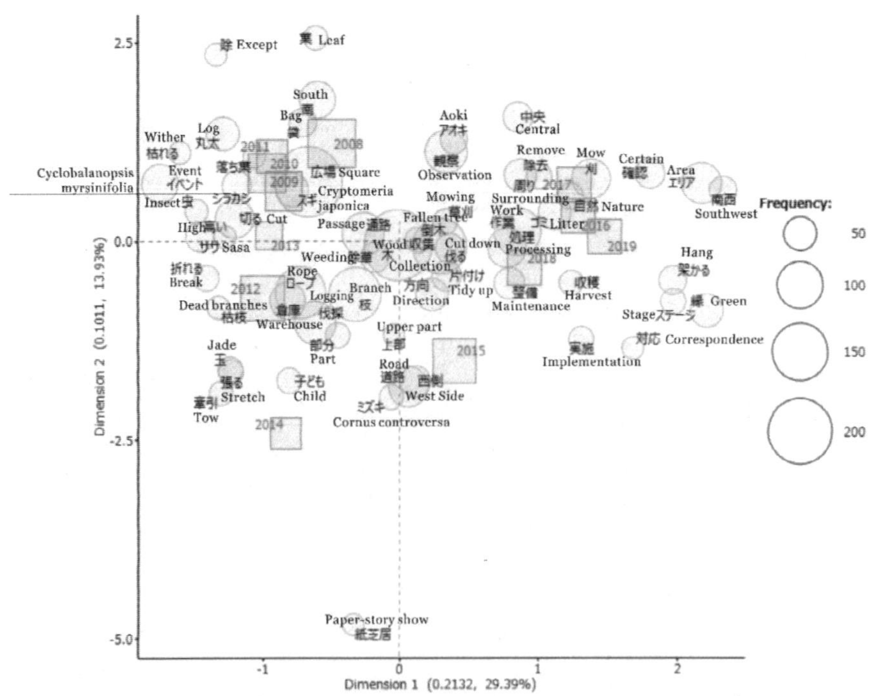

Figure 5 Correspondence analysis of Satoyama activity by year without words and frequency

The words of 2012 were related to "Dead branches", "Warehouse" and "Rope" that these are still in the preparation stage and conform to the main components between 2008 and 2013. In 2014, "Child" and other words indicated the environmental education and children's games carried out in the middle of the activity preparation. "Picture-story show" was also for children. But it was very special for Satoyama's routine activities, far away from the whole, which can be explained as Satoyama activities are not only related to nature but also the culture between human beings' heritage culture. "Westside", "road", "*Cornus controversa*", etc. appeared in 2015 like a transition phase. There were the new venues and roads as preparations to connect with activities between 2016 and 2019, including fields for harvesting food, observation of nature, concerts, and so on.

4. Conclusion

Firstly, among the most frequent 100 words, KH coder3 found that the words related to people's activities occupied the main part, indicating that the maintenance of Satoyama required a lot of labor. These activities included the construction and felling of trees in response to disasters, and the restrained development of forests to preserve natural diversity. Secondly, we learned from co-occurrence that volunteers created various facilities for children in limited venues. In order to complete these facilities, they also invested a lot of preparation labor and time. Also, their works included the treatment of garbage, maintaining a good ecological environment, turning the forest into a beautiful environment that can be used by people. Finally, according to the year classification of correspondence analysis, the overall development process of the "forest around the mountains" activity group is obtained: open up Satoyama activity site (2008-2013) - excessive children's playground (2014) - development of Satoyama (2015) - formation of diversified Satoyama activities (2016-2019).

This study is to describe the activities of Satoyama from the volunteers' perspective and find out the significance that they give to Satoyama activities. For the sustainable development and utilization of suburban green space, their activities have a great reference. For children's education, urban residents can consider carrying out environmental education in Satoyama, teaching with entertainment. And there are more places to look forward to using Satoyama diversified.

References

[1] Wang L, Omrani H, Zhao Z, Francomano D, Li K, Pijanowski B 2019 Analysis on urban densification dynamics and future modes in southeastern Wisconsin, USA *PLoS One* Volume **14**, Issue **3**, 0211964.

[2] Paul H *et al* 2020 Measuring landscape change, lot by lot: Greening activity in response to a vacant land reuse program *Landscape and Urban Planning* Volume **196**, 103729.

[3] *Ryan C et al* 2014 Biophilic design patterns: Emerging nature-based parameters for health and well-being in the built environment *Archnet-IJAR* Volume **8**, Issue **2**, **pp**62-76.

[4] *Collins R et al* 2020 A systematic map of research exploring the effect of greenspace on mental health *Landscape and Urban Planning* Volume **201**, 103823.

[5] Xiang Q, Furuya K 2020 *Physiological and Psychological Effects of Walking in Urban Parks and Its Imagery in Different Seasons in Middle-Aged and Older Adults: Evidence from Matsudo City, Japan Sustainability, MDPI* Volume **12,** Issue **10**, pp1-25.

[6] Hunter M, Askarinejad A 2015 Designer's approach for scene selection in tests of preference and restoration along a continuum of natural to manmade environments *Front Psychol.* Volume **19**, 01228.

[7] Macintyre V *et al* 2019 "I Would Never Come Here Because I've Got My Own Garden": Older Adults' Perceptions of Small Urban Green Spaces *Int. J. Environ. Res. Public Health* Volume **16,** Issue **11**, 1611994.

[8] Dewa H, Broad-leaf bambootani Y 2003 Study on management and practical use of the forest park by the citizen *Environmental Systems Research* Volume **31**, pp1-7.

[9] Shidei T 1993 learning in forests Nature protection from ecology (森に学ぶエ コロ ジか ら自然保護へ) Kaikosha, Tokyo, 241p.

[10] Hattori T and *et al*, 1995 A study on the Actual Conditions of Satoyama (Rural forests) and their Management *Humans and Nature*, No. **6**, pp1-32.
[11] Nakayama K and Matsumura N 2018 Ability Development Measured by the Fundamental Competencies for Working Persons and Effectiveness of Forest Volunteer Activities: Case Study of Forest Volunteer Activities of University Students in Niimi City, Okayama Prefecture *J. For. Plan.* Volume **22**, Issue **2**, pp63–75.
[12] Ministry of Environment, Historical transition of Satochi Satoyama *Material 4*, pp1-3. https://www.env.go.jp/nature/satoyama/conf_pu/03/mat04.pdf
[13] Nagano K, Tabata A 2017 Efforts for Conservation of Biodiversity in Satochi-satoyama and its Development in Japan and Abroad *Journal of rural planning* Volume **35**, No. **4**, pp465-468.
[14] Yoshioka A, Kadoya T, Imai J, Washitani I 2013 Overview of land-use patterns in the Japanese Archipelago using biodiversity-conscious land-use classifications and a Satoyama index *Japanese Journal of Conservation Ecology* Volume **18**, pp141-156.
[15] Kagawa T 2004 How to concern with Satoyama in future *Forest Science* Volume **42**, pp46-50.
[16] Yokoyama K, Oku H, Fukamachi K 2004 A relationship between consciousness of landscape conservation, the attributes and the Satoyama conservation activities of residents in urban fringe Satoyama area *Journal of rural planning* Volume **23**, No. suppl, pp91-96.
[17] Yoshino A 2015 Urban residents' consciousness toward Satoyama-Focusing on experience in nature play-*Environmental marketing* Working paper: 20140301.
[18] Higuchi K 2015 Quantitative Text Analysis for Social Survey-Toward Succession and Development of Content Analysis (社会調査のための計量テキスト分析—内容分析の継承と発展を目指して) printed in Japan. ISBN**978-4-7795-0803-5** C3036.

IOP Conf. Series: Earth and Environmental Science **879** (2021) 012027 doi:10.1088/1755-1315/879/1/012027

Developing green tourism to create the sustainable landscape: *evidence from Community-based Coffee Tourism (CbCT) in Puncak, Bogor, Indonesia*

Y Prihayati[1] and T O Veriasa[1]

[1] Researcher, Division of Planning and Community Development, Center for Regional Systems Analysis, Planning and Development (Crestpent), IPB University, Indonesia, 16127

E-mail: yunipri@apps.ipb.ac.id, thomas_veriasa@apps.ipb.ac.id

Abstract. Indonesia went through 3.5% growth in 2019, over the last five years in the travel and tourism sector, while the global economy increased by 2.5%. Currently, this condition encourages the creation of new jobs. Surprisingly, nowadays, there is increasing awareness of people contributing to the triple bottom line of sustainable development through initiating green tourism. Community-based Coffee Tourism (CbCT) has been running in the Puncak area for about five past years as green tourism has drawn people to come. Over time there are some significant changes in an area relating to several aspects of the three bottom line of sustainability. Through this study, we are going to 1) identify the impact of green tourism on the environment/landscape through the ecological approaches using the Social Ecological System (SES) framework and 2) to identify the contribution green tourism to the local economy using the Basic Income analysis and Benefit and Cost analysis methods. Based on the SES framework, the results show that there are some aspects from the SES framework that will be effective in the context of the case study, if there is a combination between the conservation activities and economics, which gives a direct impact on the local income. This study indicated that green tourism through CbCT has contributed to the increasing individual community income from 13.19% or IDR 2,375,000 /year/person to 76% or IDR 13,680,000/year/person. Based on B/C ratio analysis, it shows that green tourism is financially reliable to be developed at 5% and 10% of growth level, even at zero growth of the business. Hence, CbCT should be continuously developed as green tourism practices that contribute significantly to the community's income and ecological improvement to support a sustainable landscape management strategy.
Keywords: Community-based coffee tourism, Ecological approaches, Green tourism, Social-ecological system, Sustainable landscape

1. Introduction

Over the past five years, Indonesia experienced a growth of 3.6% in 2019 in the travel and tourism sector while the global economy increased by 2.5% [1]. Tourism creates new jobs, generating employment and making an ample opportunity to make money [2] but at the same time has a negative influence on the environment and social aspect.

Nowadays, the awareness of people to nature is increasing since it has promoted the quality of life through health, relaxation, and benefit from environmental services [4]. People utilize natural resources for livelihood, looking for food, and other benefits [5]. Conflicts might happen regarding the use of natural resources. There should be an excellent exit to resolve the conflicts which consider environmental aspects by conserving it [6], economic by generating some economic benefits [7] and social by encouraging the stakeholders' participation [8]. Cibulao's people, who live in the mountain area, employ natural resources as a center to create tourism attractions. Understanding how the system worked within the community is significant to ensure the sustainability of the environment and community. Cibulao's people offered green tourism named Community-based Coffee Tourism (CbCT) as their appreciation for the extraordinary nature they have.

2. Literature review

2.1. Landscape, green tourism, and sustainable tourism

Landscape is seen as a mosaic of interacting ecosystems. This definition is strengthened by [9], who clarified landscape as a heterogeneous land area composed of a cluster of interacting ecosystems. How people meet their needs in many aspects can contribute to the landscape alteration. The landscape provides many natural features to be used as tourist attractions like experience of landscape such as mountains, forests, rivers, lakes, lawns, waterfalls, etc. According to UNEP and UNWTO [10], sustainable tourism is tourism that considers economic, social, and environmental impacts, in the present and future. Green tourism is one of the sustainable tourism, which nowadays, is seen as a way to conserve the destination from a negative impact on the sustainable landscape.

2.2. Coffee, Rural tourism and tourism impacts

According to WTO [11], rural tourism has become potential because it experienced around a 6% expected growth rate, which is faster than the overall growth rate for international tourism arrivals. Coffee is one commodity that is primarily planted in rural areas and recently started to attract urban people to taste the coffee on site. The coffee tour expressed the identity economy that has grown exponentially in the past decades [12]. Gunn [13] argued that tourism would have a more social impact on a community than any other form of economic development because tourism will bring people outside as visitors or tourists and also as investors to develop the tourism industry into the host communities. He added that when the number of visitors outpaces the number of locals, there will be some consequences like some social response, both negative and positive. Tourism can create significant pressure on local resources that somehow may already be in short supply, such as biodiversity, energy, land, forest, and water. Tourism also exposes an opportunity in a change of local identity of local value.

2.3. Concept of the Social-Ecological Systems (SES)

Great concern on tourism is the negative impacts that might exist on the natural environment [2, 3, 14] and the local community [13, 15]. In some areas, tourism highly relies on landscape character and landscape quality and resulted in high-pressure conditions of the natural landscape, and the surrounding environment like exploitation due to the tourism development. According to Kirchhoff *et al.* [16] and Levin *et al.*, [17] there is a system called social-ecological to understand synergies between tourism and landscape because this viewpoint suggests that tourism and landscape are an integrated system. A Social-ecological system (SES) is a set of critical resources (natural, socio-economic, and cultural) whose flow and use are regulated by a combination of ecological and social systems [18].

3. Research methodology

3.1. Study area

The study was conducted at Cibulao Hamlet, for one year through 2018. Geographically, Cibulao Hamlet is located in Tugu Utara Village, Cisarua District, Bogor Regency, West java (Puncak Area) inside a tea plantation area and is surrounded by a forest area managed by Perhutani (state-owned forestry enterprise). Tea plantation has been established since the Dutch colonial era (1811), known as the Ciliwung tea plantation. Cibulao Hamlet has existed since 1992, with an initial population of 12 families. Currently, the population has increased to 139 families[19].

People at Cibulao Hamlet all work as labor at PT Sumber Sari Bumi Pakuan (SSBP) as infrastructure workers and tea pickers. They work for about 6 hours and spend the rest of their time farming and taking care of the small area of the coffee plantation next to their neighborhood. Endowment with a great forest, people started planting coffee in 1998 to take advantage of previously eroded forests. This effort is deemed illegal by the authority, while many people cut trees to make the land for vegetable farming. Until 2009, this attempt continued, some people who are members of the Cibulao Hijau Forest Farmers Group (FFG) have managed to plant coffee with an intercropping system in the forest and have obtained permission from the Perhutani officials with a cooperation certificate. A large area of the forest nowadays has been covered by the coffee plot. Besides, it has attracted people to have Cibulao Hamlet as a tourism destination. Nowadays, the Cibulao community has developed the CbCT (Community-based Coffee Tourism).

3.2. Data collection

The study uses a case study method, which allows researchers to understand the contextual conditions with in-depth analysis of the case [20]. The focus of SESs was the tourism systems of Cibulao Hamlet. Data collection was carried out through primary sources using field observation and interviews. Secondary data collected included annual reports and minutes of meetings. The targeted respondents in the semi-structured interview were tourism stakeholders, including tourism entrepreneurs, NGOs, Community-Based Organizations (CBO), and residents.

Selected participants were determined by the researchers through the close relationship during the community facilitation over the years. The interview process continues until the required information is reached [21]. All interviews were conducted in Cibulao Hamlet and Tugu Utara village for a year. Interview guides have been prepared in advance to establish that the interview process takes place with a focus on the relevant topic [21,22].

3.3. Data Analysis

3.3.1. The impact of tourism on environment/landscape Analysis

The SES framework, known as Ostrom framework was initially modeled for the application of common-pool resource management circumstances. Eight variables will be indicated by the SES framework to understand the relationship among them. The eight variables referred to as first-tier variables: resource systems (RS), resource units (RU), governance systems (GS), actors (A), interactions (I), outcomes (O), social, economic, and political settings (S), and related ecosystems (ECO).[23-25] .

Table 1 The second-tier variables of the SES framework for Cibulao Hamlet tourism systems

Social, economic and political settings (S) S1 Population S1.1. Size S2 Economic Development S2.1 Livelihood options S3.Markets S3.1. Tourism market S3.2. Competition S4. Political situation	Actors or stakeholders (A) A1. Number of relevant actors A1.1. group size A1.2. Type of actors A2. Socioeconomic aspects A3. Background of tourism management A4. entrepreneurship	Resource Unit (RU) RU1 Resource unit mobility RU2 Growth rate RU3 Interaction among resource units RU4 Economic value RU5 Number of units

Related ecosystems (ECO) ECO 1 Geography ECO 2 Climate ECO 3 Biodiversity	Governance Systems (GS) GS1 Government Organisations GS2 Nongovernment Organizations GS3 Property-rights system GS4 Monitoring GS4.1. Conflict resolution	Resource Systems (RS) RS1 Sector RS2 Clarity of system boundaries RS3 Size of resource boundaries RS4Human-constructed facilities RS4.1. Attractions RS4.2. Infrastructure
Outcomes (O) O1 Social performance measures O1.1. Social capital O2.1. Heritage conservation O2 Ecological performance measures O2.1 forest management O2.2. wildlife management O3 Economic performance measures O3.1 Sustainable livelihoods O3.2. Infrastructure O3.3. Capacity building	Interactions among actors (I) I1 Tourism product I2 Conflicts I3 Investment activities I3.1. Natural and cultural resources I3.2. Infrastructure I4 Self-organizing activities I5 Networking activities I6 Monitoring and evaluation	

Source: Data 2020 and modification.

In the framework there are resource units (RU) as part of a resource system (RS). Resource systems will set conditions for focal action situations (with interactions among actors to yield outcomes). Resource units are inputs for focal action situations. Vice versa, focal action situations will indirectly influence resource systems and resource units. Another variable is governance systems (GS) to define and set rules for an Actors (A) and determine procedures and set conditions for Focal Action situations related ecosystems (ECO), and in social, economic and political settings (S). Actors (A) are an essensial part of participating in focal action situations in the forms of interactions and outcomes. In the centre of this framework, there are action situations where the actors interact and yield outcomes. They all are illustrated below. This study used this Ostrom framework [24] because it is known as a tool to analyze the SES of the community since it equipped organizational clarity, and hierarchical structure.

After setting the eight first-tier variables, researcher constructed the second-tier variables of the social-ecological system framework for tourism systems. Each first-tier variable has some second–tier variables reasonable to the context of Cibulao Hamlet study area. The second-tier data could be seen in Table 1.

3.3.2. The contribution of green tourism to the local economy

The contribution of green tourism is analyzed by calculating CbCT income obtained by community groups and individuals. The community group income was calculated by using the basic income equation according to Angelsen and Lund [26] to analyze income from self-employment, such as the tourism business. The equation used is as follows:

$$I = \sum_{i=1}^{n} p_i y_i - \sum_{j=1}^{m} q_j v_j \tag{1}$$

Where, I—Nett CbCT income; n – types of CbCT packages; p_i— CbCT package price; y_i— a quantity of CbCT package sales; m— Types of CbCT packages Input; q_j— Price of CbCT packages input; v_j— number of tourists.

Furthermore, individual income from CbCT is analyzed by calculating revenue and profits from several works in CbCT earned by each community for a year. To analyze the CbCT contribution to increasing individual income, the primary income of the Cibulao community (tea plantation laborers) was used as a baseline, i.e., IDR 1,500,000 per month or IDR 18,000,000 per year per person [19].

The benefit and cost analysis is used to analyze the financial feasibility of CbCT and determine whether a CbCT will result in a net profit [27]. The benefit and cost analysis, i.e., Nett Present Value (NPV), Internal Rate Return (IRR), Nett Benefits/Costs Ratio (Nett B/C Ratio), is used with several sensitivity variables to analyze the CbCT business dynamics. Sensitivity analysis is done by comparing the growth of business turnover, and operational costs at 0%, 5% and 10% level per year for five years of business economic life.

4. Analysis and problem solving

4.1. The impact of CbCT on environment/landscape through the ecological approaches using the SES framework

4.1.1. Social, Political, Economic and Ecological Contexts of the CbCT (S)

Cibulao Hamlet is located in Puncak Area with an altitude ranging from 650 to 1200 m (ECO1). Cibulao has 06.68340°S Latitude and 106.92804°E Longitude with an elevation of 921. The altitudinal is included in the tropical area. The annual rainfall is about 2178 mm (ECO2). The geographical and climatic in the tropic climate has produced diverse flora and fauna. Cibulao is located next to the Ciliwung Upstream, the most important river in the Jabodetabek area (ECO3) [19].

Cibulao Hamlet has a population of 138 families (S1.1). Most people work as labor in tea plantations as infrastructure workers and tea pickers. Few people spend their leisure time after work by farming in the forest (planting coffee and vegetables). With the development and high appreciation of the quality of Cibulao coffee, some members of the community focused on the coffee business and offered tourism from the coffee attraction (S2.1). These economic activities then developed become Community-based Coffee tourism and have specific markets from diverse societies ranging from student, researcher, bikers, hikers, nature lovers coming from various regions and countries (S3.1). Since the Puncak area is considered a tourism destinations, so there are many tourism areas that existed in Puncak which potentially become a competitor for CbCT of Cibulao. Fortunately, Cibulao has specific attractions that are unique and authentic. It is proven from the risen numbers of tourists who visited Cibulao CbCT over the years starting 2017 to 2019, after winning competition of Specialty Coffee (S3.2).

Some potential resources have driven communities to elaborate it become tourism destinations. Cibulao Hamlet, located next to the forest, belongs to Perhutani and is laid on the buffer zone of Taman Nasional Gunung Gede Pangrango (RS1). Before 2009, people in communities cannot freely access the forest to open the farm for vegetables or the coffee field. After years of attempting to propose cooperation to Perhutani, finally, this institution let the communities (Cibulao Hijau FFG) access the 4 hectares of the forest to open the coffee field, and now it has to enlarge become 34 hectares (RS2 and RS3). There is some specific area they plotted to organize various of the tourism they offer. For the bike park area, they have a bike track along 16.700 meters. For coffee tourism, they have 36 hectares area of agroforestry coffee. For hiking and camping, they explore the next forest around the neighborhood within the buffer zone conservation forest of Taman Nasional Gunung Gede

Pangrango, and part of the forest belongs to Perhutani (RS2 and RS3). For each kind of tourism, they have human-constructed boundaries. People in the communities built bike tracks within the forest belong to Perhutani and regularly maintain the area for various types of bike mountain class from the beginning to advance class (RS4.2). For coffee tourism, they have a small coffee café next to the coffee field, some gazebo, and camping area (RS4.2). For the attractions, the communities offer the bike attraction, coffee tourism, camping, and hiking. For the specific of coffee tourism, they offered some attractions such as harvesting the coffee, hiking, camping, brewing and cupping (tasting the coffee), and learning about the coffee plants (RS4.1).

Coffee is the RU since it is the focus of the destination. The coffee field has dynamically increased since the rising awareness from various institutions that have same interest in developing coffee while conserving the forest (RU1). We have no specific number about the growth rate of coffee, yet there are significant increasing numbers of coffee planted, and the coverage of area grew coffee plants (RU2).

The interaction between resource units can be negligible, since coffee plants stay in the same place (RU3). The total net income of the community group from coffee tourism is IDR 22,800,000 per year, this calculation does not include the amount of profits from the sales of adoption tree seedlings from nurseries developed by the group (RU4). The area planted with coffee has been expanded to 36 hectares and people can harvest 5 tons of coffee beans (RU5).

There are some government institutions have an interest in coffee elaboration. Perhutani is one of the government institutions that allow the community access to the forest through scheme of social forestry [(*Pengakuan dan Perlindungan Kemitraan Kehutanan* (KULIN-KK)]. It created Cibulao Hijau FFG to ease the assistance. Another institution is the Agriculture agency that created a farmer group (*Kelompok Tani*-POKTAN) to let the people access the aid from this institution (GS1). There are some non-government organizations have the same interest in coffee namely Rumah Kopi Ranin, The Consortium of Save Puncak and some individual coffee entrepreneurs (GS2). For property-right system, there is cooperation between Perhutani and Cibulao Hijau FFG to an agreement to manage the area to be grown by coffee through the scheme of social forestry (GS3). This scheme is prevailed not the private interest but for the communities, especially the Cibulao Hijau FFG. Since there is agreement on land use from Perhutani, there is responsibility for the community (Cibulao Hijau FFG) to use the forest wisely with the agroforestry system while conserving the forest (GS4). There are potential conflicts that took place since there are two identities for one community to access the help from government institutions, namely the Cibulao Hijau FFG and the POKTAN group, which are each of them are precisely the same person/members (GS4.1). Nevertheless, it finally depends on the member itself to manage the two groups that are the same members since it is merely the administrative matters.

The main actor (A1) is the Cibulao Hijau FFG. Since 2009, there is an agreement between Perhutani and Cibulao Hijau FFG to use the forest area as a *Pengelolaan Hutan Bersama Masyarakat* (PHBM) Program (Joint the Community Forest Management) for the coffee field[15]. The area has 4 hectares large to grow coffee plants and has now expanded to become 36 hectares. The members of Cibulao Hijau FFG had 12 people at the beginning and now become 38 members in 2019 (A1.1). There are at least 7 types of actors that have a role in coffee cultivating development in the social forestry area in Cibulao Hamlet. (A1.2). There are internal agencies and external agencies based on their respective role. The first role manages land access mechanism (Ministry of Environment and Forestry, Perum Perhutani, LMDH (forest village community agency) Puncak Lestari); the second role is in community empowerment and development (The Consortium of Save Puncak, P4W LPPM IPB, Rumah Kopi Ranin) and the third role is in promotion of the coffee product (Bogor Regency Agriculture and Plantation Service (A1.2). Cibulao Hijau FFG is considered as the main actor that has a strong spirit of conserving the forest and has no capital to expand their coffee field. The opportunity from Perhutani, and capacity building from external agencies have helped the community to escalate their capacity and capability (A2). From the beginning, only three people actively ran the tourism in Cibulao Hamlet. After the establishment of Cibulao Hijau FFG, the management of tourism (especially in coffee tourism) has been handled by Cibulao Hijau FFG members (nowadays have 38 members) (A3). The initial intention to maintain the forest from degradation has been developed to become something that has an economic value from the environment that they have preserved. Planting coffee to maintain the forest from erosion has generated high-quality coffee products which are popular and attract tourists to visit and explore coffee tourism (A4).

There is some potential of tourism that can be developed by the communities. Before developing coffee tourism, people had established a bike park area for mountain bikers for years and had loyal customers to visit regularly every week (I1). After Cibulao coffee became famous, they develop coffee tourism (I1). Bestowed by the beautiful scene and forest, the community also elaborated similar kinds of tourism that explore the beauty of nature, such as hiking, camping, and tea walk (I1). Through the experience this community has, there are some conflicts that arose both internally and externally. Cooperative members have established democratic, and accountable deliberative processes, and both internal and external conflicts are resolved in the assembly (I2). Cibulao Hijau FFG has actively built and expanded its tourism infrastructure, such as pathway, gazebo, toilet, café. (I3.2). Some of them are attempted self-sufficient, and others by joining with other institutions to build a café. (I3.2). Cibulao Hijau FFG keeps growing the new coffee plants in the Perhutani area with assistance from Bogor Regency Agriculture and Plantation Service and P4W LPPM IPB for the seeds and knowledge about growing coffee and making a good harvest (I3.1). The leader of Cibulao Hijau FFG, along with their members, has participated actively in many programs about increasing coffee quality products and their skill in serving coffee (I4). Frequently, cooperative leaders establish commercial networks with local café in Bogor and Puncak and other markets from other cities (I5). Cibulao Hijau FFG has some buyers for their coffee from café in Bogor, so that they have to maintain their quality. Some café like Rumah Kopi Ranin is one of the customers that always gives observation about the quality management of Cibulao coffee. Besides, Cibulao Hijau FFG members regularly attended the training to escalate their capability to run the coffee business. Cibulao Hijau FFG has its regular meeting with the community to engage their cooperation and discuss problems in the field and evaluate their business progress (I6).

The social performance of their activity is high. Community participation in CBCT (labor) has 52 people (34 men, 18 women) who were involved in CbCT. The frequency of group meetings increased to 2 times a month from previously irregular and unscheduled. Forest monitoring/patrol activities are carried out by community groups once a week, following by maintenance activities for coffee tourism trails and adoption trees that have been planted (O1).The 3800 multi-purpose tree species (MPTS) trees have been planted by CBCT activities. The 12.7 hectares area have been enriched and improved by the community through CbCT. Zero illegal activity in the surrounding Cibulao area in 2019 (O2). The economic performance of the CbCT has provided financial benefits that vary to an individual. However, higher economic benefits are felt more collectively, especially in group income (22,800,000 per year), employment (52 people), increased skills (tour guides, tourism organizers, etc.), and strengthening social capital in the relationship among members of CbCT (O3).

4.1.2. The contribution of CbCT to the local economy
The results showed 3,040 local tourists and 260 international tourists coming to Cibulao in groups with 15-120 members in 2019. Following the equation (1), the total net income of the community group is IDR 22,800,000 per year, this calculation does not include the amount

of profits from the sales of adoption tree seedlings from nurseries developed by the group. A total of 52 people were involved in CbCT activities and benefited economically from various types of supporting activities at CbCT such as tour guides, cooking laborers, photographers, etc. (Table 2).

In Table 3, we can see that there is an increasing income from 13.19% or IDR 2,375,000/year/person to 76% or IDR 13,680,000/year/person. Of the total types of CbCT activities that have involved in the community, tour guides have become the most significant revenue contributors for the individuals involved. Whereas the smallest revenue derived from the income of homestays rental. In the Cibulao context, the types of works in CbCT that require unique expertise such as tour guides, coffee roasted bean sellers, photographers, maintenance workers of edutour tracks, and adoption trees have contributed directly to a significant increase in individual income. Even, in work as tour guides, the community has earned more than their primary income (tea plantation laborers).

Table 2 Contribution of CbCT to the community group income in 2019 (IDR/year)

Production Inputs				Production Sales				Net Income of CbCT (I)
m	v_j	q_j (IDR 000)	q_jv_j (IDR 000)	n	y_i	p_i (IDR 000)	p_iy_i (IDR 000)	p_iy_i - q_jv_j (IDR 000)
1. Entrance Ticket	3,800	10	38,000	1 Day	3,040	180	547,200	
				2 Days 1 Night	760	390	296,400	
2. Tour Guide								
1 Day Trip	3,040	30	91,200					
2 Days 1 Night Trip	760	60	45,600					
3. Meals								
1 Day Trip	3,040	50	152,000					
2 Days 1 Night Trip	760	155	117,000					
4. Coffee Cupping	3,800	40	152,000					
5. First Aids	3,800	2	7,600					
6. Edutour Tracks maintenance	3,800	3	11,400					
7. Tree Adoption	3,800	40	152,000					
8. Homestay	760	25	19,000					
9. Fish pond rental	760	15	11,400					
10. Photograph	760	30	22,800					
Total			820,800				843,600	22,800

Source: Primary data, 2019.
n — types of CbCT packages ; p_i— CbCT package price; y_i— quantity of CbCT package sales; m— Types of CbCT packages Input; q_j— Price of CbCT packages input; v_j— number of tourist

Table 3 Direct contribution of CbCT the individual community revenue (in IDR/year)

No	Types of Income	Number of people	Amount	Mean	Shared (%)
1.	Tour Guides (man)	10	136,800,000	13,680,000	34.94
2.	Cooking laborers (women)	15	38,000,000	2,533,333	06.47
3.	Maintenance workers of Edutour tracks and adoption trees.	5	30,400,000	6,080,000	15.53
4.	Homestay rental	8	19,000,000	2,375,000	06.07
5.	Fish pond rental	3	11,400,000	3,800,000	09.71
6.	Photographers	5	22,800,000	4,560,000	11.65
7.	Roasted bean sale profits	6	36,731,750	6,121,958	15.64
	Total	52	295,131,750	39,150,291	100

Source: Primary Data, 2019

In analyzing the financial feasibility of CbCT, we used the latest historical financial data in 2019. The initial investment is IDR 50,000,000. A total of IDR 45,000.0000 is used for working capital to develop a nursery garden, buy a roasting machine, and coffee brewing tools package as well as IDR 5.000.000 for CbCT operational costs. The discount rate is 7%, and the economic life of the investment is estimated to be five years, following with the national regulation on the interest rate for small business loans (*Kredit Usaha Rakyat*) in 2019. The national tax rate for small businesses is 1% about the total turnover in a year. The CbCT turnover in 2019 is IDR 843,600,000, and tourism activity costs are estimated at IDR 820,800,000. The detailed information of CbCT business turnover and costs can be seen in Table 2.

The results of B/C ratio analysis have shown that CbCT is financially feasible in all levels of sensitivity analysis (Table 4). In moderate analysis at 5% growth level, during its economic life (5 years), CbCT still provides the business profit, i.e. NPV=IDR 11.014.645; the interest rate return (14.23%) of CbCT can cover capital interest (7%) at the end of the economic life of the business; the CbCT business is profitable due to the benefits are much higher than the costs (Net B/C ratio=1.22).

Moreover, at the zero level of growth, which means the turnover and costs are stable every year following the data 2019, CbCT is still beneficial to be developed by the Cibulao community.

Table 4 B/C ratio analysis of CbCT

Analysis Tools	Variable of Sensitivity Analysis[a]			Criteria	Results
	0%	5%	10%		
NPV[b]	1,957,702	11,014,645	20,967,426	NPV=Positive, Accepted	CbCT is financially feasible at 5% and 10% growth, even at zero growth
IRR[b]	8.43%	14.23%	18.59%	IRR > prevailing discount rate[c], Accepted	
Nett B/C Ratio[b]	1.04	1.22	1.42	Nett B/C Ratio >1, Accepted	

[a] The sensitivity variable used is the growth of turnover and operational costs per year for five years.
[b] Nett Present Value (NPV), Internal Rate Return (IRR), Nett Benefits/Costs Ratio (Nett B/C Ratio).
[c] Discount Rate 7%

4.2. The CbCT in the context of economic improvement, social capital and sustainable landscape management.

Through this study, we identified that CbCT (Community based Coffee Tourism) in Cibulao Hamlet has their own governance system like self-enforcement management to support the sustainable landscape. It shows that there is an implicit code of conduct the community upholds from the main actor of the Cibulao Hijau FFG to keep the spirit in maintaining their environment as a vision in developing many activities. This self-enforcement becomes a distinctive role in keeping their measurements and development on their right track. Besides they have mutually beneficial cooperation between farmers and Perhutani to guarantee their sustainable coffee field by agroforestry system to maintain the forest from devastation that is become their essence for CbCT (GS1, GS3). The community has their facilitator to help in monitoring and evaluating the effort to conserve the landscape while developing green tourism. In economic, CbCT was able to increase individual people's income from 13.1 9% -76%. CbCT also has good business financial feasibility at various levels of sensitivity of the B/C analysis. The economic value of the CbCT will incentivize the local actors to make efforts to conserve the resources.[28, 29]

In social aspect, most of the people in Cibulao are relatives and neighbors so they have close relationship each other. In Cibulao, they have local rules of collective-choice which represent the democratic value and equity among them to hold the coexistence and cohesion of the local community network. The system prevails and remains resilient through many challenges the community face for years. This become social capital when they have system to establish the social resilience.

The existence of environmental, economic and social benefit in Cibulao hamlet is a proven that through CbCT, the sustainable landscape will be achieved with some requirements. This study found that the main menaces to the sustainability of the landscape of forest/buffer zone next to Cibulao Hamlet, in the natural system are heavy rainfall resulted in landslide and land erosion. Related to human system the threats are 1) the breaking up of cooperation (agreement) between community and the authority (Perhutani) that will restrict the forest accessibility; 2) the extended growth of the interest for tourist services; 3) the ecological impact from other tourism destination and attraction development around the village that has no relation with quality environment improvement; and 4) illegal tree cutting.

5. Conclusions

Based on the study, through the SES framework, the researcher found some positive impacts of tourism on the Cibulao landscape/environment. There are good interactions between actors and resources and the local governance system as well, which ended up in good social and economic performance as an outcomes. The economic performance represented with increasing income is an crucial element to attain the sustainable use of resources guided by effective local governance rule. The social performance is reflected by the capability of the community to make a decision at the local level by establishing collective rules that allow an equitable distribution of resources, which support cohesion to the system. This social performance created the community-based governance that will establish their social resilience to pursue their sustainable tourism and sustainable landscape/environment since their sense of appropriation of the resources of Cibulao well-formed. This study implies the central government for establish the general rules for people next to the forest to access the forest and extract the resources through the social forestry scheme.

There is a recommendation that CbCT should be continuously developed as green tourism practices that contribute significantly to the community's income and ecological improvement to support a sustainable landscape management strategy. Local governance systems succeed in managing their tourism system because there is the main actor who leads the vision of tourism management to protect the forest and environment. The cooperation with other organizations and institutions that have the same interest in guarding the forest, has a good impact on monitoring and evaluation of the tourism system to help protect the forest from devastation (including illegal tree cutting and natural disasters).

Acknowledgments
Thanks to Cibulao people who have given contribution and important information for this study. Kang Yono, Kang Jumpono, and others who cannot be mentioned one by one.

References

[1] World Travel and Tourism Council (WTTC) 2020 *Travel and tourism: Global economic impact & trends 2020* Retrieved at https://wttcweb.on.uat.co/Research/Economic-Impact
[2] Khan A, Bibi A, Lorenzo A, Lyu J and Babar ZU 2020 Tourism and development in developing economies: A policy implication perspective *Sustainability* **12** 1618 pp 1-19
[3] Río-Vázquez MESd, Rodríguez-Rad CJ and Revilla-Camacho MÁ 2019 Relevance of social, economic, and environmental impacts on residents' satisfaction with the public administration of tourism *Sustainability* **11** 6380 pp 1-15
[4] World Economic Forum 2019 *The travel & tourism competitiveness report 2019* Retrieved at http://www3.weforum.org/docs/WEF_TTCR_2019.pdf
[5] Alston, L.J.; Libecap, G.D.; Mueller, B. Land reform policies, the sources of violent conflict and implications for deforestation in the Brazilian Amazon. *J. Environ. Econ. Manag.* **2000**, *39*, 162–188.
[6] Crowder, L.; Norse, E. Essential ecological insights for marine ecosystem-based management and marine spatial planning. *Mar. Policy* **2008** 32 pp 772–778
[7] Weninger, Q.; Waters, J.R. Economic benefits of management reform in the northern Gulf of Mexico reef fish fishery. *J. Environ. Econ. Manag.* **2003**, *46*, 207–230
[8] Lebel, L.; Anderies, J.M.; Campbell, B.; Folke, C.; Hatfield-Dodds, S.; Hughes, T.P.; Wilson, J. Governance and the capacity to manage resilience in regional social-ecological systems. *Ecol. Soc.* **2006**, *11*, 19.
[9] Forman RTT and Godron M. 1986. *Landscape ecology* (New York: John Wiley)
[10] The United Nations Environment Programme (UNEP) and The World Tourism Organization (UNWTO) 2012 *Tourism in the green economy*
[11] World Trade Organization (WTO) 2009 Press Release. URL: (November 23, 2009) Retrieved at https://www.unwto.org/international-tourism-growth-continues-to-outpace-the-economy
[12] Comaroff J and Comaroff J 2009 *Ethnicity inc* (Chicago: University of Chicago Press)
[13] Gunn CA 1988 *Tourism planning* - second edition (New York: Tylor &. Francis)
[14] Saarinen J 2006 Traditions of sustainability in tourism studies. *Ann. Tour. Res.* **33** pp 1121–1140
[15] McCombes, L.; Vanclay, F.; Evers, Y. Putting social impact assessment to the test as a method for implementing responsible tourism practice. *Environ. Impact Assess. Rev.* **2015**, *55*, 156–168
[16] Kirchhoff T, Brand FS, Hoheisel D, Grimm V 2010 The one-sidedness and cultural bias of the resilience approach. *Gaia* **19** pp 25–31
[17] Levin S, Xepapadeas T, Crépin AS, Norberg J, deZeeuw A, Folke C, Hughes T, Arrow K, Barrett S, Daily G, Ehrlich P 2012 Social-ecological systems as complex adaptive systems: Modeling and policy implications *Environ. Dev. Econ* **18** pp 111–132
[18] Redman C, Grove MJ, Kuby L 2004 Integrating social science into the long term ecological research (LTER) network: Social dimensions of ecological change and ecological dimensions of social change. *Ecosystems* **7** pp 161–171
[19] Veriasa TO, Rustiadi E, Kinseng RA 2020 The impact of joint community forest management (PHBM) on local community income in upstream of Ciliwung Watershed, Bogor Regency-Indonesia. *IOP Conf. Ser.: Earth Environ. Sci* **556** 012009 pp 1-12
[20] Creswell JW 2016 *Research design: Pendekatan metode kualitatif, kuantitatif dan campuran* (Yogyakarta: Penerbit Pustaka Pelajar)
[21] Padgett DK 2008 *Qualitative methods in social work research* (2nd ed.) (Thousand Oaks, CA: Sage)
[22] Ostrom E 2007 A diagnostic approach for going beyond panaceas *Proceedings of the National Academy of Sciences* **104** 39 pp 15181-15187
[23] McGinnis MD and Ostrom E 2014 Social-ecological system framework: initial changes and continuing challenges *Ecology and Society* **19** 2 p 30.
[24] Ostrom E 2009 A general framework for analyzing sustainability of social-ecological systems *Science* **325** 5939 pp 419–422

[25] Basurto X, Gelcich S, Ostrom E 2013 The social-ecological system framework as a knowledge classificatory system for benthic small-scale fisheries *Global Environmental Change* **23** 6 pp 1366–1380
[26] Angelsen A and Lund JF 2011 *Measuring livelihoods and environmental dependence: Methods for research and fieldwork designing the household questionnaire* eds A Angelsen, H O Larsen, J F Lund, C Smith-Hall, S Wunder (Washington DC: Earthscan).
[27] Berk J and DeMarzo P 2014 *Corporate finance* - third edition (Boston: Pearson Education Inc)
[28] Gutiérrez NL, Hilborn R, Defeo O 2011 Leadership, social capital and incentives promote successful fisheries *Nature* **470** 7334 pp 386–389
[29] Libosada CM 2009 Business or leisure? Economic development and resource protection-concepts and practices in sustainable ecotourism. *Ocean Coast. Manag* **52** pp 390–394

Mapping the village forest of Pattaneteang through drone participatory mapping

Firnawati[1], Kaswanto[2] and S Sjaf[3]
[1]Graduate Student, Department of Landscape Architecture, Faculty of Agriculture, IPB University
[2]Lecturer, Department of Landscape Architecture, Faculty of Agriculture, IPB University
[3]Lecturer, Department of Communication Science and Community Development, Faculty of Human Ecology, IPB University

Email: firnawati05@gmail.com

Abstract. The Government of Bantaeng Regency, South Sulawesi Province is one of the first districts established policies to support community-based forest management and to adopt the Ministry of Forestry (Kemenhut) policy on Community Forests (CF – Hutan Kemasyarakatan HKM) and Village Forests (VF – Hutan Desa HD. There is one VF located in Pattaneteang, which has characteristics and great resource potential. In the VF management process, several problems arise in conflicts between villages over boundaries that increase with reducing the landscape as a resource. The utilization of a geographic information system (GIS) in landscape management is expected to provide complete information for managing the form of VF areas integrated with each other. This study aims to produce a precise map of the VF area based on the image map of the drone results. The latest methods are used to produce accurate delineation of the region through drone participatory mapping (DPM) technology involving several related parties. DPM is a group-based research method that gives participants the freedom to make discussions on topics with minimal intervention from researchers. Simply put, it is an activity in which villagers choose to location referencing and other features related to the topic given on the map. The participatory monitoring method was carried out through a Focus Group Discussion (FGD). The results showed that Pattaneteang Village which, has an area previously about 1303.81 ha after the village was carried out, participatory mapping reducing became 1107.79 ha, then participatory mapping was carried out based on the result of the drone, Pattanetang area was obtained to be 1109.92. While the extensive VF based on Decree of the Minister of Forestry No.55 / Menhut-II / 2010 dated January 21, 2010, Pattaneteang VF forest area was 339 ha after drone based-participatory mapping (DPM) to 429,71 ha related to three types of forest, namely Core Forests, Utilization Forests, and Dispute Forests are covering 48.15 ha, 198.12 ha. and 181.43 ha, respectively. The results of this study can be used as a reference in planning sustainable village forest management areas.
Keywords: geographic information system, precision maps, village forests

1. Introduction

Decentralization has revived a new bottom-up approach to forest governance in Southeast Asia. In Indonesia, social forestry and customary rights (customary forests) are among the leading forms of forest management [1]. Forests are a source of landscape services that are important for the sustainability of ecosystems. Well, preserved forests are very important to biodiversity conservation and provisioning landscape services [2]. Tropical forests provide ecosystem services directly or indirectly derived from ecosystem functions [3] [4].

The Ministry of Forestry has provided access to village communities to manage forest areas legally, one of which is the village forest (VF-Hutan Desa, HD) scheme with management access rights for 35 years [5]. Village forest development can contribute to the development of livelihood security for people dependent on forest resources. VF management, in principle, is how to involve the community around the forest to benefit from the existence of the forest without changing the function and status of the forest area.

In recent years, the community forestry model has become a popular approach to forest management globally [6]. Community Forest Management (CF-*Hutan Kemasyrakatan*, HKM) is increasingly recognized as a potentially effective means of protecting forests [7]. Community forestry (CF) development such as village forest (VF), has been considered as a successful national strategy to improve rural livelihoods and environmental protection, where local people protect and manage forest resources to increase forest cover and improve forest conditions to provide forest products such as fuelwood, wood, animal feed, leaf litter and other ecosystem services (ES) [8].

Participatory mapping is a type of public participation that includes creating and using spatial information for various purposes. Participatory mapping differs significantly in design and implementation, for example, *public participatory* GIS (PPGIS), *participatory* GIS (PGIS), and *Volunteered Geographic Information* (VGI) [9]. Community assessments of land use are important for planning and allocation, especially in areas where conflicts over limited resources have occurred historically or are likely to arise [10].

The application of participatory mapping in land use planning can be regarded as a planning support system (PSS), emphasizing the participatory components of the system. As a form of public participation, participatory surveying and mapping have many ideal goals for public participation, such as increasing trust, reducing conflict, providing information and education to the public, incorporating public values into decision-making, and improving the quality and legitimacy of decision-making [11]. Specifically, this study aims to map the village forest area (VF) of Pattaneteang Village based on drone images to produce a precise map of the VF area and check the VF spatial distribution based on the participatory mapping.

2. Data and Methods

The research was carried out in Pattaneteang Village, Tompobulu District, Bulukumba Regency, South Sulawesi, Indonesia. Geographically, Pattaneteang Village is located between 119° 57' 53" and 120° 2' 9" East Longitude (EL) and 5° 22' 31" and 5° 24' 40" South Latitude (SL). This village is 155 Km from the capital city of South Sulawesi Province. Pattaneteang Village is located in the ecosystem at the foot of the Bawakaraeng-Lompobattang Mountains, with the topography at an altitude between 650-1760 masl. The research data consisted of primary data and secondary data. This research uses primary data in the form of drone image maps, existing data on village land use, and types of commodities developed by the village community. Secondary data is in the form of village administrative boundary data set by the government and village administrative boundary data resulting from the participation of villagers.

Four main methodological steps were taken to carry out participatory mapping of the Pattaneteang Village Forest. It begins with taking drone images of Pattaneteang Village based on the village administrative map. Unmanned Aerial Vehicle (UAV) technology or drones are currently gaining popularity. In addition to increased sensor capabilities, drones also produce aerial images that can be used to map accurately, easily understand to managing environmental landscapes [12]. Second, making an orthophoto map by combining drone photos to form a mosaic image of the entire study area. Third, the implementation of participatory mapping based on drone image maps. Fourth, delineating the boundaries of the Pattaneteang VF area as well as comparing village administrative boundaries based on the results of drone based-participatory mapping, administrative boundaries of participatory village maps, and administrative boundaries by the government.

2.1 Taking Aerial Photographs using Drones

To take aerial photos using drone DJI Mavic Pro to get the village aerial photo data Pattaneteang. Preparation begins with a flight path planning, which includes the polygon design of the research area based on a village map with a buffer as far as 50 meters from the village administrative boundary, determining the size of the photos that are overlapping (sidelap and overlap), planning starting location take-off, landing, and fly high. The applications used are Pix4D Capture and Ctrl + Dji. The flight path created includes all research areas, which are divided into 35 polygons. Sidelap and overlap are used by 70%, the take-off and landing locations are in open areas with a flying height of 200 meters above ground level at the take-off locations.

Determination of the Ground Control Point (GCP) for geometry correction by installing markers at 30 points at the research location, measured by GPS Garmin, this correction is made if the coordinates of the drone do not match the truth, but if it is appropriate, then no correction is made. Camera calibration and drone GPS calibration are carried out with the DJI Go 4 application so that the resulting aerial photos can match the correct coordinates. Calibration is done before drone take-off.

2.2 Combining Drone Photo results

The raw images of each flight were compiled using the photogrammetric software Agisoft Metashape. The principle is that each photo has coordinated and overlaps with another photo next to it. The ortho mosaic results are then combined to form an image mosaic of the entire study area (orthophoto map) using the Arc Gis 10.5 software.

2.3 Drone Participatory Mapping(DPM)

After generating an orthophoto map of Pattaneteang Village based on the drone results, the map was then printed for the purpose of participatory mapping (PM) based on drone images. In implementing participatory mapping (PM) is combined with Focus Group Discussion (FGD). Use a combination of PM and FGD as these methods are well established in the field of development studies and planning. Using this method is an efficient way to know a group perspective. Participatory discussions make an important contribution to social knowledge and learning about forest management models [13]. The FGDs, in turn, helps to generate more descriptions of the topics at issue as well as a deeper understanding of local history, perceptions, and meanings of the region itself [14].

The Participatory Mapping (PM) based on orthophoto maps, is printed at a scale of 1: 10,000 (size A0). Mapping was carried out together with 15 community members representing community specialists as informants (Forestry extension, village government and NGO members, farmers, and community leaders). In community-based forest management, organizations outside the community play an important role in achieving various social and ecological goals [15]. The PM's main objective is to accurately delineate the Bantaeng District Village Forest Area's boundaries and extent.

The PM first introduces the topic and collects background information about the informant (such as age, family details, main livelihood, education, personal knowledge of the landscape), and then locates it on the aerial photo map. First, informants mark their houses on the map, and then indicators for each different landscape service are mapped one by one using colored paper for each marked indicator. Allow the information provider to draw as many locations as possible for each indicator. Each indicator is fully mapped with descriptive statements to add relevant information, such as which plants are grown, how to use medicinal plants, and why certain places are considered beautiful or important. PM to determine boundaries area and area sizes were carried out together with selected community groups who then delineated village boundaries and village forest boundaries, determining boundaries marked with attributes that could be confirmed in the field. The colored paper markers are manually copied onto the A3 paper copy of the aerial photo map. At the end of each interview, the original image map and marked data were also taken for verification.

2.4 Overlay of Drones Participatory Mapping (DPM) Results

DPM results are then delineated using Arc GIS 10.5 and then overlaid with the base map used are village administration map by the government and the participatory map that has been carried out by the village government. The aim is to compare the mapping results using drones with the available maps.

3. Result and Discussion

3.1 *Orthophoto Map of Pattaneteang Village*

An orthophoto is a reproduction of a photo that has been corrected for errors by plane tilt, relief, and lens distortion [16]. An orthophoto is a result of the process of aerial photogrammetry. An orthophoto is a type of aerial photograph that has been corrected for geometric errors using DEM data and GCP data to be used for mapping purposes without any scale inconsistencies throughout the scope of the photo [17]. Orthophoto can be created after the Dense Point Clouds, Mesh, and DEM stages are complete. Orthophoto maps (drone images) are used to obtain high accuracy and make it easier to determine different land cover boundaries. Sjaf (2015) research results reveal that used drones can make it easier for rural communities to prepare careful planning with a substantial drone innovations database. The Orthophoto map used is the result of photographs from drones combined to form a mosaic image of the entire study area. The principle is each photo has coordinated and overlaps with another photo next to it, processed using Agisoft Metashape (figure 1).

3.2 *Drone Participative Map Pattaeneteang Village*

Pattaneteang Village was a division of Labbo Village in 1980 [18]. Participatory mapping (PM) is a type of public participation that includes creating and using spatial information for various purposes [19]. Based on government administration, Pattaneteang Village has an area of 1309.9 ha (Figure 2). After participatory mapping (PM) was carried out by the village government in 2018 through tracking the Pattaneteang Village Mapping Team, it was found that the area of Pattanetang Village was 1107.79 ha. Participatory mapping by the Pattaneteang Village government was carried out for six months. On the Bantaeng Regency map, the administrative area of Pattaneteang Village boundaries is in the Labbo Village area of 196.9 ha. Develop local participation as a strategy to protect and maintain the biological and environmental resources and historical value of cultural landscapes [20]. Then, after participatory mapping based on the drone results, the final area of Pattanetang Village was 1109.92 ha. This area is not much different from the participatory results that have been implemented by the village government. The Pattaneteang Village Government made the river flow as the ecological boundary between villages, and these rivers are the Bialo River and the Kulepang River. PM may include objective assessments (for example, the distribution of species based on community science or mapping paths for entertainment) and/or subjective data (for example, the perceived quality of the landscape or the location of intangible environmental services)[21]. Community assessment of land use is very important for planning and distribution, especially in areas where conflicts of limited resources have occurred or may occur in history can arise [10].

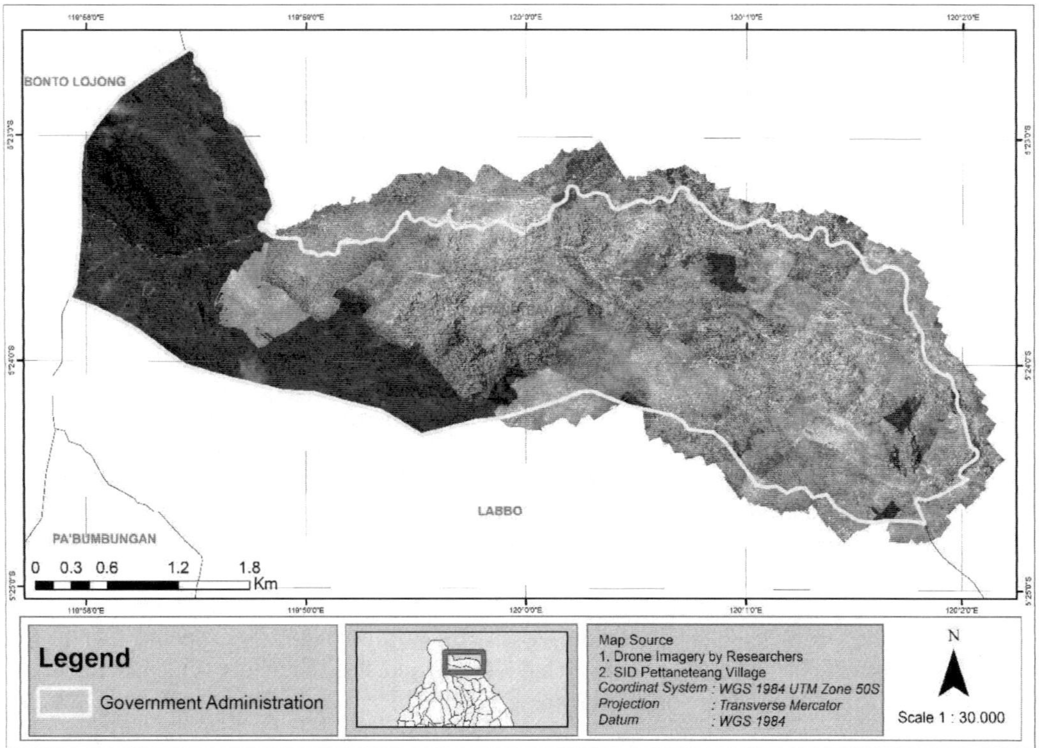

Figure 1 Orthophoto map of Pattaneteang Village, Tompobulu District, Bantaeng Regency

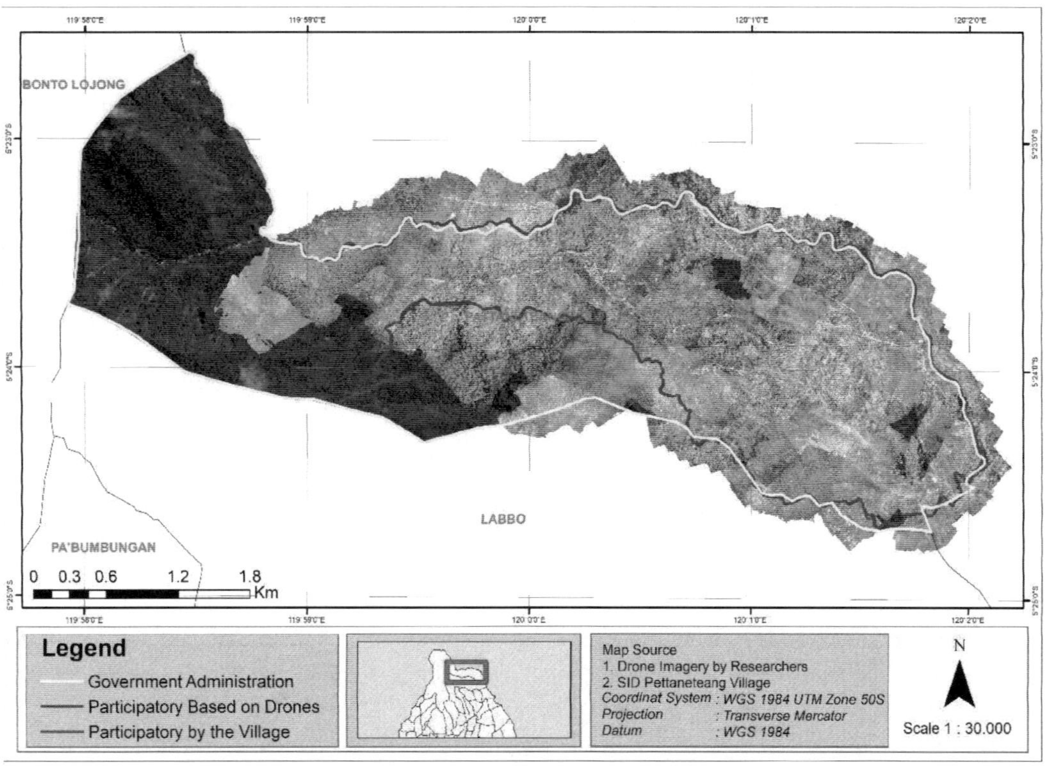

Figure 2 Administrative boundaries map of Pattaneteang Village

IOP Conf. Series: Earth and Environmental Science **879** (2021) 012028 doi:10.1088/1755-1315/879/1/012028

3.3 *Pattaneteang Village Forest Area by Drone Participatory Map*

Drone based-participatory mapping was also carried out to determine the boundaries and area of the Village Forest area managed by the Pattaneteang community. Village Forest mechanism is a form of forest management. Collective forest management has a positive impact on the provision of environmental services that are beneficial to local, regional, national, and international users [8].

Based on the results of DPM, the area of Pattaneteang Village Forest (VF) is 427.70 ha, which is categorized into three types and forest status according to the results of the PM that has been carried out by the village government (Figure 3). In the final stage, the VF area is then compared with the area from each of the three different data sources (Table 1). Village Forest area previously determined by the government in the Decree of the Minister of Forestry No.55 / Menhut-II / 2010 dated January 21, 2010, Pattaneteang Village covering an area of 339 ha. After carrying out PM by the village government, the area of VF is 429,71 ha, which is divided into Core Village Forest covering 48,38 ha, Utilization Village Forest covering an area of 199,05 ha, and Administrative Dispute Village Forest covering an area of 182,28 ha. However, after carrying out DPM, the total area of VF Pattaneteang was 427,70 ha, consisting of Core Village Forest covering an area of 48,15 ha, Utilization Village Forest covering an area of 198,12 ha, and Administrative Disputing Village Forest covering an area of 181,43 ha.

Utilization VF is the forest that is intensively managed by the community to cultivate coffee and cloves. Core Village Forest is part of the Lompobattang Protected Forest area, which the community only uses for their daily needs in non-timber forest products (NTFP). Not only as a producer of wood and forest NTFPs, it also functions as a provider of environmental services, reducing natural disasters while protecting nature [22]. Meanwhile, administrative dispute VF is an area that also has the status of Lompobattang Protected Forest. The area is administratively part of the administration of Pattaneteang Village, Tompobulu District, Bantaeng Regency, but the population in this area are residents of Borong Rappoa Village, Kindang District, Bulukumba Regency. Participatory methods allow decision-makers to understand the views and issues of local communities and determine strategies to support agriculture, and public participation can also be used as a landscape planning tool [20].

Table 1 Comparison of the area and type of Pattaneteang Village Forest based on data sources

No	Data Sources	Area and Type of Forest			Total area (Ha)	Difference Area: DPM (%)
		Core Village Forest (Ha)	Utilization Village Forest (Ha)	Village Forest Administrative Dispute (Ha)		
1	Government	-	-	-	339	20,74
2	Participatory mapping by Village	48,38	199,05	182,28	430	0,47
3	Drone Participatory Mapping	48,15	198,12	181,43	428	

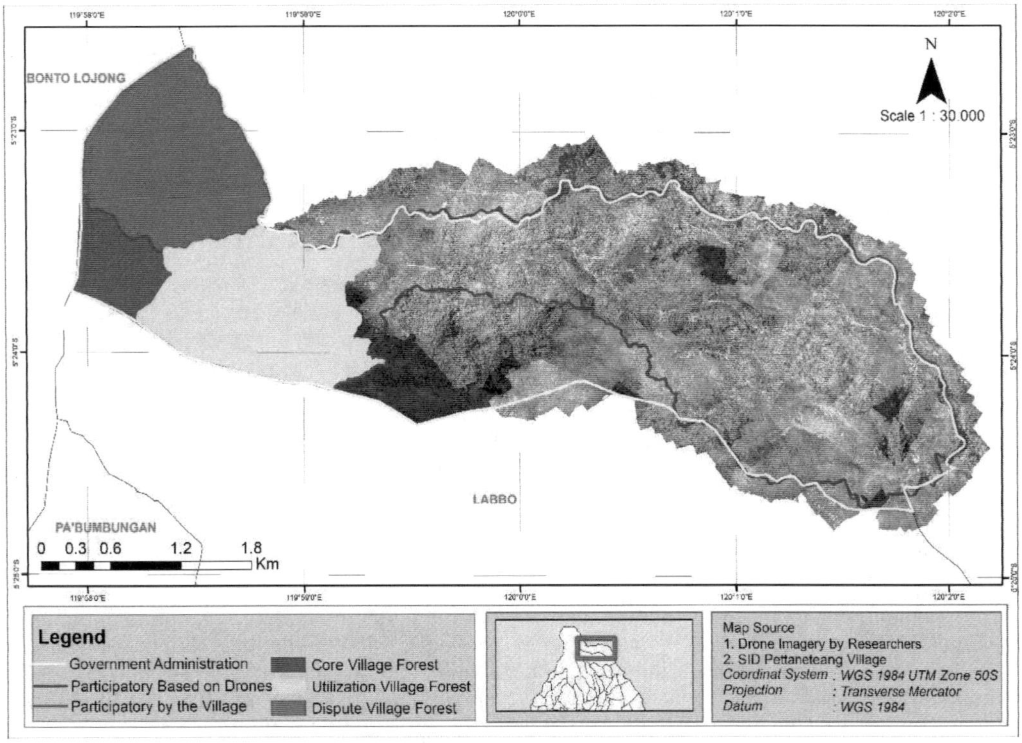

Figure 3 Map of the Pattaneteang Village forest area

4. Conclusion

The results showed that through drone based-participatory mapping, accurate data would be obtained regarding the extent and boundaries of an area. The area obtained in DPM is not much different from the PM that has been carried out by the village government. However, the important difference between the use of participatory and tracking methods is that this will take a very long time. The study also shows that a participatory approach, integrated with precise satellite imagery (drone imagery), is an appropriate approach to involve local communities in discussions about forest landscape management. A participatory approach can increase efforts to integrate perceived public values into land and resource management decision-making processes, especially in data-poor rural areas.

References

[1] Fisher M R, Workman T, Mulyana A, Moeliono M, Yuliani E L, Colfer C J P and Adam U E F B 2017 Striving for PAR excellence in land use planning: Multi-stakeholder collaboration on customary forest recognition in Bulukumba, South Sulawesi *Land use policy* 0–1

[2] de Freitas D S, de Oliveira T E and de Oliveira J M 2019 Sustainability in the Brazilian Pampa biome: A composite index to integrate beef production, social equity, and ecosystem conservation *Ecol. Indic.* **98** 317–26

[3] Delgado-Aguilar M J, Konold W and Schmitt C B 2017 Community mapping of ecosystem services in tropical rainforest of Ecuador *Ecol. Indic.* **73** 460–71

[4] Delgado-Aguilar M J, Hinojosa L and Schmitt C B 2019 Combining remote sensing techniques and participatory mapping to understand the relations between forest degradation and ecosystem services in a tropical rainforest *Appl. Geogr.* **104** 65–74

[5] Batari A, Jusuf Y and Sahide M A K 2017 Analisis tingkat keaktifan pengelolaan hutan Desa Labbo *J. Hutan dan Masyarakat.* **9** 54

[6] Purnomo H, Sekar Arum G, Achdiawan R, and Irawati R H 2012 Rights and Wellbeing: an analytical approach to global case comparison of community forestry *J. Sustain. Dev.* **5** 35–48

[7] Wood A, Tolera M, Snell M, O'Hara P and Hailu A 2019 Community forest management (CFM) in south-west Ethiopia: Maintaining forests, biodiversity and carbon stocks to support wild coffee conservation

Glob. Environ. Chang. **59** 101980
[8] Paudyal K, Baral H, Burkhard B, Bhandari S P and Keenan R J 2015 Participatory assessment and mapping of ecosystem services in a data-poor region: Case study of community-managed forests in central Nepal *Ecosyst. Serv.* **13** 81–92
[9] Brown G and Kyttä M 2018 Key issues and priorities in participatory mapping: Toward integration or increased specialization *Appl. Geogr.* **95** 1–8
[10] Pearson A L, Rzotkiewicz A, Mwita E, Lopez MC, Zwickle A and Richardson R B 2017 Participatory mapping of environmental resources: A comparison of a Tanzanian pastoral community over time *Land use policy* **69** 259–65
[11] Brown G, Sanders S and Reed P 2018 Using public participatory mapping to inform general land use planning and zoning *Landsc. Urban Plan.* **177** 64–74
[12] Hamylton S M, Morris R H, Carvalho R C, Roder N, Barlow P, Mills K and Wang L 2020 Evaluating techniques for mapping island vegetation from unmanned aerial vehicle (UAV) images: Pixel classification, visual interpretation and machine learning approaches *Int. J. Appl. Earth Obs. Geoinf.* **89** 102085
[13] Marques M, Oliveira M and Borges J G 2020 An approach to assess actors preferences and social learning to enhance participatory forest management planning *Trees, For. People* **2** 100026
[14] Ramirez-Gomez S O I, Torres-Vitolas C A, Schreckenberg K, Honzák M, Cruz-Garcia G S, Willcock S, Palacios E, Pérez-Miñana E, Verweij P A and Poppy G M 2015 Analysis of ecosystem services provision in the Colombian Amazon using participatory research and mapping techniques *Ecosyst. Serv.* **13** 93–107
[15] Friedman R S, Guerrero A M, McAllister R R J, Rhodes J R, Santika T, Budiharta S, Indrawan T, Hutabarat J A, Kusworo A, Yogaswara H, Meijaard E, St. John F A V, Struebig M J and Wilson K A 2020 Beyond the community in participatory forest management: A governance network perspective *Land use policy* **97** 104738
[16] Husna S, Subiyanto S and ah H 2016 Penggunaan parameter orientasi eksternal (Eo) untuk optimalisasi digital triangulasi fotogrametri untuk keperluan ortofoto *J. Geod. Undip* **5** 178–87
[17] Anurogo W, Lubis M Z, Khoirunnisa H, Pamungkas D S, Hanafi A, Rizki F, Surya G, Situmorang A D L, Timbang D, Sihombing P N, Lukitasari C A and Dewanti N A 2017 A simple aerial photogrammetric mapping system overview and image acquisition using unmanned aerial vehicles (UAVs) *J. Appl. Geospatial Inf.* **1** 11–8
[18] Pemerintah Desa Pattaneteang 2019 *Pattaneteang Desa Untuk Indonesia 2019* vol 1, ed M Rahim (Bantaeng)
[19] Brown G, Rhodes J and Dade M 2018 An evaluation of participatory mapping methods to assess urban park benefits *Landsc. Urban Plan.* **178** 18–31
[20] Gullino P, Devecchi M and Larcher F 2018 How can different stakeholders contribute to rural landscape planning policy? The case study of Pralormo municipality (Italy) *J. Rural Stud.* **57** 99–109
[21] De Vreese R, Leys M, Fontaine C M and Dendoncker N 2016 Social mapping of perceived ecosystem services supply-The role of social landscape metrics and social hotspots for integrated ecosystem services assessment, landscape planning and management *Ecol. Indic.* **66** 517–33
[22] Bösch M, Elsasser P, Franz K, Lorenz M, Moning C, Olschewski R, Rödl A, Schneider H, Schröppel B and Weller P 2018 Forest ecosystem services in rural areas of Germany: Insights from the national TEEB study *Ecosyst. Serv.* **31** 77–83

IOP Conf. Series: Earth and Environmental Science **879** (2021) 012029 doi:10.1088/1755-1315/879/1/012029

Landscape design of Ciliwung Riverbank based on ecological aspect in Pejaten Timur Settlement Area, South Jakarta

A N Amelia[1] and F N H Utami[1]

[1] Department of Landscape Architecture, Faculty of Agriculture, IPB University, Bogor, Indonesia

Email: ayunovitaamelia15@gmail.com

Abstract. Jakarta is a city that has a reasonably dense population and continues to grow every year. The rapid increase in the number of residents has caused a lack of harmony between the need of land for housing and existing land availability. This condition has led to a change in the function of riverbank in the area of settlements which happened in Pejaten Timur, South Jakarta. The existing settlements area on the riverbank caused environmental problems such as river width change and pollution. These conditions caused this area is often flooded every rainy season. Therefore, this research was conducted to design the Ciliwung Riverbank and restoring the function of its riverbank which was supposed to be a green open space. Green open space is designed in the form of an urban community park that can be used as a water catchment area, improving the ecological function of the riverbanks, and as a place to relax for the citizens around the site. The design of the park is carried out by the design process method, these design process included preparation, inventory, analysis, synthesis, and site design. In the analysis section, the writer determines the potential and site constraints, and then some appropriate site development plans can be produced. The applied concept is called "Eco-relaxing riverside park", which is an urban park on the riverbank that is environmentally friendly and has the primary function of a place to relax for visitors. The result of the urban community park design is expected to increase the aesthetic and functional value of the Ciliwung Riverbank in Pejaten Timur, South Jakarta. The final result of this research is the site plan, overall perspective, spot perspective, visible section images, detailed drawings, and explanations of the images.
Keywords: Ciliwung Riverbank, riverbank settlement, urban green space

1. Introduction

The increase in population in Jakarta every year causes the problem of decreasing land for settlements. Based on Statistics Indonesia (BPS) [1], the total population of Greater Jakarta in 2016 was 10.28 million people and increased to 10.37 million people in 2017. The lack of land for settlements area creates other problems in the form of land use that is not supposed to be used as residential lands, such as the emergence of settlements built on riverbanks. Based on Statistics Indonesia (BPS) [2], there are 267 settlements located along the riverbanks in Greater Jakarta. The riverbanks are ideally used as conservation areas that can prevent, maintain and protect river bodies from landslides and erosion due to natural disasters or human behavior, and as water catchment areas when water flows exceed normal [3].

The river is one of the environmental components which is very influential for human life. The river has an essential function as a water source, fisheries, livestock, agriculture, and business in urban areas. Ciliwung River is one of the 13 rivers flowing in Jakarta. This river has the most significant impact because it passes through the middle of Greater Jakarta and crosses many settlement areas, dense housing, and slum areas. Pejaten Timur is one of the sub-districts whose settlements are crossed by the Ciliwung River. One of the consequences of settlements along the river is causing river pollution. The Ciliwung River is entirely polluted due to the garbage and household waste dumped into the river and drainage around the site. This problem is also one of the triggers for flood when the existing waste has clogged the drainage and caused water to overflow when the rainfall is high.

Pejaten Timur, South Jakarta, is one of the areas that was flooded in February 2018. One of the government's efforts to prevent the flood was doing the Ciliwung River normalization project. This project aims to restore the condition of the river width by expecting the river water holding capacity. The riverbanks of the Ciliwung River in the residential area of Pejaten Timur, which is used as the location of this research, has been acquired by the Jakarta National Land Agency for the implementation of the Ciliwung River normalization project since 2016.

Meanwhile, the area of green open space in Greater Jakarta only amounts to 9.98% of the total area. Based on Jakarta's Department of Parks and Cemeteries [4], this number is still far from the 30% owned by the Greater Jakarta. Therefore, the implementation of the Ciliwung River normalization project needs to be balanced with some efforts to increase green open space by building some parks. The park, which will be built in the settlement area of Pejaten Timur, proposes to restore the function of the riverbanks as green open space is such a kind of an ecological-based community park. This community park is expected to create a sense of togetherness, bonding between residents, restore the buffer area between the river and land ecosystems, so that river functions and human activities are not disturbed, and improve environmental sustainability.

The objectives of this study are as follows:

1. identifying and analyzing the characteristics of the landscape and use of the riverbank of the Ciliwung River in the settlement area of Pejaten Timur, South Jakarta,
2. analyzing the opportunities and constraints on the riverbank of the Ciliwung River in the settlement area of Pejaten Timur, South Jakarta,
3. designing an ecological-based urban community park on the riverbank of the Ciliwung River in the settlement area of Pejaten Timur, South Jakarta.

2. Methods

2.1. Research location and time

This research was conducted on the Ciliwung River riverbank, it is located in residents of community unit number 03, Pejaten Timur settlement area, South Jakarta (Figure 1). This research took place from July to October 2019.

2.2. Research methods

This research was conducted with a design process, according to Booth [5], including preparation, site inventory, analysis, synthesis, and site design

2.2.1. Preparation

At this stage, the problem formulation is carried out. To determine the objectives for landscape design, conducting literature studies, compiling research proposals, arranging permits for related parties, and gathering initial information from various sources or related agencies.

2.2.2. Site inventory
Data collection activities were carried out by direct surveys (collecting primary data) and literature studies (collecting secondary data). Related data can be obtained through field observations, interviews and questionnaires, and literature studies.

2.2.3. Analysis
The analysis was carried out to process the data that had been obtained from the site inventory process. The collected data were analyzed in four ways, including qualitative descriptive analysis, quantitative descriptive analysis, social analysis, and spatial analysis.

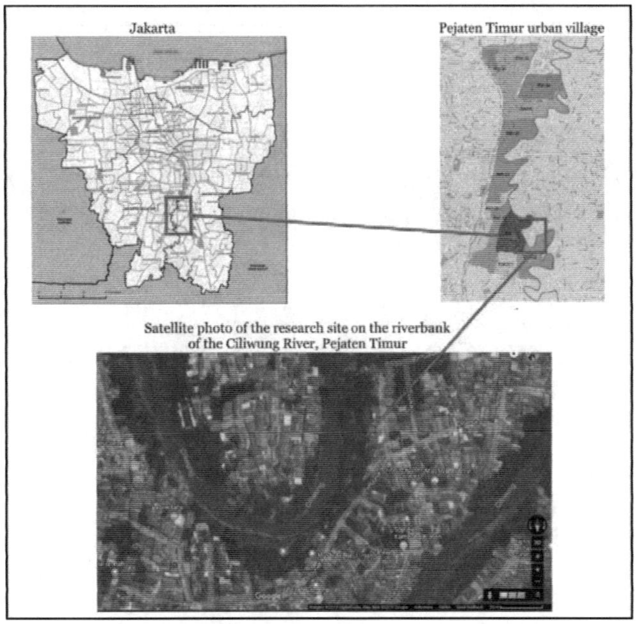

Figure 1 Research location on the map

2.2.4. Synthesis
Analysis results are grouped based on potential and constraints. The existing potential is maintained and developed, while the existing constraints are sought for solutions. These solutions then produce a site synthesis that is used as the basis for designing the landscape. This process results in site zoning, specifically determining the appropriate space for its utilization purposes, which is then carried out by the development of basic concepts and design concepts. The output of this synthesis process is in the form of the development concepts and a block plan.

2.2.5. Design
This stage is using design principles, which include unity, balance, rhythm, emphasis, harmony, scale, and proportion. The development concept that has been created in the synthesis process can be processed into a block plan, which is then used to make design development. The design development is carried out in various aspects under the ecological concept. This stage produces output in design drawings, site plan, section drawings, detailed engineering drawings, planting plans, overall perspectives, and spot perspectives.

3. Results and Discussions

3.1. Site inventory

The site is located in an open area on the riverbank of the Ciliwung River, which is administratively located in RW 03 and RW 09, Pejaten Timur subdistrict, Pasar Minggu, South Jakarta. The site has an area of 1.0159 ha, which is directly next to the north by the Ciliwung River, to the south by Jalan Gunuk Raya, to the east with the RW 3 settlement area, and to the west with the RW 09 settlement area. The site is an empty land along the Ciliwung river that is flooded almost every year. The densely populated settlements around the site are also the cause of the large amount of waste dumped into the Ciliwung River. Therefore, it is increasing the possibility of flooding due to overflowing river water. The area around the site is quite busy with the activities of residents who live around the riverbank. The image of the site inventory can be seen in Figure 2.

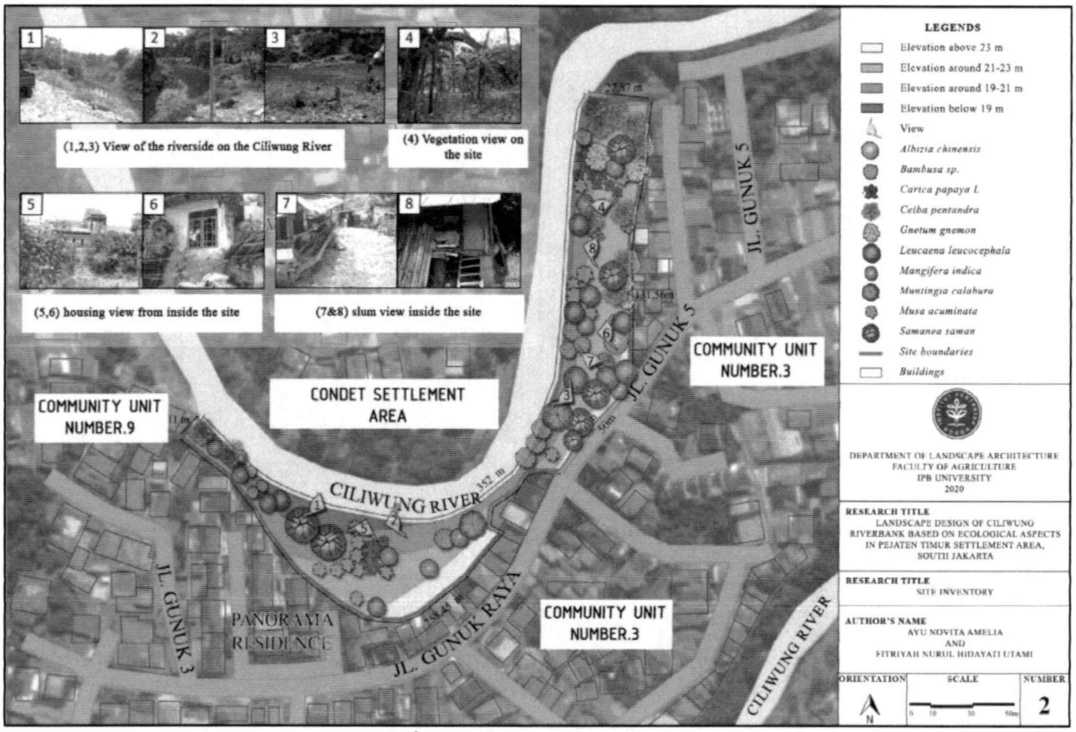

Figure 2 Site inventory

3.2. Site analysis and synthesis

This site is an empty land and large area. It is located in the middle of a settlement with no green open space, so it has the potential to be developed. Based on the Regulation of the Minister of Public Works Number 5 of 2008 [6], according to the area of land available, the public space that can be developed on the site is a sub-district community park with a minimum area of 9000 m^2 or 0.9 ha.

The soil condition on the site is alluvial soil with better productivity than the topsoil because of deposition. Based on the soil erodibility class, according to USDA-SCS [7], alluvial soil types are included in the soil category which has high erosion sensitivity. This condition is not possible for medium to large scale landscape construction. The park can be built by carrying out various site engineering, especially the special treatment of soil. Consideration of the landscape construction load needs to be paid attention to maintain soil stability.

IOP Conf. Series: Earth and Environmental Science **879** (2021) 012029 doi:10.1088/1755-1315/879/1/012029

The entire area has an altitude of 19-23 m asl, with a slope ranging from 8.1-22.2%. According to the Minister of Agriculture Decree number. 837/Kpts/Um/11/1980 [8], the slope on the site enters the rather steep category. To create a community park with an ideal slope, site engineering needs to be done in the form of grading or cut and fill at several areas, so that the site has a flatter topographic condition and has a safe height from flooding, and does not endanger park visitors when carrying out various activities in it. Barrier vegetations are planted to add more safety in the leveled area, and bio-retaining walls are provided to support the ecological concept.

In general, the climatic conditions on the site are included in the climatic conditions of the City of South Jakarta, which is a tropical climate. According to the Meteorological, Climatological, and Geophysical Agency: BMKG [9], the climatic condition on the site has an average air temperature of 28.76°C per year. There are shady areas under the trees and also open space areas exposed to direct sunlight. The average rainfall in the last year in the research period was 134mm per year, it is related to the potential for flooding in Jakarta, which usually increases at the end of the year.

Access into the site can be reached through Jalan Gunuk Raya and Jalan Gunuk V, the main roads along the research site area. The facilities and utilities on the site are minimal, without pedestrian walk or lighting. Facilities and utilities still need to be added according to the needs of visitors and activities carried out. The choice of materials for facilities and utilities that are strong and waterproof is also critical because it is still potentially flooded. However, the materials used must also be environmentally friendly to support the ecological concept.

Designing vegetation arrangement is very important on the site to increase the aesthetic value of the area. The selection of vegetation on the site must also meet the criteria, because the site is still potentially flooded. These criteria include vegetation that is resistant to excess water conditions and has strong roots so that it is not easily wiped away when exposed to flood water flows.

The lack of public green space in this residential area is one reason why residents feel essential to build or present a community park on this site. So that residents can carry out the desired activities in this community park. Based on the questionnaire obtained, the most results regarding the activities that people want are to relax and exercise. However, other activity spaces were also added with an appropriate division of area. Based on the questionnaire, the community walked to the site by walk and rode a motorbike. That can be fulfilled by adding paths for pedestrians in the park and parking lots for motorbikes or bicycles in the welcome area. For the people's preference for ecological design, the dominant community considers that the ecological design in the park is made with local materials and is environmentally friendly. However, the application of community preferences for an ecological design still considers site conditions. It applies various plants, conserves energy use, uses a recycling system (reduce, reuse and recycle), dominantly uses local plants, and practices stormwater management.

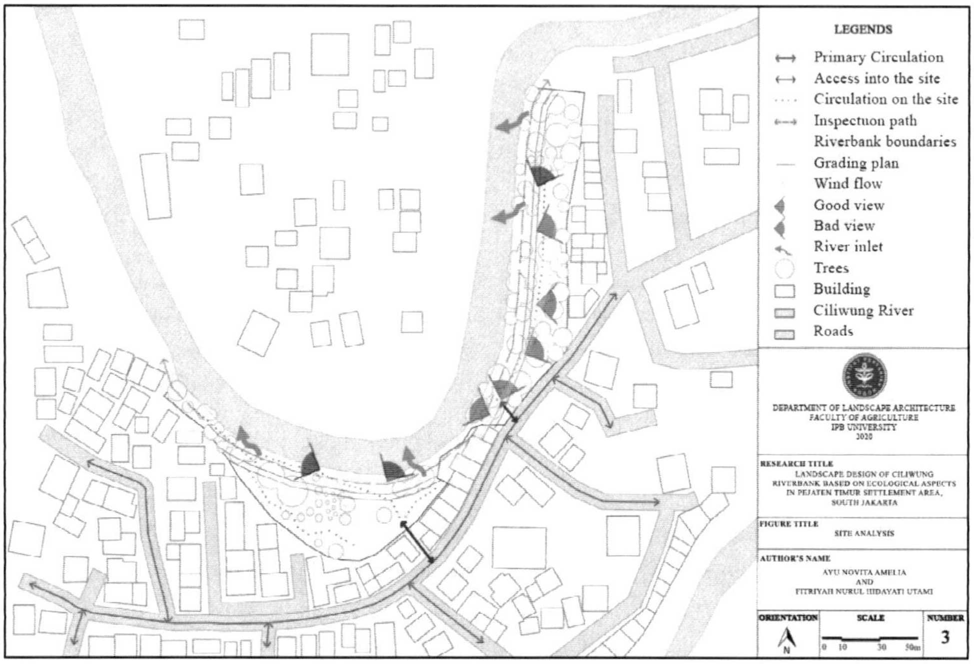

Figure 3 Site analysis

3.3. Concept

The main concept in this research is "Eco-Relaxing Riverside Park". "Eco-Relaxing" refers to the initial concept of an ecology-based environmental park. This park will become a public space on the riverbank that has the primary function of relaxation for its users. The environment and activities in this park refer to the user's goal of going to the site, to relax and relieve fatigue or pressure caused by work on a busy day. These objectives can be achieved by presenting an ecological and green atmosphere to the site.

The design concept in this community park is based on the site's location on the Ciliwung River's edge and is closely related to water. The shape and color of the water itself are created in several elements of the park design. The form of flowing water like the Ciliwung River itself is transformed and connected to form circulation paths in this park. These shapes are created with modern designs and materials under the residents' preferences from filling out the questionnaire that has been done. Modern materials and designs are also applied but still combined with environmentally friendly materials that make the park still seem ecological.

The ecological-based concept applied to the site focuses on water, to support and maintain the primary function of the site as a rainwater catchment area and hold the flow of the Ciliwung River from overflowing into residential areas. The concept of hydrology by using bio-retention ponds is essential to facilitate the passage of rainwater as a run-off to the Ciliwung River and filter wastewater from the settlement area so that the water becomes cleaner.

Moreover, the concept of circulation by using permeable pavement also makes it easier for users to do activities and helps the soil absorb water properly. The permeable pavement itself is a pavement system that can flow rainwater into the soil and aggregate storage areas [10]. The overflow water is stored and absorbed to the ground or flew directly into the drainage system. The use of energy-efficient facilities and utilities also increases the ecological value of the site, the application of waste separation can make it easier to recycle waste with the implementation of 3R (Reduce, Recycle, Reuse), and the use of various

IOP Conf. Series: Earth and Environmental Science **879** (2021) 012029 doi:10.1088/1755-1315/879/1/012029

environmental friendly materials is also applied to this park to provide adequate benefits for the user and the environment.

3.4. Block plan

The combined results of the site analysis, development concepts, and community preferences form a block plan. The areas in this community park are expected to facilitate the community around the site to carry out various activities such as gathering, relaxing, exercising, also educating the public about the environment from the ecological concept applied to the site. The development concept that is combined into a block plan can be seen in Figure 4.

3.5. Landscape design

Combining primary concepts, design concepts, and development concepts produces a landscape design with a site plan drawing product. The Ciliwung Riverbank, which is designed as a green public space, is implemented after considering the Ciliwung River normalization project, which will be carried out by river widening by 10-15 m. This community park design provides a public space to relax and socialize for its users, especially residents around the site. This riverbank is designed to be a community park that provides river inspection path facilities next to the Ciliwung River, and it has three main areas such as a welcome area, recreation area, and conservation area. The site plan can be seen in Figure 5, the overall perspective can be seen in Figure 6.

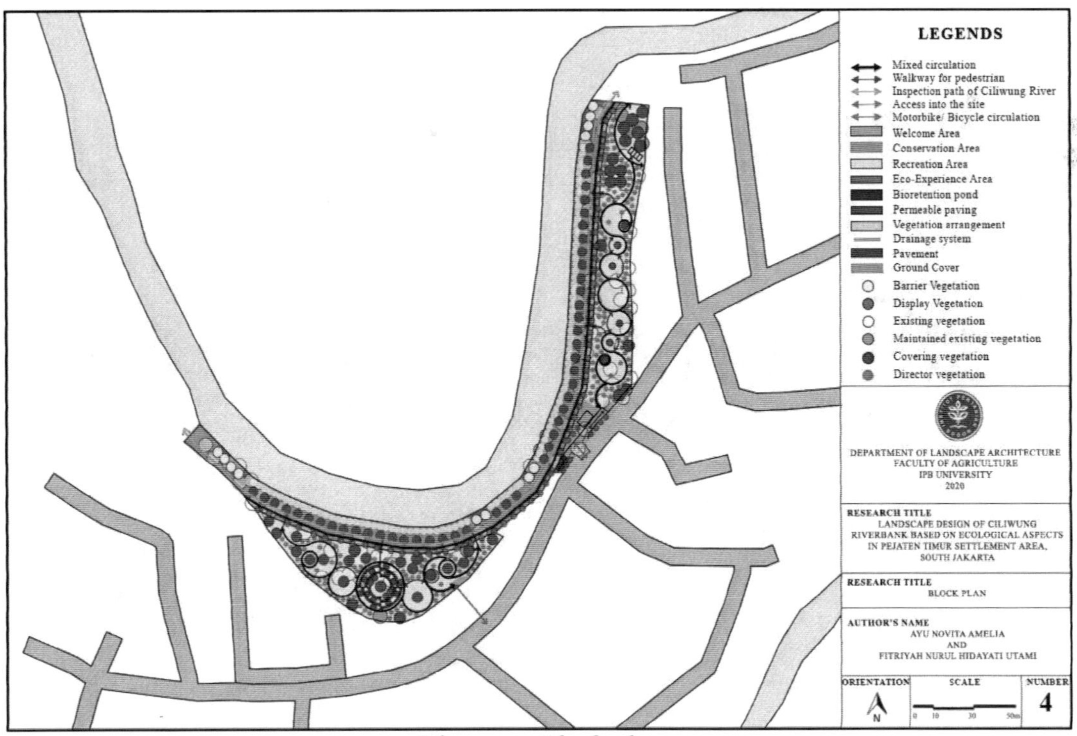

Figure 4 Block plan

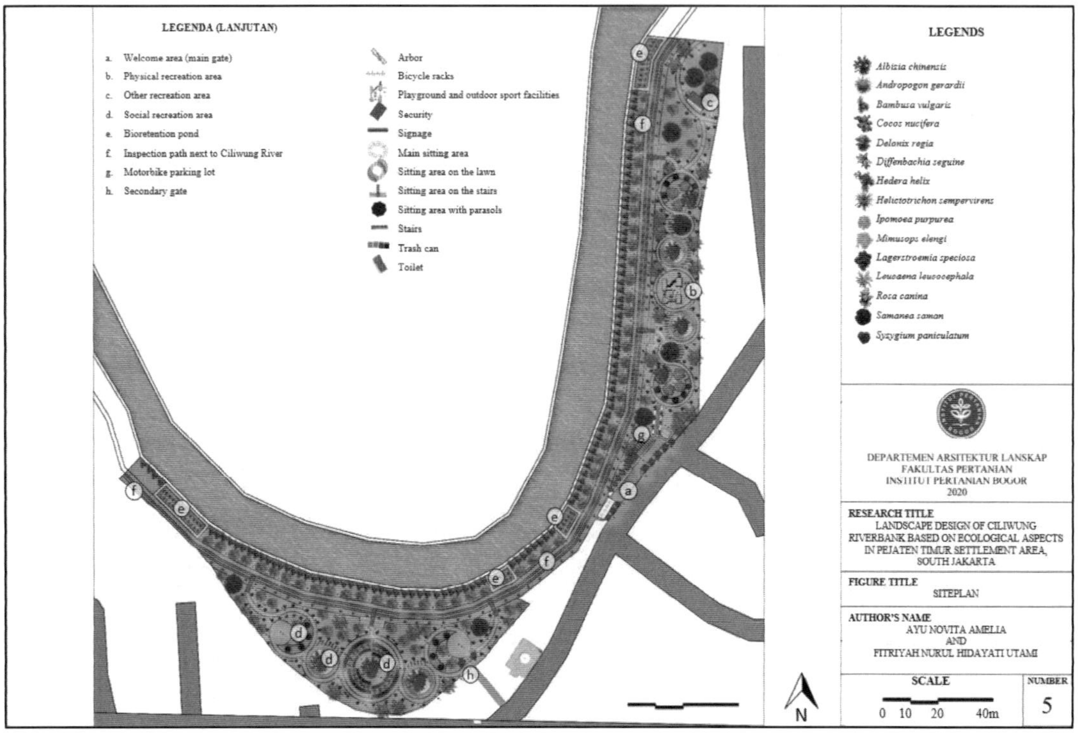

Figure 5 Site plan

Figure 6 Overall perspective

4. Conclusions

Pejaten Timur is currently a middle-class settlement area which is quite dense and has caused pollution problems in the Ciliwung River. The functions of the riverbank itself are currently disturbed, and the inadequate absorption of rainwater on the site causes the area to flood every rainy season frequently. One of the efforts to prevent the flood problems on the site is to restore the Ciliwung Riverbank's function in the settlement area of Pejaten Timur to become a more organized and more optimal green open space in absorbing rainwater.

The development of this site is based on the concept of "Eco-relaxing riverside park," which is an ecology-based community park on the riverbank that can be used to relax as its primary function. The application of these basic concepts is supported by an organic design concept because it is closely related to water and the Ciliwung River itself. The ecological concept is also applied to various environmentally-friendly facilities on the site to meet user needs and increase the ecological, aesthetic, and functional values of the riverbanks. The spatial plan applied to this park consists of two welcome areas, recreation area, and conservation area. The welcome area planted with display plants to attract visitors. The recreation area for various activities, such as relaxations, sports, and more. The conservation area used for river inspection track planted with various vegetations to prevent landslide or erosion. It is hoped that the development of this landscape design can anticipate floods on the site.

Acknowledgment

The authors would like to pay special regards to the staff of the Ciliwung Cisadane River Region Center (BBWSCC), Headman of the Pejaten Timur sub-district, and residents that participated in the questionnaire.

References

[1] [BPS] Statistics Indonesia 2017 Total population and population growth rate by regency/ city in DKI Jakarta province
[2] [BPS] Statistics Indonesia 2015 Number of villages/ sub-district according to the existence of settlements on riverbanks, under extra high voltage airways (SUTET), and slum settlements
[3] Salmah S 2010 Arrangement of river banks in terms of environmental aspects (Jakarta, Indonesia: Trans Info Media)
[4] [Department of Parks and Cemeteries] Data of recap area of green open space per municipality in DKI Jakarta [internet] 2017 [cited 2018 Aug 27] Available from: https://data.jakarta.go.id/dataset/rekapluasruangterbukahijauperkotamadyadidkijakarta
[5] Booth N K 1983 Basic element of landscape architectural design. (Illinois, US: Waveland Press, Inc)
[6] Regulation of the Minister of Public Works 2008 Number 5 of 2008 about Guidelines for the Provision and Utilization of Green Open Space (RTH) in Urban Areas
[7] El-Swaify and Dangler D W 1976 Erodibilities of selected tropical soils in relation to structural and hydrologic parameters in soil erosion: prediction and control. (Iowa, US: Soil Conservation Society of America)
[8] Decree of the Minister of Agriculture 1980 Number 837/Kpts/Um/11/1980 about Criteria and Procedures for Designating Protected Forest Area
[9] [BMKG] Meteorological, Climatological, and Geophysical Agency 2019 Climate Data for South Jakarta City, 2018-2019
[10] [UTTIPEC] 2009 Pedestrian design guidelines (New Delhi: Delhi Development Authority)

Fore Karst: *A Biotope-focused planning approach for the ecological management of Susong Dalaga, Quirino*

D Lavilla[1], N Navarra[2] and N Palarca[3]

[1] B Landscape Architecture Student, University of the Philippines Diliman, Quezon City, PH
[2] Professor, University of the Philippines Diliman, Quezon City, PH
[3] Professor, Adviser, University of the Philippines Diliman, Quezon City, PH

E-mail: davelavilla@gmail.com

Abstract. Conventional planning approaches that prioritize built-up areas and rapid urbanization for economic growth and social security, instead of the environment and the ecological services they provide have consistently been problematic in maintaining sustainability. Because of this, there is a growing need to shift from the conventional to a greener approach. The research argues that an ecological management approach is a viable choice of approach in protecting the ecological integrity of the landscape, especially in biotopes where ecological services are sought-after, such as the karst system Susong Dalaga found in Cabarroguis, Quirino Province. This approach is done by identifying the different biotopes to find the existing relations, services, and underlying issues of the place. In assessing the items mentioned earlier, the research finds that the natural systems are slowly dwindling due to the expansion of agriculture and built-up areas. Each biotope is then assigned with an appropriate approach, and with the consultation of the stakeholders, policies are then strategized for specific functions. It is hoped that the study would help push the application and integration of ecological management approach, and green approaches and principles in general, to future urban planning projects in the Philippines.
Keywords: biotope, ecology, ecological infrastructure, environment, urban planning

1. Introduction

According to the Philippine Statistics Authority [14], the rate of urbanization from 2010 to 2015 is recorded to be at 4.6 percent. This is reflected as the percentage of people who live in urban areas, comparing 2010 and 2015, had a remarkable leap of 45.3 percent to 51.2 percent. This data reflects of the reality that the Philippines is being urbanized and developed at a swift rate. This is to be augmented by the government's focus on infrastructure. In a report from Manila Bulletin in 2017, the National Economic and Development Authority stated that the government plans to spend a staggering 18.7 billion US Dollars on infrastructure under the "Build, Build, Build" Program, which focuses on transportation and social structures.

Situated in the Cagayan Valley as the youngest province, Quirino encompasses vast vegetative farmland covers, with rice and corn being the major crops, in addition to other fruit and vegetable produce. The province is also part of the Sierra Madre mountain range, which is home to virgin forests and complex riparian systems of the Cagayan, Addalam, and Ganano Rivers. This mountain range is currently secured by the Quirino Protected Landscape, which covers 175,943.62 hectares (around 67 percent of the total land area of the province). These natural systems offer excellent ecological services (such as protection from floods, landslides, and strong winds) and provide an economic and cultural livelihood to the residents.

Despite rural qualities, Quirino is proliferating – its population, for instance, is growing at a rate of 2-3 percent annually since the eighties [12]. From 148,575 in the 2000s, Quirino stands at 176,786 people as of 2010, increasing by 19 percent over the past decade. Also, households recorded in 2010 were at 39,770, which is higher by 9,189 from the past decade. This rapid development calls for the province to use its rich and untapped potential resources for infrastructural development.

Ultimately, its local government is confronted with the challenge of implementing the appropriate approach to urban planning and handle urbanization as a whole. The failure of this challenge invites future threats and fosters unsustainability to Quirino Province. This failure can be often brought about by conventional planning approaches that prioritize built-up areas and rapid urbanization in the name of economic growth and social security, rather than protecting the environment and the ecological services that it provides. These types of urban planning approaches have been proven by time to be ineffective in maintaining sustainability.

As the problem of negligence towards environmental issues increases and spreads worldwide, so does the concern and the need to mitigate and solve these arising issues. Ever since the 1980s, there has been a growing demand for sustainable development. This demand peaked at the release of the United Nations World Commission on Environment and Development's 1987 report entitled "Our Common Future," or simply and more commonly known as the Brundtland Report, where one of the widely recognized definitions of the topic can be described as a "development that meets the needs of the present without compromising the ability of future generations to meet their own needs."

With the growth of sustainable development comes the innovation of approaches that suits its ideals, goals, and priorities. Since its emergence and development in the 1980s, landscape ecology – a systems science that deals with the relationships between ecological services in habitats, became a relevant and valid field of study that serves bases for urban and natural growth developments as the science explores the different ecological factors and determinants in an area (Lin & Zheng, 2012). The global movement of green infrastructure brought about by the call for sustainable development made way for the use of green, renewable sources of energy, the consideration and prioritization of natural habitats and ecosystems over urbanization and infrastructure building, and the never-ending research to create more sustainable materials from either green or recycled sources – among others.

There is a need to shift to green methods of planning to prevent Quirino, a province full of natural resources, from suffering the same fate if conventional planning techniques where built infrastructure remains a priority over the environment are to be applied. New methods should focus more on sustainable goals and priorities.

2. Problem Setting

One of the areas in Quirino that can be potentially threatened by conventional urban planning is the sub marginal forest resting on top of a hill jutting out in the middle of the agricultural lands of Barangays Gundaway, Zamora, and Villamor. The locals call this hill Susong Dalaga, named after its panoramic breast-like shape.

Susong Dalaga is one of the many karst landscapes found in the province. Karst landscapes, formed when the underlying rock is soluble, are characterized by unique and highly specialized ecosystems that house specific flora and fauna and give off an abundance of ecological services. Karst landscapes provide an abundant source of drinkable freshwater via its aquifers; influence the microclimate due to how water filtration acts differently than regular soils and rocks; contain the collection of minerals in these landscapes; house not only specific and rare flora and fauna, but also act as a shelter to our indigenous people; and preserve any prehistoric data for multiple millennia that may indicate and increase our knowledge about the culture of ancient people. These stress the need to protect these systems.

For as much as it is a valuable resource, karst landscapes are very fragile as well. Even the slightest disturbance by human activity can disrupt and degrade these services as harm can quickly induce significant effects on the whole system. Human disturbances are activities done within or near the karst landscape, such as agriculture, which might cause soil degradation, soil erosion due to deforestation and overgrazing; urban development and land use, which might lead to the infilling of sinkholes, poor management of sinkholes and caves, water pollution especially in streams and aquifers; tourism which might lead to the destruction of rock formations, vandalism, and infrastructure; and so on [7][18].

3. Methodology

Patterned after Danao's ecological infrastructure planning approach [3], the research focuses on the study of different biotopes (habitat). Careful and meticulous on-site visits and mapping are done to identify the different biotopes in the area, and identify its aesthetic, biodiversity, culture, and risks. Identifying and consolidating the date will create a biotope map, which will be the foundation of the assessment of ecological values and the formulation of the ecological management plan (preservation, conversion, improvement, or conversion). This ecological management plan will then be used to create and enact the different strategies and policies for each biotope, ultimately protecting the area's ecological integrity (in this case, Susong Dalaga).

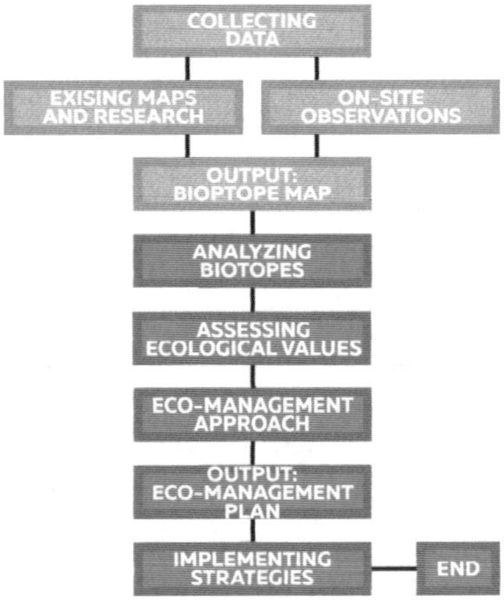

Figure 1 The Methodology of The Study

3.1 Collection, Analyses, and Interpretation of Existing Maps and Data

Data provided by the LGU of Quirino include: statistics and word documents of the three barangays involved in the research, and remote sensing and geographic information system technology files of the Quirino province. The data will be analyzed to determine the existing conditions of the immediate surroundings of Susong Dalaga.

3.2 On-site Observation

Site visits are conducted in and surrounding Susong Dalaga to crosscheck the existing conditions of the site and the different biotopes present in and surrounding the karst landscape. Observation of the different biotopes will help determine the best approach to each biotope in terms of the dimensions of aesthetic, biodiversity, and culture in ecological infrastructure.

3.3 Discussion of Biotopes

Biotopes are studied and analyzed to determine the current situation of the area. Identification of characteristics, distribution, and the following dimensions of ecological infrastructure – biodiversity, aesthetic, culture, and risk – will be enumerated.

Biodiversity pertains to the variety of flora and fauna found. A biotope is considered to have zero biodiversity if the data finds no sign of life, and the amount of biodiversity increases as the amount of plant and animal life increases. High biodiversity is considered when there is a wide variety of plant life, which encourages animal life to thrive.

Aesthetic deals with the visual interest and beauty of the biotope. Aesthetic can be the scenic views of the natural and manmade landscapes (or lack thereof), such as the rivers and creeks flowing in the site, the vast agricultural land, the presence of greenery, etc.

Culture encompasses all interactions, activities, and traditions that take place in the biotope. It can be as simple as day to day life activities or even as grand culturally and historically as festivals and celebrations. Culture can also encompass how much human traffic is in one place.

Risk looks into the underlying threats that loom in certain conditions. As the site is near creeks and rivers, this aspect will take into consideration the flooding and erosion aspect of the site. If the biotope is prone to the dangers mentioned before, it is considered to be at high risk and should be taken into account the amount of damage it sustains. Risk also encompasses any threat that a biotope may encounter in such the case that improper management of other biotopes was to take place.

3.4 Assessment of Ecological Values

Ecological values are assigned to each biotope after its identification, study, and analysis. Since the study focuses on the ecological integrity of Susong Dalaga's ecological services, ecological values are heavily determined on its current environmental conditions.

An assessment of ecological values will help determine which biotopes currently have excellent conditions of biodiversity and which biotopes do not. Identifying the different values will be used in simulating changes, predicting or projecting future issues, threats, opportunities and possibilities. The assessment will also help prioritize which biotopes must be protected and which biotopes should be regulated, as guided by the ecological management approach.

3.5 Ecological Management Approach

Ecological management approaches and strategies are adapted to each biotope after having assessed the ecological values. These approaches are the following: preservation, conservation, improvement, and conversion.

Preservation is the highest form of protection - biotopes under preservation are to remain intact, with zero intervention, before they get damaged by change or development. These are usually nature and wildlife preserves, national parks, and large patches of intact vegetation.

Next is conservation, which aims to protect the existing landscape elements by implementing action. These areas are usually suffering from impending pressure, such as development.

Improvement comes after conservation. Improvement involves implementing actions, management, and reconfiguration of structures, so that the existing manmade elements do not destroy the existing landscape. These are usually built-up areas that threaten nearby natural biotopes.

The last approach, which has the most amount of intervention, is conversion. Conversion recognizes that the biotope needs to be changed so that its ecological functions will be restored, or so that adjacent biotopes will not be threatened. Determining these approaches will help create and implement different strategies for each biotope.

IOP Conf. Series: Earth and Environmental Science **879** (2021) 012030 doi:10.1088/1755-1315/879/1/012030

4. Findings, Results, Discussion

4.1 Discussion of Biotopes

Below is a thematic map that showcases the current distribution of the natural, semi-natural, and built-up areas of the three barangays, along with a table showing a comparison of the different biotopes in terms of total area. The table also shows the different categories of biotopes: the densely vegetated natural systems, the manmade but still vegetated semi-natural systems, and the urbanized and developed built-up systems. It also shows the appropriate ecological management approach for each biotope based on the study's findings (to be discussed further in 4.4. Ecological Management Plan).

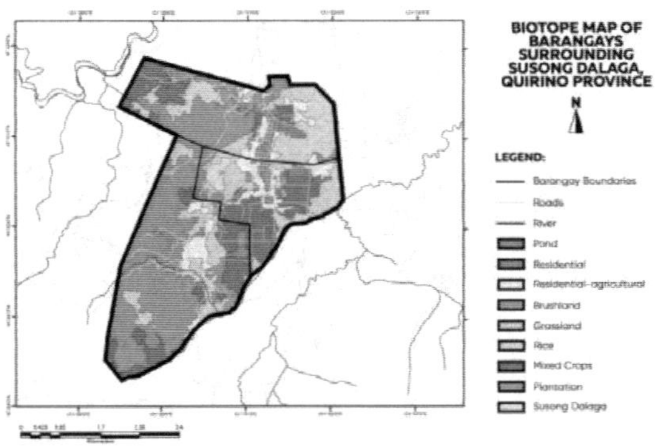

Figure 2 Biotope Map of Barangays Surrounding Susong Dalaga

Tabel 1 Table Listing the Areas of Biotopes Found in and Surrounding Susong Dalaga

BIOTOPES		AREA (HA)	PLANNING APPROACH
SEMI-NATURAL SYSTEMS		**806.9324**	
Agricultural Land	Rice Paddy	409.7351	Conservation, Improvement
	Mixed Crops	379.6083	Conservation
	Plantation	17.5891	Improvement, Conversion
NATURAL SYSTEMS		**902.5666**	
Grassland		104.699	Preservation
Brushland		769.581	Preservation
Sub Marginal Forest		23.4902	Preservation
Pond		4.796	Preservation
BUILT-UP SYSTEMS		**224.4132**	Improvement, Conversion

Source: Cabarroguis' Comprehensive Land Use Plan (CLUP) [1]

As seen in the table, natural biotopes still dominate in terms of the amount of area that encompasses the barangays, as seen in the east of the area, surpassing the manmade agricultural lands by around a hundred hectares, which can be found at the west. Build-up areas are still sparse, at around 11.6 percent (compare to 46.7 percent of the total area of natural biotopes, and 41.7 percent of agricultural lands). These build-up areas are situated in the middle of the semi-natural biotopes.

The following map reflects these systems (with aquatic systems to differentiate them from the terrestrial natural systems) about Susong Dalaga.

IOP Conf. Series: Earth and Environmental Science **879** (2021) 012030 doi:10.1088/1755-1315/879/1/012030

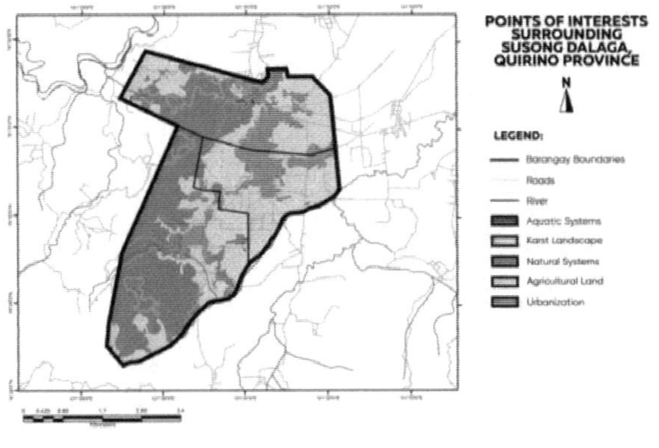

Figure 3 Map Showing Key Points

Based on the site's land classification map, alienable and disposable lands are to the east of the map while a strip of land on the west remains forested. The data shows that there is currently a build-up of urbanization happening in the east side of Gundaway, as well as the majority of Zamora, all located on the east side of Susong Dalaga.

Two patches of built-up areas are visibly seen from the map: one from Gundaway-Zamora, and a smaller one at Villamor. The two patches are slowly connected by a slim strip of urbanization that barely connects the two via the road that connects the barangays together. Alarmingly, to the south, there is also a small patch forming, and considering what surrounds Susong Dalaga is alienable land, it is a possibility for the rice fields that surround the karst landscape be converted into a built-up area. This possibility has been attempted before, like the subdivision attempt along Tourist Road a few years back.

Meanwhile, the undisturbed, forested west of Susong Dalaga acts as a long corridor of shrublands and grasslands that connect the karst landscape upward to Capitol Hill and downward to Villamor Hill. However, during the researcher's site visits, some of the natural systems have been slowly converted into rice and cornfields. They are trickling along roads and rivers, fragmenting the shrublands and grasslands and further isolating Susong Dalaga. Considering the topography and slope of the barangays, converting too many woodlands into agriculture may not bode well for the constituents in the long run. The creation of more agricultural land to the west also invites for more agricultural residences. This possible influx of residences may turn into patches of urbanized areas, further isolating Susong Dalaga.

4.2 Assessment of Ecological Values

Presented in Figure 4 is a map that shows the different ecological values of the biotopes, ranging from high, average, to low ecological value.

A biotope of high ecological value means that based on the dimensions of ecological infrastructure, they have immensely sought-after qualities. In terms of biodiversity, a plethora of different species of plants and animals are present. These biotopes are encouraged to grow and be sustained. Included in this category are the shrublands, the grasslands, the aquatic systems, and the sub marginal forest, where Susong Dalaga is located.

Meanwhile, a biotope of average ecological value based on the area's biodiversity might mean a vastly vegetated area, dominated by a handful of species. While these biotopes benefit from being sustained, they do not necessarily need to expand, especially not at the expense of high ecological valued biotopes, where several species are found. Included in this category are the rice paddies, mixed crops, and residential-agricultural lands.

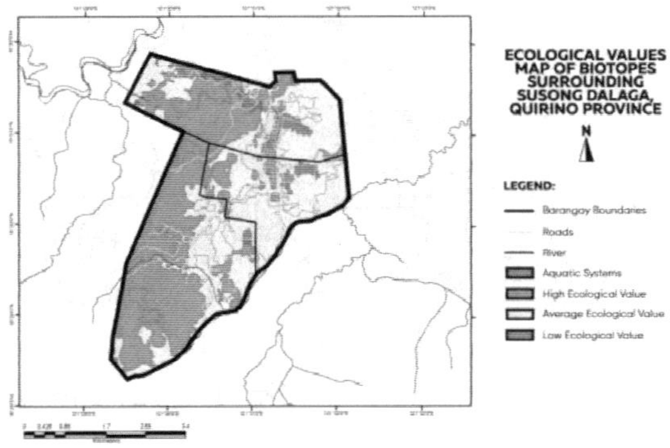

Figure 4 Map Showing The Differing Ecological Values of Biotopes

Lastly, a biotope of low ecological value means that based on the dimensions of ecological infrastructure, they have lesser qualities as than the other two categories. While these biotopes may have high culture, their built-up nature means less area for wildlife. They are encouraged to be sustainable, which might mean shifting these biotopes to cultivate qualities that will help grow their ecological qualities. Included in this category are the plantations, and the residential lands.

With the ecological values set in place, theoretical simulations are created to present any hypothetical and possible benefits, dangers, and issues in map form. Simulations can present the best possible outcome, as well as the worst. These projections can help formulate the ecological management plan, and subsequently, create and enact policies to prevent issues in the future and foster sustainability.

4.3 Simulation of Ecological Values

Based on the data from Cabarroguis' CLUP for 2013-2023, current trends, such as agricultural fields following the streams, the trickling down of built-up areas, are assumed and projected by spreading these areas out. Assuming that the people's dereliction and negligence to Susong Dalaga and the other natural systems allowed the spread and growth of biotopes with average and low ecological value, this trend leads to the conversion of high to average/low and average to low, as seen in the map (Figure 5).

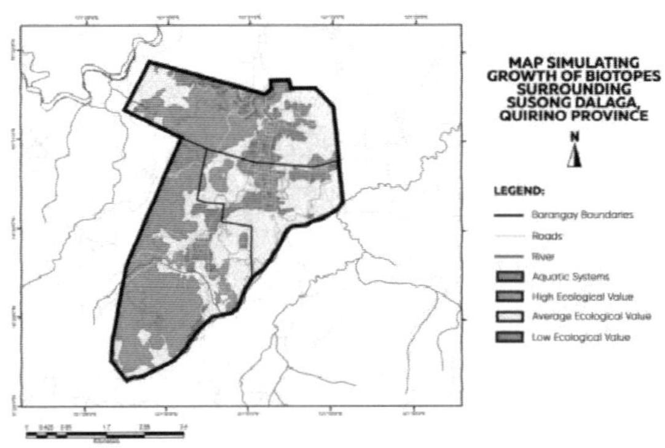

Figure 5 A Map Showing A Simulation That Portrays The Negligence of Biotopes with High Ecological Value in Favor of Biotopes with Lesser Ecological Value

Agricultural fields have continued to fragment the shrublands of Barangay Villamor, which also continue to sever the corridor that allows Susong Dalaga to connect to the Villamor Hill (which in itself is also suffering from the expansion of agriculture) located below. This is likely caused by two factors:

The first factor would be the fast growth of Quirino Province as a whole. In the 2000s, Quirino was expected to double its population by 2027. In a later report by the Philippine Statistics Authority, the population in Quirino in 2013 – 176, 786, is expected to double in 2053. While an increase in population would not necessarily reflect heavily on the three barangays' population, it might mean more demand for food production in the province, which may be a factor in expanding agricultural land. This expansion of agriculture, and the doubling of population soon, would mean the shift to residential-agricultural lands spreading along with these newly created fields, and in turn, expanding the residential hub.

The second factor would be the increasing number of makeshift homes and illegal settlements in the barangays, particularly in the natural systems. They are trying to make a living by clearing out natural systems to convert to agricultural lands. As mentioned briefly in the findings, there are currently over a hundred people under this category. This societal issue will only increase as more people will illegally settle over the years, and those who already have been in the past will continue to grow as families exponentially grow per generation. In twenty to forty years, the existing population will double or triple, and in fifty to seventy years it will grow even fivefold, creating patches of built-up and semi-natural systems if neglected.

The overall degradation of ecological value of the biotopes will result in the diminishing of the services that it provides. This will cause a decline in production, a lesser observable biodiversity, etc.

To avoid this scenario (among many other circumstances), an ecological management plan must be created to plan for a scenario that favors the landscape as a whole. The plan must also be beneficial not only for the environment, but also for its constituents.

4.4 Ecological Management Plan

By assessing the current situation, the researcher decided upon the action needed for each biotope based on the four approaches of ecological management approach (preservation, conservation, improvement, and conversion). The researcher was then able to produce an ecological management plan based on these decisions:

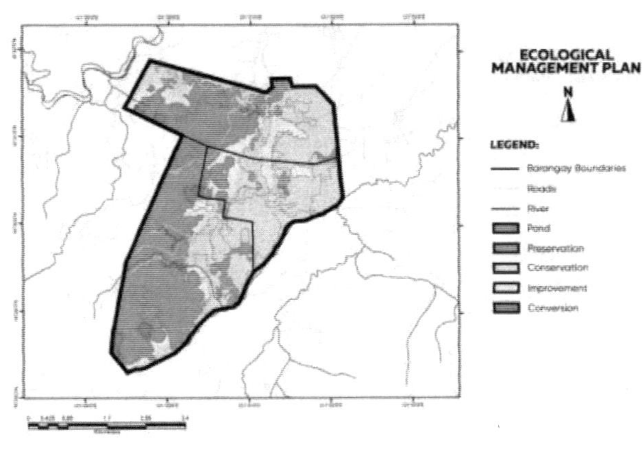

Figure 6 Ecological Management Plan

The plan was based on the decision that the researcher felt most justified with the study's aims and goals:

Biotopes of high ecological value should be preserved to maximize the ecological network, biodiversity, and sustainability, and to prevent Susong Dalaga from being fragmented and separated from its other natural biotopes.

Next, the agricultural land that surrounds Susong Dalaga and envelops the preservation areas will act as buffers and shields to the ever-expanding urbanization happening in the east. Meanwhile, the agricultural-residential lands will act as places of bustling culture and activity that will protect these farmlands. Therefore, most of these areas should be conserved. Exceptions are fields that are encroaching and permeating on biotopes of high ecological value. In this case, they would have to be converted instead.

Finally, there needs to be an improvement on the built-up areas of barangay Zamora and Gundaway, where commercial activity happens – biotopes of low ecological value can be slowly restructured back into agricultural villages. They can serve as agro-business hubs to further flourish the agricultural economy that Quirino boasts.

5. Conclusion

Despite being about one percent of the total area of the scale of the study, Susong Dalaga contributes so many services to its constituents. Whether it is of economic, social, environmental, cultural, or ecological significance, there is enough justification for protecting the karst from existing and potential threats.

Analyzing the biotopes, it is clear that while Susong Dalaga's constituents thrive from these services, existing issues in the area arise from anthropogenic pressures. Future issues will also spread if they are neglected, and will cause a steady decline not only for Susong Dalaga's ecosystem itself, but also to those who benefit from it.

By assessing the present conditions and ecological values of these biotopes and then assigning them with the appropriate ecological management approach, the result presents policies that, in theory, maximize the protection of natural systems, which will benefit society in the long run.

In addition to this, there is of grave importance to contribute to the study of one of the rarest, if not the rarest ecosystem/landscapes in the earth, especially studies conducted in the Philippines. The Philippines has a lot of karstic environments, some of them are lesser known, such as Susong Dalaga. Karst landscapes provide services that benefit society, and studying them will advance what we already know about karsts, and the mysteries that these landscapes keep, such as the bountiful flora and fauna, the culture and history, among others.

Planning is a long-term process that deals with the organization and selection of choices based from the information gathered. With these objectives accomplished, it is concluded that undergoing a biotope analysis, assessment, and an ecological management approach is a viable way of urban planning, especially now that there is a need and emphasis for the protection of the environment.

As of the writing of this study, biotope analysis has never been considered in comprehensive land-use plans. This study sheds new light on one of the many greener options of urban planning. Hopefully, in the years to come, ecological management approach, as well as the principles and strategies akin to the ones done in this study, can be integrated to many more urban planning approaches in the future.

References:

[1] Cabarroguis, Local Government Unit of (2013). Comprehensive Land Use Plan 2013-2023.
[2] Dalton, G., & Lewis, T. (2011). Metrics for measuring job creation by renewable energy technologies, using Ireland as a case study. Renewable and Sustainable Energy Reviews, 15(4), 2123–2133. doi: 10.1016/j.rser.2011.01.015
[3] Danao, M. (2014). Ecological infrastructure planning: corridor resource landscape management approach of Angat River basin (Master's thesis). University of the Philippines Diliman, Quezon City, Philippines.

[4] Evidente, M. R. (2009). Present opportunities and future crises: managing tourism development at the chocolate hills, Bohol, the Philippines. Yale University Tropical Resources Bulletin, 28, 17-22.
[5] Gurung, D. B., & Scholz, R. W. (2008). Community-based ecotourism in Bhutan: Expert evaluation of stakeholder-based scenarios. International Journal of Sustainable Development & World Ecology, 15(5), 397–411. doi: 10.3843/susdev.15.5:1
[6] Ignatieva, M., Meurk, C., & Newell, C. (2000). Urban biotopes: the typical and unique habitats of city environments and their natural analogues. Urban Biodiversity and Ecology as a Basis for Holistic Planning and Design.
[7] Kiss, M., Tanács, E., & Ilona, B. (2011). Ecosystem Services in Hungarian Karst Areas. Acta Climatologica et Horologica. 44-45. 41-49.
[8] Leyco, C. (2017). Gov't unveils P3.6 trillion three-year infrastructure plan. Manila Bulletin. Retrieved May 10, 2019, from https://business.mb.com.ph/2017/04/18/govt-unveils-p3-6-trillion-three-year-infrastructure-plan
[9] Lin, W., & Zheng, X. (2012). Ecological Exploration of Urban Planning. 2012 International Conference on Biomedical Engineering and Biotechnology. doi:10.1109/icbeb.2012.163
[10] Government Unit of Cabarroguis (2018). Community Based Monitoring System (CBMS) for Barangay Gundaway, Villamor, and Zamora.
[11] Löfvenhaft, K., Björn, C., & Ihse, M. (2002). Biotope patterns in urban areas: A conceptual model integrating biodiversity issues in spatial planning. Landscape and Urban Planning, 58(2-4), 223-240. doi:10.1016/s0169-2046(01)00223-7
[12] Philippine Statistics Authority (2001). Quirino's Population Expected to Double in Twenty Six Years. Retrieved May 10, 2019, from https://psa.gov.ph/content/quirinos-population-expected-double-twenty-six-years
[13] Philippine Statistics Authority (2015). Region II (Cagayan Valley). Total Population by Province, City, Municipality and Barangay. Census of Population.
[14] Philippine Statistics Authority (2019). Urban Population in the Philippines (Results of the 2015 Census of Population). Retrieved May 10, 2019, from https://psa.gov.ph/content/urban-population-philippines-results-2015-census-population
[15] Rodríguez, S., Roman, M., Sturhahn, S., & Terry, E. (2002). Sustainability assessment and reporting for the university of Michigan's Ann Arbor campus.
[16] Salomon, J. (2012).A Mysterious Karst: the "Chocolate Hills" of Bohol (Philippines). Acta Carsologica.
[17] Urich, P., Day, M., & Lynagh, F. (2002). Policy and practice in karst landscape protection: Bohol, the Philippines. The Geographical Journal, 167(4), 305-323. 10.1111/1475-4959.00028.
[18] van Beynen, P., Brinkmann, R., & van Beynen, K. (2012). A sustainability index for karst environments. Journal of Cave and Karst Studies, 74(2), 221–234. doi: 10.4311/2011ss0217
[19] Vatalis, K. I., Manoliadis, O. G., & Charalampides, G. (2011). Assessment of the economic benefits from sustainable construction in Greece. International Journal of Sustainable Development & World Ecology, 18(5), 377–383. doi: 10.1080/13504509.2011.561003
[20] Yu, K., Wang, S., & Li, D. (2011). The negative approach to urban growth planning of Beijing, China. Journal of Environmental Planning and Management, 54(9), 1209-1236. doi:10.1080/09640568.2011.564488

IOP Conf. Series: Earth and Environmental Science **879** (2021) 012031 doi:10.1088/1755-1315/879/1/012031

Continuity of heritage urban landscape: the case of *Jeron Beteng* Yogyakarta

D H Rahmi
Associate Professor, Universitas Gadjah Mada, Yogyakarta, Indonesia
E-mail: dwitahr@ugm.ac.id

Abstract. The urban landscape, which is in a dynamic environment, continues to change, meaning it has transformed the relationship between humans and the environment. Many changes have occurred since the early formation of the city until now. Nevertheless, many heritage urban landscape that has been formed hundreds of years ago are continuing to date with their various elements remain the same. Jeron Beteng is a part of the Yogyakarta Sultanate Palace, which has existed since the early construction of the palace. This traditional area was once a residential area for the *abdi dalem* (people who served the Sultan and Palace) and the sultanate princes or brothers of the Sultan. Along with the developments of Yogyakarta, Jeron Beteng has developed into a dense urban settlement. Various changes have occurred, both the function of the area, architecture building, and land use. As a heritage area, it is interesting to know whether Jeron Beteng can adapt to the changes and continue as the cultural identity of the Yogyakarta Palace. This paper is based on research that intends to identify the urban landscape characters of Jeron Beteng and examine the continuity of its physical landscape characters as the cultural identity of the Yogyakarta Palace. A qualitative method is used with two approaches: interpretation of the history of Jeron Beteng and qualitative approach based on the contemporary empirical phenomenon. This research found that Jeron Beteng today is the culmination of hundreds of years of physical and social management. It is not just a collection of heritage buildings, events, and precincts, but the whole landscapes, that have unique characters. The character of Jeron Beteng is mostly formed by its physical elements that have heritage values, mainly the Palace, Alun-alun, and the *beteng* wall. Changes in the spatial and physical environment have happened, however, the essence or core of the physical landscape character of Jeron Beteng is continuing and contributes to the cultural identity of the Yogyakarta Palace and the city.
Keywords: change, continuity, heritage urban landscapes, Jeron Beteng

1. Introduction

The term landscape was first introduced by Alexander von Humboldt, a German geographer, and biologist, in the early 19th century: "landscape is the total character of region of the earth". This definition is still accepted and used today [1]. Furthermore, there are several definitions of landscape, including [2], who define a landscape as an area with topographic configuration, vegetation closure, land use, and settlement patterns with its natural and cultural processes and activities. Landscape is a relationship between social and environmental processes [3], and landscape is the result of a relationship between natural factors and human activity [4]. In this context, the landscape is closely related to land that has characters due to the relationship between nature and human culture.

In the establishment and development of a traditional city, the landscape is one of the considerations incorporated into planning and design. Traditional cities generally have urban landscapes with old assets or heritage assets as its landscape elements. The urban heritage landscape of the traditional cities is the achievement of our predecessors in forming a city, so what is seen today is an expression of its altered heritage landscape, the interaction of nature and culture shaped over decades and centuries. The heritage urban landscape, then, contributes to the shape and character of the city [5]. O'Donnel also mentions some premises for the historic landscape of the heritage cities, which are:

- *The urban landscape has heritage values, comparable to urban architecture*
- *Urban historic landscapes are worthy of conservation as expressions of the spirit of place*
- *Unique landscape identity and character of a place expres heritage and spirit*
- *Shaped through time, the spirit of the landscape is the combined work of humanity and nature*
- *Processes of continuity and change are present in the historic urban landscape*
- *Multiple tangible and intangible values are imbedded in historic urban landscapes*
- *Recognition, documentation and analysis of the complexity of the urban cultural landscape is a doorway to the retention of values and spirit of place*

Those premises shows that a historic urban landscape can also be called a heritage urban landscape because of its heritage values; it relates with time and space; shapes unique identity and character of a place; and experiences change and continuity. The character of a heritage urban landscape has been formed in a long period of time and is always associated with the history and uniqueness of the city [6]. The unique spatial and architecture forms the city identity, as well as the socio-culture of the community and all the combination of natural and cultural values [6].

Since a heritage city is in a dynamic environment, many changes have occurred in its landscape elements. Changes as the impact of the increasing number of people, urbanization, and urban development. Nevertheless, there are also elements of the heritage urban landscape that are unchanged or continuing until now, both the physical or cultural elements. Old assets need to be continued for fulfilling the present and future needs [7]. The heritage urban landscape needs a balance between its continuity and change of all the resources and cultures. The sustainability of heritage areas with history, landscape, and culture needs to be continuous in accepting change and development, between maintaining the continued of old assets and fulfilling the needs of present and future life [8].

As a traditional city, Yogyakarta has its heritage urban landscape in the city center, in which existence still colors the city to this day. One area of urban landscape of Yogyakarta is Jeron Beteng where the Sultanate Palace of Yogyakarta is located. Jeron Beteng is an area that has a unique, well-planned residential area and was built by Sri Sultan Hamengku Buwono I, the first king of the Yogyakarta Sultanate. The *beteng* (wall) structure [9] is a part of the most recent palace built-in 1782, after the construction of the palace was completed in 1757. The many cultural heritage and historical values contained in it, Jeron Beteng is considered to be able to reflect the image of Yogyakarta as a cultural heritage city.

Jeron Beteng is considered as the first residential area in Yogyakarta and still continues to function today with historical traces of heritage landscape elements that can be found. Amid the city's ongoing development, many changes are experienced by Jeron Beteng settlement as the heritage urban landscape, and two questions arise: how many changes have occurred and how many remain? Is Jeron Beteng heritage landscape still continuing? This paper aims to identify the urban landscape character of Jeron Beteng and examine the continuity of its physical landscape character as the cultural identity of the Yogyakarta Palace.

2. Methods

This paper is based on qualitative research that explores Jeron Beteng area, a unique heritage urban landscape in the heart of Yogyakarta city. Primary data was collected through a field survey on the physical environment of the Jeron Beteng to get an understanding and data on the spatial pattern, land use, spatial functions, and landscape elements. Identification of the landscape elements, which area relic of the past, and the new one were undertaken during the survey by spatial mapping, photo, and notes taking.

Data was analyzed by 1) Spatial analysis to know the spatial structure and pattern of the area based on the survey result; 2) Explanation building, that is constructing and evaluating the organized data and information into a narration. A narrative explanation is used to examine the history and values of Jeron Beteng. Supported data, such as maps and photos are needed to be interpreted.

3. Result and discussion

3.1. Jeron Beteng

Jeron Beteng area is the government center of Yogyakarta Sultanate and is the beginning of the Yogyakarta city development. Physically, Jeron Beteng is an area of Yogyakarta Palace bordered by a wall fence called Beteng Baluwerti. Inside there is a palace complex, prince's houses, Tamansari Water Castle complex, and settlements. Beteng Baluwerti has 1,200 meters long and 940 meters wide in the West and 1140 meters in the East, while the wall thickness is 3 meters and the height is 3-4 meters. Beteng Baluwerti has four wall corners and five gates called *plengkung* (arch), including Plengkung Tarunasura (Wijilan), Plengkung Jagasura, Plengkung Jagabaya, Plengkung Madyasura, and Plengkung Gadhing (Nirboyo). Plengkung that still exists today, is Plengkung Wijilan, located in the East, and Plengkung Nirboyo in the South. Other *plengkungs* have been dismantled into ordinary gates. Above the arch is an open court called Gapura Panggung. Each arch is equipped with a guard post or bastion in the corner of the court (Pojok Beteng), which used to be a place to snoop on enemies and lay canons. On the outer side of the *beteng* is a *jagang* or moat along the edge of the *beteng* and planted with *gayam* trees, but at this time, the *jagang* no longer exists.

The *beteng* wall and space inside seem to have been planned and very well designed by Sultan Hamengku Buwono I. The primary function components in the *beteng* are the Sultan Palace, surrounded by Beteng Cepuri as a barrier, and the Alun-alun Utara and Alun-alun Selatan (North Square and South Square). Outside Beteng Cepuri is the settlement of Jeron Beteng, which was initially the residence of the nobles, who are the Sultan's brothers. The Sultan built residences for the princes and nobles with a traditional Javanese architecture called Dalem Pangeran, and each was confined to a high wall and equipped with a gate as the entrance. There are 13 (thirteen) Dalem Pangeran scattered within the Beteng Baluwerti area, while the other Dalem Pangeran were built outside the beteng [10]. In the *dalem* complex, there are also *magersari* or *ngindung* (shelterers) who are in charge of serving the princes. The existence of the Dalem-dalem Pangeran has undergone many changes, both in terms of the ownership and its main function as the residence of the royal with *magersari* relatives in it. In the *beteng* area also, in 1758 Sultan built the Tamansari complex, which serves as a recreation area, princess baths, and vegetable and spice gardens for the palace's needs. Although some areas have been lost, Tamansari still is functioned as a tourist attraction of the city. Figure 1 shows the position of the Jeron Beteng area in Yogyakarta city, and Figure 2 shows the Jeron Beteng area with its physical elements, and the spatial pattern formed by the road network.

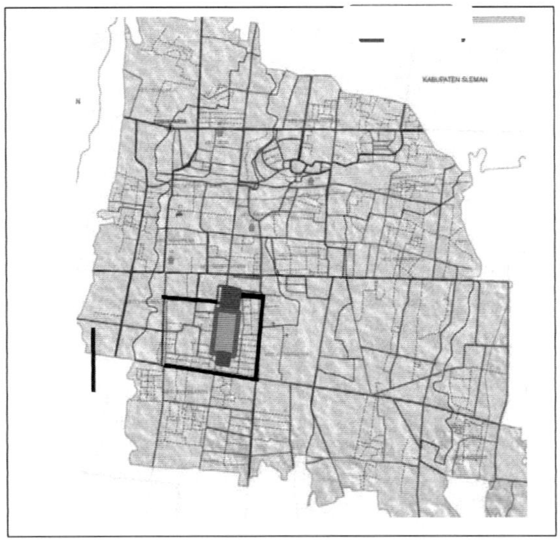

Figure 1 Jeron Beteng area in Yogyakarta City

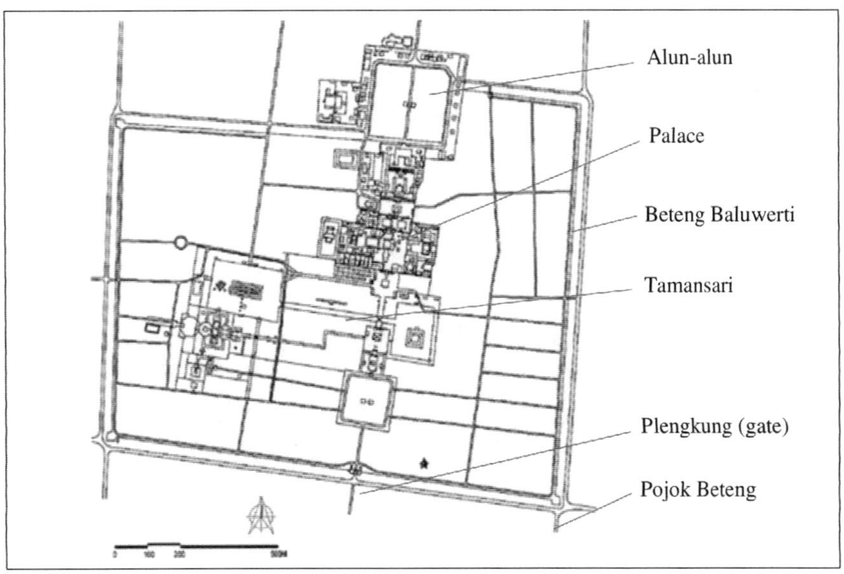

Figure 2 Physical elements and spatial pattern of Jeron Beteng
Source: RIK Budaya Perkotaan Yogyakarta 2014-2034, 2016 [12]

Furthermore, in *beteng* lived the *abdi dalem* who served the Sultan Palace and were granted the right of *lungguh* (residence) by the Sultan. The settlement of the *abdi dalem* is the oldest in Yogyakarta [11]. Regarding the right to live and land ownership in Jeron Beteng, it currently uses Law No.13 of 2012 on The Privilege of Yogyakarta as a legal basis. The area in Jeron Beteng is divided by the Sultan into several *kampungs* (housing areas) for the *abdi dalem* based on the types of their work. At this time, the division of the *kampungs* has been lost, but replaced by the naming of the roads according to the naming of the *abdi dalem kampung*. Here are the names of the *kampung* of *abdi dalem* in Jeron Beteng [10]:

Table 1 Names of *kampung of abdi dalem* in Jeron Beteng

No.	Kampung's name	Meanings
1	Musikanan	*Kampung* of *abdi dalem:* string and wind musical instrument players
2	Ngrambutan	*Kampung* of *abdi dalem:* hairstylist for the Sultan's family
3	Kenekan	*Kampung* of *abdi dalem kenek* (co-driver of the Palace's horse carts)
4	Bludiran	*Kampung* of *abdi dalem:* clothes embroidery makers for the Palace
5	Kemitbumen	*Kampung* of *abdi dalem kemitbumi:* persons who maintained the cleanliness of the Palace yard.
6	Gebulen	*Kampung* of *abdi dalem* who prepared a fire for the cooking
7	Sekullanggen	*Kampung* of *abdi dalem* who prepared the rice (*sekul langgi*).
8	Pandean	*Kampung* of *abdi dalem pande:* blacksmith (iron makers)
9	Mantrigawen	*Kampung* of *abdi dalem*: head of the Palace's staffs
10	Pesindenan	*Kampung* of *abdi dalem*: Javanese traditional singers (*sinden*)
11	Gamelan	*Kampung* of *abdi dalem gamel* who in charge of caring the Sultan's horses
12	Namburan	*Kampung* of *abdi dalem:* drum players.
13	Siliran	*Kampung* of *abdi dalem silir* who in charge of managing the Palace lamps
14	Patehan	*Kampung* of *abdi dalem* who in charge of managing drinks for the Palace
15	Palawijan	*Kampung* of *abdi dalem palawija*, who have physical deficiencies such as dwarfs and albinos
16	Rotowijayan	*Kampung* of *abdi dalem sais*: Palace horse cart drivers and makers
17	Langenastran	*Kampung* of *abdi dalem* warrior guard Langenastra.
18	Langenarjan	*Kampung* of *abdi dalem* warrior guard Langenarja.

Source: Gupta et al., 2017

With the division of *kampung abdi dalem* according to their duties above, Kraton and the two *Alun-alun*, the space in Jeron Beteng is divided into patterns of common space, which can still be found today. Jeron Beteng, in the past, was indeed the center of the Yogyakarta kingdom, with the king or Sultan as the power.

3.2. Landscape character

The characteristic of the Jeron Beteng heritage landscape is influenced mainly by the physical elements and the social and cultural life of its community. Physically, the character of Jeron Beteng area is different from other areas in Yogyakarta because of the distinctive elements of the landscape, which have long existed and are not found anywhere else. With the traditional Javanese athmosphere, the uniqueness of the Jeron Beteng landscape is also influenced by the Palace's rules as the landowner. These rules, include:

a. Palace gives the right to the building, usage rights, *ngindung* right, and *magersari* right;
b. The place of worship that can be established is only *musholla* and mosque;
c. Lodging or homestay is not allowed;
d. Foreign/Chinese residents are not allowed;
e. The building is only allowed 1 (one) floor only.

The application of the Palace's rules is seen physically in Jeron Beteng from the density and regularity of the buildings. Buildings with the right to use are usually built-in an orderly and low-density manner. Meanwhile, buildings with *ngindung* and *magersari* rights grow irregularly and have high-density.

Jeron Beteng has grown into a dense settlement, and at this time, not only the *abdi dalem* and princes who live there, but many of the descendants of the *abdi dalem* and others from outside the *beteng* wall. Even Dalem Pangeran has changed its function even though it retains its architectural form. For example, Dalem Mangkubumen is now Widya Mataram University; Dalem Mantrigawen currently serves as an inn, although this is breaking the Palace's rules; and Dalem Pakuningratan which is functioned as a restaurant. The character-forming elements of Jeron Beteng heritage landscape are shown in Table 2 below.

Table 2 Physical elements of Jeron Beteng landscape

No.	Physical elements	Pictures	Notes
1	The Palace		The Palace with its Alun-alun (square) is the main landscape elements that physically remain the same (no changes). Both become the center of the area, and the Palace is as the center of the life in Jeron Beteng.
2	*Beteng Baluwerti*		Walls that border the Palace and the area outside it, where there are also 13 *Dalem Pangeran* and the houses of *abdi dalem*. The wall along the north side is no longer exists, so is the moat/*jagang* around the *beteng*.
3	*Plengkung* (Gate)		The five gates are called Plengkung. Plengkung Tarunasura (Wijilan) in the East and Plengkung Nirboyo (Gading) in the South. Plengkung Jagasura, Plengkung Jagabaya, and Plengkung Madyasura have been dismantled into ordinary gates.
4	*Pojok beteng* (*Beteng* corner)		The corner of *beteng* is located in all four corners of the *beteng* wall with the unique shape of a bastion that has not changed since long ago.
5	*Dalem Pangeran*		13 *Dalem Pangeran* buildings still exist even though some of their functions have been changed into schools and restaurants. The architecture of the building persists with traditional Javanese architecture.
6	*Tamansari* complex		Tamansari is currently one of the tourist attractions in the Palace area. The location of the bathing pool, Gumuling Well, barrier wall, and some old building sites are still there. Nevertheless, vacant land has been filled with residential homes making it a dense settlement.
7	Traditional houses/ kampung		Many houses with traditional Javanese architecture are still found, providing a robust traditional atmosphere.

No.	Physical elements	Pictures	Notes
8	Spatial pattern		The spatial pattern formed by space zoning and road patterns remains the same as they used to be. The existing road patterns composed the zones of the Palace, Alun-alun, Tamansari, Dalem-dalem Pangeran, and the *abdi dalem* houses.
9	Toponym of roads		The roads still use traditional *kampung* names that have a meaning following the work of the *abdi dalem* in the past.
10	*Ngindung/ magersari* houses		The houses of the inhabitants who occupy the land in the courtyard of Dalem Pangeran (they were once the prince's *abdi dalem*); remaining land along the inner *beteng* wall; land in Tamansari complex; and some other remaining land.
11	Commercial houses		They are residential buildings that change functions into shops or stalls.
12	Commercial areas		Commercial activities develop in some road corridors with a variety of sale materials, such as *gudeg* food, clothing, and souvenir.
13	Buildings with modern architecture		Traditional buildings have been transformed into buildings with modern architecture. Also new buildings with modern architecture.

Source: analysis, 2020

Table 2 shows that Jeron Beteng has many landscape elements that have existed since the beginning of its construction. The elements of the *beteng* wall and entrance gates (*Plengkung*) with their large shape become the main physical identity of Jeron Beteng. All those heritage elements highly contribute to form the landscape's characters, and even new commercial activities have given character variations, balancing the traditional character of the existing settlements.

Along with the development of the city, the Jeron Beteng area that used to function as a historical residence, began to experience building functions, building architecture, and the intensity of buildings triggered by the increasing economic needs of the community. Nevertheless, the character of the landscape physically still shows the uniqueness of the area as a traditional gated settlement with the Palace as the center of the area.

3.3. Continuity of Jeron Beteng heritage urban landscape

The urban landscape that is in a dynamic environment indeed continues to change. The physical and cultural changes of the landscape occur due to the dynamics of development and changes of the community activities. This has undoubtedly also affected the existence of landscape elements, and the quality of the landscape as a whole. The elements of the Jeron Beteng landscape have rich past values that continue until today. This means that these elements are still functioning properly today. An assessment of the continuity of the physical landscape elements of Jeron Beteng is done, using some criteria:1) past and present values, 2) physical form in the past and at present, 3) well function in the past and present, 4) the number of changes at present (Table 3).

Table 3 Assessment of the continuity of the landscape elements

No	Landscape elements	Criteria					
		1a	**1b**	**2a**	**2b**	**3a**	**3b**
1	Palace dan *Alun-alun*	vvv	vvv	vvv	vvv	vvv	vvv
2	*Beteng Baluwerti*	vvv	vvv	vvv	v	vvv	vvv
3	*Plengkung* (Gate)	vvv	vvv	vvv	v	vvv	vvv
4	*Pojok beteng* (*Beteng* corner)	vvv	vvv	vvv	vvv	vvv	vvv
5	*Dalem Pangeran*	vvv	vvv	vvv	v	vvv	vvv
6	*Tamansari* complex	vvv	vvv	vvv	v	vvv	vvv
7	Traditional houses/ *kampung*	vvv	vvv	vvv	v	vvv	vvv
8	Spatial pattern	vvv	vvv	vvv	vvv	vvv	vvv
9	Toponym of roads	vvv	vvv	vvv	vvv	vvv	vvv
10	*Ngindung/magersari* houses	v	vvv	v	vvv	-	vvv
11	Commercial houses	-	vvv	-	-	-	vvv
12	Commercial areas	-	vvv	-	-	-	vvv
13	Buildings with modern architecture	-	vvv	-	-	-	vvv

Source: analysis, 2020

Note:
1a : past values
1b : present values
2a : past physical form
2b : present physical form
3a : past function
3b : present function

vvv high
v low

The assessment result shows that all elements of the Jeron Beteng landscape still have essential values, both past values, and values for today's life. So is the physical existence of these elements, which still exist at present. Functionally, all elements of the landscape still work well, despite there are changes. Some of the changes found are seen in Table 4 below.

Table 4 Assessment of changes of the landscape elements

No.	Landscape elements	Scale	Changes
1	Palace and *Alun-alun*	v	Physically and the function of Palae and Alun-alun does not change much.
2	*Beteng Baluwerti*	vv	The wall along the north side no longer exists and is transformed into a row of *magersari* houses. The moat (*jagang*) around the beteng are gone.
3	*Plengkung* (Gate)	vv	There are 3 gates: Plengkung Jagasura, Plengkung Jagabaya, and Plengkung Madyasura which has been dismantled into regular gates.
4	*Pojok beteng*	v	All four *beteng* corners only have minor physical changes.

No.	Landscape elements	Scale	Changes
5	*Dalem Pangeran*	vv	Some functions of the 13 Dalem Pangeran have been transformed into schools, restaurants, etc. The architecture of the buildings remain or slightly changes.
6	*Tamansari* complex	vv	No longer serves as a bathing and garden area, but is currently a residential area that occupies empty land around the relics.
7	Traditional houses/ *kampung*	vv	*Kampung abdi dalem* has been transformed into an ordinary settlement, due to changes in land ownership and the current function of *abdi dalem* that has been different from the past
8	Spatial pattern	v	Almost no physical changes
9	Toponym of roads	v	Road names use *kampung* names
10	*Ngindung/magersari*	vv	The number of houses are growing but still under control.
11	Commercial houses	vvv	Many changes in house function into shops and stalls as the increase of economic needs.
12	Commercial areas	vvv	The development of commercial activities along several roadsides marked by converting old houses into shops and the growth of new buildings with commercial functions (shops, stalls, arts hops, cafés, restaurants, etc.).
13	Buildings with modern architecture	vvv	Many traditional buildings have been transformed into buildings with modern architecture. So is the development of new buildings with modern architecture.

Source: analysis, 2020

Note: vvv big/significant changes vv medium changes v small changes

Table 4 indicates that all elements of the landscape have changed, although many are at least different. Most changes occurred particularly in the use of houses and new buildings for commercial activities. Also, changes in the residents' activities and the emerge of modern houses, due to the residents' desire to change their traditional houses into houses with modern architectural style. Nevertheless, the physical landscape elements built from the beginning of the *beteng* construction can still be found and seen today. Those elements also still function well as their functions in the past.

Change has certainly been ongoing and will continue to affect the continuity of the Jeron Beteng heritage landscape. Nevertheless, the study result shows that many changes have not yet impacted to the decrease of the existence and function of the landscape. The changes that exist today are still expected to be a result of the ongoing dynamics of development, which are inevitable. Most elements of Jeron Beteng heritage landscape existence are still continuous from the past until now. The form of elements of the Jeron Beteng landscape, which are the Beteng Baluwerti, Palace, Alun-alun, Dalem Pangeran, spatial pattern, and part of Tamansari, is still survive as it existed for hundreds of years ago. The landscape is continuing in the middle of the development pressures. The balance between changing and unchanging is the key to the continuity of the heritage urban landscape. Jeron Beteng's historical, spatial, and cultural landscapes need to be sustainable in accepting change and development, between maintaining continuity of old assets and fulfilling current and future living needs.

4. Conclusion

Jeron Beteng, with its landscape character, is the heritage of Yogyakarta city, which is not found anywhere else. The genius of Sri Sultan Hamengku Bowono I as a planner and the Sultanate Palace's power made the Jeron Beteng residential area continue amid the rapid development of the city. The physical elements of the landscape built from the beginning to the present still exist and function properly. The heritage elements include the Palace, Alun-alun, Tamansari complex, *beteng* wall, Plengkung (gate), spatial pattern, and traditional houses. Despite the changes, its existence dominates the formation of the landscape character of Jeron Beteng.

The continuity of the Jeron Beteng heritage landscape is achieved because of the strong power of the Sultanate Palace of Yogyakarta. It also means that the Palace's rules have been appropriately implemented. Nevertheless, the pressure of urban life continues to be experienced in the Jeron Beteng area. Therefore, good management and public awareness of the importance of Jeron Beteng conservation is indispensable. Jeron Beteng heritage landscape is continuing to be the identity of Yogyakarta Sultanate and the city of Yogyakarta.

References

[1] Haber W 1995 Concept, Origin and Meaning of "Landscape," in von Droste B, Plachter H, and Rossler M *Cultural Landscapes of Universal Value*, Gustav Fisher Verlag, New York
Ndubisi F 1991 Landscape ecological planning, in Steiner F *The Living Landscape: An Ecological Approach to Landscape Planning,* McGraw-Hill, Inc., New York
[2] Green B H, Simmons E A and Woltjer I 1996 *Landscape Conservation. Some Steps Towards Developing a New Conservation Dimension,* Draft Report IUCN-CESP Landscape Conservation Working Group, Department of Agriculture, Horticulture and Environment, Wye College, Ashford, Kent
[3] Steiner F 1991 The living landscape: an ecological approach to landscape planning, McGraw-Hill, Inc., New York
[4] Ndubisi F 1991 Landscape Ecological Planning, dalam Steiner F *The Living Landscape: An Ecological Approach to Landscape Planning,* McGraw-Hill, Inc., New York
[5] O'Donnell P M 2008 Urban cultural landscapes and the spirit of place In: 16th ICOMOS General Assembly and International Symposium: 'Finding the spirit of place – between the tangible and the intangible', 29 sept – 4 oct 2008, Quebec, Canada. Conference or Workshop Item] http://openarchive.icomos.org/67/1/77-hFCw-13.pdf
[6] Yaldız E, Aydın D and Sıramkaya S B 2014 Loss of city identities in the process of change: the City of Konya-Turkey
[7] Adishakti L T 1999 Pengantar konservasi lingkungan kota berejarah Writing Project- Directorate of Higher Education, Republic of Indonesia, Jakarta
[8] Rahmi D H 2015 The Cultural Landscape of Borobudur: Borobudur Villages – continuity and changes, in Kanki K, Adishakti L, Fatimah T (editors) *Borobudur as Cultural Landscape. Local Community Initiatives for the Evolutive Conservation of Pusaka Saujana Borobudur*, Kyoto University Press, Kyoto
[9] Dinas Kebudayaan (Kundha Kabudayan) Daerah Istimewa Yogyakarta 2019 *Kraton Yogyakarta Pusat Budaya Jawa*, Yogykarta
[10] Gupta D, Handayani T, Hamoko D and Yuliani P 2007 *Toponim Kota Yogyakarta*, Yogyakarta: Dinas Pariwisata, Seni, dan Budaya, Kota Yogyakarta
[11] Santosa J 2008 *Arsitektur-Kota Jawa: Kosmos, Kultur & Kuasa*, Centropolis: Universitas Tarumanegara, Jakarta
[12] Pemerintah Daerah Daerah Istimewa Yogyakarta Dinas Pekerjaan Umum, Perumahan, Energi dan Sumber Daya Mineral 2016 *Rencana Induk Kawasan Budaya Perkotaan Yogyakarta 2014-2034*, Yogyakarta

Re-designing of the bottleneck on Suryanata Street Samarinda City East Kalimantan

E P Raharjo[1], A M Rahayu[2] and V O Hartini[3]

1 2 3 Indonesian Land Transportation Polytechnic - STTD, Jl Raya Setu no 89 Cibuntu Cibitung Bekasi 17510

E-mail: efendhisttd@gmail.com

Abstract. Some indicators of traffic problems like congestion, pollution and accident become the benchmark to evaluate the infrastructure performance. One kind of problem that often happened are the Bottleneck problems. In this kind of condition, the capacity of one side of the road is lower than before condition. On the other hand, in the process of landscape design of bottleneck deciding only by the availability of the space and those kinds of condition are not getting research by some specialist. The purpose of this research is to analyze the design parameter to enhance the capacity so case we can decreasing the number of congestion in a significant way. This research has taken place in Suryanata Street, Samarinda City.This research uses the model of Greenshield Linear, Greenberg Logarithmic, and Exponential Underwood to state the relationship between the performance of Traffic, Volume, Speed, and Density so we can decide the best model to analyze the shock wave that is using to design the parameter that effecting the design of bottleneck road area. Meanwhile, the pollution that can identify it consist of: CO (in the part of Per Million / PMM), NOX (calculate in Cubic meter / M3) and level of smoke (calculate in Cubic Meter / M3) with using the model of Hobbs (1979). Re-designing will be implemented to get the optimal performance infrastructure in the research location, and evaluate those kinds of terms.The result of the research got three design parameters in the Bottleneck area. There are track width, length of the road, and traffic volume. From the result analysis, the researcher is doing the Re-designing in the bottleneck area of the research location by using the parameter performance of congestion, specifically on delayed and pollution, like CO, NOX and level of smoke. Based on the re-designing, we can be decreasing the delay in the amount of 29.31% decreasing the number of NOX by 25.03% on CO we can be decreasing in the amount of 14.88% and also decreasing the level of smoke by 11.94%.
Keywords: Bottleneck, CO, Delayed, Delayed, Level of Smoke, NOX, Shock Wave

1. Background

In general, the problem of road narrowing (bottleneck) is a traffic problem in urban areas [1]. The road narrowing results from changes in road dimensions, both width and number of lanes [2]. No exception to the road narrowing on Suryanata Street, Samarinda City, performance traffic problems arise due to road narrowing, a section of the road with the road type 4/2 D becomes 2/2 UD, in other words the condition of the traffic capacity afterwards is smaller than the entrance section (before) so that there will be a decrease in traffic performance. Indicators of traffic problems in road narrowing (Bottlenecks) include: congestion, pollution and accidents used as benchmarks in evaluating infrastructure performance. Meanwhile, in terms of the design (landscape) road narrowing has not been studied much, it arises because of the lack of land, while the optimization of existing infrastructure has never been studied and applied in its application [3].

The purpose of this study is to determine the design parameters to increase service capacity so that it can significantly reduce congestion by using the analysis of the relationship between traffic flow performance on roads and the effect of road narrowing, as well as implementing the results of research on narrowing on Suryanata Street Samarinda City.

The benefit of research is availability of road narrowing design parameters from the results of the analysis of the traffic flow performance relationship that can be used in overcoming congestion caused by road narrowing, as well as the implementation of the results of the design planning analysis as a solution to the problem of road narrowing at the study location.

2. Research methodologi

The methods of collecting data are divided into 2 (two), namely:

a) Secondary Data,

b) Primary Data, including [4]:

1) Segment geometric data,
2) Traffic volume data,
3) Speed data and vehicle travel time,
4) Vehicle Queue Data.

In road narrowing, design variables that can be used as candidate variables include variables from road characteristics and traffic characteristics. The characteristics of the road include: The initial road length and the final road width or the number of initial and final lanes, and the length of the road connecting the initial and final road segments. While the road characteristics variable.

Here is a few variables which affects the design on the Bottleneck:

1) Lane Width
2) Traffic Flow Performance
3) Length of the road on Suryanata Street Segment 2

The method for analyzing the relationship between volume, velocity, and density uses a data distribution chart. Furthermore, the method used to analyze the mathematical relationship between the three parameters of traffic performance parameters using three types of models, namely the Greenshield Model, Greenberg Model, and Underwood Model [5]. Shock wave analysis using the results of the selected model. Analysis of vehicle exhaust emissions using the Hobbs air pollution regression model [6].

3. Research result

Road geometric that affects road capacity and performance, namely the type of road that determines the difference in traffic loading, traffic lane width, which can affect the value of free-flow speed and capacity. The geometric characteristics of the road are from the road conditions (4/2 T) to (2/2 T).

3.1. Traffic performance analysis

The goal is to measure traffic performance and identify problems that occur on roads and traffic space usage. The traffic flow performance analyzed are volume, speed and density [7].

IOP Conf. Series: Earth and Environmental Science **879** (2021) 012032 doi:10.1088/1755-1315/879/1/012032

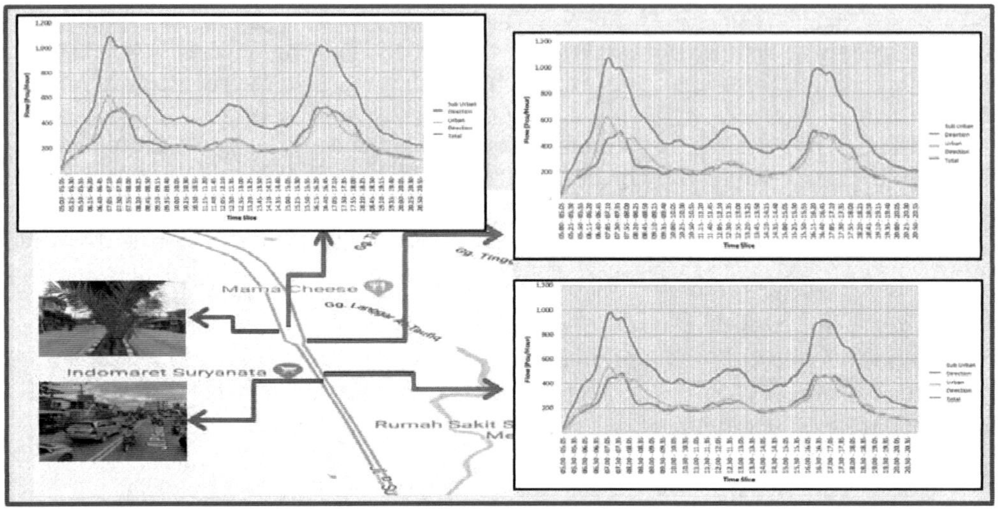

Figure 1 The fluctuation of traffic volume on Suryanata Street road segment between standard and narrow

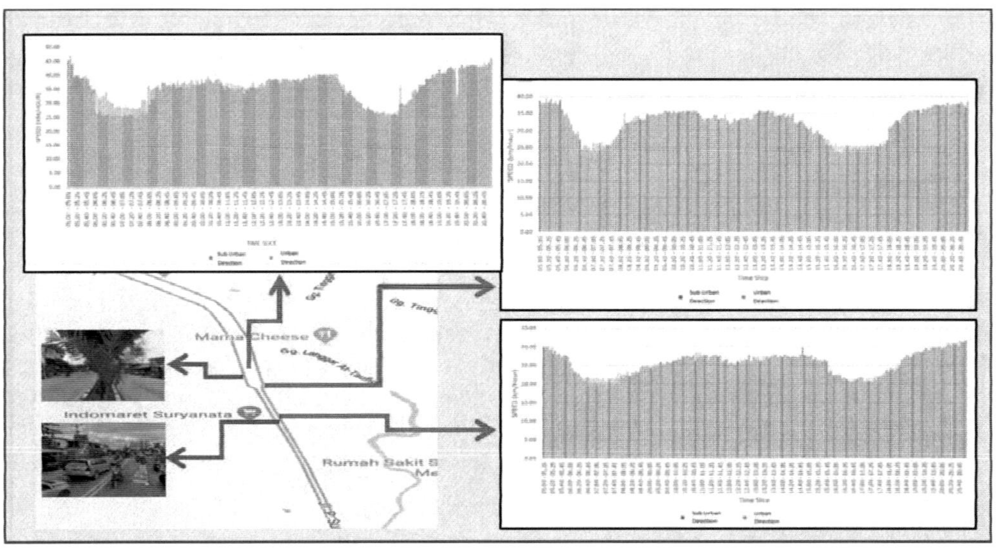

Figure 2 The fluctuation of traffic speed on Suryanata Street road segment between standard and narrow

IOP Conf. Series: Earth and Environmental Science **879** (2021) 012032 doi:10.1088/1755-1315/879/1/012032

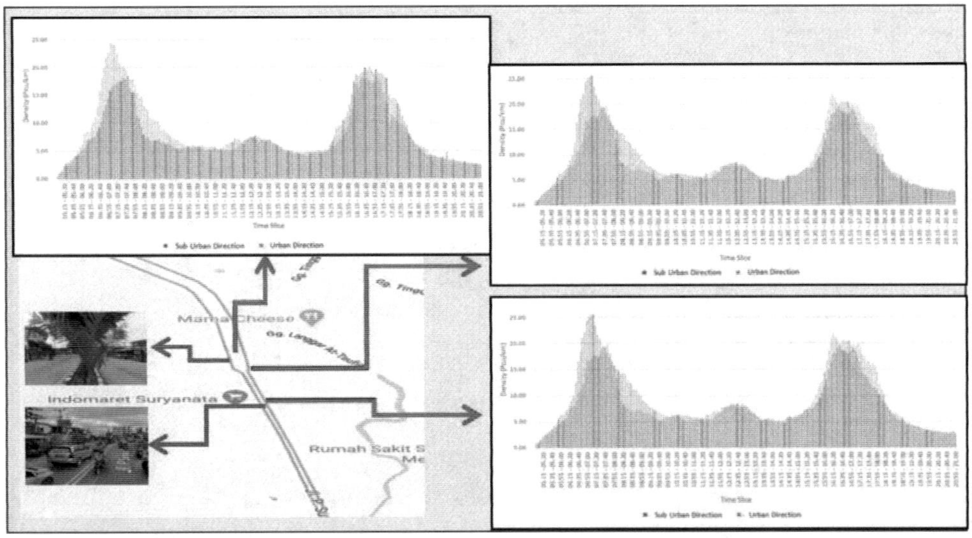

Figure 3 The fluctuation of traffic density changes on Suryanata Street, the road segment between standard and narrow roads

3.2. Analysis of the Relationship of Speed, Density and Volume

The choice of the traffic volume (V) and speed (S) data relationship model using the Greenshield, Greenberg and Underwood methods. From the field data, the relationship between volume and speed is obtained as in Table 1.

Table 1 The value of R2 Suryanata Street Determination for the three segments in the three models

Condition	Relationship	*Greenshileds*	*Greenberg*	*Underwood*
Normal road direction out of town	SD	0.952	0.869	0.876
	VD	0.986	0.986	0.988
	VS	0.629	0.740	0.727
	Average	0.856	0.865	0.864
Normal road direction to enter the city	SD	0.910	0.842	0.845
	VD	0.962	0.975	0.971
	VS	0.583	0.560	0.642
	Average	0.818	0.793	0.819
The road between normal and narrow roads out of town	SD	0.950	0.825	0.877
	VD	0.979	0.981	0.982
	VS	0.714	0.568	0.747
	Average	0.881	0.791	0.869
The road between normal and narrow roads in the direction of entering the city	SD	0.942	0.774	0.797
	VD	0.972	0.977	0.976
	VS	0.596	0.471	0.622
	Average	0.837	0.740	0.798
The road between normal and narrow roads out of town	SD	0.927	0.850	0.857
	VD	0.985	0.991	0.988
	VS	0.631	0.711	0.681
	Average	0.848	0.851	0.842
The road between normal and narrow roads in the direction of entering the city	SD	0.966	0.852	0.911
	VD	0.987	0.989	0.990
	VS	0.782	0.677	0.818
	Average	0.912	0.839	0.906

Based on the largest coefficient of determination, the selected model is the *Greenshield* model.

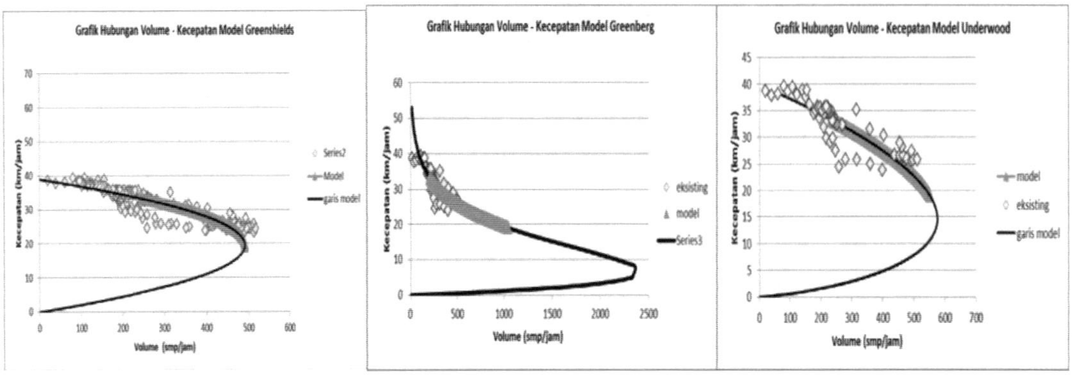

Figure 4 Relationship model using the Greenshield, Greenberg and Underwood methods.

3.3. Shock wave analysis

Shock waves can be described as movements in traffic volume due to traffic density changes and volume values. The shock wave on the road occurred due to the incident's impact so that the traffic volume was disturbed, resulting in the blocking of part or all of the road lanes [8].

By using the shock wave method, the values are obtained:

a. The shock wave (ω) is calculated using the formula, the value ω = -1.38 km/hour is obtained. A negative value indicates a shock wave backward forming in the direction of the vehicle's arrival.
b. According to the formula, the number of vehicles in the queue will get N = 260.25 *pcu*.
c. The time it took for the last vehicle to leave the last queue was 20.83 minutes.

3.4. Selection of variables that affect the performance of the Bottleneck area

There are three candidate variables observed, namely road width, bottleneck area length and traffic volume. The method used to assess the effect is by using the variable sensitivity method on traffic performance [9].

Graph of the Sensitivity to Change the Length of the Bottleneck area (Figure A) | Traffic Volume Change Sensitivity Graph (Figure B) | Sensitivity Graph to Change Road Width (Figure C)

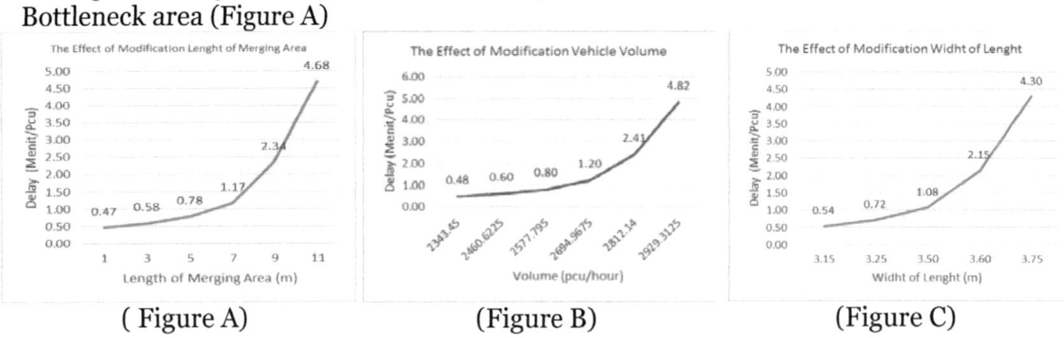

(Figure A) (Figure B) (Figure C)

Figure 5 Method tests the change in the value of each candidate variable using a trial-error method with the assumption that the value of the other candidate variables is fixed

From the sensitivity graph, it is obtained the relationship between candidate variables that affect traffic performance in the form of delays. So that the three candidate variables affect on traffic performance in the bottleneck area.

4. Design planning

From the results of the analysis of variable selection obtained variables: road lane width, bottleneck area length, and traffic volume. These three variables are used as a proposed improvement by changing the geometric shape of the road length in the narrowing area so that a geometric repair proposal is obtained for the narrowing area as follows:

The results of field observations (existing conditions) obtained data on the length of the bottleneck area, namely 12 m, with an angle α of 19°. as in the picture beside.

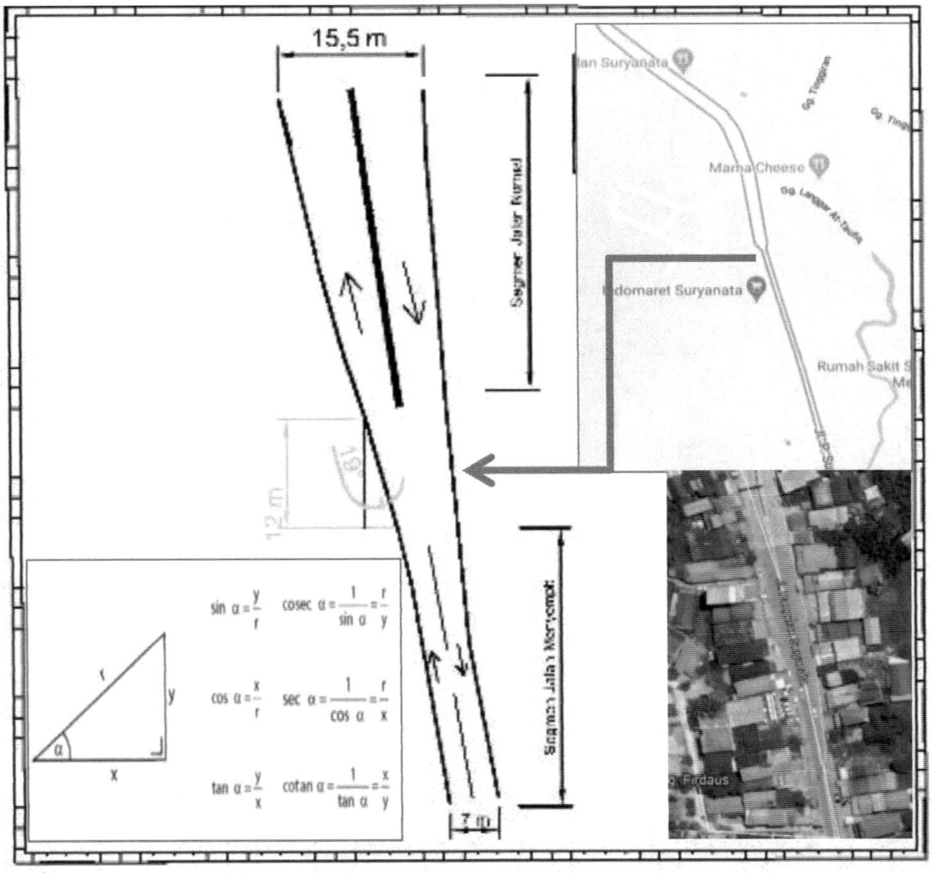

Figure 6 Existing and Road Area Planning Narrowing

To redesign the bottleneck on Suryanata Street by looking at the density of land use and traffic conditions, it is not possible to change the road width variable, and it is also difficult to control the amount of traffic volume so that the proposed change in the length of the bottleneck area becomes a mishandling of the design. The method used to propose changes in the length of the bottleneck area that can be proposed is by changing the length of the bottleneck area by using trial-error and measuring the traffic performance. The bottleneck area length assumption is accepted if the speed and density in the bottleneck area are close to the traffic flow on the standard road segment (without narrowing).

From the trial-error result, it is obtained that the bottleneck area length is 22 m with angle α is 90. It is sufficient to say that the speed and density of the bottleneck area and the standard road area are the same.

Design the narrowing area after the proposed improvements are as follows:

IOP Conf. Series: Earth and Environmental Science **879** (2021) 012032 doi:10.1088/1755-1315/879/1/012032

Existing	Redesign
Length of Bottleneck Area is 12 M, Angle α is 19°	Length of Bottleneck Area is 22 M, Angle α is 9°
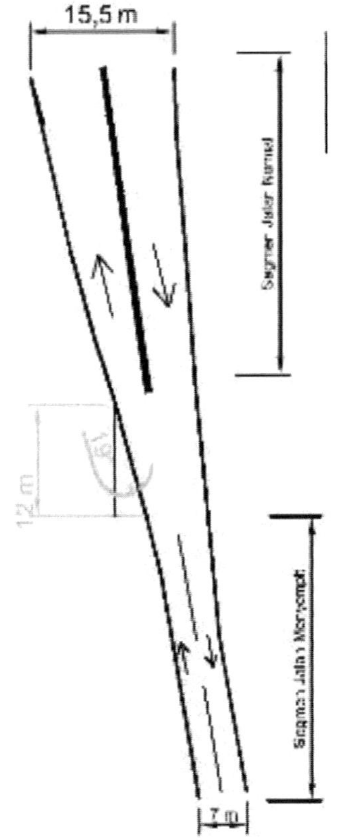	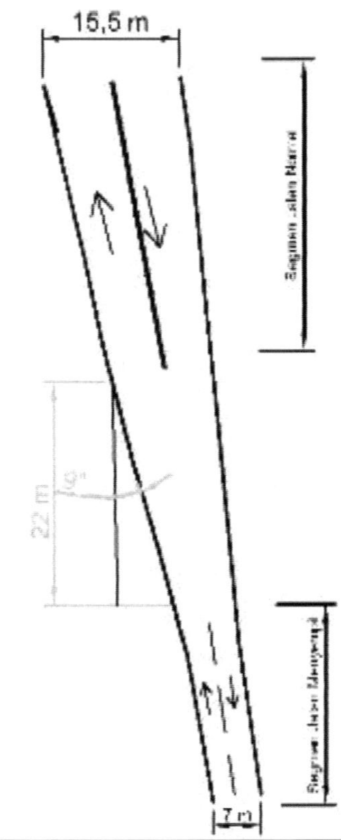

Figure 7 Existing and Redesign the narrowing area

5. Delay analysis

Delays are obtained from the calculation results by comparing the travel time required by vehicles to cross narrowed roads and normal roads. The following is the result of the calculation of delay analysis [8].

Table 2 The result of the existing delay calculation

	Existing	
Delay	T3-T2 (minutes)	T4-T2 (minutes)
6.31 (minutes / Pcu)	13.57	7,26
378.6 (sec / Pcu)	814.2	435.6

	Redesign	
Delay	T3-T2 (minutes)	T4-T2 (minutes)
4.46 (minutes / Pcu)	9.89	5.43
267.6 (sec / Pcu)	593.4	325.8

The amount of delay will be obtained from the travel time on narrow roads reduced by travel time on normal roads calculated from the beginning of the narrowing to the end of the narrowing, while the travel time on normal roads is calculated on roads that do not experience narrowing.

6. Emission calculation analysis

From the results of the redesign of the design in the constriction area, there were several changes in the results of the analysis of vehicle exhaust emissions [10], to determine the relationship between traffic volume and air pollution, among others, presented by *Hobbs* (1979) which are summarized in the table below:

Table 3 Calculation results with the Air Pollution Regression Model [6]

Existing

Types of Pollutants	Regression Formulas	Information	Survey Results	Survey Time	Traffic Volume (Pcu)	Concentration of Gas
CO concentration (ppm)	C = 2.96 + 0.00032V + 0.0000005V ^ 2	V = Volume of the vehicle for 3 hours	Out of Town Directions	07.00 - 10.00	10455.1	60.96
			Directions to Enter the City	14.00 - 17.00	15107.7	121.92
Nox Concentration (m3)	N = 46.9-0.036T + 0.00004T ^ 2	T = Vehicle Volume for 1 hour	Out of Town Directions	07.00 - 08.00	5379.4	1010.76
			Directions to Enter the City	16.00 - 17.00	6111.3	1320.81
Smoke Rate (m3)	S = 9.49 + 0.022 V	T = Vehicle Volume for 1 hour	Out of Town Directions	07.00 - 08.00	5379.4	127.84
			Directions to Enter the City	16.00 - 17.00	6111.3	143.94

Proposal

Types of Pollutants	Regression Formulas	Information	Survey Results	Survey Time	Traffic Volume (Pcu)	Concentration of Gas
CO concentration (ppm)	C = 2.96 + 0.00032V + 0.0000005V ^ 2	V = Volume of the vehicle for 3 hours	Out of Town Directions	07.00 - 10.00	9608.6	52.20
			Directions to Enter the City	14.00 - 17.00	13840.1	103.16
Nox Concentration (m3)	N = 46.9-0.036T + 0.00004T ^ 2	T = Vehicle Volume for 1 hour	Out of Town Directions	07.00 - 08.00	4615.4	732.82
			Directions to Enter the City	16.00 - 17.00	5410.2	1022.94
Smoke Rate (m3)	S = 9.49 + 0.022 V	T = Vehicle Volume for 1 hour	Out of Town Directions	07.00 - 08.00	4615.4	111.03
			Directions to Enter the City	16.00 - 17.00	5410.2	128.51

The reduction of pollution with a change in the length of the bottleneck area from the existing to the standard conditions can be seen as follows:

Table 4 Calculation results the reduction of pollution

Types of Pollutants		Decline	Percentage reduction
CO concentration (ppm)	Out of Town Directions	8.76	14.37%
	Directions to Enter the City	18.75	15.38%
	Total	**27.51**	**14.88%**
Nox Concentration (m^3)	Out of Town Directions	277.94	27.50%
	Directions to Enter the City	297.87	22.55%
	Total	**575.81**	**25.03%**
Smoke Rate (m^3)	Out of Town Directions	16.81	13.15%
	Directions to Enter the City	15.42	10.72%
	Total	**32.23**	**11.94%**

7. Conclusion

The variables that affect traffic performance in the bottleneck area are variable: road lane width, bottleneck area length, and traffic volume. In the study area, the geometric road improvements were carried out by changing the length of the bottleneck area. The length of the bottleneck area in the existing 12 meters was changed to 22 meters. From the results of the redesign of the narrowing area in the study location with performance parameters in the form of a decrease in performance delay from 378.6 sec/pcu in the existing condition to 267.6 sec/pcu for the proposed condition or by 29.31% and a decrease in NoX of 25.03%. Co is 14.88% and the smoke level is 11.94%.

8. Suggestion

For the further studies related to the addition of other candidate variables that affect the performance of the bottleneck area, especially those related to the percentage of heavy vehicles and motorbikes.

References

[1] Fitriani A 2012 Pengaruh Penyempitan Jalan terhadap Karakteristik lalu lintas jalan (Studi kasus: Jl.P.Kemerdekaan dekat Mitos Jembatan Tello). Universitas Hasanudin
[2] Ariadi 2016 Pengaruh Penyempitan Jalan Pada Jembatan (Bottleneck) terhadap karakteristik lalu lintas (Studi kasus : Jembatan lamyong, jalan teuku nyak arief banda aceh) (Universitas Syiah Kuala)
[3] Tamin O Z 2008 *Perencanaan. Pemodelan. & Rekayasa Transportasi* (Bandung: ITB)
[4] Indrajaya Y 2003 Pengaruh Penyempitan Jalan Terhadap Karakteristik Lalu Lintas
[5] Tamin O Z 1992 Hubungan Volume, Kecepatan, dan Kepadatan Lalu Lintas Di Ruas Jalan H.R Rasuna Said," *J. Tek. Sipil*, vol. 5
[6] Hoobs F D 1979 *Traffic Planning and Engineering 2ed.* London: Pergamon Press
[7] Arthur L K 2013 Analisis Kinerja Jaringan Jalan Dalam Kampus Universitas Sam Ratulangi
[8] Imammul M M 2014 Hubungan Tundaan dan Panjang Antrian Terhadap Konsumsi Bahan Bakar Akibat Penyempitan Jalan (Bottleneck) pada pembangunan Flyover Palur (Studi kasus : jalan Raya Palur Km 7.5)
[9] Harinaldi 2005 *Prinsip-prinsip Statistik untuk Teknik dan Sains*. Jakarta: Erlangga
[10] Ruktiningsing R 2014 Kajian Hubungan Volume Lalu Lintas Terhadap Emisi Gas Buang Kendaraan di Ruas Jalan Majapahit Semarang (Studi Kasus : Kadar CO dan PM10), Universitas Katolik Soegijapranata, Semarang

The role and importance of plant purification systems in the ecological approach to landscape design

I R Dika[1] and M Rimac[2]

[1] Assistant professor, University of Zagreb, Faculty of Agriculture, Department of ornamental plants, landscape architecture and garden art / School of Landscape Architecture, Svetošimunska cesta 25, 10000 Zagreb, Croatia

[2] MLA, Pilatušćak 3, 10000 Zagreb, Croatia

E-mail: [1] irechner@agr.hr; [2] rimac.martina@gmail.com

Abstract. The continuous city expansion and use of various impermeable pavement materials greatly hinder the rainwater surface runoff by conventional drainage systems, especially during heavy and extreme precipitation, which is occurring more frequently. Also, mostly as a rule, natural watercourse discontinuity and/or their canalization results in more frequent floods in cities. An additional problem is rainwater pollution, which, with conventional drainage systems, ends up in natural watercourses, rivers, lakes and seas often without any or even minimal treatment. These are some of the main reasons why rainwater management is becoming an increasingly more important and indispensable segment of contemporary sustainable city development. Among many other approaches, it also includes conventional rainwater drainage systems replacement and the application of biological rainwater treatment systems as much as possible. The positive impact of sustainable and environmentally friendly rainwater treatment systems has already been widely recognized. However, their application is still slightly limited compared to the problems that occur in modern cities. The aim of this paper is to present biological rainwater treatment systems (basic categories and characteristics) and to analyze their implementation possibilities in urban green areas. Their ecological and aesthetic values will be considered, and how users generally perceive structures of 'natural' characteristics within designed landscapes will be discussed. The analysis of fourteen recently constructed parks will be carried out to determine the ways of integrating biological purifier systems in the park, the possibilities of their use, and the degree of biological purification systems influence on the overall park design concept. Criteria or guidelines that should be taken into account when designing biological rainwater purification systems will be proposed in order to simultaneously meet their basic function and other roles of such structures within the city park.

Keywords: parks, sustainable water management, wetlands

1. Introduction

The continuous city expansion and use of various impermeable pavement materials greatly hinder the rainwater surface runoff by conventional drainage systems, especially during heavy and extreme precipitation, which is occurring more frequently. Also, mostly as a rule, natural watercourse discontinuity and/or their canalization results in more frequent floods in cities. An additional problem is rainwater pollution, which, with conventional drainage systems, ends up in natural watercourses, rivers, lakes and seas often without any or even minimal treatment.

These are some of the main reasons why rainwater management is becoming increasingly more important and indispensable segment of contemporary sustainable city development. Sustainable rainwater management involves different approaches and systems, and depending on the requirements can be applied in combination or independently. Today, sustainable stormwater drainage systems are increasingly being used, and they gradually supplement or replace conventional drainage systems, especially in urban areas.

Depending on the function they perform in the rainwater management process, the systems can be divided into six basic categories: infiltration systems, filtration systems, constructed wetland systems, retention systems and flow control structures and detention systems [1]. Infiltration systems enable the slow flow and filtration of rainwater and its drainage to another area (most often natural watercourses) and thus mitigate urbanization impact on the natural water balance in a particular area. They can also contribute to the protection against erosion due to the large amounts of rainwater influx and flood protection. Filtration systems also collect water, filter and remove the contaminants, but are primarily used to collect rainwater from roads and other non-impermeable areas [2]. Although there are several different filtration systems, only rain gardens in which plants play the most important role in the water treatment mechanism are mentioned. Rain gardens are usually designed as narrow linear elements that collect water, infiltrating it gradually into the soil, collecting at the bottom of the rain garden, or discharging it into a natural watercourse [1]. The successful functioning of these systems does not require the constant presence of water. That is a basic difference compared to wetland systems that must be underwater throughout the year. Constructed wetlands are designed shallow depressions with dense plantations of aquatic plants used to maximize pollution removal from rainwater, and include designed wetlands and wet ponds [1]. These systems are the most important plant purification systems and provide various applications in the design of open green areas that will be discussed in detail later in the paper. Retention systems are ponds, designed/created using natural depressions that are constantly underwater, with landscaped banks to mitigate surface impact and erosion due to heavy rainfall [3]. The last category is detention systems (includes dry ponds and dry swales) and their role is to retain rainwater during heavy rainfall in order to reduce the pressure on other rainwater collecting systems and reduce the possibility of flooding [1].

Even though the positive impact of all sustainable and environmentally friendly rainwater treatment systems has already been widely recognized, their application is still quite limited considering the advantages that they have over conventional systems. This statement is partially, in the context of a theoretical approach to modern sustainable/ecological approach to landscape design, confirmed by a research which identified sustainable rainwater management and biological (plant purification) water treatment as some of the basic principles associated with ecological system imitation [4]. Although, according to the analysed theoretical works, these two principles should be generally applicable as a gold standard in contemporary landscape design, the analysis of recent park projects shows that the principle of rainwater management is used in 60% and the principle of biological water treatment in 37% of parks [4]. Despite all the limitations and possible shortcomings of that research the results indicate that approaches to sustainable rainwater management and biological water treatment are not applied in the design of open green areas to the extent expected, given the global importance of the sustainability paradigm in contemporary society.

That is why the aim of this paper is to present biological rainwater treatment systems (basic categories and characteristics) and to analyze the possibilities of their application in urban green areas.

2. Materials and methods

A literature review and analysis of scientific and professional literature from several different fields (sustainable rainwater management, wetland/plant purification/systems, ecology, environmental psychology) was conducted to determine the main characteristics of plant purification systems from different aspects.

Also, an analysis and comparison of fourteen recent parks (Table 1) with a system of plant purifiers as a water element in the park was made. Parks selection was based on these criteria: realization within the last 15 years, location within the urban structure, public park and plant purification system as an integral part of the park. The main source of data about the parks were the park authors' descriptions published on the web.

Analysis and comparison of parks refers to the relationship of plant purification systems with other functions in the park, the design of plant purification system and the impact of the application of plant devices on the overall design concept of the park.

Table 1 List of analyzed parks

	Park Name	Location	Year	Previous Use	Area	Authors
1	Shanghai Houtan park	Shanghai, China	2010	Industrial	14 ha	Turenscape
2	Qunli park	Qunli New District, Haerbin, China	2011	Wetland area	34,2 ha	Turenscape
3	Tianjin Qiaoyuan park	Tianjin, China	2008	(shooting range) used as a landfill	22 ha	Turenscape
4	Minghu wetland park	Liupanshui, Guizhou, China	2012	A canalized river in the city	90 ha	Turenscape
5	Weiliu wetland park	Xianyang, China	2017	River area between two cities	152 ha	Yifang Ecoscape
6	Sydney park	Sydney, Australia	2015	Industrial	44 ha	Turf Design Studio i Environmental partnership
7	Le Pare du Chemin de l'Ile	Nanterre, France	2006	River bank	14 ha	Mutabilis Paysage et Urbanisme
8	Yichang Yunhe park	China Yichang, Hubei, China	2015	Former fish pond	12 ha	Turenscape
9	St Jacques ecological park	Ille & Vilaine (35), France	2013	Outside of the city along the agricultural land	40 ha	Atelier des Paysages Bruel-Delmar
10	Grorudparken	Oslo, Norway	2013	One of 4 parks by the river	12 ha	LINK Arkitektur
11	Pirrama park	Sydney, Australia	2010	Port	1,8 ha	ASPECT Studios
12	Magnuson park	Seattle, Washington, USA	2009	Airport takeoff	142 ha	The Berger Partnership and Sheldon Associates
13	Northala Fields park	London, Great Britain	2008		27,5 ha	By FoRM Associates
14	Barry Curtis park	Auckland, Australia	2009		94 ha	Isthmus

3. Wetland (plant purification) systems

Plant purification systems include designed systems that imitate natural wetland systems [5]. Their goal is to create conditions for enabling the treatment of wastewater flowing through the system thanks to various physical, biological and chemical processes [6]. In addition to purifying rainwater in cities, these systems can also successfully be used for treating gray household water and even wastewater generated as a product in various types of industrial activity.

3.1. Types of plant purification systems

Plant purification systems can have different design, construction and functional characteristics, and can be classified according to: plant type, water flow regime, types of plant purification systems configuration (hybrid, single-stage or multistage systems), type of treated wastewater, level of wastewater treatment, type of pre-treatment , type of substrate (gravel, sand, etc.) and the method of wastewater distribution [7]. The water flow regime has become established as the basic criterion for the typology of plant purification systems, according to which surface and subsurface plant purification systems are distinguished. The main difference between them is that in the surface plant purification systems the surface of the water passing through the system is exposed to the atmosphere while in subsurface plant purification systems, wastewater passes through the substrate (gravel or soil) in which the plants are planted and the water face is not in contact with the atmosphere [8]. Hybrid systems that consist of a combination of one or more surface and subsurface plant purification systems are also mentioned in literature [7].

A plant purification system with a surface water flow consists of a shallow pool of water, soil, or another medium that serves to grow rooted wetland plants, looks and functions like a natural wetland, and is a habitat for many plant and animal species. Depending on the type of plants, they can be divided into systems with plants whose parts protrude above the water surface level, systems in which plant parts float on the water surface and systems in which plants are submerged below the water surface level [5]. A subsurface plant purification system consists of one, or more, interconnected, embedded pools filled with gravel, sand or some similar medium substrate of different granulation. The same authors state that the surface can be dry and wet, and is regularly planted with selected plants whose roots penetrate the substrate and serve as a medium for developing microorganisms.

Although the hydrological requirements of plant purification systems are similar to natural wetland systems, plant purification systems, due to their large water surface and shallow depth, are highly dependent on meteorological conditions (precipitation and high temperatures) that lead to excessive evaporation.

3.2. Criteria for preparation and construction of plant purification systems

Although in this paper plant purification systems are considered primarily from the aspect of landscape design, it is important to note that in order to achieve a functional system, it is crucial to involve an interdisciplinary team of experts for their design and implementation.

For successful planning and construction of a plant purification system, it is necessary to research and analyze the area climatic characteristics, terrain topography, soil composition, groundwater characteristics (if any) and socioeconomic factors of the local community [5]. It is also necessary to ensure watertightness and gravitational water flow towards the system by the natural terrain topography (reduces the construction energy consumption) or, if necessary, to remodel the terrain.

An important factor in designing a plant purification system is certainly the quantity (including seasonal, weekly and daily variations) and quality (concentration of pollution) of wastewater and the requirement for the quality of treated water, which all affects the determination of capacity or size of the plant purification system. It is necessary to stress the importance of ensuring an adequate volume of one, or more, rainwater reception and retention basins long enough for the purification process to be completed [9].

Also, it is significant to select appropriate plant species. Various authors emphasize the importance of using local and indigenous wetland and aquatic plant species precisely because indigenous species are best adapted to the local climate thus increasing the probability of successful wastewater treatment [6,10,11]. In this context five planting zones of certain plant species and communities (depending on the depth of water and flooding frequency) are distinguished.

3.3. Advantages and disadvantages of plant purification systems

Among the advantages of plant purification systems, firstly their ability to remove contaminants from water should be mentioned. Various authors list five basic groups of substances that need to be removed: suspended solids, organic matter, nitrogen and nitrogen compounds, phosphorus and phosphorus compounds, and microorganisms [5,7,12]. Heavy metals and pathogenic substances are also mentioned by some authors [13].

The overall system cost-effectiveness, lower construction costs compared to conventional systems, low costs and energy consumption for system maintenance (carried out periodically), and recycling and reuse of water facilitation are also emphasized as their advantages [11]. According to the same author, the main disadvantages of plant purification systems are: significantly larger area required for construction compared to conventional systems (in the context of space as a resource and land economic value), the sensitivity of biological components to toxic compounds (e.g., ammonia and pesticides), but also clogging or poor functioning of a system that can disrupt the flow of water.

Regular monitoring, maintenance (a period of several months to several years depending on the part of the plant purification system) and remediation are the key factors for the successful long-term functioning of the plant purifier system. The most common problems due to poor management are: sediment accumulation (reduced storage volume), waste accumulation (blocks water supply and drainage), damage to pipes and pumps, the too rapid spread of invasive plant species that clog the rest of the wetland vegetation, vegetation loss on slopes and erosions and the collapse of the structure of embankments or basins [14].

3.4. Ecological and aesthetic values of plant purification systems

When it comes to ecological benefits, it should be said that natural wetland systems are considered to be one of the most biologically productive ecosystems on Earth and as such are characterized by exceptional biodiversity. The goal of plant purification systems is to imitate natural wetland systems [5]. Their design and construction create habitats suitable for the development of numerous plant and animal species, which ultimately has a positive effect on the biological diversity of the area. The approach of plant purification system design additionally, to a certain extent, contributes to this goal and should strive to form irregular edges of plant purification system with complex bottom topography, achieve changes in hydrological conditions (permanently and occasionally flooded zones) and alternating groups of plants and open water face [9]. Although, essentially the functioning of plant purification systems (wetlands) is not related to the way they are designed (geometrically or close to nature), it can still be concluded that from the aspect of biodiversity or landscape ecology, 'naturalistic' design might achieve better results.

This paper discusses the possibilities of application and the role of wetland systems in urban public open green areas, so it is important to mention their aesthetic role in the context of user perception. Water is unquestionably one of the elements that generally positively affects the attractiveness of the landscape, although, according to some research, wetlands are much less positively assessed than other aquatic phenomena (e.g., lakes and rivers) [15,16]. Also, people generally find scenes that look natural more attractive [17,18,19]. However, what is perceived as a natural landscape may differ from ecological naturalness, which is one of the basic wetland characteristics [20]. In this context, it should be mentioned that respondents rate landscapes of extreme biological (ecological) diversity as untidy and non-maintained, and ecologically restored landscapes were not necessarily attractive to them.[21] This perception can be associated with a sense of security in a particular space (extremely important to its users), which is the lowest in more closed spaces with a dense lower vegetation, and the highest in open spaces [22].

Despite the different results of numerous analyzed studies naturalness, spaciousness and coherence are some of the landscape characteristics that contribute to a more positive assessment.[4] The same author also points out that it is not possible to draw general conclusions about peoples' preferences for visual landscape features because they are diverse and depend on several different spatial, social, and even economic criteria and factors. It can be concluded that people in their environment want diverse landscapes [23] that will satisfy the needs and preferences of, primarily, local users.

4. Analysis of wetland (plant purification) systems in park design

Fourteen public parks were considered in which wetlands, i.e., plant purification systems, are applied in different ways, located in several countries on four continents - six parks in China, four in Europe, three in Australia and one in the USA. They are mostly situated within the densely built-up urban tissue along the river or sea, in a wetland, and almost all examples are brownfield sites (former industry, port, fishing ponds, landfill, etc.).

Probably this is the reason why the basic goal of almost all parks was the regeneration or revitalization of the degraded area as well as the water purification and/or soil purification. Reconstruction of the bank or riverbed, flood protection, creating a protective green belt between cities, etc., are also mentioned. In addition to the importance of a positive environmental impact, particular emphasis is placed on the educational role of parks in the context of learning about polluted water treatment technology and wetland habitats.

Since these are all public parks, they are characterized by various facilities and structures, however, the focus is mainly on pedestrian and bicycle pathways, various pavilions, canopies, benches, etc., and in several parks, there are other facilities such as a children's playground, a park for dogs, etc. In any case, the primary function of all parks is spending time in nature, leisure and recreation. The presence of a natural water element largely defines the character of all considered areas (except in three parks). Probably this is the reason why all parks have one feature in common, and that is providing access to water/wetlands to park users, which is most often achieved by placing a plateau above the water and a network of pedestrian pathways. This approach encourages spending time in an ecologically diverse area, which on the one hand, enriches the diversity of experiences in the park, and on the other hand, allows direct observation and education about natural processes.

The process of water purification by its nature presupposes gradualness, which most often results in a series of interconnected wetland elements, which are ultimately, at the end of the process, connected with some natural water element (river, lake, sea). In design, this often results in elongated, linear formations, and less frequently, and in examples of small-format wetlands, the elements are grouped into a field (e.g., Tianjin Qiaoyuan Park). As stated in the previous discussion, there is a tendency for a design that mimics nature, especially in the design of larger wetlands. Still, even in these cases, often some of the edges bordering the existing geometrically shaped structures (coast, road, etc.) are geometrically defined. Several characteristic examples showing these tendencies are considered in this paper.

Houtan Park (Figure 1) is a longitudinal spatial composition that stretches between the indented river bank and the city road. The main formal element of the park is a designed wetland that stretches along the entire length of the park and divides it into two parts with significantly different design characteristics. Part of the park towards the river is characterized by close to nature (curvilinear) design, which contrasts with the geometrically designed edge from previous industrial use.

Quinly Park (Figure 2) is located in the existing wetland area, which remains the main, central part of the park retaining its natural forms. All newly introduced facilities and elements are located at the edge of the park along the road. The peripheral belt forms a linear series of interconnected surface plant purification elements and mounds of irregularly broken geometric shapes connected by two pedestrian pathway systems designed in the same manner.

Figure 1 Houtan Park (www.landezine.com)

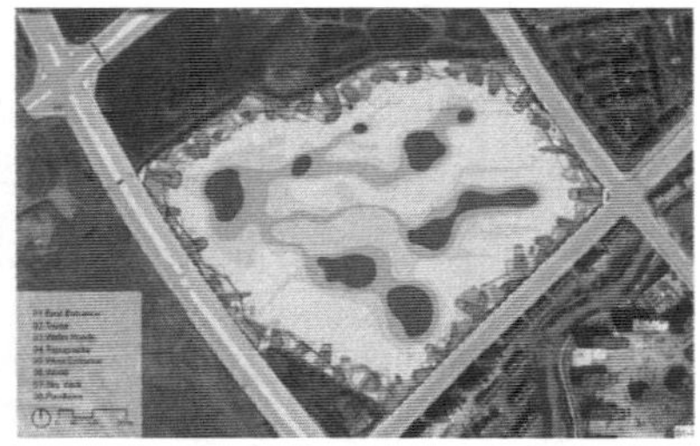

Figure 2 Quinly Park (https://www.archdaily.com/446025/qunli-stormwater-wetland-park-turenscape)

The spatial composition of Tianjin Qiaoyuan Park (Figure 3) consists of two contrastingly designed units, one of which is dominated by design close to nature, and the other of regular geometric shapes, separated from each other by a large L-shaped water surface of the same characteristics. The dominant part consists of egg-shaped swamps and modelled elevations of exact edges. Secondary elements - a pathway network in an irregular geometric grid is formed in their interspaces as a distinct contrast. The network and the wetlands system are framed by a curvilinearly shaped main promenade and separated from, on the one hand, by high vegetation-rich space leaning on the road, and on the other hand, from the contact surface to the water. The second part of the park is characterized by regular geometric shapes.

Unlike previous examples in Minghu Wetland Park (Figure 4), the central motif is a lake enriched with new architectural and park elements of a predominantly curvilinear character along its entire northern and southern shore. Existing streams and wetland ecosystem are integrated into the ecological rainwater wetland system, parallel to the lake shore that is formed by a series of elongated terraces connected by an intertwined dense network of pedestrian communications. The most pronounced element is the circular organic form bridge that connects the facilities on the northern shore of the lake with the wetland system.

Weiliu Park (Figure 5) also has a longitudinal form with edges of contrasting characteristics - curvilinear towards the river and a linear edge along the road. The central axis of the park is a linear promenade, both the main pedestrian and bicycle corridor stretch from the southwestern to the northeastern edge of the park, with many other facilities arranged next to it.. The promenade divides the park into two parts, both of design close to nature, but on the south side there is a system of plant purification elements, and on the north side are located all other facilities.

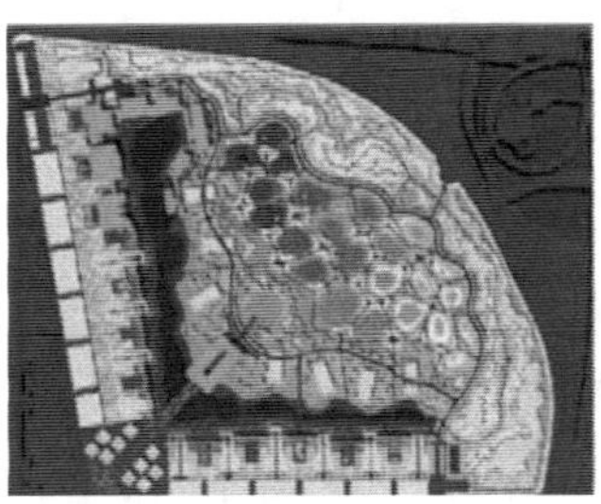

Figure 3 Tianjin Qiaoyuan Park (www.landezine.com/index.php/2011/03/tianjin-qiaoyuan-park-by-turenscape-landscape-architecture/)

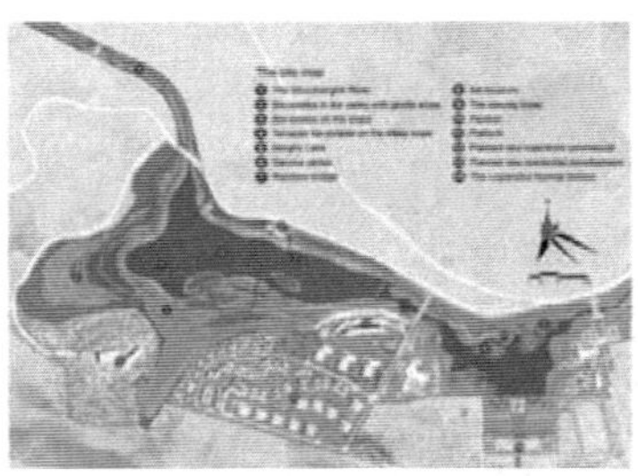

Figure 4 Minghu Wetland Park (www.dezeen.com/2015/01/11/turenscape-the-slow-down-liupanshui-minghu-wetland-park-meandering-causeways-landscape-architecture/)

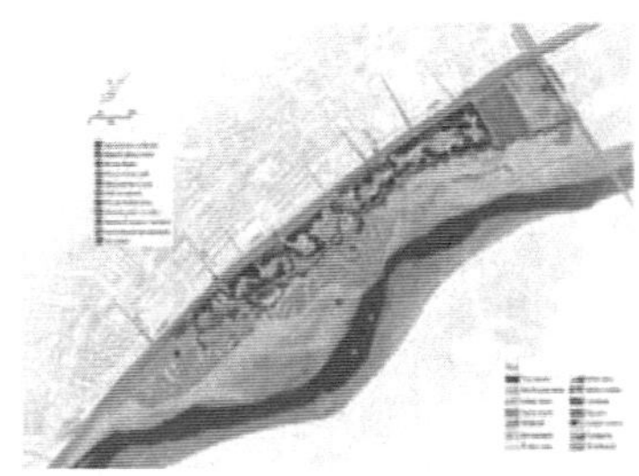

Figure 5 Weiliu Wetland Park (www.landezine.com/index.php/2019/01/weiliu-wetland-park-by-yifangecoscape/)

In Sydney Park (Figure 6), the central motif of the park is a wetland system composed of four bioretention water bodies of nature like character. Water flows through a series of cascades from the north to the south of the wetland system. Here, too, all the facilities are located on the periphery, and are connected by a network of curvilinear pedestrian pathways.

Northala fields park (Figure 7) is divided by the main promenade into two differently designed parts. The main design elements of one part are the geometric structures of four modelled regular cones, and the other part a series of interconnected and watercourse-connected lakes. The shape of the lakes varies, so those located along the main promenade, are almost regular geometric shapes. In contrast, the more distant ones have more and more curvilinear, close to nature design elements. The paths are of a geometric character (rectilinear, segments of a circle and a spiral).

Figure 6 Sydney Park (www.archdaily.com/793523/sydney-park-water-re-use-project-turf-design-studio-plus-environmental-partnership-alluvium-turpin-plus-crawford-dragonfly-and-partridge?ad_medium=gallery)

Figure 7 Northala Fields Park (www.landzine.com)

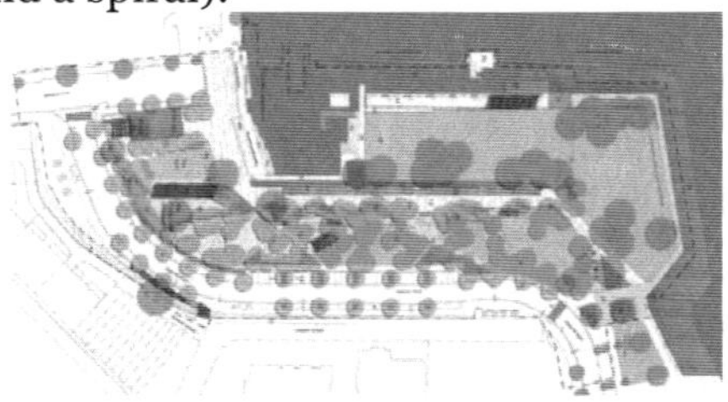

Figure 8 Pirrama Park (www.cityofsydney.nsw.gov.au)

The Pirrama park (Figure 8) design backbone is the main promenade (linear element), which also divides the park into two completely different parts. The part of the park towards the coast is a unique area of lawn plane on which smaller irregular groups of high vegetation have been introduced along the perimeter. The contact with the sea is formed by the existing sequence of wooden platforms, supplemented by a geometrically articulated bay coast that gradually descends towards the sea. The second part of the park is fragmented by a series of angled straight paths and retaining walls in the same geometry. High vegetation in numerous groups is freely superimposed on the geometric base. Biological water treatment pools are accompanying elements and occur as a narrow linear sequence along the promenade.

5. Gudelines and conclusion

The recent ecological design paradigm (water purification, etc.) is not experiencing its full potential in newer park designs and it is only expected to be more widely used outside the brownfield area. When introducing wetlands into urban areas, it is important to pay attention to the design that users will perceive as safe and desirable to use.

The design of these close to nature ecosystems implies creating complex spaces and forms, because the spaces designed in this way encourage the emergence of diverse habitats, which ultimately encourages biodiversity in a particular area. The experience of previously built parks shows that in intensive water treatment, wetlands usually form a separate spatial unit relatively separate from other facilities, and trails and plateaus for leisure and observation in these parts are the predominant facilities. The educational component of wetlands is vital for the overall understanding of ecological processes and raising awareness of their importance, and could be used as a demonstration site on a smaller scale and in smaller spatial units.

References

[1] Metropolitan Council and Barr Engineering Company 2001 Minnesota urban small sites bmp manual, stormwater best management practices for cold climates metropolitan Council Environmental Services St. Paul, MN
[2] Philadelphia water department 2018 The Philadelphia stormwater management guidance manual, chapter 4 - Stormwater management practice guidance, 4.9 Media Filters
[3] European Commission 2013 Natural water retention measures, individual nwrm retention ponds http://nwrm.eu/measure/retention-ponds
[4] Rechner D I 2012 Ekološko oblikovanje kao paradigma suvremenog krajobraznog oblikovanja Sveučilište u Zagrebu Agronomski fakultet, doktorski rad, Zagreb, p 222
[5] Ružinski N and Anić Vučinić A 2010 Obrada otpadnih voda biljnim uređajima (Hrvatska sveučilišna naklada ISBN 978-953-169-201-4)
[6] Malus D and Vouk D 2012 Priručnik za učinkovitu primjenu biljnih uređaja za pročišćavanje sanitarnih otpadnih voda (Sveučilište u Zagrebu, Građevinski fakultet) p 75
[7] Stanković D 2017 Constructed wetlands for rainwater treatment *Građevinar* Vol **69** No 8 pp 639-652
[8] Šperac M, Kaluđer J and Šreng Ž 2013 Biljni uređaji za pročišćavanje otpadnih voda *Electronic Journal of the Faculty of Civil Engineering Osijek-e-GFOS* Vol **4** No 7 pp 76-86
[9] Davis L 1998 A guide to creating wetlands for agricultural wastewater, domestic wastewater, coal mine drainage and stormwater in the Mid-Atlantic Region *Volume 5: stormwater* (Natural Resources Conservation Service United States Environmental Protection Agency.Region III. Pennsylvania (USA) Dept. of Environmental Resources)
[10] Stanić Z, Baić L and Kraš V 2016 Mogućnost uporabe samoniklih močvarnih biljnih vrsta u biljnim uređajima za pročišćavanje otpadnih voda *Inženjerstvo okoliša* Vol **3** No 1 p 47-62
[11] Davis L 1995 A guide to creating wetlands for agricultural wastewater, domestic wastewater, coal mine drainage and stormwater in the Mid-Atlantic Region *Volume 1: general considerations* (Natural Resources Conservation Service United States Environmental Protection Agency.Region III. Pennsylvania (USA) Dept. of Environmental Resources)
[12] Kayombo S, Mbwette T S A, Katima J H Y, Ladegaard N and Jørgensen S E 2004 Waste stabilization ponds and constructed wetlands design manual (UNEP-IETC with the Danish International Development Agency /Danida) p 59
[13] Peluso V F and Marshall A 2002 Best management practices for south florida urban stormwater management systems (South Florida Water Management District) p 26
[14] Environmental protection agency and Tetra Tech inc. 2009 Stormwater Wet Pond and Wetland Management Guidebook p 26
[15] Herzog T R 1985 A cognitive analysis of preference for waterscapes *Journal of Environmental Psychology* Vol **5** No 3 pp 225-241
[16] Howley P 2011 Landscape aesthetics: Assessing the general publics' preferences towards rural landscape *Ecological Economics* Vol **72 pp** 161-169
[17] Kaplan R and Kaplan S 1989 *The experience of nature: a psychological perspective (*Cambridge, Cambridge University Press)
[18] Cifrić I and Trako T 2008 Usporedba percepcije prirodnog i kulturnog krajobraza u Hrvatskoj, Primjena metode semantičkog diferencijala *Socijalna ekologija* Vol **17** No 4 pp 379-403
[19] Swanwick C 2009 Society's attitudes to and preferences for land and landscape Land *Use Policy* Vol **26** pp 62–75
[20] Tveit M S 2009 Indicators of visual scale as predictors of landscape preference; a comparison between groups *Journal of Environmental Management* Vol **90** No9 pp 2882–2888
[21] Hands D E and Brown D 2002 Enhancing visual preference of ecological rehabilitation sites *Landscape and Urban Planning* Vol **58** *No 1* pp. 57-70
[22] Jorgensen A, Hitchmough J and Calvert T 2002 Woodland spaces and edges: their impact on perception of safety and preference *Landscape and Urban Planning* Vol **60** No 3 pp 135-150
[23] De Groot W T and Van den Born R J G 2003 Visions of nature and landscape type preferences: an exploration in The Netherlands. *Landscape and Urban Planning* Vol. **63** No. *3* pp. 127-138

PERMACUL(SU+RE): Designing a Sustainable and Regenerative Agricultural Eco-park through Permaculture

J S Diquit[1] and H T Palarca[2]
[1] Student, Bachelor of Landscape Architecture, College of Architecture, University of the Philippines, Quezon City, Philippines
[2] Lecturer, Landscape Architecture Program, College of Architecture, University of the Philippines, Quezon City, Philippines

Email: [1] jsdiquit@up.edu.ph; [2] noritpalarca@gmail.com

Abstract. To protect the beauty and diversity of Quirino from the impacts of natural and human-made disasters, the call for sustainable and regenerative developments strengthens. With the introduction of permaculture in the province, it can serve as a framework for these developments. As such, the study seeks to showcase the potentials of permaculture by integrating its principles in the landscape design of Faraon Integrated Farm, as it aims to be a sustainable and regenerative agricultural eco-park. To achieve its goal, a series of literature studies, analyses, and conceptualization was conducted to identify the areas where permaculture can be incorporated. As a result, a permaculture-based landscape design of a sustainable and regenerative agricultural eco-park with a closed-loop system is developed. The eco-park was divided into six (6) permaculture zones, arranged according to the amount of attention it needs, consisting of areas chosen and designed based on the twelve (12) permaculture principles. To improve its sustainability and regenerative capacity, various technologies, practices, and techniques were also incorporated into the design, such as intercropping, use of natural farming inputs, vermicomposting, rainwater harvester & catchment, and utilization of renewable sources of energy. By imitating the Earth's natural processes and relationships, permaculture can create sustainable and regenerative developments that could preserve and enhance the natural environment.
Keywords: Agriculture, Permaculture, Regeneration, Sustainability

1. Introduction

Known as the Philippines's "Unexplored Paradise of the North," Quirino has numerous picturesque and breathtaking spots waiting to be seen and experienced. [5] As an agricultural province, it is susceptible to the effects of typhoons, drought, soil degeneration, and degradation. With that, it is time for Quirino to invest in sustainable and regenerative solutions to preserve its natural beauty and diversity while remaining economically sound. Sustainability takes into account environmental, social and economic aspects in finding solutions to world problems within the limits of the Earth [1]. At the same time, regenerative developments aim to close all the open loops that cause environmental damages, and consequently, direct these resources back to the system where they are needed. [4]

This is where permaculture comes in. Permaculture, according to David Holmgren, is "consciously designed landscapes which mimic the patterns and relationships found in nature, while yielding an abundance of food, fibre and energy for provision of local needs." [11] It is an ecological design approach [7] that creates systems that can heal the planet for future generations. And to realize its potential benefits, the study aims to integrate the permaculture principles in the landscape design of a sustainable and regenerative agricultural eco-park.

2. Methodology

2.1. Study Site
As per recommendation by the local government of Quirino, a farm in Brgy. Villa Pagaduan, Aglipay was selected as the study site, the Faraon Integrated Farm. The three-year-old farm is the only registered farm tourism site in Quirino, making it a priority for development.

Figure 1 Faraon Integrated Farm [A] Location Map, [B] Site Map, [C-F] Site Photos

2.2. Research Methods

2.2.1. Data Collection
Primary data were collected through Rapid Rural Appraisal (RRA), maps and plans, and census and statistics to understand the daily activities in the farm, its current condition, its profile, and land characteristics. RRA includes direct observation, transect walk, and key informant interviews [3] with the owner, organic agriculture expert, and the provincial agriculture and tourism offices. On the other hand, secondary data were collected through journals, reports, theses, books, manuals, brochures, online courses, and documentaries.

2.2.2. Data Analysis
For a more in-depth analysis of data, the study used Permaculture Decision Making Matrix and Market Analysis. The Permaculture Decision Making Matrix provides the designer with information on how elements are related and arranged to create a permaculture system. This matrix is composed of (1) Topography, (2) Sectors, (3) Permaculture Zones, and (4) Permaculture Principles. The topography and shape of the land tell how the water and soil behave in a specific environment. With that, the permaculture landscape can optimize the water as it moves through the site. On the other hand, sectors pertain to all forces that permeate the site from outside, such as sun direction, prevailing winds, and noise. By understanding these, the landscape can be designed as a direct response to them. Next, zones are based on the way elements are placed and arranged in the system (i.e., how near or far are these elements to the center of human activity). Lastly, the permaculture design principles provide the designer a guide on how the different elements in the system would relate to each other, giving a theoretical backbone for the design. [10] On the other hand, market analysis is the quantitative and qualitative valuation of a market to help in planning and decision making. [8] For this study, the researcher conducted a qualitative market analysis to show the marketability of developing farm tourism sites in the province. However, due to constraints, a quantitative market analysis was not conducted.

2.2.3. Design Development
After analyzing the collected data, a permaculture-based landscape design was developed to directly respond to the strengths, weaknesses, opportunities, and threats found during the analysis.

3. Findings

3.1. General Condition

Faraon Integrated Farm is a relatively isolated farm envisioned to serve as a haven for people who want to get away from their busy lives and to provide the owner with most of his daily needs for a long time. For two years, the owner has been developing it by planting various crops and building recreational areas. In 2019, it became an accredited learning site of the Agricultural Training Institute, where the owner trains and teaches people on farming. With that, the farm has also started transitioning to organic agriculture.

3.2. Physical and Biophysical Aspects

Site and Site Boundaries. Faraon Integrated Farm is composed of five lots with a total land area of 60 201.8 square meters. It is located near the barangay and a Communal Irrigation System (CIS) canal line. The farm is surrounded by forest and agricultural lands on the north, east, and south, while a fish farm is situated on its west side.

Circulation. In terms of circulation, the farm's only access from the provincial road is through a service road because most of the site has steep slopes. On the other hand, the pedestrian circulation inside the farm is mostly free. From the service road, guests must climb up the customized stairs on the hill to reach the other parts of the farm.

Structure. With most of the existing structures in the farm being temporary (e.g., nipa huts, chicken coops), they can easily be transferred and reused in the proposed design. On the other hand, the permanent structures include a training hall and a groundwater pump. The training hall is used for training, storing equipment, and resting, while the groundwater pump serves as a source of water for irrigation and daily activities (e.g., bathing, washing). All these structures are located near the house, making it easier to perform daily activities.

Vegetation. The farm is mostly comprised of agricultural vegetation as it is its primary purpose. It is composed of high-value crops, cash crops, fruit trees, and nitrogen-fixing trees. On the other hand, more than a hectare of the farm are forested. The landscaped areas are where recreational and educational activities usually happen, while the wetland on the lowest part of the site contains water all year round.

Agro-climate. Aglipay has a Type III climate which is characterized by a short period of the dry season and no definite period of the rain season. The average temperature in Aglipay is 26.70 degrees Celsius. Being located in the moist zone, Faraon Integrated Farm receives an average rainfall of more than 1 500 millimeters. [13] Areas in the moist zone are agriculturally more productive than those located in the dry zone as it receives more rainfall.

Soil. Most of the farm's lands used for agricultural purposes are Alaminos clay loam, while the forest lands and wetland are composed of Cauayan sandy loam. According to PhilRice [12], these soil types are both acidic, with a medium level of N, a low level of P and K, and a medium level of organic matter, constraining crop growth. It is essential to treat and prepare these soils before usage by applying agrochemicals to enhance their quality.

Topography & Slope. The farm is located within the 0-18 percent sloped area, making it prone to slight to moderate erosion. [13] Based on the transect walk, the farm has a rolling hills terrain. With this, the topography concludes that the lowest part of the site is a valley, creating a wetland. Lastly, the site has many high areas where people can get a nice view of the farm and its surroundings.

4. Analysis

4.1. Permaculture Decision Making Matrix

4.1.1. Topography

Hydrology Analysis. The hydrology analysis shows how water moves in the site to design according to it. As shown in Figure 2, the areas where the arrows meet can collect water and (re)use it before going out of the system. They can store rainwater and irrigate crops that need more water, like *palay*. Also, the areas with blue arrows are more prone to erosion and landslide, so the usage of trees and crops can help control these hazards.

Slope Analysis. The slope analysis shows the different elevations in the site and their potential hazards and opportunities. Figure 3 shows that relatively flat areas can accommodate recreational and agricultural activities, while low areas can act as catch basins for rainwater, and relatively steeper areas can be planted with trees and crops. Lastly, high areas can contain recreational activities that utilize the view.

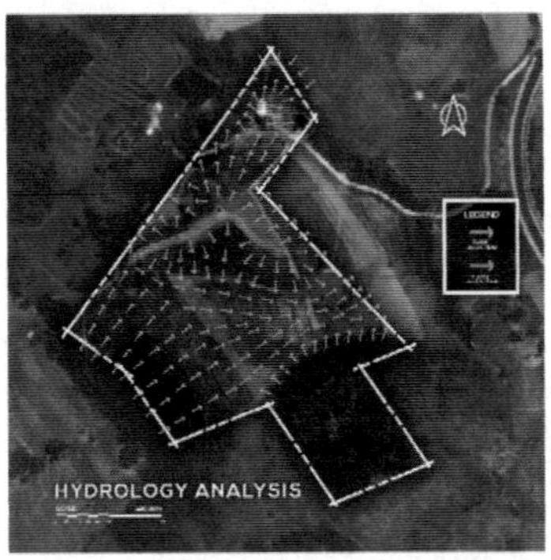

Figure 2 Hydrology Analysis

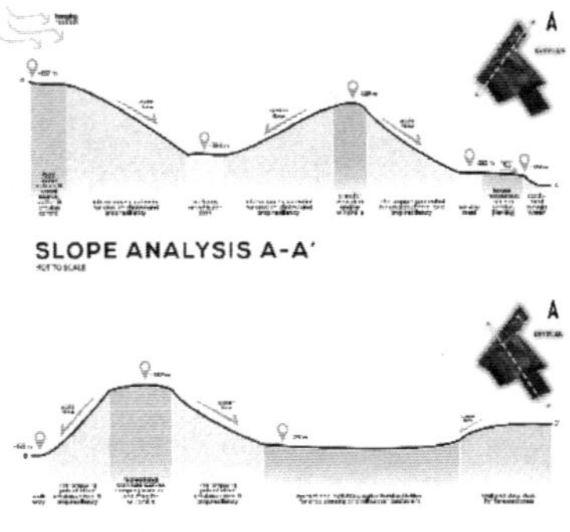

Figure 3 Slope Analyses

4.1.2. Sectors

Figure 4 shows that the farm sectors include the sun, wind, water, wildlife, landslide, and view. The sun sector follows the natural sun path while the wind sector represents the Northeast and Southwest Monsoons. Based on the wind's direction and the farm's topography, the northeast and southwest sides are prone to landslide. On the other hand, the wildlife sector comes from all sides except for the side near a fish farm. The farm is also surrounded by hills, forests, and croplands, resulting in pleasant and desirable views from all directions. More importantly, the compass shows that most of the sectors come from the southwest, making it the most susceptible area of the farm.

4.1.3. Permaculture Zones

In Figure 5, Zone 0 are areas for farm management, including the owner's home. For Zone 1, this includes areas that require regular maintenance, such as accommodation spaces. In the proposed zoning, two Zone 0s and 1s are created for more efficient management and more user-friendly circulation since these areas have different elevations. As for Zones 2 and 3, these areas need less attention than Zone 1, such as the croplands, orchard, greenhouse, and camping area. Zone 4 contains wood and food forests, and lastly, Zone 5 includes the unmanaged part of the site.

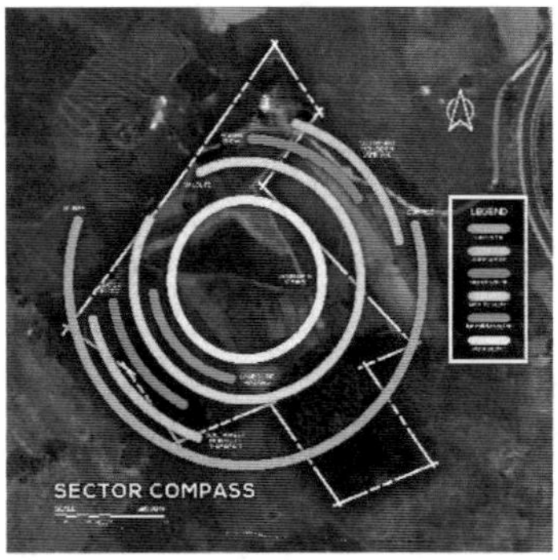

Figure 4 Sector Compass

Figure 5 Permaculture Zones

4.1.4. Permaculture Principles

Given the farm's existing conditions and how waste and energy flow in the site, Table 1 shows the potential applications of the permaculture principles in the proposed agricultural eco-park to make it sustainable and regenerative.

Table 1 Potential Applications of Permaculture Principles.

Principles [7]	Potential Applications
Observe and Interact	1. Data collection and analysis 2. Management and maintenance of the farm
Catch and Store Energy	1. Using the ponds and rainwater harvesters to store water for farm usage.
Obtain a Yield	1. Having enough croplands that would provide the farm and its guests with food 2. Unprocessed and processed crops and fruits can be an additional source of income
Apply Self-Regulation and Accept Feedback	1. Management and maintenance of the farm
Use and Value Renewable Resources and Services	1. Using solar panels and building of windmills as sources of electricity 2. Planting and using bamboos as primary building materials
Produce No Waste	1. Reusing wastes as agrochemicals for the farm
Design from Patterns to Details	1. Designing the farm based on the patterns naturally found in nature 2. Placement of catch basin to areas where the water goes, and planting based on the terrain
Integrate Rather than Segregate	1. Using pond for catching water, irrigation, aquaculture, and recreational purposes. 2. Using wetland for catching water and for recreational activities 3. Using the training hall as a learning center and as alfresco 4. Using forest lands as homes for wildlife and as recreational areas 5. Developing the farm as an agricultural, educational, and recreational site
Use Small and Slow Solutions	1. Management and maintenance of the farm
Use and Value Diversity	1. Providing different purposes for the farm, including agricultural, educational, and recreational. 2. Using intercropping to control pests naturally and to provide the farm with different crops.
Use Edges and Value the Marginal	1. Planting bamboo and fruit-bearing trees along edges to serve as wood and food source, as a buffer, and as shade 2. Planting ornamentals that can be used to beautify the farm and to serve as agrochemicals
Creatively Use and Respond to Change	1. Management and maintenance of the farm

4.2. Market Analysis

Farm tourism has been growing in the Philippines, boosting the country's agri-tourism. In 2019, the country recorded 174 DOT-accredited farm tourism sites [2], including the Yamang Bukid Farm in Palawan that drew over 254 000 visitors in the same year. [15]

In developing Faraon Integrated Farm as an agricultural eco-park, the target market is the local and international tourists, farmers, students, and visitors who want to learn about agriculture and permaculture. As the only accredited farm tourism site in Quirino, the farm has a comparative advantage over other farms in the province regarding funding, marketability, and government support. Faraon Integrated Farm would become the pioneering permaculture farm in Quirino. But more importantly, it can spark the growth of the province's agri-tourism industry.

5. Design Development

5.1. Design Concept

Sanglad is an Ilocano word that means "to anchor." As the design concept, it aims to showcase how permaculture creates a design that symbiotically integrates people and nature, synergistically connects all elements, and harmoniously incorporates different functions.

To embody the design concept and incorporate permaculture principles, such as design from pattern to detail and integrate rather than segregate, the vernacular style, as the design theme, uses locally available materials and knowledge in building structures and creating spaces. [6] The design used bamboo as the primary construction material to make it more environmentally and economically sustainable as it lessens construction costs. At the same time, native and multifunctional plants are used for softscape materials.

To visually represent Sanglad in the design, the researcher uses roots as the form concept. Like how permaculture anchors people and nature, the roots anchor a plant to its environment to create a harmonious and mutually beneficial relationship. The design uses the characteristics of roots in terms of hierarchy for effective circulation, connectivity for integration of areas, and organic shape for space optimization and rurality preservation.

5.2. Landscape Design

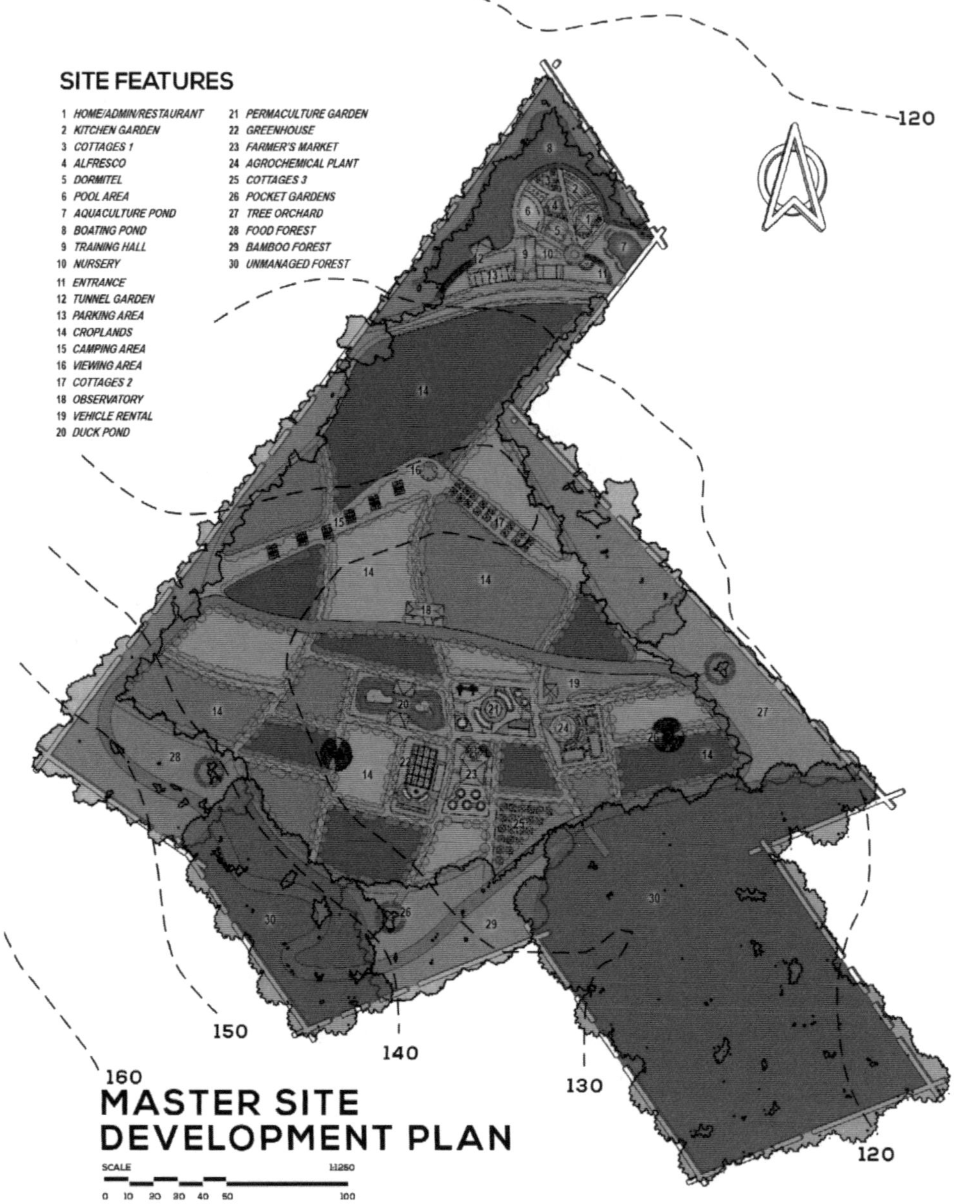

Figure 6 The Proposed Permaculture-based Agricultural Eco-park

Given its hilly terrain, the site's topography plays an essential role in laying out and designing the permaculture-based agricultural eco-park (see Figure 6), as it dictates where to place what and what to place where. In the proposed design, the farm is divided into three (3) major areas according to topography, namely, the lower area, hilltop area, and upper area.

Lower Area (1-13). The lower area has the lowest elevation and a relatively flat area. This area comprises home/admin/restaurant, kitchen garden, cottages 1, alfresco, dormitel, biological pool area, aquaculture pond, boating pond, training hall, nursery, entrance, tunnel garden, and parking area which mainly function to accommodate and entertain guests. The home/admin/restaurant primarily manages the whole development. On the other hand, the kitchen garden, ponds, and alfresco are placed near the restaurant for easier access. Although most of the features in the lower area are for recreation, they also function as rainwater catchment (e.g., ponds, biological pool), as a food source (e.g., entrance, gardens, ponds), and as a learning area (e.g., training hall, nursery). And to help with electricity, solar panels are also placed in all roofed structures in the eco-park.

Hilltop Area (15-17). On the other hand, the hilltop area has the highest elevation but limited flatlands. This functions primarily for recreation, which includes the camping area, viewing area, and cottages 2. The camping area is for people who want to stay overnight and experience the eco-park at night. With an average wind speed of 4 m/s, the agricultural eco-park is a perfect site to run wind turbines [14] placed in this area to maximize its elevation. Lastly, view deck and cottages are found in the area for guests to appreciate the totality of the eco-park and the beauty of Quirino.

Upper Area (18-26). Lastly, the upper area has an elevation between the lower and hilltop area but relatively flat terrain. This area includes the observatory, vehicle rental, duck pond, permaculture garden, greenhouse, farmer's market, agrochemical plant, cottages 3, and pocket gardens. The observatory, together with the home/admin/restaurant, manages the upper area given its location in the eco-park. On the other hand, the greenhouse and agrochemical plant serve as the eco-park's source of composts, fertilizers, and soil improvers to the eco-park. The duck pond serves as a rainwater catchment in the upper area for irrigation. On the other hand, the permaculture garden features an outdoor museum to teach permaculture and its application in the eco-park. Moreover, the farmer's market is where processed products are made and bought.

Aside from these areas, some areas can be found all over the site – the croplands and the buffer area.

Croplands (14). The croplands are the primary source of income for the agricultural eco-park. The proposed design applies different organic farming techniques, making it sustainable and regenerative. Various intercropping practices are integrated (e.g., crop rotation, contour farming, companion planting) to increase crop resiliency, soil productivity, and erosion control. And to help lower costs in terms of chemical usage, nitrogen-fixing leguminous trees are planted around the croplands.

Buffer Area (27-30). The buffer area serves as the park's barrier, which aids in controlling erosion and water flow, including the tree orchard, the food forest, the bamboo forest, and the unmanaged forest. The tree orchard is filled with fruit trees that can be harvested by guests to lessen labor costs. Also, the integration of chicken coops in the orchard would help fertilize the trees and lessen feeding costs. On the other hand, the food forest serves as an additional food source with its diverse selection of fruit trees and shrubs, while the bamboo forest serves as a wood source and added source of income as a high-value crop. Lastly, forest trails provided in the unmanaged forest are used to meditate or to connect with nature.

Figure 7 Perspectives of the Different Features in the Agricultural Eco-park

5.2.1. Circulation

As shown in Figure 8-A, the paths in the lower and hilltop areas follow the farm's contour, as taught by the principle of designing from pattern to detail. On the other hand, the upper area's circular path is a one-way ATV trail chosen based on the topography, the view, and the user experience it can give. Lastly, to increase field efficiency and lessen inefficiency costs, regular and rectangular paths are used as field shape in the upper area. [8]

5.2.2. Accessibility

Figure 8-B shows the public areas or those that can be accessed by anyone with or without paying fees, the semi-private areas or those with both public and private features, and the private areas or those accessible only to a particular person or group. Aside from a few areas, the design is publicly accessible. The public areas do not have entrance fees; however, the services offered in these areas can only be availed by paying for them so that the farm remains profitable.

5.2.3. Classification

As mapped in Figure 8-C, the different classifications of site features include agricultural areas that aid in crop and livestock production, educational areas that help teach and train people, and recreational areas that improve people's social and mental health. Since agricultural production is the park's primary function and income source, this area occupies about 75 percent of the eco-park. At the same time, most areas fall under two or more classifications, aligned to the permaculture principle of integrating rather than segregating.

5.2.4. Su+Re (Sustaining + Regenerating)

Lastly, Figure 8-D maps out areas for sustaining, which aid maintains environmental, social, and economic health while preserving the natural systems, and areas for regenerating, which aid returns wastes into the permaculture system. Based on the plan, most areas in the agricultural eco-park both function to sustain and regenerate. These areas are mostly composed of agricultural spaces, being the primary producers and reusers of wastes. Moreover, these areas sustain the farm economically for its agricultural production and environmentally for practices that preserve soil quality. In addition to these, the eco-park promotes social sustainability by providing jobs to the people of Quirino, resulting in a healthier and more equitable community. Being a province that primarily generates income from the agriculture and tourism industry, its people have more experience in these areas; thus, cultivating a culture of sharing among people. At the same time, it offers local and international guests space where they can relax mentally and physically by interacting with other people and with nature.

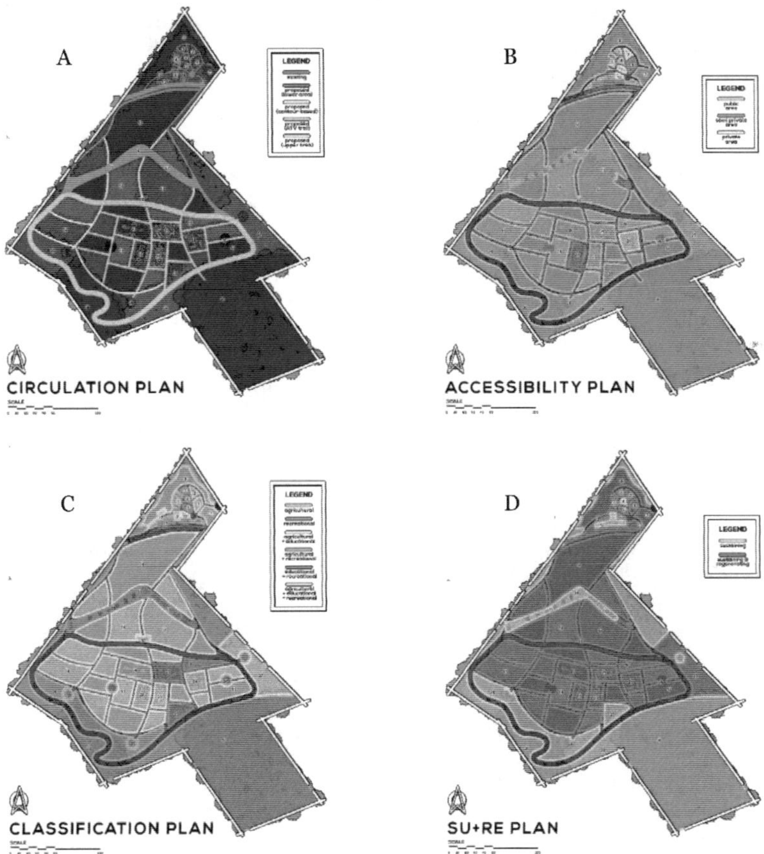

Figure 8 Thematic Plans [A] Circulation, [B] Accessibility, [C] Classification, [D] Su+Re

6. Conclusion

To alleviate the impacts of natural and human-made disasters, Quirino has to start building sustainable developments and regenerating its environment. Through permaculture, a design theory that mimics the relationships and patterns naturally found in nature and incorporates them into the landscape, Faraon Integrated Farm in Quirino became a self-sustaining agricultural eco-park. With the Permaculture Decision Making Matrix, the potential applications of permaculture in the farm were identified, resulting in the design concept *Sanglad,* which harmoniously integrates people and nature to design a sustainable and regenerative agricultural eco-park. With this, the landscape design features elements that strengthen organic farming, utilize various renewable energy sources, and reuses wastes to create a closed-loop system. With these design interventions, the cost of running the agricultural eco-park decreases, while its ecological quality improves over time. At the same time, it creates a more cooperative and livable community by providing people with space to relax, interact and work. With the people and environment working synergistically together, the development would remain sustainable and regenerative for a long time. The agricultural eco-park proved that with proper integration of permaculture in the layout and design, sustainable and regenerative developments could be created.

However, the study on permaculture does not stop from the development of a sustainable and regenerative agricultural eco-park. With that, for those who wish to continue the study, perhaps on a different site, the researcher recommends developing a farming schedule that reflects the crops to be planted, the sequence of the crops, the time when to make and apply natural farming inputs, and other related methods to improve the agricultural aspect of the development. In addition to this, the researcher recommends quantifying and estimating the permaculture farm's economic feasibility, in the short run and the long run. As such, a comparison with other kinds of farms can also be made. Lastly, the researcher recommends applying permaculture in urban planning and design, which would be an excellent opportunity to see how it works in a landscape with more human elements.

References

[1] ACCIONA, What is sustainability? [Internet]. Activesustainability.com. [cited 2019 May 20]. Available from: https://www.activesustainability.com/sustainable-development/what-is-sustainability/
[2] Domingo R, Farm tourism growing in PH [Internet]. INQUIRER.net. 2019 [cited 2020 June 09]. Available from: https://business.inquirer.net/280507/farm-tourism-growing-in-ph
[3] Freudenberger K, Rapid rural response appraisal and participatory rural appraisal manual. Baltimore, Md.: Catholic Relief Services; 1999.
[4] Gabel M, Regenerative Development: Going Beyond Sustainability [Internet]. Kosmosjournal.org. 2015 [cited 2020 June 10]. Available from: https://www.kosmosjournal.org/article/regenerative-development-going-beyond-sustainability/
[5] Gorospe-Ibuan J, PROTECTING THE 'UNEXPLORED PARADISE OF THE NORTH' [Internet]. Newsflash.org. 2004 [cited 2019 May 17]. Available from: http://www.newsflash.org/2004/02/si/si001805.htm
[6] Gross R, Sustainable Design: Lessons We Can Learn From Vernacular Housing [Internet]. Houzz. 2017 [cited 2020 June 09]. Available from: https://www.houzz.com.au/magazine/sustainable-design-lessons-we-can-learn-from-vernacular-housing-stsetivw-vs~88142199
[7] Holmgren D, About Permaculture - Holmgren Permaculture Design [Internet]. Holmgren Permaculture Design. [cited 2019 May 18]. Available from: https://holmgren.com.au/about-permaculture/
[8] How to do a market analysis for a business plan [Internet]. The Business Plan Shop. [cited 2020 June 10]. Available from: https://www.thebusinessplanshop.com/blog/en/entry/market_analysis_for_business_plan
[9] Hunt D, Farm Power and Machinery Management. 10th ed. Illinois: Waveland Press; 2017.
[10] Millison A, Introduction to Permaculture [Internet]. Oregon: Pressbooks; [cited 2019 May 18]. 251 p. Available from: https://open.oregonstate.education/permaculture/
[11] Moore K, Bradley L, North Carolina Extension Gardener Handbook. North Carolina: NC State Extension; 2018.
[12] Philippine Rice Research Institute. Simplified Keys to Soil Series: Isabela. Isabela; p. 28-29, 34, 37-38.
[13] Province of Quirino. Comprehensive Land Use Plan Aglipay 2012-2022. Quirino; 2012.
[14] U.S. Department of Energy. Small Wind Guidebook. WINDExchange; p. 14. Available from: https://windexchange.energy.gov/small-wind-guidebook
[15] Yamang Bukid Healthy Products Inc. Yamang Bukid Farm draws over 254,000 visitors in 2019 [Internet]. Yamangbukid.com. 2020 [cited 2020 June 10]. Available from: https://www.yamangbukid.com/farm/yamang-bukid-farm-draws-over-254,000-visitors-in-2019

Analysis of ecological and visual quality impact on urban community activities in Bogor City

S Anwar[1] and R L Kaswanto[2]

[1] Department of Landscape Architecture, Faculty of Agriculture, IPB University, Indonesia
[2] Department of Landscape Architecture, Faculty of Agriculture, IPB University, Indonesia

E-mail: [1] salsabila_2004@apps.ipb.ac.id

Abstract. Bogor City's implication as a satellite city is directly related to carrying capacity and its influence on environmental quality, specifically ecological and visual quality. Urban communities naturally tend to properly utilize open space in their daily activities, such as in urban park along Ciliwung River riparian. Psychologically, the individual's environmental situation has more role in influencing a person's actions than the individual's characteristics. Therefore, as a socio-cultural function, public space can reflect people's lives in the city. This study analyses the ecological and visual quality impact on park visitors' activity in Bogor City to determine that environmental factors can influence community activities. Sempur Park has the highest visual quality value according to respondents' perceptions. This result is the same as the result of ecological quality that places Sempur Park as the highest diversity index. The high value for these two qualities is in line with the increased activity carried out in the park, which is 41 activities. As a result, the connection between ecological and visual quality shows a strong and positive correlation with the urban park's amount of activity. This study's output is recommendations for managing the urban park landscape to increase the level of activities by improving the urban park's visual and ecological quality.
Keywords: ecological quality, park visitor's activity, visual quality

1. Introduction

Economic progress and the increased population in Bogor City as a buffer city for DKI Jakarta have stimulated the increasing demand for land, transportation, industrial activities, economic activities and spaces for community social activities. Human development or activities in exploiting the potential of waterfront ecosystems called riparian landscapes often cause various environmental problems, including ecological and visual aspects [1]. Riparian ecosystems have a crucial hydrological function and need special attention to their management systems [2].

One of the non-physical factors that affect a city's development is the city activity factor, which reflects the urban life [3]. Activities in the city, such as park visitor activities, are generally associated with its users' various needs, beauty, and interests. Parks are useful in architectural, social, economic and ecological terms and are also useful in increasing various users' accommodation. Psychologically, people, objects and meaningful events around the individual build an atmosphere or environmental situation in a place. Environmental conditions that are experienced directly have a role in influencing the form of a person's behaviour/actions rather than individual characteristics [4]. Meanwhile, the Bogor City Government has focused on increasing Green Open Space (GOS) by building parks to improve environmental quality. So that the existence of city parks also has the potential to increase the value of city-scale biodiversity.

IOP Conf. Series: Earth and Environmental Science **879** (2021) 012035 doi:10.1088/1755-1315/879/1/012035

This study aims to analyze the impact of ecological and visual quality on city park visitors' activity and find their correlation and figure out the environmental factors that affected the level of community activities in Bogor City. Besides, to produce park management recommendations with high ecological and visual quality and high community activity level to form harmony between human interests and natural sustainability.

2. Research Methods

2.1. Study site and period

This research was conducted in the Ciliwung river riparian at Bogor City, West Java (Figure 1) for seven months.

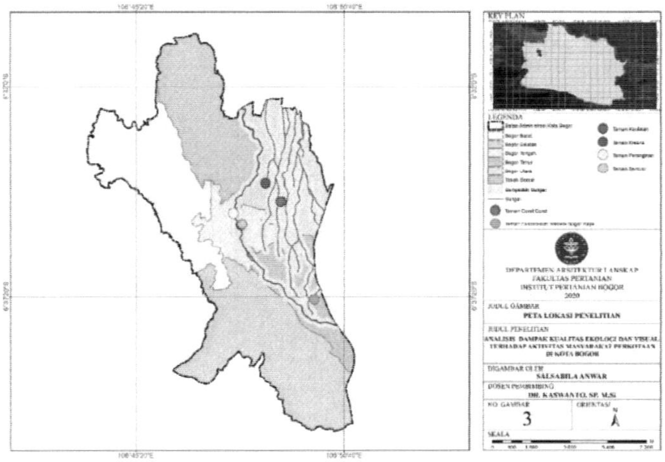

Figure 1 Study Site

2.2. Research Methods

Preparation: This step includes preparing the research proposal and determining the site location (Table 1).

Table 1 Site Location

No.	Territory	District location	Area of Green Open Space (m²)	Park Name	Symbol	Area (m²)	Distance(m)
1	Top	East Bogor	53.395,70	Mutiara Bogor Raya	MBR	1200	172
2	Middle	Central Bogor	163.792.96	Sempur	LS	13.632	30
				Kaulinan	K	600	4
				Peranginan	P	1.699	78
3	Bottom	North Bogor	46.755,17	Kresna	LP	8.547	85
				Corat Coret	CC	571	158

Inventory: The inventory processes collect data on ecological quality and visual quality and community activities—retrieval of ecological quality data by identifying all existing vegetation in six sample parks. According to their function in forming and filling space, vegetation is ground cover plants, wall forming plants, roofing or shade plants and plants as ornaments and space fillers. Wall-forming plants are divided into the boundary, directional, windbreaker and screen-forming plants, while plants as space-filling ornaments are often called ornamental plants [5]. Plant inventory results will be grouped based on their function to see plants' dominance in Bogor City parks. The method of taking visual quality images is in the form of photographs representing the four cardinal directions (n=24) which will be assessed by expert respondents, namely the Landscape Architecture Students of IPB University (n=60)—collecting park visitors' activities data using a questionnaire method which was distributed to 10 visitors per park (n = 60).

Analysis: The analysis phase was carried out in four steps. First, this study measures species biodiversity's value as a representation of the landscape's ecological character using a vegetation approach in a predetermined park sample. Biodiversity analysis as a representation of one of the ecological qualities [6]. Species biodiversity was observed using various analysis techniques. The indexes used in this study were the Shannon-Wiener Species Diversity Index, the Species Evenness Index and the Margalef Species Richness Index. The following is the formula for each index:

Species Diversity Index

$$H' = -\sum Pi \ln(Pi); \quad Pi = (ni/N) \qquad (1)$$

Information:
H' = Shannon-Wienner diversity index
ni = number of individuals type-i
N = number of individuals of all types
The criteria for the Shannon - Wiener (H ') diversity index value are as follows:
$H' < 1$: low diversity
$1 < H' < 3$: moderate diversity
$H' > 3$: high diversity

Species Evenness Index

$$E = \frac{H}{\ln S} \qquad (2)$$

Information:
E = Species evenness index
H' = Shannon-Wienner diversity index
S = The number of individuals of all species
The evenness index value ranges from 0 to 1; if the value is 0, it shows that the evenness level of plant species in the community is very uneven. In contrast, if the value is close to 1, almost all species that exist have the same abundance [7].

Species Richness Index

$$DMg = \frac{S-1}{\ln N} \qquad (3)$$

Information:
DMg = Margalef species richness index
S = Number of species
N = The number of individuals of all species
The criteria for the Margalef type wealth index (DMg) are as follows:
DMg < 3,5 : low species richness
3,5 <DMg<5 : moderate species richness
DMg>5 : high species richness

Second, the visual quality is an experience or influence that is felt on each person, and it can only exist when there is someone in it so that people need to be able to assess it. Every visual quality assessment method required people explicitly or implicitly, directly or indirectly[8]. Visual quality analysis method uses Semantic Differential (SD) by asking the Landscape Architecture student as the expert respondents. The method is done by presenting some photos or images to measure the connotative meaning of concepts by the respondents, with can be expressed as a point in semantic space [9] The respondent will assess ten pairs of criteria and resulting in a graph of attributes that shows an average value of the six parks. The third analysis of park visitors' activities, seen from the type and level of park visitors' activity, produces a map of park visitors' activity level. Fourth, the process combines the three analysis results to create a management recommendation for the Bogor City landscape suitable for community activities and environmental quality. This step is generated by looking at the relationship between ecological quality, visitors park activities, and visual quality and visitors park activities using Biplot analysis.

Recommendation: for landscape management with high ecological quality, visual quality and community activities so that harmony is formed between human interests and natural sustainability for the sustainability of a city's landscape.

3. Result and Discussion

3.1. General Condition of Bogor City

Bogor city has 11,850 hectares, divided into six urban village and 68 sub-districts and located in West Java Province. It has a strategic location with a distance of ± 56 km from DKI Jakarta. It is one of the satellite cities that support the development and economy growth and services of the State Capital. The population growth rate reaches 1,67% per year, with a total population of 1.1 million people in 2019 [10]. This lower population growth rate than other satellite cities such as Depok, Bekasi and Tangerang makes Bogor City has moderate of city landscape quality from a combination of building structures, land use and the existence of urban infrastructure and green infrastructure [11]

Residents living in urban areas are happier than those who live in rural areas as seen from the West Java Happiness Index of 68.91. The Happiness Index is measured using three dimensions: the dimensions of life satisfaction, feelings, and the meaning of life from the result. The second-highest indicator is the environment's state, which includes the dimensions of life satisfaction of 75.65 [12]. This indicates the importance of environmental quality for urban communities, especially in Bogor City, which has the potential for parks as public green open spaces. In line with the Bogor City Government's efforts in spatial planning and strengthening Bogor City's image's identity.

3.2. Existing Condition of Bogor City Park

Green Open Space in Bogor City, which is included in public green open space, is generally managed by the Bogor City government, through the Landscaping and Sanitation Service sub-sector of landscaping. Bogor City Government divides Green Open Space into nine types with a total area of 41.8 ha at the beginning of 2019, which has increased by 1.77% from the previous year. Public GOS in Bogor City should be at least 20% based on Law Number 26 of 2007 concerning Spatial Planning. So, it is necessary to maximize the use of existing public green open space to improve the quality of the urban landscape, especially in the riparian areas of the Ciliwung River, which in essence has a higher ecological value. The largest green open space area was in the Central Bogor District of 16.38 ha, and the smallest was in the South Bogor District of 2.57 ha.

The existing conditions in the six parks in Bogor City are overall good. The visual quality of the Peranginan Park is considered beautiful on a seating spot facing the Ciliwung River. The impression of being closed, shady and cool is felt by 50% of visitors to this park. Meanwhile, for ecological quality, 60% of respondents rated Peranginan Park have moderate variations of vegetations. As many as 80% of visitors to Sempur Park rate that the visual quality is beautiful and the park's impression is cool, clean, and open, while 70% of respondents think that the vegetation varies. The Sempur signage is a spot that is considered beautiful by the visitors, according to the remaining 60%, that the seating area on the south and north side and the field area is a good view spot. The questionnaire results show that 93% of park visitors had visited other parks than one they were visited when the questionnaire was given to them, thus indicating a trend towards community activity in the park.

4. Analysis

4.1. Ecological Quality Analysis

The inventory results were 122 species included in 47 families, 50% of which were dominated by ten families can be seen in Table 2. The highest percentage of families were Asparagaceae at 11.47%, followed by Fabaceae with 5.73% and Moraceae and Myrtaceae with the same percentage of 4.91%. Species belonging to the Asparagaceae family are *Canna* Sp, *Dracaena sanderiana, Dracena loureiri, Dracaena fragrans, Dracaena dremensis, Cordyline* Sp and *Sanseviera trifasciata*. This family was found in many parks as ornamental plants. Meanwhile, the Fabaceae family was dominated by plants with a shade function, which is the element forming the roof. It is used as a microclimate regulator, including *Samanea saman, Delonix regia* and *Pterocarpus indicus*.

The results showed that the percentage of ornamental plants had a dominant function, namely 38.88%, while 24.60% shade plants or roofing plants became the second-highest dominant function. The results show that plants in Bogor City parks have the function of increasing visual beauty and controlling the microclimate according to [13], open space must give a pleasant impression to the user from every movement so that there are harmony and unity with the characteristics of the landscape. exist functionally, both physically and visually. The next function is that boundary plants and direction plants have a percentage of 11.90% and 9.52% followed by a percentage of functions of ground cover plants (9.52%), windbreakers (3.96%) and sight forming plants, and other functions have the same percentage of (0.79%).

Table 2 Percentage of 10 Highest Families

No	Family	Number of Species	Percentage (%)	Function
1	Asparagaceae	14	11,47	Barrier, Direction, Display
2	Fabaceae	7	5,73	Ground Cover, Wind Breaker, Shade, Display
3	Moraceae	6	4,91	Ground cover, shade, Display
4	Myrtaceae	6	4,91	Barrier, Shade, Display
5	Acanthaceae	5	4,09	Ground Cover, Barrier, Direction, Display
6	Euphorbiaceae	5	4,09	Barrier, Display
7	Apocynaceae	5	4,09	Barrier, Direction, Display, Shade
8	Araliaceae	5	4,09	Ground cover, barrier, Display
9	Aracaceae	4	3,27	Director, Shade
10	Meliaceae	4	3,27	Shade

The species diversity of plants in the six parks in Bogor City showed that Sempur Park is the only one park classified as high diversity level category, with an index value (H ') of 3.14. This park has 43 species, with total individuals of all species reaching 2809. Meanwhile, five other parks in the category of moderate diversity, including Kresna Park with a value of 2.24 with 40 species and 296 individuals of all species, Kaulinan and Peranginan Park have the same index value 2.08. Kaulinan Park has 12 species with 730 individuals of all species, while Peranginan Park has 26 species and 1207 individuals of all species. Mutiara Bogor Raya Park at 1.87 with 21 species totalling 412 individuals. The last, Corat-Coret Park, with a value index of 1.71, has ten species and 173 individuals of all species. The value of the plant diversity index in the three regions, namely the upper, middle and lower areas, have including in the medium diversity class. The middle area has the highest index value compared to others, and this is influenced by the area of green open space in Central Bogor District, the largest of all districts.

The species evenness index results show that the six parks have high evenness values, which shows the more evenly distributed species and the more stable the vegetation in each park. The highest evenness index value was Sempur Park and Kaulinan Park at 0.83, followed by Kresna Park and Corat-Coret Park, which received the second-highest evenness index value 0.74. Meanwhile, Peranginan Park was next with a value of 0.63 and Mutiara Bogor Raya Park with the lowest level of the evenness index value of 0.62. The results of the evenness index value can be seen in Table 3.

Table 3 Value of Diversity Index, Evenness Index and Species Richness Index

No	Territory	City Park in Bogor City	H'	E	DMg
1	Top	Mutiara Bogor Raya	1,87	0,62	3,32
2	Middle	Peranginan	2,08	0,63	3,52
3		Kaulinan	2,08	0,83	1,60
4		Sempur Park	3,14	0,83	5,28
5	Bottom	Kresna Park	2,24	0,74	3,33
6		Corat-Coret	1,71	0,74	1,74

The evenness index value for the three regions, namely the upper, middle and lower areas, shows the high evenness of the park's distribution of vegetation species. The highest index value is in the middle region with a value of 0.76, including Sempur Park, Kaulinan Park and Peranginan Park. The bottom Area is the second-highest index value, which is 0.74, this value is only 0.02 different from the Central Area, the lower area includes Kresna Park and Corat-Coret Park. Meanwhile, the upper area has Mutiara Bogor Raya Park to be the lowest evenness index value, which is 0.62. The calculation of the vegetation species richness index's value shows different results from low to high richness classifications. The Sempur Park has the highest vegetation species richness index value, classified as a high wealth classification of 5.28 and is the only park among the other five parks that shows a relatively high species richness. Likewise, Peranginan Park is the only park that shows species richness in the medium category with a value of 3.52. Meanwhile, the other four parks, namely Mutiara Bogor Raya Park, Kresna Park, Corat-Coret Park and Kaulinan Park, are included in the low species richness value. The species richness index value results in the three sample areas are included in the low index value because the resulting value is less than 3.5. The lower area is the area with the lowest index value among the other two areas, namely 2.54, Corat-Coret Park and Kresna Park.

4.2. Visual Quality Analysis

Visual quality is a beauty that can be felt through the interpretation of images based on the observer's experience to recognize a meaningful form [14]. The assessment in this study was carried out through a questionnaire with a google form. Respondents consisted of 38 women and 22 men who rated 24 photos using the SD method on ten pairs of adjectives as criteria, resulting in a perceived visual quality (Figure 2).

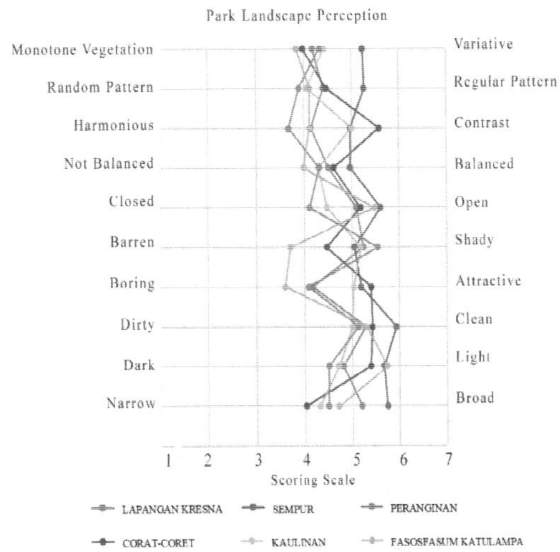

Figure 2 Semantic Differential Analysis Results for Visual Quality of Parks in Bogor City

The urban landscape has a good visual variation within the range of values, and the best landscape has values that are neither the highest nor the lowest [15]. The results obtained show the six parks in good visual quality because they range from values 1-7 (Table 4).

Table 4 Average Value of Visual Quality Using Semantic Differential

	Sempur	Corat-Coret	Kresna	Kaulinan	Peranginan	MBR
Narrow-Broad	5,75	4,02	5,19	4,32	4,49	4,71
Dark-Light	5,67	5,38	4,81	4,71	4,51	5,74
Dirty-Clean	5,93	5,42	5,27	5,00	5,38	5,30
Boring-Attractive	5,18	5,40	4,17	5,03	4,18	3,60
Barren-Shady	5,05	4,47	5,24	5,18	5,54	3,71
Closed-Open	5,60	5,18	5,09	4,48	4,11	5,26
Not balanced-Balanced	4,97	4,62	4,51	4,31	4,32	3,99
Harmonious-Contrast	4,97	5,58	4,14	5,00	3,84	4,07
Random Pattern - Regular Pattern	5,26	4,47	4,41	4,07	3,90	4,12
Monotone-Variative Vegetation	5,23	3,98	4,18	4,42	4,33	3,84
Average value	5,36	4,85	4,70	4,65	4,46	4,43

Sempur Park is a park with the best visual quality, which has an average scale of 5.36 with the highest score on clean, bright and wide, according to respondents' perceptions for the lowest attributes are balanced and contrast. This result is the same as the result of ecological quality that places Sempur Park as the highest diversity. Meanwhile, the ecological quality results place Corat-Coret Park as the park with the lowest diversity value, but the visual quality analysis results place this park in second place with an average value of 4.85. The majority of respondents considered this park to have attractive attributes and contrast among other parks. This is in accordance with the name, visitors who visit this park are given facilities in the form of a row of walls as an image medium whose results can be seen by other visitors. Located between road intersections, this park has contrasting conditions with its surroundings. Respondents consider the existing vegetation to be monotonous not to disturb the focus of the view. Also, the limited land area affecting the existing vegetation is limited. There are only three Jacaranda trees as shade trees with branch canopy that are not too tight, making the impression of this park is bright because sunlight can reach any part of the park.

4.3. Analysis *of Park Visitors Activities*

Collecting data using a questionnaire to 60 park visitor respondents covering the respondent's characteristics, the respondent's preferences and the respondent's perception of the conditions directly in the park. The respondents' age range ranged from 13-56 years, with the percentage of female 58%.

Occupation is considered to influence one's perception of a park. Also, work background affects the activities carried out by a person. The biggest occupational background for park visitors is students with 20% or as many as 12 people predominantly in Corat Coret Park and Sempur Park. Based on the data collected, respondents' occupational backgrounds with the most "other" categories in Sempur Park and Kaulinan are housewives.

Overall, 47% of the respondents were park visitors who lived more than 3 km from the park they visited when the questionnaire was administered. Meanwhile, respondents who have a place to live less than 500 m from the park are in the lowest position, 10% of the total respondents. It can be concluded that distance does not affect people's propensity to visit the park.

The frequency of visits is listed in this questionnaire so that the respondent's visit patterns in the park can be identified and the respondents' perceptions. As many as 59% of the total respondents visited the park with an uncertain frequency, 13% of the respondents routinely visited the park every day. Mutiara Bogor Raya Park has the highest number of respondents who visit this park every day. This is because the park's location is in a residential area to pass by and stop by this park. The accumulated results of all responses are known that weekend is the choice of the most days chosen by the respondent to visit the park.

The activities of the respondents in this questionnaire are categorized into seven options. Namely, Photography, Gathering, Playing, Walking, Sports, Sitting Relax, and Others. Respondents can choose Other if they carry out activities outside the existing activity choices. Respondents can choose more than one category of activities carried out in the park. The results showed that the two biggest activities carried out by park visitors were sitting relaxed and taking a leisurely walk by 29% and 20%, this indicates that the majority of respondents visited the park to enjoy the park atmosphere in their spare time or when they took a break from their daily routine. Meanwhile, the three categories of activities such as gathering, playing and sports have almost the same percentage of 14%, 13%, and 12%. Photography activity was 9% while for other activities 3% each included eating.

All respondents chose Kresna Park and Peranginan Park in the park to carry out sitting and relaxing activities. The facilities at Kresna Park support visitors who want just to sit back and relax. There are sculptures in the shape of footprints covered with synthetic grass that visitors can use to sit or gather activities in this park. Meanwhile, various sports activities are carried out, such as jogging, playing volleyball, gymnastics, and playing badminton. Sempur Park is a park that can accommodate various activities of respondents in the park. Respondents mostly choose sports activities and leisure walks in Sempur Park because the sports facilities on the park's side support sports activities. Also, in Sempur Park, a Jogging Track coated with rubber so that visitors can use it for jogging or walking comfortably.

This study shows that the highest total activity in Sempur Park is 41 activities followed by Kresna Park with 27 activities and four other parks, namely Kaulinan Park, Mutiara Bogor Raya Park, Peranginan and Corat-Coret, respectively 23, 22, 18 and 16 activities.

4.4. Correlation between Ecological, Visual Quality and Park Visitors Activities

At this step, the correlation between ecological quality, visual quality and park visitors' activities shows positive and strong results in both relationships (Figure 3).

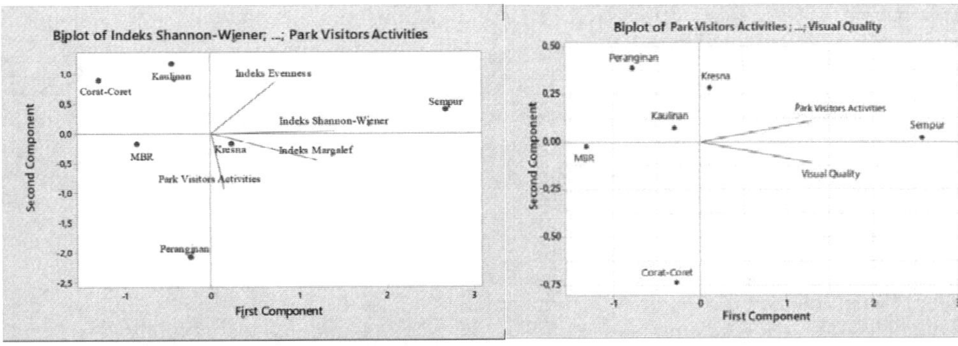

Figure 3 Biplot Graph: (a) Visitors Activities and Ecological Quality (b) Visitors Activities and Visual Quality

The result of biplot graph interpretation shows that the correlation between the level park visitors' activity and the ecological quality variable shows a positive and strong correlation based on the angle formed between vectors of ecological quality (Shannon-Wiener Index and Margalef Index) with park visitors activity variable. Meanwhile, the Biplot graph between the park visitors' activity and the visual quality variables shows a positive and strong correlation. The angle formed between the two variable vectors is 9 °. This is greater compared to the ecological quality and park visitors' activity correlation graphs. The relatively same vector length indicates the same variability of these two variables from the two Biplot charts. Through the results obtained, the ecological and visual quality can affect community activity in the park. The higher the park's ecological and visual quality, the higher the park's level of community activity. Vice versa, if the ecological and visual quality is low, the level of community activity is also low. In line with the Bogor City Government's efforts in providing GOS as a space for community activities, this study's results can be used as a reference in managing the landscape of Bogor City.

5. Recommendation

The landscape of a city is closely related to its activities that form the city structure [16]. Psychologically, the environmental situation that is experienced directly has more role in influencing the form of a person's behaviour or actions rather than individual characteristics [4]. In this study, the environmental situation was seen from its ecological and visual quality aspects. Based on the results of the analysis, several recommendations were made for the management of park landscapes in Bogor City :

5.1. Selection of multifunctional plants with various levels of species

Based on the analysis of ecological quality, parks in Bogor City are moderate to high diversity conditions. Increasing vegetation diversity, it is necessary to arrange planting by maximizing plants' function and aesthetics. The vegetation planted in the park must also be able to adapt ecologically to the site location. Several plant functions can be used in city parks, including ground cover plants, boundaries, guides, windbreakers, viewers, shade, decoration, fruit producers, and pollutant absorbers. Planting plants that have more than one function (multi-function) has a higher value in the park so that it has the potential to increase the diversity of plant functions that can increase biodiversity. The diversity of species in a park is influenced by the richness and evenness of species in the park, the number of different species planted in the park can also increase the richness of plant

IOP Conf. Series: Earth and Environmental Science **879** (2021) 012035 doi:10.1088/1755-1315/879/1/012035

species by being balanced with the even distribution of species in a park which will increase the stability and sustainability of the park in its role as an ecosystem. Namely, a species with a high level of stability has a greater chance of maintaining its species' sustainability [6].

The high vegetation biodiversity allows for high variations in the shape, texture, and colour of flowers and canopy. This condition supports the formation of landscapes with attractive views because the variations that exist can reduce the monotony of the scenery, thereby increasing the value of a park's visual quality [17]. This vegetation planting can be done in a multi-layered manner, namely planting plant layers vertically by utilizing differences in plant height so that it gives a good and beautiful visual impression as well as being able to maximize the diversity of vegetation in the park, an example of applying plant layers can be seen in Figure 4.

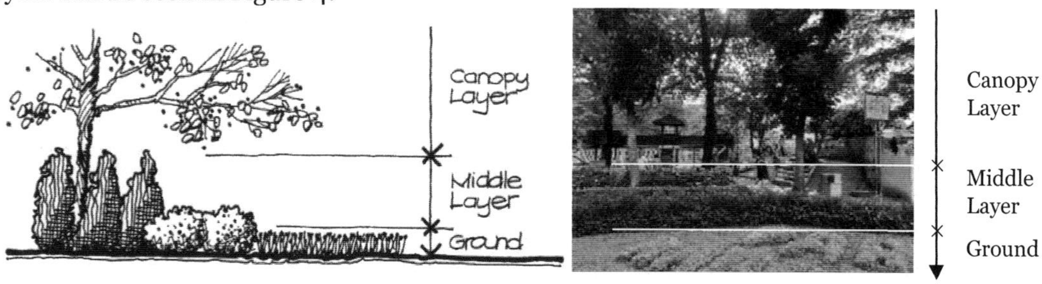

Figure 4 Application of crop layers: (a) literature and (b) Kaulinan park
Sources: (a) [18] and (b) Personal Documentation

5.2. Park Facilities Improvement
The park's ecological and visual qualities can increase community activity in the park, but park visitors' types of activities also depend on the park facilities. Therefore, this recommendation is aimed at developing or managing parks in Bogor City in the future facilities to accommodate these four activities need to be considered, because, from this research, the visitors of parks that have high ecological and visual quality and parks with low ecological and visual quality mostly will do these four activities. However, activities have a strong relationship with ecology and visuals. Therefore, even though many sitting facilities are provided, there will not be many visitors that use the facilities if there are not supported by improvements in ecological and visual quality.

6. Conclusions and Suggestion

6.1. Conclusions
The ecological quality analysis resulted in a vegetation diversity index value, including moderate to high diversity categories in six parks in Bogor City. High diversity is in Sempur Park with an index value of 3.14, and the other five parks are in the medium diversity category. The Species Evenness Index value is in the high species evenness class for the whole park, and the Species Richness Index value in the park varies from low to high. While the visual quality analysis results showed good visual quality in the six parks because they were between the 1 and 7 value ranges, the highest average overall attribute value was in Sempur Park, namely 5.36. The ecological and visual quality analysis results show the similarity with visitor activity in the park. Sempur Park also has the highest total activity value of the other five parks of 41 activities.

This proves that the correlation between ecological, visual quality, and community activity level in Bogor City has a positive and strong correlation so that the ecological and visual quality affects the level of community activity. If ecological and visual quality is high, the level of community activity is also high. Also, park management recommendations include applying design principles, improving supporting facilities based on the top five activities in the park, and education for the community.

6.2. Suggestion

Ecological and visual quality should be enhanced and maintained in urban landscapes at microscales such as environmental parks and macro scales. Collaboration between the community and stakeholders such as city governments and community organizations is needed to achieve harmony between human interests and the environment. In future research, it is suggested that other environmental quality variables are necessary to complete this research.

Acknowledgements

This paper was supported by Penelitian Terapan Unggulan Perguruan Tinggi (PTUPT) of Ministry of Research Technology/National Research and Innovation Agency Indonesia with the title of 'Model Manajemen Lanskap Agroforestri sebagai Strategi Peningkatan Jasa Lanskap Mewujudkan Lanskap Rendah Karbon Tahun 2020.

References

[1] Arifin H S, Kaswanto RL, Nurhayati. Manajemen lanskap riparian Sungai Ciliwung berbasis pemberdayaan masyarakat dalam mengatasi banjir di wilayah hilir. Seminar Hasil-hasil Penelitian PPM IPB. Bogor (ID); 2014: 10 hal.
[2] Konsorsium Revisi HCV Toolkit Indonesia. Panduan Identifikasi Kawasan Bernilai Konservasi Tinggi di Indonesia. Jakarta(ID) :Tropenbos International Indonesia Programme; 2008.
[3] Jamaludin A N.Sosiologi Perkotaan. Bandung (ID):CV Pustaka Bandung; 2017.
[4] Bechtel R B, Churchman A. *Handbook of Environmental Psychology*. New York (US): John Willey and Sons; 2002.
[5] Departemen Pekerjaan Umum. Tata Cara Perencanaan Teknik Lanskap Jalan. Jakarta (ID): DirektoratJendral Bina Marga; 1996.
[6] Odum E P, Barret G W. *Fundamentals of Ecology*. California(US) : Cengage Learning; 2005.
[7] Magurran A E. Ecological Diversity and Its Measurement. New Jersey (US): Princeton University Press; 1988.
[8] Bruce H R, Revell G R B 1989 Issues in sampling landscapes for visual quality assessments Landscape and Urban Planning 17 4 pp 323–330.
[9] Hewstone M and Stroebe M 2006 Social psychology: Modern textbook of social psychology Oxford: BPS Blackwell
[10] Badan Pusat Statistik Kota Bogor. Bogor dalam Angka. Bogor (ID): BPS Kota Bogor; 2019
[11] Rahmawati E F. Evaluasi lanskap perkotaan untuk meningkatkan quality of life masyarakat di Kota Bogor [skripsi]. Bogor (ID): Institut Pertanian Bogor; 2018.
[12] Badan Pusat Statistik Kota Bogor. Kompilasi Indikator Terkini Provinsi Jawa Barat (ID): BPS Kota Bogor; 2020
[13] Simonds J O dan Starke B W. *Landscape Architecture*. New York (US): McGraw-Hill Book Co; 2006.
[14] Ching, Francis D K. *A Visual Dictionary of Architecture*. New York (US): Van Nostrand Reinhold Company; 1995
[15] Rapoport A. Human Aspects of Urban Form: Towards A Man Environmental Approach to Urban Form and Design. New York (US): Pergamon Press; 1977.
[16] Crow T, Brown T, De Young R. *The Riverside and Berwyn experience: Contrasts in landscape structure, perceptions of the urban landscape and their effects on people*. Landscape and Urban Planning; 2006.75(1):282-299.
[17] Yulianto D. Persepsi Kualitas Estetika dan Ekologi pada Jalur Wisata Alam Taman Nasional Gede Pangrango. [Skripsi]. Bogor (ID): Institut Pertanian Bogor; 2006.
[18] Booth N K, and Hiss J E. *Residential Landscape Architecture*. Ohio (US): Prentice-Hall; 2012.

Response of bird community to gradual landscape transformation in Darmaga campus, IPB University, Bogor, Indonesia

S P A Nugroho[1], **A Mardiastuti**[2], **Y A Mulyani**[3] **and D A Rahman**[4]

[1] Tropical Biodiversity Conservation Study Program, Faculty of Forestry and Environment, IPB University, Bogor, Indonesia

[2] Department of Forest Resources Conservation and Ecotourism, Faculty of Forestry and Environment, IPB University, Bogor, Indonesia

E-mail: septianputra@apps.ipb.ac.id

Abstract. When a landscape gradually transforms, the bird community that inhabits the landscape will also adapt and change accordingly. Long-term data on the bird-habitat relationship, however, is still lacking in the sub-urban tropical areas. The objective of this paper was to analyze the response of bird community along with the gradual transformation of the landscape. Darmaga Campus of IPB University (± 2.67 km2) was selected as the study area due to its landscape transformation for campus development/construction, and the availability of bird monitoring data. Data on landscape transformation were gathered from the IPB building construction book published in 2017, and bird community (1982-2020) were gathered from published reports and papers. During the observed period, Darmaga Campus that initially consisted mostly of old rubber plantation has been gradually transformed into various academic-related uses, such as some areas for buildings dan field laboratories. From 1982 to 2020, the cumulative bird species in the study area at least was 124 species. During the observed years, the bird community was highly dynamics (bird species number in 1982 = 41 species, 1985 = 39 species, 1986 = 68 species, 1991 = 68 species, 2001 = 39 species, 2003 = 72 species, 2013 = 52 species, 2020 = 99 species). New incoming species were Chestnut-winged Cuckoo, Asian Brown Flycatcher, Tiger Shrike, and babblers (e.g., Black-capped Babbler); Brown Shrike, White-rumped Shama, Oriental Magpie-Robin and Green Junglefowl have not been observed or have become increasingly rare in the last ten years. Meanwhile, Red-breasted Parakeet, Fulvous-breasted Woodpecker, Sunda Pygmy Woodpecker, Black Drongo, and Sooty-headed Bulbul are always present species. Bird species such as Orange-headed Thrush, White-rumped Shama, White-browed Shortwing, and Velvet-fronted Nuthatch are species that entered to rarely found species. Gradual landscape transformation, from an old rubber plantation into a mosaic of small patches of human-made ornamental plants, shrub, and tree plantation, turns out to be beneficial for some bird species, but could also disadvantageous for others, causing a dynamic bird community composition.
Keywords: feeding guilds, rubber plantation, species composition, sub-urban

1. Introduction

The human population has continued to increase from time to time, and this increase has also been followed by an increase in land requirements [1]. The use of land by humans is carried out to fulfil all interests to maintain life, such as to produce food, use it for shelter, and various other purposes. The increasing human need for land has transformed the shape of a landscape a lot.

Transformations in a landscape have occurred in many areas, e.g., in the Darmaga Campus, IPB University, Bogor. In the past, the Darmaga Campus was founded in 1963, most of its area was an old rubber plantation, and there were only a few buildings. Then to fulfil academic needs, the campus area has gradually continued to develop until now. The

Dramaga Campus landscape has undergone many transformations gradually due to area development by humans.

The development of an area can influence the existence of animals, especially birds. Birds are one of the animals that are very sensitive to changes in their environment. If the landscape undergoes transformations or disturbances that affect the bird habitat condition, the bird species composition in the landscape will also change [2,3], hence birds can be used as a bio-indicator to see changes in conditions environment [4] because the bird diversity in a place is closely related to their habitat conditions.

Birds can respond in different ways to the landscape transformations that occurred. There are bird species that can adapt and survive from transformations in their habitat, but there are also bird species that are lost or become extinct due to habitat transformations. This causes the bird community to be dynamic due to transformations in the landscape. This paper aims to determine how birds respond to gradual landscape transformation at the Darmaga Campus.

2. Methods

This paper tries to analyze the shapes of landscape transformation on the Darmaga Campus in several periods and tries to relate to the bird species present in the area whose data is obtained from several research. Darmaga Campus of IPB University (± 2.67 km2) was selected as the study area due to its landscape transformation for campus development/construction, and the availability of bird monitoring data. Landscape transformation data were obtained from the IPB building construction book written by [5] in 2017, then the bird data in this paper were taken from the research results of [6–13]. The data on the bird species presence obtained from several previous research were then grouped into several groups based on presence criteria (Table 1) to find out which bird species can survive from changes in habitat, rarely found on Darmaga Campus, newcomer, and that cannot be found anymore in the last ten years.

Table 1 Criterions of bird species presence

Always present	Found in 7 to 8 survey period
Intermittently present	Found at least 2 to 6 survey period
Rarely found	Found only once
Newcomer	Found in the last 10 years
Extirpated	Cannot be found anymore in the last 10 years

3. Results

3.1 Landscape transformation in Darmaga Campus, IPB University

The Darmaga Campus landscape has undergone many transformations since it was used as a campus until now. In 1963, when it was built as a campus, the Darmaga Campus area was an old rubber plantation from the Dutch colonial era. During the period 1963 - 2017, many building constructions were carried out for academic, laboratory and residential purposes. Previously, at the beginning of the campus, almost the entire of Darmaga Campus area was the green area, namely forest or garden with very few buildings. As time goes by, to fulfil all academic needs, more and more developments are being made at the Darmaga Campus and further reducing the green area. The number of people is increasing and academic needs are increasing, the more building construction is being carried out. The following is a list of building construction carried out at Darmaga Campus, IPB University from 1963 - 2017 (Table 2).

Table 2 The buildings built from 1963 - 2017 in Darmaga Campus, IPB University [5]

Period	Building
1963 - 1970	• Main building of forestry faculty and forest product department (2.419 m²) • Silviculture laboratory building (2.629 m²) • Sylvasari and sylvalestari dormitory building (2.256 m²) • Housing for lecturers and employees (± 60 unit) • Al-Hurriyah mosque (332 m²) • The building of fisheries faculty (3.016 m²) Total: 10.652 m² +
1971 - 1978	• Gymnasium (1.854 m²) • Aerial portrait building (1.019 m²) • Forest engineering laboratory (902 m²) and wood engineering laboratory (154 m²) • Facilities of Food Technology Development Center (FTDC) (4.370 m²) • Area of Agricultural Product Processing Plant (AP4) (1.754 m²) Total: 10.053 m²
1979 - 1987	• The central building of agricultural technology faculty (23.189 m²) • The building of information resources institute (12.361 m²) dan center for environmental research (PPLH) (5.133 m²) • Sylvapinus dormitory (3.856 m²) and Darmaga female dormitory (667 m²) • 3 Faperta greenhouse units • Seed lecture room (252 m²) • I and II analytical laboratory building (585 m²) • Agricultural cultivation department building (652 m2) • Kornita senior high school (252 m²) Total: 46.947 m² +
1988 - 1998	• Inter-university central building (9.000 m²) • Graha Widya Wisuda (GWW) building (4.195 m²) • Fisheries and marine science faculty building (39.971 m²) • The building of animal husbandry faculty (47.682 m²) • Sawmill laboratory (613 m²) and forest product harvesting laboratory (1.176 m²) • Lecture room of forestry faculty (216 m²) and forest resources conservation building (2.768 m²) • Household section building-A (655 m²) and household section building-B (675 m²) Total: 106.951 m²
1999 - 2010	• Student dormitory (6.504 m²) • Sylva Pertamina auditorium (400 m²) • Animal closed house building (1.869 m²)

Period	Building
1999 - 2010	• A2 female dormitory (4.058 m²), A3 female dormitory (3.802 m²), male dormitory (9.767 m²), international dormitory (396 m²), and dormitory secretariat building (329 m²) • Agriculture faculty building (50.545 m²) • The central building of veterinary medicine faculty (20.254 m²) and animal hospital (7.720 m²) • Public electronic agrimedia building (600 m²) Total: 106.244 m²
2011 - 2017	• Biotropics, bioscience, and technology library building (3.641 m²) • Common Class Room (6.750 m²) and teaching laboratory building (3.216 m²) • Forestry faculty building (4.500 m²) and Tanoto building • Chemistry building of mathematics and natural science faculty (2.400 m²) • Greenhouse laboratory of forestry faculty (300 m²), drying and preserving wood laboratory of forest products department (608 m²) and, PPM creata laboratory • Student cooperative building of female dormitory (128,56 m²) • Primate animal cages (1,076 m²) • Outdoor tennis court (175 m²) • Agriananda kindergaten (758 m²) • Alih jenis building • Management economics faculty building (FEM) (4.980 m²) • Biology building (7.073 m²) and mathematics and natural science faculty auditorium (1.193 m²) Total: 35.723, 636 m² +

3.2 Bird communities in Darmaga Campus, IPB University based on several previous research

There are at least 124 bird species founded in the study location based on several previous research results, which is relatively high for a tropical urban periphery. The number of bird species from several previous research results at Darmaga Campus can be seen in Figure 1. All of the species founded, then are grouped into several groups based on presence criteria (Table 3). Prescribed presence criteria are always present, intermittently present, rarely found, newcomer, and extirpated. There are 26 species included in the always present, 58 species intermittently present, 39 species rarely found, 103 newcomer and 23 extirpated species.

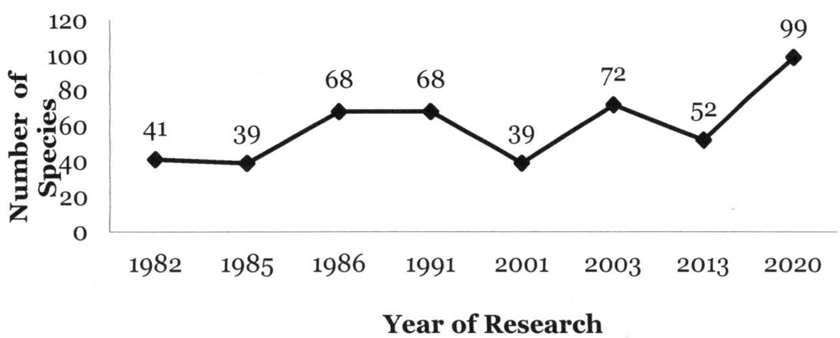

Figure 1 Bird species number found from several previous research from 1982 - 2020 in Darmaga Campus, IPB University [6-13]

Table 3 Bird species presence from several previous research [6-13] which has been grouped based on several presence criteria (bird name and sequence based on [14]

	Common Name
Always present	White-breasted Waterhen, Pink-necked Green Pigeon, Spotted Dove, Red-breasted Parakeet, Lesser Coucal, Javan Kingfisher, Collared Kingfisher, Fulvous-breasted Woodpecker, Sunda Pygmy Woodpecker, Pied Triller, Small Minivet, Common Iora, Sooty-headed Bulbul, Long-tailed Shrike, Bar-winged Prinia, Common Tailorbird, Great Tit, Scarlet-headed Flowerpecker, Brown-throated Sunbird, Olive-backed Sunbird, Oriental White-eye, Javan Munia, Scaly-breasted Munia, Black-naped Oriole, Black Drongo, White-breasted Woodswallow
Intermittently present	Javan Pond Heron, Black-crowned Night Heron, Cinnamon Bittern, Crested Honey Buzzard, Crested Serpent Eagle, Crested Goshawk, Besra, Spotted Kestrel, Green Junglefowl, Barred Buttonquail, Island Collared Dove, Zebra Dove, Chestnut-winged Cuckoo, Banded Bay Cuckoo, Plaintive Cuckoo, Rusty-breasted Cuckoo, Asian Drongo-Cuckoo, Greater Coucal, Barn Owl, Indian Scops-Owl, Collared Scops Owl, Savanna Nightjar, Edible-nest Swiftlet, Cave Swiftlet, House Swift, Asian Palm-Swift, Blue-eared Kingfisher, White-throated Kingfisher, Chestnut-headed Bee-eater, Barn Swallow, Pacific Swallow, Striated Swallow, Black-winged Flycatcher-shrike, Yellow-vented Bulbul, Brown Shrike, Oriental Magpie-Robin, Black-capped Babbler, Horsfield's Babbler, Striped Tit-Babbler, Striated Grassbird, Brown Prinia, Olive-backed Tailorbird, Arctic Warbler, Yellow-bellied Warbler, Asian Brown Flycatcher, Hill Blue Flycatcher, Golden-bellied Geryone, Pied Fantail, Ruby-cheeked Sunbird, Sunbird, Little Spiderhunter, Java Sparrow, Eurasian Tree Sparrow, Black-winged Starling, White-vented Myna, Ashy Drongo, Slender-billed Crow, Large-billed Crow
Rarely found	Grey Heron, Purple Heron, Great Egret, Little Egret, Cattle Egret, Brahminy Kite, Chinese Sparowhawk, Japanese Sparrowhawk, Black Eagle, Crested Hawk-Eagle, Slaty-legged Crake, Oriental Pratincole, Imperial Pigeon, Brown Cuckoo Dove, Common Emerald Dove, Yellow-crested Cockatoo, Long-tailed Parakeet, Yellow-throated Hanging Parrot, Moustached Hawk-Cuckoo, Chestnut-breasted Malkoha, Barred Eagle-Owl, Spotted Wood Owl, Sunda Frogmouth, Large-tailed Nightjar, Black-nest Swiftlet, Black-banded Barbet, Grey Wagtail, Tiger Shrike, White-browed Shortwing, White-rumped Shama, Sunda Robin, Orange-headed Thrush, Black-capped Babbler, Velvet-fronted Nuthatch, Plain Flowerpecker, Mountain White-eye, Asian Glossy Starling, Pied Myna, Common Hill Myna
Newcomer	Grey Heron, Purple Heron, Great Egret, Little Egret, Cattle Egret, Javan Pond Heron, Black-crowned Night Heron, Cinnamon Bittern, Crested Honey Buzzard, Brahminy Kite, Crested Serpent Eagle, Crested Goshawk, Chinese Sparowhawk, Japanese Sparrowhawk, Besra, Black Eagle, Crested Hawk-Eagle, Spotted Kestrel, Barred Buttonquail, Slaty-legged Crake, White-breasted Waterhen, Pink-necked Green Pigeon, Island Collared Dove

	Common Name
Newcomer	Spotted Dove, Zebra Dove, Common Emerald Dove, Red-breasted Parakeet, Long-tailed Parakeet, Yellow-throated Hanging Parrot, Chestnut-winged Cuckoo, Moustached Hawk-Cuckoo, Banded Bay Cuckoo, Plaintive Cuckoo, Rusty-breasted Cuckoo, Asian Drongo-Cuckoo, Chestnut-breasted Malkoha, Greater Coucal, Lesser Coucal, Barn Owl, Collared Scops Owl, Barred Eagle-Owl, Spotted Wood Owl, Sunda Frogmouth, Large-tailed Nightjar, Savanna Nightjar, Edible-nest Swiftlet, Black-nest Swiftlet, Cave Swiftlet, House Swift, Asian Palm Swift, Blue-eared Kingfisher, White-throated Kingfisher, Javan Kingfisher, Collared Kingfisher, Chestnut-headed Bee-eater, Black-banded Barbet, Fulvous-breasted Woodpecker, Sunda Pygmy Woodpecker, Barn Swallow, Pacific Swallow, Striated Swallow, Grey Wagtail, Pied Triller, Small Minivet, Black-winged Flycatcher-shrike, Common Iora, Sooty-headed Bulbul, Yellow-vented Bulbul, Tiger Shrike, Long-tailed Shrike, Black-capped Babbler, Horsfield's Babbler, Striped Tit-Babbler, Striated Grassbird, Brown Prinia, Bar-winged Prinia, Common Tailorbird, Olive-backed Tailorbird, Arctic Warbler, Yellow-bellied Warbler, Asian Brown Flycatcher, Yellow-rumped Flycatcher, Hill Blue Flycatcher, Golden-bellied Geryone, Pied Fantail, Great Tit, Velvet-fronted Nuthatch, Plain Flowerpecker, Scarlet-headed Flowerpecker, Brown-throated Sunbird, Ruby-cheeked Sunbird, Olive-backed Sunbird, Crimson Sunbird, Little Spiderhunter, Oriental White-eye, Javan Munia, Scaly-breasted Munia, Eurasian Tree Sparrow, Black-naped Oriole, Black Drongo, Ashy Drongo, White-breasted Woodswallow, Slender-billed Crow
Extirpated	Green Junglefowl, Oriental Pratincole, Imperial-pigeon, Brown Cuckoo Dove, Yellow-crested Cockatoo, Indian Scops-Owl, Brown Shrike, White-browed Shortwing, Oriental Magpie-Robin, White-rumped Shama, Sunda Robin, Orange-headed Thrush, Plain Flowerpecker, Sunbird, Crimson Sunbird, Mountain White-eye, Java Sparrow, Asian Glossy Starling, Pied Myna, Black-winged Starling, White-vented Myna, Common Hill Myna, Large-billed Crow

4. Discussion

The Dramaga Campus landscape has changed gradually and can affect the bird communities in the landscape. In the past, the Darmaga Campus landscape was initially built by a campus as an old rubber plantation. Gradually, the Darmaga Campus has undergone a lot of development until now to fulfill academic needs. Transformations in a landscape can impact the animals in it, especially birds, because they are sensitive to changes in their habitat.

Regional development gradually will make the shape of landscape mosaic and heterogeneous. The development of the area can create threats in the form of fragmentation and habitat loss for animals, which forces them to determine whether they will emigrate to the new places, more suitable habitats, or survive and adapt to new conditions [15,16]. Birds will respond differently to any development in their habitat. There are bird species that can survive changes in their habitat, and others choose to avoid or disappear.

IOP Conf. Series: Earth and Environmental Science **879** (2021) 012036 doi:10.1088/1755-1315/879/1/012036

Some shy bird species will require large patches of habitat and forest-resembled habitat (e.g., thick canopy, dense shrub), some bird species that can live in open forests or areas with humans will be able to adapt to changes in their habitat. Bird responses can vary according to the bird species' characteristics and how they use the resources around them.

Most of the bird species that are included in the always present group like open habitats and habitats that are close to the presence of humans. Bird species such as the Red-breasted Parakeet, Fulvous-breasted Woodpecker, Sunda Pygmy Woodpecker, Black Drongo, and Sooty-headed Bulbul like open areas or open forests [17]. Besides, the Javan Munia species are seed-eating, usually founded in rice fields, and the Javan Kingfisher species are always founded in the Darmaga Campus landscape because this landscape can provide habitat for them in the form of rice fields and wet areas.

Meanwhile, in the rarely found group, species encountered such as Orange-headed Thrush, White-rumped Shama, White-browed Shortwing, and Velvet-fronted Nuthatch are shy bird species to human presence and require forested habitats [17]. Therefore, the increasing number of developments and the more human activities will have an impact on this type of bird.

In the newcomer group or new species that enter, there are bird species such as the Chestnut-winged Cuckoo, Asian Brown Flycatcher, Tiger Shrike, and babblers (e.g., Black-capped Babbler). Meanwhile, species such as the Brown Shrike, White-rumped Shama, Oriental Magpie-Robin, and Green Junglefowl have not been observed or have become increasingly rare in the last ten years. The existence of bird species in a place is closely related to the conditions of the place to support birdlife. According to [18] that there are several factors determine the presence of birds in a place, namely the availability of food, a place to rest, play, breed, nest, perch, and take shelter. If bird habitat is disturbed, newcomer species may disappear or no longer be found, such as extirpated species.

The existence of area development carried out at the Darmaga Campus should be balanced with the fulfilment of bird habitat components to support birdlife. Some activities that can be carried out are protecting forested habitats for birds, intensifying the planting of bird feed sources, and planting trees.

5. Conclusion

The conclusion of this paper is the gradual landscape transformation that occurred at Darmaga Campus, IPB University, from previously forest and old rubber plantations to many building constructions, turns out to be beneficial for some bird species, but can also be detrimental for others, causing a dynamic bird community composition. There are several recommendations for area development activities so that the development can also consider the birdlife needs, such as protecting forested habitats for birds, intensifying the planting of bird feed sources, and planting trees.

Acknowledgment

The authors would like to thank the Deputy for Strengthening Research and Development, Ministry of Research and Technology-National Research and Innovation Agency to provide funds so that can carry out research and publication. Similar remarks are also addressed to people who have helped the author to compile this paper. Moreover, the authors are also grateful to the Department of Landscape Architecture, Faculty of Agriculture, IPB University, which has arranged The 5th International Symposium for The Sustainable Landscape Development 2020.

References

[1] Kusrini, Suharyadi and Hardoyo S R 2011 Perubahan penggunaan lahan dan faktor yang mempengaruhinya di Kecamatan Gunungpati Kota Semarang *Maj. Geogr. Indones.* **25**(1):24–40

[2] Canterbury G E, Martin T E, Petit L J and Bradford D F 2000 Bird communities and habitat as ecological indicator of forest condition in regional monitoring *Conserv. Biol.* **14**(2):544–58
[3] Lee T M, Soh M C K, Sodhi N, Koh L P and Lim S L H 2005 Effects of habitat disturbance on mixed species bird flocks in a tropical sub-montane rainforest *Biol. Conserv.* **122**(2):193–204
[4] Schultze C H, Waltert M, Kessler P J A, Pitopang R, Shahabuddin, Veddeler D et al 2004 Biodiversity indicator groups of tropical land use systems: comparing plants, birds, and insects *Ecol. Appl.* **14**(5):1321–33
[5] Nugroho N, Mansjoer S, Djajanegara O, Arsyad S, Suryokusumo S, Sutarahardja S et al 2017 Pembangunan gedung-gedung IPB Bogor: IPB Press
[6] Putro H R 1982 Keragaman jenis burung di lingkungan kampus IPB Darmaga Institut Pertanian Bogor
[7] Mulyani Y A 1985 Studi keanekaragaman jenis burung di lingkungan Kampus IPB-Darmaga Institut Pertanian Bogor
[8] van Balen S, Hernowo J B, Mulyani Y A and Putro H R 1986 The birds of Darmaga *Media Konserv.* **1**(2):1–5
[9] Hernowo J B, Soekmadi R and Ekarelawan 1991 Kajian pelestarian satwaliar di Kampus IPB Darmaga *Media Konserv.* **3**:43–65
[10] Mulyani 2001 Keragaman jenis burung di Kampus IPB Darmaga Bogor Institut Pertanian Bogor
[11] Kurnia I 2003 Studi keanekaragaman jenis burung untuk pengembangan wisata birdwatching di Kampus IPB Darmaga Institut Pertanian Bogor
[12] Mulyani Y A, Ulfah M and Sutopo 2013 Bird use of several habitat types in an academic campus of Institut Pertanian Bogor in Darmaga, Bogor, West Java *Media Konserv.* **18**(1):18–27
[13] Mustari A H 2020 Biodiversitas Kampus IPB, Bogor Bogor: IPB Press
[14] Sukmantoro W, Irham M, Novarino W, Hasudungan F, Kemp N and Muchtar M 2007 Daftar Burung Indonesia No.2 Bogor: Indonesian Ornithologists' Union
[15] Marzluff J M and K E 2001 Restoration of fragmented landscapes for the conservation of birds: a general framework and specific recommendations for urbanizing landscapes *Restor. Ecol.* **9**(3):280–92
[16] McKinney M L 2002 Urbanization, biodiversity, and conservation *Bioscience* **52**(10):883–90
[17] MacKinnon J, Phillips K and van Balen S 2010 Seri Panduan Lapangan Burung-Burung di Sumatera, Jawa, Bali, dan Kalimantan Bogor: Burung Indonesia
[18] Ontario J, Hernowo J B, Haryanto and Ekarelawan 1990 Pola pembinaan habitat burung di kawasan pemukiman terutama di perkotaan *Media Konserv.* **3**(1):15–28

Landscape transformation and wildlife: road mortality in Rawa Aopa Watumohai National Park, Southeast Sulawesi, Indonesia

T Purnomo[1], M D Kusrini[1], and A Mardiastuti[1]
[1]Department of Forest Resources Conservation & Ecotourism,
Faculty of Forestry and Environment, IPB University, Dramaga Campus
Bogor 16680, Indonesia

E-mail: teguh_purnomo@apps.ipb.ac.id

Abstract. The development of high way that crosses wildlife habitat will inevitably cause wildlife casualties. The purpose of this study was to identify wildlife road mortality in Rawa Aopa Watumohai National Park, Southeast Sulawesi (mostly consisted of savanna). Road kill data was collected by driving along the 22.4 km highway in February and March 2020 (n=28). Traffic volumes were recorded by taking traffic samples in 5 different times of the day, an hour duration each (totaling 70 h). There were 37 species being killed (totaling 529 individuals, of which 49.3% happened along km 0-10), consisted of 19 bird species, 3 small mammal species, 10 reptile species, and 5 amphibian species. Based on taxa, the highest number of road mortality was amphibians (dominated by *Ingerophrynus biporcatus*), followed by birds (mostly *Lonchura malacca*), mammals (dominated by *Rattus argentiventer*), and reptiles (dominated by *Varanus salvator*). Among mammals, no big mammal (booted macaque *Macaca ochreata*, wild boar *Sus scrofa*) was killed. Landscape transformation clearly has impacted wildlife in the study area. To reduce the occurrence of road kills, it is recommended to create corridors for wildlife crossings, limiting vehicle speed in the hot spots locations, create speed bumps, as well as adding more signage and road signs along the highway.
Keywords: amphibians, road kills, savanna, traffic, wildlife crossing.

1. Introduction

Economic development has driven changes of landscape in various regions. To improve access in remote areas, roads and highways are constructed and often crosses natural landscapes. The existence of the road has created a problem for wildlife that usually moves among various habitat, as the habitat become fragmented, and might cause mortality or injured while trying to cross the road. Wildlife road mortality has become a leading cause of population decline in some areas, i.e. in East Africa, India and Taiwan [1], [2], [3].

Rawa Aopa Watumohai National Park in Southeast Sulawesi (Indonesia) has been known as its tropical savanna habitat. In 2009, a provincial highway funded by the Australian project “The Eastern Indonesia National Road Improvement Project (EINRIP)” was built, which separated the northern and southern part of the savanna of the Park [4]. Considering that there are at least 321 species of fauna consisting of 28 species of mammals, 218 species of aves [5], 10 species of amphibians and 32 species of reptiles [6], the existence of this highway might become a problem for wildlife. The discovery of carcasses of Amboina box turtle (*Cuora amboinensis*) along the highway indicates the possibility that the road could affects the population of this turtle [7]. Although the species is not protected, it is categorized as Vulnerable by IUCN Red List [8] and listed in CITES Appendix II.

Report on the impact of highway to wildlife in the Indonesia is scarce. Only one report in grey publication on roadkill available, which refers to road kill between Sanggi – Bengkunat

IOP Conf. Series: Earth and Environmental Science **879** (2021) 012037 doi:10.1088/1755-1315/879/1/012037

at Bukit Barisan Selatan National Park in Sumatera [9]. Thus, research on road mortality in Rawa Aopa Watumohai National Park is important. The purpose of this study is to identify the wildlife road mortality in Rawa Aopa Watumohai National Park.

2. Methods

The Rawa Aopa Watumohai National Park (105.194 ha) was established on 17 December 1990 and lays in four regencies: East Kolaka Regency, Konawe Regency, South Konawe Regency and Bombana Regency [5] (Figure 1). Surveys in the Park were carried out along the 22.4 km provincial highway, starting from the entrance gate of Konawe Selatan to the exit gate of Bombana. The road passed through several ecosystems, i.e, savanna, lowland forest, swamp, and mangrove, with savanna (22.963 ha) as the dominant ecosystems (Figure 2).

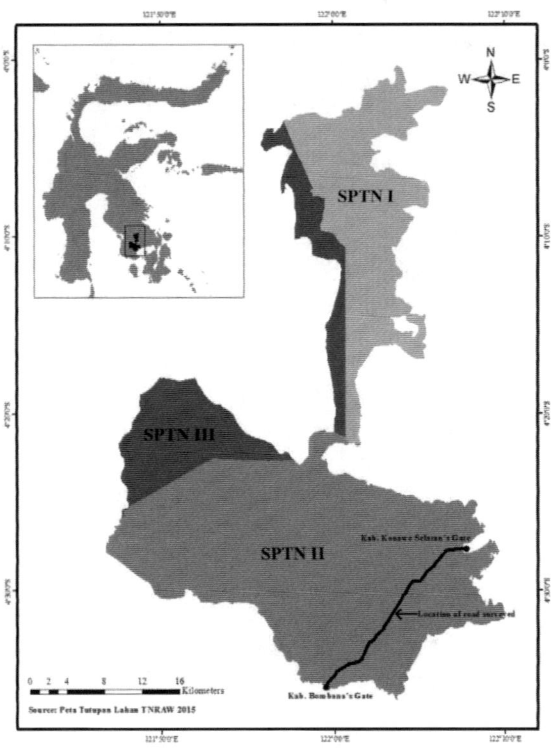

Figure 1 Study area in Rawa Aopa Watumohai National Park (Southeast Sulawesi, Indonesia), showing position of highway surveyed

Figure 2 Ecosystem along the high road. (a) savanna ecosystem, (b) low land forest ecosystem, (c) swamp ecosystems, (d) mangrove ecosystem

Data were collected for 28 days (15-28 February; 15-28 March 2020) by driving along the 22.4 km road as the main transect where observer travel at a constant speed [10] similar to research elsewhere [11]. Observations were made by two-person travelling with motorbike with a constant speed of 20-40 km/hour, 2 times a day in the morning (6am local time), and the afternoon (4pm). Observers will stop on the road shoulder whenever wildlife carcasses were sighted. The starting point for the observation were either the entrance gate in Konawe Selatan Regency or the entrance gate in Bombana Regency, with the same number of repetitions, totaling 627.2 km. All carcasses spotted were recorded with *Global Positioning System* (GPS) and marked using spray paint. Carcasses were measured and recorded and removed afterward from the road using shovels and latex gloves to avoid subsequent recounts. Species were identified at least to family level. Wildlife roadkill diversity was estimated as the number of species found dead on the surveyed.

3. Results and Discussions

3.1 Roadkill diversity

Roadkill during the study period was 529 cases from 37 spesies, with only 453 cases identified to genus and species, as some could not be identified due to the deterioted conditions (Figure 3). The number of roadkill consisted of 263 amphibians (51%), 120 aves (23%), 63 mammals (12%), and 69 reptiles (14%). The daily average of roadkill in this study was 37.8 cases with a range of 18–69 cases. The number of roadkill in this study was quite high and diverse. The annual roadkill was estimated to reach 13,797 (6,570-25,185). No previous report on the number of roadkill in this Park and the high number of road mortality in this study showed that road mortality in this Park has been overlooked. Overall from the above data, many roadkills occur near the main gate of Bombana Regency.

In term of diversity, the number of species killed included at least 19 species of aves (Table 1), 3 species of mammals (Table 2), 10 species of reptiles and 5 species of amphibians (Table 3). The number of species was higher compared to report by [9] which only reported 4 species of mammals, 1 species of birds, and 6 species of reptiles. Other research outside Indonesia mostly reported specific classes of animals. For example, [12] recorded roadkill of 18 species of mammals from 10 families on major roads in Brazil. Meanwhile, [13] reported roadkill of 8 herpetofauna species on the road, consisting of 180 amphibians and 72 reptiles. During 6 years of observation from 2010-2015, [14] recorded 102 (80.3%) *Cuora amboinensis* and 25 (19.7%) *Cyclemys dentata* roadkill in Palawan, Philippines.

For bird (aves), *Lonchura malacca* was killed the most compared to other birds (Figure 2b). This bird species lives mostly in lowland ecosystems including savanna [15],often seen flying low and crossing the highway from one side to the other, which caused this species vulnerable to being hit.

Roadkill mammals are dominated by rats (*Rattus argentiventer*) and civets (*Viverra tangalunga;* Figure 2c) which have been known to adapt with human settlements [16], [17]. There were no incidents of road kill of big mammals such as *Macaca ochreata* and *Sus scrofa* at the time of the study, although both species were often seen around the highway.

The highest roadkill among wildlife were amphibians (n = 263) with 143 cases of crested toad *Ingerophrynus biporcatus*, of which 99% of cases occurred near the savanna. Savannas are one of the main habitats of this species [6] and during the survey, the toad were often found hopping in the asphalt in the middle of the road at night, increasing the possibility of getting hit by vehicles (Figure 2). Although the toad is endemic to Indonesia, its original distribution confines to Sumatra, Java, Madura, Bali, Lombok, and Kalimantan. The toad has been introduced to Sulawesi and able to lives in lowland forests, degraded forests, and nearby settlements [18], [19].

As expected, the Amboina box turtle was also recorded as road kill. Six species of snake were recorded as roadkill with the highest number of snake killed was *Hypsiscopus matannensis,* an endemic species of Southeast Sulawesi. *Varanus salvator* was the highest number of reptile killed on the road (Table 3, Figure 3d). This species was often found basking on the side of the road during afternoon and late afternoon. The higher number of this species killed in the road might be caused of its feeding habits as scavengers [20]. The high number of carcasses in the road might attract the monitor to eat the carrion and increase the possibility of getting hit by a vehicle while eating.

Table 1 Diversity of avian road kill and mortality rate in Rawa Aopa Watumohai National Park during February-March 2020, listed based on alphabetical order

No	Species	Family	Number	Mean mortality/10 km
	Aves		**120**	**53.6**
1	*Aerodramus fuciphagus*	Apodidae	2	0.9
2	*Amaurornis phoenicurus*	Rallidae	1	0.4
3	*Caprimulgus affinis*	Caprimulgidae	1	0.4
4	*Centropus bengalensis*	Cuculidae	13	5.8
5	*Cisticola exilis*	Passeridae	5	2.2
6	*Cisticola juncidis*	Passeridae	13	5.8
7	*Cisticola sp.*	Passeridae	10	4.5
8	*Gallirallus torquatus*	Rallidae	4	1.8
9	*Hirundo tahitica*	Hirundinidae	1	0.4
10	*Ixobrychus sinensis*	Ardeidae	2	0.9

No	Species	Family	Number	Mean mortality/10 km
11	*Lalage sueurii*	Campephagidae	2	0.9
12	*Lewinia striata*	Rallidae	1	0.4
13	*Lonchura malacca*	Estrildidae	16	7.1
14	*Passer montanus*	Passeridae	2	0.9
15	*Pycnonotus aurigaster*	Pycnonotidae	5	2.2
16	*Saxicola caprata*	Muscicapidae	1	0.4
17	*Synoicus chinensis*	Phasianidae	4	1.8
18	*Turnix suscitator*	Turnicidae	1	0.4
19	*Zosterops chloris*	Zosteropidae	6	2.7
20	Unidentified bird	-	30	13.4

Table 2 Diversity of mammal road-kill and mortality rate in Rawa Aopa Watumohai National Park during February-March 2020, listed based on alphabetical order

No	Species	Family	Number	Mean mortality/10 km
	Mammals		**63**	**28.1**
1	*Rattus argentiventer*	Muridae	33	14.7
2	*Rattus sp.*	Muridae	22	9.8
3	*Viverra tangalunga*	Viverridae	2	0.9
4	Unidentified	-	6	2.7

Table 3 Diversity of herpetofauna road-kill and mortality rate in Rawa Aopa Watumohai National Park during February-March 2020, listed based on alphabetical order

No	Species	Family	Number	Mean mortality/10 km
	Reptiles		**69**	**30.8**
1	*Calamaria nuchalis*	Colubridae	1	0.4
2	*Cuora amboinensis*	Goemydidae	3	1.3
3	*Dendrelaphis pictus*	Colubridae	8	3.6
4	*Gekko gecko*	Gekkonidae	1	0.4
5	*Hypsiscopus matannensis*	Homalopsidae	17	7.6
6	*Lycodon capucinus*	Colubridae	10	4.5
7	*Ophiophagus hannah*	Elapidae	1	0.4
8	*Ramphlotyphlops braminus*	Typhlopidae	1	0.4
9	*Varanus salvator*	Varanidae	19	8.5
10	*Xenopeltis unicolor*	Xenopeltidae	1	0.4
11	Unidentified reptile	-	7	3.1
	Amphibian		**263**	**117.4**
1	*Fejervarya cancrivora*	Dicroglossidae	7	3.1
2	*Fejervarya limnocharis*	Dicroglossidae	28	12.5
3	*Fejervarya sp.*	Dicroglossidae	49	21.9
4	*Ingerophrynus biporcatus*	Bufonidae	143	63.8
5	*Polypedates iskandari*	Rhacophoridae	17	7.6
6	Unidentified amphibian	-	19	8.5

Figure 3 Species most killed during research in Rawa Aopa Watumohai National Park (a) *Ingerophrynus biporcatus;* (b) *Lonchura malacca;* (c) *Rattus argentiventer;* (d) *Varanus salvator*

Conservation implication

The development of highway in the middle of a national park area is a major problem for the existence of wildlife in this area. The construction of road infrastructure in forest areas might cause several negative impacts, such as physical changes, pollution, invasion of alien species and the roadkill [21]. Lack of signs stating animal crossing areas and maximum speed limit of 80 km/hour increases the likelihood of an animal being hit. Even though about 150 m from the direction of South Konawe Regency there are animals crossing sign, but these signs are not clearly visible (Figure 4).

Figure 4 Road signs in the Rawa Aopa Watumohai National Park. Left: 80 km limits; Right: a nondescript animal crossing sign

There is another project within the Park area in progress, namely the 150 kV Andolo-Kasipute Sutt Electricity Infrastructure Acceleration Project. Geothermal power plants is an environmentally friendly energy [22], however there might be negative impact of the project, especially during the construction phase with high mobility of construction's vehicle. There are several steps to reduce roadkill by construction of corridors for wildlife crossings, reducing speed limit of vehicles in "hot spots" area, and adding more signboards and road signs.

Corridors have an important function as connecting routes between a different type of land cover and neighbors on both sides [23]. The highway has divided the savanna into two parts, and ponds are found in both sides. These ponds are very important and wildlife will move to puddles on other sides when ponds dried up. To enable this wildlife species, especially amphibian and reptiles, to cross the road safely, there is a need to develop a tunnel as a corridor for the species' movement. Corridor construction can be carried out near the Bombana's Gate, but must spend high funds. Previous study showed suitable types of tunnels for frogs are the one mostly made by concrete but lined with soil [24].

Speed restrictions need to be put in place so that animals have the opportunity to run when vehicles approach. At high speed, drivers and animals have limited time to avoid each other [25]. Lowering speeds not only reduce collisions but also lowering noise intensity. Tall and dense roadside trees can increase the likelihood of collisions with animals, so tall and medium vegetation should be avoided [26].

It is necessary to install notification boards or road signs. Road sign along the highway are mostly in poor condition and the number is too few. Signs must be made every 1 km to enable drivers to read the signs. Creating speed bumps in hot spot area might be needed to reduce speed, especially at km 1-10 of Bombana Gate.

With the development of other infrastructure in the area, it is necessary to ensure that an environmental impact analysis has been carried out to identify the potential effects of this development on the ecology and wildlife in the national park. Local and international stakeholders as well as the government need to take strategic steps to protect the Rawa Aopa Watumohai National Park and its wildlife.

4. Conclusions

Based on the research, the highway of Rawa Aopa Watumohai National Park might negatively affect animal that live in the park. There were 529 cases of roadkill consisted of 37 species. The animals most frequently hit are amphibians (*Ingerophrynus biporcatus*), aves (*Lonchura malacca* and *Centropus bengalensis*), mammals (*Rattus argentiventer*), and reptiles (*Varanus salvator*). It is necessary to consider the construction of corridors in the form of flyovers and improvement of several tunnels along the road, renewing and adding road signs on animal crossing areas, and reduce maximum speed limit to reduce roadkill.

Acknowledgments

This study was supported by Asosiasi Pengusaha Eksportir Kura-kura dan Labi-labi Indonesia (APEKLI). We are grateful for the help of Maraden Purba from APEKLI and Amir Hamidy from PHI. We thank the Rawa Aopa National Park for asisstance, especially head of National Park (Ali Bahri, SSos Msi), and Head of SPTN II (Benny Ermmyadi Purnama, SHut), National Park staff and others (Aris, Ana, Acho, Ikhsan Fajrin, Brigdal Karhut, and Harnum Nurazizah). This research was conducted with permit Surat Izin Masuk Kawasan Konservasi (SIMAKSI) Nomor SI.03/T.22/TU/KSDAE/01/2020 issued on 27 January 2020.

References

[1] Baskaran N and Boominathan D 2010 Road kill of animals by highway traffic in the tropical forests of Mudumalai Tiger Reserve, southern India *Threatened Taxa* **2** 753-759
[2] Kioko J, Kiffner C, Jenkins N, Collinson WJ 2015 Wildlife roadkill patterns on a major highway in northern Tanzania *African Zoologyi* 1–6
[3] Yue S, Bonebrake TC, Gibson L 2019 Informing snake roadkill mitigation strategies in Taiwan using citizen science *Wildlife Management* **83** 80–88
[4] Pratiwi D 2015 *Perubahan penutupan lahan dan faktor penyebabnya di Taman Nasional Rawa Aopa Watumohai, Provinsi Sulawesi Tenggara* skripsi (Bogor: Fakultas Kehutanan, Institut Pertanian Bogor)
[5] [BTNRAW] Balai Taman Nasional Rawa Aopa Watumohai. 2018. *Buku Informasi Kawasan Taman Nasional Rawa Aopa Watumohai* (Konawe Selatan: Balai Taman Nasional Rawa Aopa Watumohai)
[6] Ridha MA 2018 *Herpetofauna pada Empat Tipe Habitat di Taman Nasional Rawa Aopa Watumohai, Provinsi Sulawesi Tenggara* skripsi (Bogor: Fakultas Kehutanan, Institut Pertanian Bogor)
[7] Aini Y 2018 *Pemodelan spasial kesesuaian habitat kura-kura ambon (Cuora amboinensis) di Taman Nasional Rawa Aopa Watumohai* skripsi (Bogor: Fakultas Kehutanan, Institut Pertanian Bogor)
[8] [IUCN] International Union for Conservation of Nature and Natural Resources 2000 About Red List IUCN [Internet]. [downloaded 2020 Jun 26] Available on: https://www.iucnredlist.org/species/5958/97349919
[9] Ardiantiono A. 2016 Kematian satwa di jalan Sanggi-Bengkunat, Taman Nasional Bukit Barisan Selatan *Warta Herpetofauna* **8** 34-37
[10] Heyer WR, Donnelly MA, McDiarmid RW, Hayek LA, Foster MS 1994 *Measuring and Monitoring Biological Diversity: Standard Methods for Amphibians* (Washington: Smithsonian Institution Press) pp 364
[11] Carvalho NC, Bordignon MO, Shapiro JT 2014 Fast and furious: a look at the death of animals on the highway MS-080, Southwestern Brazil *Série Zoologia* **104** 43-49
[12] Zanzini AC, Machado FS, Oliveira JE, Oliveira EC 2018 Roadkills of medium and large-sized mammals on highway br-242, Midwest Brazil: a proposal of new indexes for evaluating animal roadkill rates *Oecologia Australis* **22** 248–25
[13] Heigl F, Horvath K, Laaha G, Zaller JG 2017 Amphibian and reptile road-kills on tertiary roads in relation to landscape structure: using a citizen science approach with open-access land cover data *BMC Ecology* **17** 24
[14] Bernardo AA 2019 Road mortality of freshwater turtles in Palawan, Philippines *The Palawan Scientist* **11** 97–111
[15] Sutarya P and Dea AA 2018 Keragaman Jenis Burung Kawasan Taman Nasional Rawa Aopa Watumohai Jilid 1 (Konawe Selatan: Balai Taman Nasional Rawa Aopa Watumohai) pp 34
[16] Sudarmaji and Herawati A 2017 Perkembangan populasi tikus sawah pada lahan sawah irigasi dalam pola indeks pertanaman padi 300 *Penelitian Pertanian Tanaman Pangan* **1** 125-13
[17] Vaughan TA, Ryan JM, Czaplewski NJ 2011 Review of Mammalogy 5th ed University of Nebraska-Lincoln *Mammalogy* **92** 478-479
[18] Iskandar DT 1998 *Amfibi Jawa dan Bali: Seri Panduan Lapangan* (Bogor: Puslitbang Biologi-LIPI)
[19] Kurniati H 2013 Distribusi kodok *Ingerophrynus biporcatus* (Gravenhorst, 1829) di Sulawesi *Warta Herpetofauna* **6** 8-9
[20] Deraniyagala P 1953 *A Colored Atlas of Some Vertebrates from Ceylon* vol 2 (Tetrapod Reptilia Colombo: The Ceylon Government Press) pp 122
[21] Laurance WF, Goosem M, Laurance SGW 2009 Impacts of roads and linear clearings on tropical forests *Trends in Ecology & Evolution* **24** 659–669
[22] Kartikasari D, Pudyatmoko S, Wawandono NB, Utami P 2018 Response of bird community to Kamojang Geothermal Power Plant, Bandung, West Java *Journal of Forest Science* **12** 156-171
[23] Forman RTT 1995 Some general principles of landscape and regional ecology *Landscape Ecology* **10** 133–142
[24] Lesbarrères D, Lodé T, Merilä J 2004 Short Communication: What type of amphibian tunnel could reduce road kills? *Oryx* **38** 220–223
[25] Lin SC 2015 Fasek lansekap dan TraffIc meningkatkan hewan kematian di jalan *Teknik Lingkungan dan Manajemen Lansekap* **24** 10-20
[26] Beben D 2016 Crossings construction as a method of animal conservation *Transportation Research Procedia* **14** 474-483

Patch dynamics in the Javan Hawk-Eagle (*Nisaetus bartelsi*) habitat of East Java

A R P Murad[1] and Syartinilia[1*]

[1] Landscape Architecture Department, Faculty of Agriculture, IPB University, Indonesia

E-mail: syartinilia@apps.ipb.ac.id

Abstract. Javan Hawk-Eagle (JHE, *Nisaetus bartelsi*) is an endemic species in Java Island and an important biological indicator of ecosystem health. The government has issued regulations to protect this species and increase the population by 10% from 2015 until 2019. East Java has the largest JHE potential habitat in Java Island based on a previous study using satellite images of 2002. Therefore, the current habitat distribution of JHE's is essential for getting knowledge about patch dynamics in JHE's habitat. This study's objective was to analyze patch dynamics of JHE's habitat from 2002 until 2015 and validate habitat distribution. Previously predicted probability map (2002) of JHE's were updated using Landsat 8 satellite images of 2015 and was validated through ground-truth checked. Results showed that the distribution of JHE's habitat after validation is 28 patches, which is covered 4766.26 km². The dynamics that occur in the JHE's patch are patch lost(1 patch), patch area decreased (5 patches), patch area increased (13 patches), new patch (4 patches), and merged patch. After validation, there are six newly identified patches, and one patch area increased. The total area increased by 2156.14 km² or 82.61% of the total area occupied by JHE's in 2002. About 39.89% of total habitat patches were located inside the protected area. This study recommends continuing monitoring activities on habitat patches, including potential habitat patches in lowland areas, and proposing conservation activities based on habitat patch dynamics that occurred from 2002 to 2015.

Keywords: habitat distribution, habitat validation, patch, Javan Hawk-Eagle

1. Introduction

Javan Hawk-Eagle (*Nisaetus bartelsi*) is an endemic raptor in Java Island, an indicator species for natural landscapes due to their role as top-level predators in the food chain [1]. JHE is one of the non-migratory birds of prey with a narrow distribution so that the function of ecosystem indicators is only on a local scale. In its natural habitat, JHE plays a role in controlling the animals' population that become its prey. This bird is susceptible to environmental changes that cause its population to be disturbed if the environment's quality is disturbed [2]. The small population size coupled with the destruction and fragmentation of natural forests and illegal trade and hunting caused JHE by the IUCN to be granted rare status [3,4]. The Director of Forest Protection and Nature Conservation (PHKA), through the Decree of the Director-General of Layoffs No. 200/IV/KKH/2015 to establish JHE as an endangered species that is prioritized to increase its population by 10% in 2015-2019 following biological conditions and habitat availability—supported by Regulation of the Minister of Forestry of the Republic of Indonesia Number: P.58/Menhut-II/2013 on Strategy and Action Plan of JHE Conservation (*Spizaetus bartelsi*) in 2013-2022.

East Java is the largest location on the island of Java, which is a habitat for JHE because of the high quality of its natural landscape [5]. Based on the probability model of habitat suitability of JHE using satellite imagery in 2002 conducted [5], there are an estimated 189 pairs of JHE in East Java. Currently, there are several locations of JHE encounters outside of the existing model. Based on several locations where there is the presence of JHE, it

IOP Conf. Series: Earth and Environmental Science **879** (2021) 012038 doi:10.1088/1755-1315/879/1/012038

assumes that the population of this species in Java is increasing, and its distribution is also widespread, especially in East Java [6]. Hence, it is necessary to evaluate the current distribution of JHE's habitat in East Java.

This study aimed to analyze the patch dynamics in the current condition of the JHE habitat distribution using Landsat 8 imagery in 2015 compared with previous research using 2002 satellite imageries [1,5]. This study also validates the patch occupancy by JHE through ground truth-checked activities to determine the actual JHE's habitat distribution. The results of current conditions, dynamics, and validation of habitat distribution are used to provide recommendations for JHE's habitat management based on patch dynamics in East Java.

2. Method of analysis

2.1. The study area

This study focused on the East Java province (figure 1), and Madura Island was excluded from this research. The ± area of 42,406 km² with a geographical location of 111,0'-114.3 BT and 7,0'-8,30'LS.

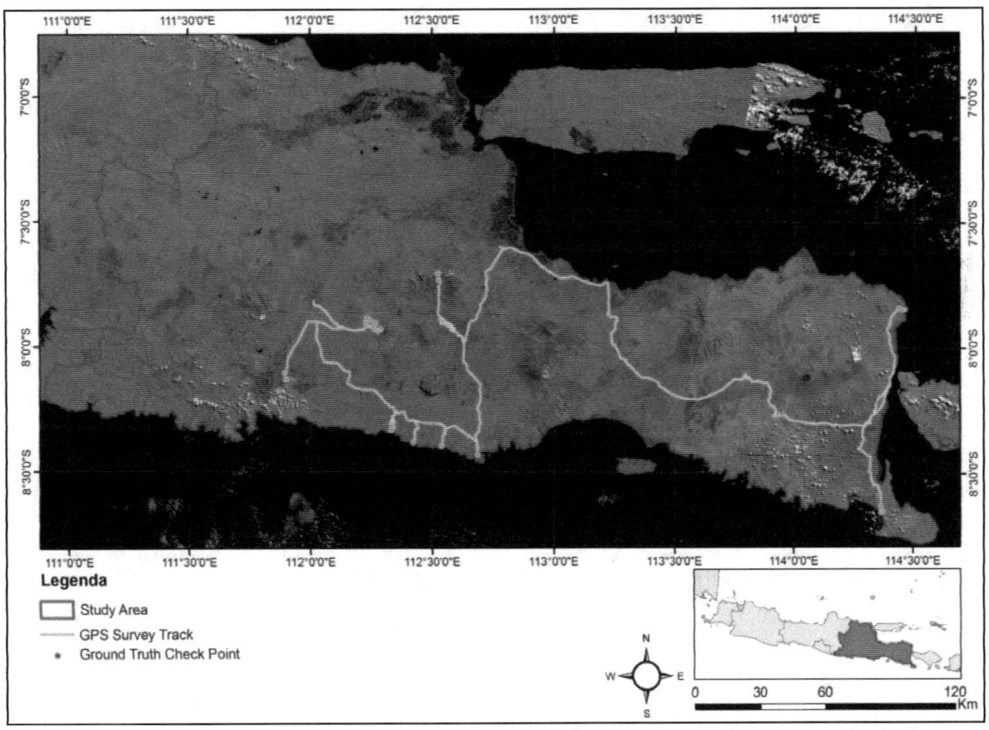

Figure 1 Study area: East Java province (Source: Landsat 8 OLI/TIRS)

2.2. Analysis

At this stage, data updates are carried out to produce the current conditions of JHE habitat distribution using the habitat suitability model of the JHE's that has been formulated previously [1] using logistic regression model as follows:

$$Pi = \frac{1}{1 + \exp[-(0.3804SLP + 0.0148ELV - 0.1229NDVI - 19.6518)]}$$

Slope (SLP) and Elevation (ELV) derived from the ASTER GDEM data of 2015. Normalized Difference Vegetation Index (NDVI) obtained from Landsat 8 OLI/TIRS scene (path/row) 119/66, 119/65, 118/66, 118/65, 117/66, 117/65 taken in May-June 2015. Furthermore, a habitat suitability map is obtained by using a modeler in ERDAS Imagine 2014 software. The habitat suitability model contains probability values ranging from 0 (not suitable) to 1 (suitable). In this process, the threshold value used is 0.5; it is intended to identify a habitat patch. This stage's result is the current habitat distribution of the JHE. All environmental variables used in this study have a spatial resolution of 30 m. The current habitat distribution has re-sampled to a resolution of 250m to compare with the previous habitat distribution of 2002.

Two types of errors identify at the validation stage, omission error, where the model is predicted as an inappropriate habitat. Still, the species can be found at that location, and a commission error in the model is predicted as suitable habitat, but the species cannot be found there [7]. Data of JHE encounters from local researchers, and related agency reports are used to validate the current habitat conditions of JHE. Then ground truth-checked activities are carried out for further validation and observed the land cover in the patch area and the habitat threats. Based on the validation results, the condition of distribution of JHE's habitat is obtained. Map of habitat distribution of JHE is then superimposed (overlay) with a map of land cover and protected areas.

3. Results and discussions

The update of the probability model of the suitability of JHE's habitat in East Java resulted in 22 patches in 2015. The patch area varies from 4.01 km^2 to 936.95 km^2, with a total patch area of JHE habitat in East Java of 3744.13 km^2. Compared with 2002, the total area of JHE's habitat patch in East Java has increased by 1134.01 km^2 (43.45%). The dynamics that occur in the JHE's patch are patch lost (1 patch), patch area decreased (5 patches), patch area increased (13 patches), new patch (4 patches), and merged patch (**Table 1**, **Figure 2**).

Table 1 The habitat patches distribution in 2002 dan 2015

Patch Number	Location Name	Patch Area (km^2)		Change (km^2)	Patch Dynamics
		2002	2015		
1	Mt. Lawu	194,32	271,22	76,90	Patch area increased
2	Mt. Liman - Wilis	221,01	347,06	126,05	Patch area increased
3	Namangan	4,70	21,36	16,66	Patch area increased
4	Badengan	9,47	6,71	-2,76	Patch area decreased
5	Slangu	4,72	0	0,00	Patch lost
6	Sambit	4,61	4,01	-0,60	Patch area decreased
7	Karangan	12,31	7,18	-5,13	Patch area decreased
8	Mt. Kelud	35,73	52,11	16,38	Patch area increased
9	Kasembon	5,92	7,73	1,81	Patch area increased
10	Mt. Kawi - Arjuna - Welirang	472,13	658,34	186,21	Patch area increased
11	Mt. Penanggungan	12,25	12,57	0,32	Patch area increased
12	Tutor*	15,75	0	0,00	Patch merged to 13
13	Bromo Tengger Semeru National Park*	663,65	936,95	273,30	Patch area increased
14	Candipuro	7,67	5,22	-2,45	Patch area decreased
15	Tiris	13,49	12,71	-0,78	Patch area decreased
16	Pakuniran	5,02	6,72	1,70	Patch area increased
17	Yang Plateau - Mt. Argapura	435,94	587,86	151,92	Patch area increased
18	Tegalampel	6,24	18,99	12,75	Patch area increased

Patch Number	Location Name	Patch Area (km²)		Change (km²)	Patch Dynamics
		2002	2015		
19	Meru Betiri National Park	5,16	19,35	14,19	Patch area increased
20	Alabang*	5,60	0	0,00	Patch merged to 22
21	Arjuna*	8,61	0	0,00	Patch merged to 22
22	Mt. Raung – Ijen*	465,82	722,19	256,37	Patch area increased
23	Bandar	0,00	15,75	15,75	New patch
24	Pule	0,00	7,26	7,26	New patch
25	Bunga	0,00	9,31	9,31	New patch
26	Mt. Baluran	0,00	13,53	13,53	New patch
Total		2610,12	3744,13		

Note: The marked location (*) is habitat merged

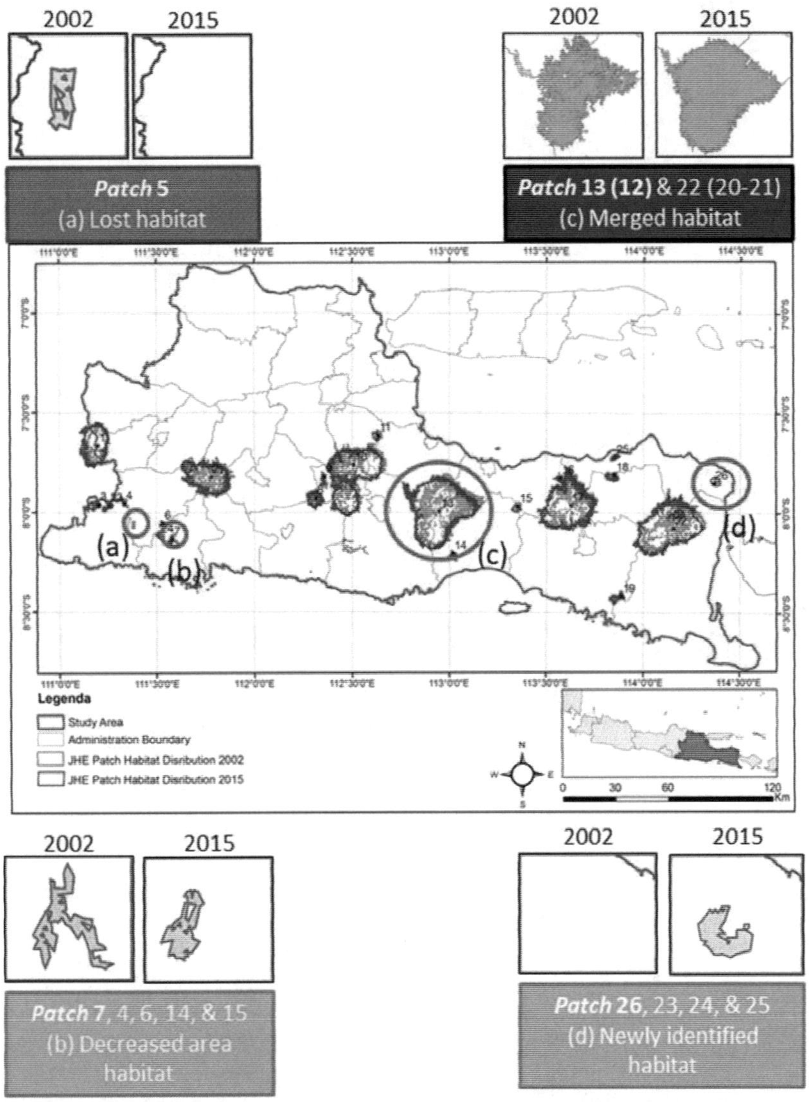

Figure 2 Patch dynamics from 2002 to 2015 in East Java

Among the three environmental variables (SLP, ELV, NDVI) in the predicted probability model, NDVI becomes a factor that tends to make the dynamic patch results. The slope and elevation tend not to change unless there is a natural disaster that can change the landform. This study's NDVI was transformed from MODIS NDVI 250 m [5] to Landsat 8, 30m resolution. Differences in the resolution of imagery are thought to be one influencer for increasing suitable habitat areas in 2015. Higher image resolution used will produce a more precise calculation of NDVI [8]. Meanwhile, environmental quality changes are suspected of causing the natural forest area to increase in Java Island, especially in protected areas.

Ground truth-check was carried out at nine locations: Mount Kelud, Mount Budeg, Jolosutro Beach, Ngeliyep Beach, Kondang Merak, Sempu Island Nature Reserve, Mount Arjuna-Welirang, Baluran National Park, and Alas Purwo National Park. The JHE was found in two patches from ground truth-checked (patch 2, patch 28), and the JHE nests were found in patches 28 and 30. All validation patches are located in the southern part of the study site (**Figure 3**).

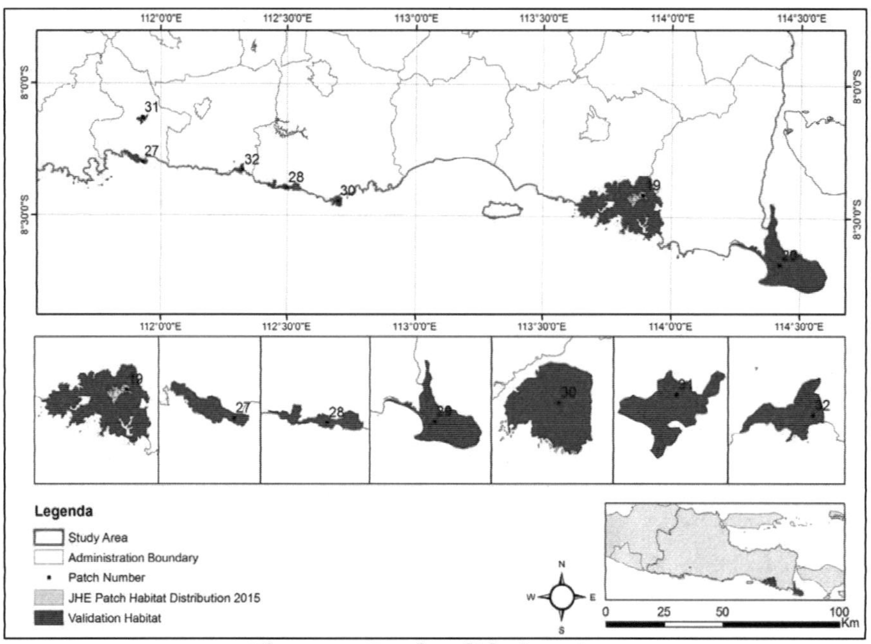

Figure 3 Patch validation sites

Based on the validation results, six newly identified patches (patch 27 to patch 32) and one patch area increased (patch 19). The total area of the habitat patches increased by 1022.13 km^2 after the validation. The total number of patches after validation became 28 patches, which is covered 4766.26 km^2. The total area increased after validation became 2156.14 km^2 or 82.61% of the total area occupied by JHE's in 2002 (**Table 2**) (**Figure 4**). At the validation stage, 15 patches (48.39%) were detected by the model, ten patches (32.26%) are commission errors, and six patches (19.35%) are omission errors. Commission error is an unavoidable error because JHE not occupies all potential habitats. Therefore, commission errors are considered patches that can become the habitat of JHE in this study. So that the patch detected correctly has the potential as a habitat of JHE is 80.65%.

JHE is known can live from 0 to 3000 meters above sea level [9]. The patch omission error sites are located at an altitude of 0-500 meters above sea level and characterize a tropical lowland rainforest [10]. JHE is also explicitly referred to as slope specialist because of its dependence on varied topography [11],[12]. This steep slope is related to protection against humans or other predators [1]. However, ground truth-check results show that the nest trees in Alas Purwo National Park (patch 29) are on a flat area, and shrubs overgrow the surroundings and are located below 100m above sea level. The previous probability model of JHE was developed based on JHE nest in Mount Gede Pangrango National Park in West Java, a highland tropical rainforest ecosystem [1]. Alas Purwo National Park (patch 29) is an omission error in the previous study [5]. The probability model of JHE's did not predict Alas Purwo National Park as a potential habitat for JHE. Therefore, the model does not correctly identify the habitat of JHE in the lowlands. This study recommends conducting intensive observations on the potential patches of JHE in lowland forests in East Java and Central Java, and West Java.

Table 2 Habitat distribution validation in 2015

Patch Number	Location Name	Area (km^2) 2015	Area (km^2) Validation	Description
1	Mt. Lawu	271.22	271.22	√
2	Mt. Liman - Wilis	347.06	347.06	√
3	Namangan	21.36	21.36	Comission Error
4	Badengan	6.71	6.71	Comission Error
5	**Slangu**	-	-	**Patch lost**
6	Sambit	4.01	4.01	Comission Error
7	Karangan	7.18	7.18	Commission Error
8	Mt. Kelud	52.11	52.11	√
9	Kasembon	7.73	7.73	Commission Error
10	Mt. Kawi - Arjuna - Welirang	658.34	658.34	√
11	Mt. Penanggungan	12.57	12.57	√
12	**Tutor**	-	-	**Merged to patch 13**
13	Bromo Tengger Semeru National	936.95	936.95	√
14	Candipuro	5.22	5.22	Comission Error
15	Tiris	12.71	12.71	Comission Error
16	Pakuniran	6.72	6.72	√
17	Yang Plateu - Mt. Argapura	587.86	587.86	√
18	Tegalampel	18.99	18.99	√
19	Meru Betiri National Park*	19.35	527.56	√
20	**Alabang**			**Merged to patch 22**
21	**Arjuna**			**Merged to patch 22**
22	Mt. Raung - Ijen	722.19	722.19	√
23	Bandar	15.75	15.75	Comission Error
24	Pule	7.26	7.26	Commission Error
25	Bunga	9.31	9.31	Comission Error
26	Mt. Baluran	13.53	13.53	√
27	Sangria Beach	-	20.72	Omission Error
28	Kondang Merak –Ngeliyep Beach	-	25.98	Omission Error
29	Alas Purwo National Park	-	437.84	Omission Error
30	Sempu Island Nature Reserve	-	9.87	Omission Error
31	Mt. Budeg	-	9.49	Omission Error
32	Jolosutro Beach	-	10.02	Omission Error
	Total	3744.13	4766.26	

Note: - The marked location (*) is a patch that has increased in one area
- (√) is a patch detected by the model

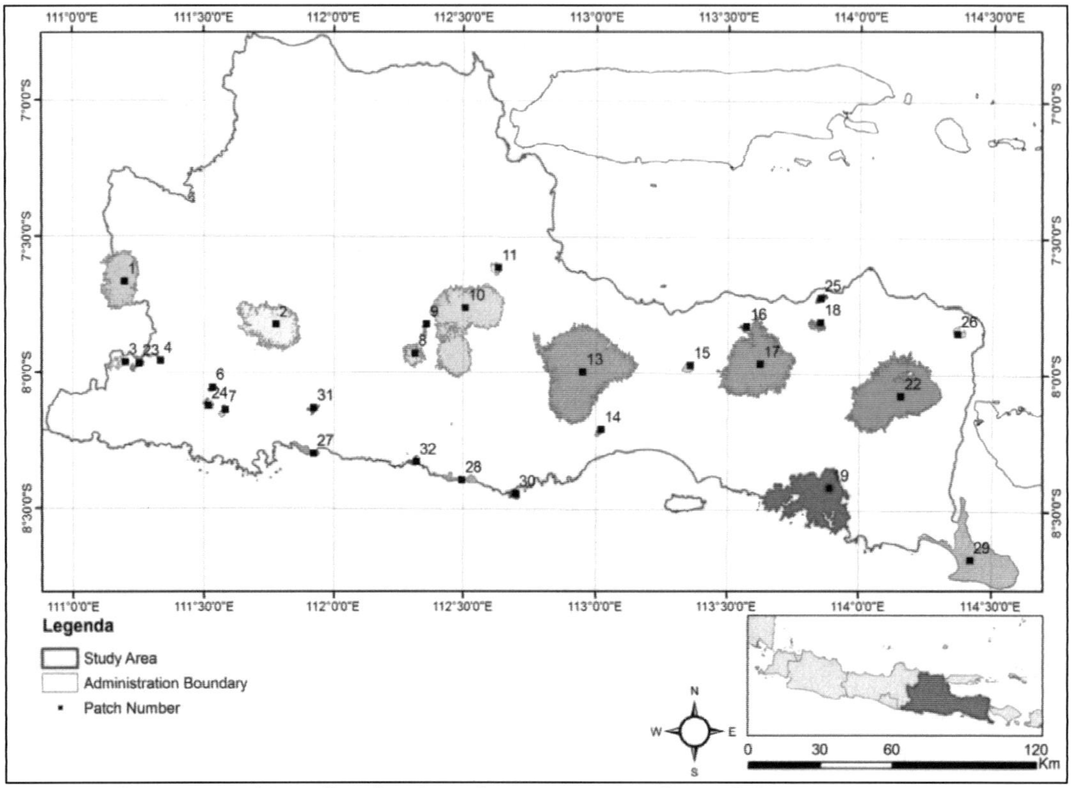

Figure 4 Habitat distribution of JHE in 2015 after validation in East Java

Based on the overlay results with a map of protected areas, there are ten patches (1901.08 km² or 39.89% of JHE area) located inside the protected areas. Ten patches of JHE's habitat in protected areas are Mount Lawu (patch 1), Mount Liman-Wilis (patch 2), Mount Kawi-Arjuna-Welirang (patch 10), Bromo Tengger Semeru National Park (patch 13), Yang Plateau-Mount Argapura (patch 1 7), Meru Betiri National Park (patch 19), Mount Raung-Ijen (patch 22), Mount Baluran (patch 26), Alas Purwo National Park (patch 29), and Sempu Island Nature Reserve (patch 30). The natural tourism area contained in JHE's habitat patch area, Mount Lawu (patch 1), Mount Kawi-Arjuna-Welirang (patch 10), Mount Penanggungan (patch 11), Bromo Tengger Semeru National Park (patch 13), and Kondang Merak-Ngeliyep Beach (patch 28). JHE's habitat patches outside the protected area as much as 18 patches with an area of 2865.18 km² or 60.11%.

The results showed that the habitat patches are dominantly located in forests (71,62%) and followed by plantations and shrubs. The forest land, especially natural forests, is the nesting place of JHE, while other land covers, i.e., plantations, shrubs, agricultural areas, become essential hunting areas for JHE. A large and healthy forest can provide a home range area for JHE [13], including natural forests and secondary forests [14]. The forest can provide the prey for JH e.g. reptiles, small mammals (bats, squirrels, rats, infant) [13]. Characteristics of JHE's nest tree in lowland forest, which is found in this study and Ujung Kulon National Park [15], have a height between 25-50 m with a diameter between 60-230 cm and have the tree architecture of "rauh." The tree canopy should be open with horizontal branches and have a broad view.

This study recommends continuing monitoring activities on habitat patches, including potential habitat patches in lowland areas, and proposing conservation activities based on habitat patch dynamics that occur especially for the patch area decrease over the 2002-2015 period. Perform landscape restoration on patch habitat lost and preserving landscapes on patches that have decreased of area.

4. Conclusion

This study successfully updated the distribution of JHE's habitat patches based on 2015 data and validated them through ground truth-checked activities in 2016. East Java has 22 patches with a total patch area of JHE's habitat in East Java of 3744.13 km^2 or an increase of 1134.01 km^2 (43.45%) of the total patch area the JHE's habitat in 2002. The dynamics that occur in the JHE's patch are patch lost(1 patch), patch area decreased (5 patches), patch area increased (13 patches), new patch (4 patches), and merged patch. After validation, the number of patches becomes 28 patches. There are six newly identified patches, and one patch has increased area (patch 19). The total area of JHE's patch habitat in 2015 after validation became 4766.26 km^2, where 2156.14 km^2 increased or 82.61% of the total area occupied by JHE's in 2002. The patch detected correctly has the potential as a habitat of JHE is 80.65%. This study's results will be useful for habitat management activities based on patch dynamic occurred from 2002-2015.

Acknowledgments

The author would like to thank The Ministry of Research, Technology, and Higher Education, Indonesia, the fiscal year of 2016–2017 (grant number 011/SP2H/LT/DRPM/VIII/2017), and the Ministry of Research and Technology/National Agency for Research and Innovation, Indonesia, the fiscal year of 2019–2020 (grant number 1/AMD/E1/KP.PTNBH/2020) for supporting this study. Thanks to Heru Cahyono for providing us the local information and guide us during the field survey.

References

[1] Syartinilia and Tsuyuki S 2008 GIS-based modelling of Javan Hawk-Eagle distribution using logistic and autologistic regression models. *Biological conservation* **141** 756-769
[2] Prawiradilaga D M 2003 *Panduan Survei Lapangan dan Pemantauan Burung- Burung Pemangsa.* (Jakarta: PT. Binamitra Megawarna)
[3] BirdLife International 2000 *Threatened Birds of the World.* (Barcelona: Lynx Edicions and BirdLife International)
[4] BirdLife International 2001 *Threatened Birds of Asia: The BirdLife International Red Data Book.* (UK: BirdLife International Cambridge)
[5] Syartinilia, Tsuyuki S and Lee J S 2009 Gis-Based Habitat Model Of Javan Hawk-Eagle (Spizaetus bartelsi) Using Inductive Approach In Java Island, Indonesia. *Wildlife: Destruction, Conservation, and Biodiversity.* (New York: Nova Science Publisher) pp 302-311
[6] Ardiansyah M, Pramudhita H E, Cahyono H, Hermadhiyanti W, Arifiantari R, Sa'adah A, Ramadhani C and Retnaningtyas R W 2015 The Sighting report of Javan Hawk-Eagle (Nisaetus bartelsi) in the Area of Kondang Merak, Malang Regency, Indonesia *Proceeding of The 3rd International Conference on Biological Science* ISSN 2413-0877 Vol. 2 530-532
[7] Boone R B and Krohn W 1999 Modeling the occurrence of bird species: are the errors predictable? *Ecological Applications* **9** 835-848
[8] Jiang Z, Huete A R, Chen, J, Chen Y, Li J, Yan G and Zhang X. 2006. Analysis of NDVI and scaled difference vegetation index retrievals of vegetation fraction. *Remote sensing of environment* **101** 366-378
[9] Setiadi AP, Rakhman Z, Nurwatha PF, Mucthar M, Raharjaningtrah W 2000 *Status, Distribusi, Populasi, Ekologi & Konservasi Elang Jawa Spizaetus bartelsi Stresemann 1924 Di Jawa Barat Bagian Selatan.* (Bandung: BP/FFI/Birdlife International/YP AL-HIMBIO UNPAD)
[10] Santoso Y 1996 Diversitas dan Tipologi Ekosistem Hutan yang Perlu Dilestarikan. *Proseding Simposium Penerapan Ekolabel di Hutan Produksi* pada tanggal 10-12 Agustus 1995. (Bogor: Kerja Sama Fakultas Kehutanan IPB dengan Yayasan Gunung Menghijau dan Yayasan Pendidikan Ambarwati)
[11] Wells DR 1985 *The forest avifauna of Western Malesia and its conservation*, in: Diamond, A.W., Lovejoy, T.E. (Eds.), Conservation of Tropical Forest Bird, ICBP Technical Publication No. 4 (London: International Council for Bird Preservation) pp 213-232

[12] van Balen S, Nijman V, Prins HHT 2000 The Javan hawk-eagle: misconceptions about rareness and threat. *Biological Conservation* **96**, 297-304
[13] Prawiradilaga DM 1999 Elang Jawa Satwa Langka: Seri Pendidikan Konservasi Keanekaragaman. (Jakarta: Pusat Penelitian dan Pengembangan Biologi [P3B], Direktorat Jendral Perlindungan dan Konservasi Alam [DJPHKA], Japan International Cooperation Agency [JICA])
[14] Nijman V, van Balen S. 2003. Wandering stars: age-related habitat use and dispersal of Javan Hawk-eagles (Spizaetus bartelsi). *Journal of Ornithology* **144** 451-458
[15] Fahmi I, Syartinilia 2020 Habitat preferences of Javan Hawk-eagle's new occurrence on lowland forest in Ujung Kulon National Park. *IOP Conf. Series: Earth and Environmental Science* 590, 012004. https://doi.org/10.1088/1755-1315/590/1/012004

Landscape planning for geodiversity conservation of Ciletuh geo area in Ciletuh Geopark using bioregional approach

M F Rifqi[1] and Q Pramukanto[2]
[1] Student of Landscape Architecture Department, Faculty of Agriculture, IPB University
[2]Lecturer of Landscape Architecture Department, Faculty of Agriculture, IPB University

E-mail: qpramukanto@apps.ipb.ac.id

Abstract. Indonesia has 15 National Geoparks, and four of them have been recognized by UNESCO as the Global Geopark Network (GGN). The Ciletuh Geopark is one of the geo-parks in Indonesia that UNESCO has recognized. With the status of GGN, the management of the Ciletuh Geopark area must be directed according to the area development criteria set by UNESCO. The form of designated area development is related to efforts to increase cooperation, increase local communities' welfare, develop infrastructure, and establish sustainable geopark areas. This Area's result must pay attention to the Area's sensitivity with geological uniqueness so as not to damage the geological diversity (geodiversity) in it. Therefore it is necessary to arrange the area with the direction of area development under the stipulations outlined. This research aims to develop a landscape plan to conserve the Ciletuh Geo Area geodiversity as part of Ciletuh Geopark using a bioregion approach. This process starts from inventory, literature study to analysis and synthesis, and planning. The research is carried out with a bioregion approach to determine the bioregion unit, landscape unit, and place the unit. Biophysical and geological characteristics are used in the preparation of bioregion classes. The synthesis was carried out by compiling the place units with significance values at 18 geological sites, making each place unit's intrinsic value. The landscape plan prepared by referring to the rules of conservation of nature reserves can be considered in the development and management of the Ciletuh Geo Area in particular and Ciletuh Geopark area in general.
Keywords: bioregion approach, Ciletuh geo area, geodiversity

1. Introduction

With its diversity of natural resources, Indonesia has 4 (four) National Geoparks, which UNESCO has recognized as the Global Geopark Network (GGN). Geopark is an area that has outstanding geology, which includes geological, ecological, and cultural values. Efforts to protect and enhance natural heritage function are carried out by involving community participation [1]. The development of the Geopark area is carried out to empower 3 (three) elements of diversity, namely geological diversity (geodiversity), biodiversity (biodiversity), and cultural diversity (cultural-diversity) so that in this Area must take Place at least 3 (three) functions, namely conservation, education, and community development. One of the Geoparks in Indonesia that UNESCO has recognized is the Ciletuh Geopark, Palabuhanratu, Sukabumi, West Java.

The Ciletuh Geopark area has a unique geological diversity and the oldest age in West Java. This Area was formed as a result of activities in the subduction zone, namely: the Eurasian Plate, which is composed of granite (acid), and the Indo Australian Plate, which is composed of basalt (alkaline), then produces deep troughs where deep-sea sedimentary rocks, metamorphic rocks, and igneous rocks alkaline to ultra-alkaline [2]. This geology's uniqueness means that the Ciletuh area must be developed comprehensively as an area with conservation, education, and sustainable development functions with comparative and competitive advantages. These three functions are aspirations in the Geopark development, as mentioned by [3] and [4].

The Ciletuh Geopark status, which UNESCO has recognized as GGN, must be maintained. Efforts to maintain this status are by fulfilling the recommendations set by UNESCO. One of the 13 recommendations given by UNESCO to the Sukabumi regional government that must be completed to maintain its status as UNESCO GGN is the development of infrastructure and empowerment of local communities in the Ciletuh Geopark. One of the efforts that can be done in developing a total area focused on preserving the geodiversity function of the Geopark area. The resulting conservation is aimed at maintaining the geological quality contained in it. The geological quality should be maintained that have impacts to sustainability of area are the uniqueness of geodiversity (special geological phenomena, landform features), significance in occurrence and distribution special ecological function, and suitable for both educational and research purposes [7]. The maintained those geological quality will have an impact on the sustainable use of the Area. The Geopark concept is focused on and considered a balance between the conservation of geological heritage, geo-tourism, and local communities' welfare. To ensure the preservation of the geopark area in a balanced manner, landscape planning is needed to maintain the Ciletuh Palabuhanratu Geopark Area's sustainability.

Bioregion, which comes from the words bio (living) and region (territorial), is a place to live (life place), which is a unique environment where the boundaries are more determined by the natural order, which can support uniqueness. The activities of the biotic community therein [5]. The limits set by the government are administrative boundaries that are not necessarily following natural boundaries, which the bioregion approach can determine. In this study, a method was made between the Area's natural characteristics with the geological distribution and the above sites' presence. By planning to conserve bioregion-based areas, it is hoped that the geodiversity of the Ciletuh Geopark area will be conserved.

2. Methods

2.1. Location and Time of Research
The research location was conducted at the Ciletuh Geo Area, which is part of the Ciletuh Geopark with an area of 128,000 Ha (1,280 km^2). Administratively, the Area is located in Sukabumi Regency, West Java Province, Indonesia (Figure 1). The research was carried out for six months, starting in February 2020, which was continued with data processing and compilation of study results until July 2020.

2.2. Preparation
At this stage, the activities include setting planning objectives, gathering initial information, drafting research proposals, and arranging permits for research. This stage is carried out before the field survey.

2.3. Inventory
Inventory is carried out by collecting data/information required in spatial analysis in the form of secondary data. The data and information required include biophysical and geological aspects. A thematic map/data was compiled based on the data and information predefined classification criteria.

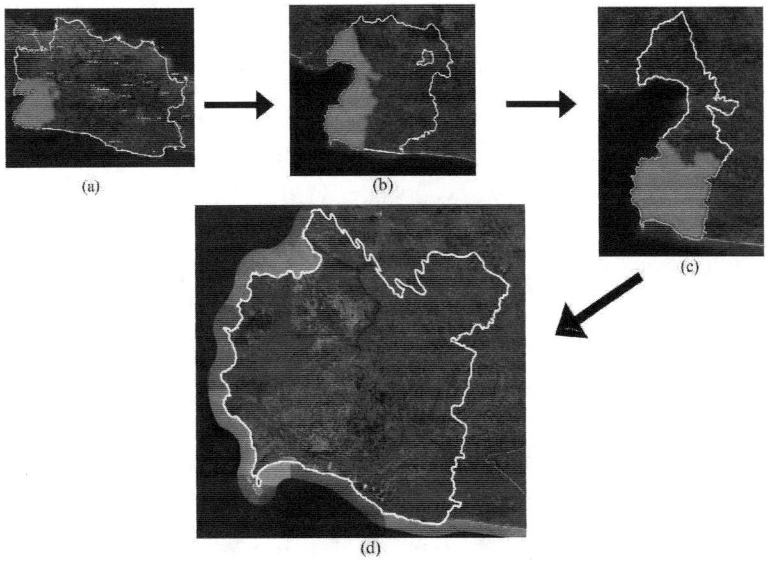

Figure 1 Research location map
(a) West Java Province (b) Sukabumi Regency (c) Ciletuh Geopark and
(d) Ciletuh Geo Area

2.4. *Bioregional Analysis*
Geodiversity analysis of the Ciletuh Geo Area was carried out based on the bioregion approach. The classification of bioregion units is arranged hierarchically based on the biophysical data and geological data of the Ciletuh Geopark with modifications to [6]. The data are classified hierarchically into three units: bioregion unit, landscape unit, and place unit.

2.4.1. *Bioregion Unit*
The Ciletuh Geopark bioregion unit classification is based on topographic characteristics outlined in a watershed to obtain watershed classification data. The bioregion unit is the highest level or the widest in the division of areas defined on the similarity of natural characteristics within which form the Ciletuh Geo Area.

2.4.2. *Landscape Unit*
The landscape unit is a subdivision of the bioregion unit, which is aggregative has homogenous characteristics in it. This unit can be identified by classifying the bioregion unit into sub-unit based on the sub-watershed, soil type, slope of the land, and the Ciletuh Geo Area's geological unit area.

2.4.3. *Place Unit*
The Place Unit is the lowest level unit in the bioregion subdivision, based on land cover in the landscape unit and evaluation of the characteristics of the geodiversity significance value based on the modification of [7].

2.5. *Synthesis*
At this stage, directives for the preservation of the Ciletuh Geo Area were drawn up. The preservation direction is based on evaluating the bioregional analysis results that form the unit of Place. Based on the bioregion analysis, each unit of Place's development's direction can be determined based on its geodiversity significance. The geodiversity significance of each unit of space forms the basis for the formulation of landscape planning.

2.5.1. Landscape Planning

2.5.1.2. Basic Concept of Landscape Planning
The concept of planning for the Ciletuh Geo Area is to make the area preserved based on the uniqueness of geodiversity. The uniqueness of geodiversity as a system guides the development of landscape layout based on the bioregion approach.

2.5.1.2. Landscape Plan Development Concept
The main development concept plan is outlined in the form of a spatial zoning concept. The spatial concept is based on the Ciletuh Geo Area system according to the significance of geological formations' distribution and formation. The idea of zoning refers to [8]. The Ciletuh Geopark area's space is divided into 3 (three) spaces: protection blocks, rehabilitation blocks, and special blocks. Meanwhile, the concept of supporting development was developed in the form of a conservation supporting infrastructure system in the form of access (circulation) connecting the sub-systems of the Ciletuh Geo Area. Besides, the concept of land designation that supports the conservation of the Area is developed.

The landscape plan is the final result for the geodiversity conservation area of Ciletuh Geo area with a bioregion approach. The program is equipped with a schematic plan drawing.

3. Result And Discussion

3.1. General Condition
The research was conducted at Ciletuh Geo area, Sukabumi Regency, West Java. Ciletuh Geo area covers four districts, namely Ciemas District, Ciracap District, Surade District, and Waluran District, with approximately 568.53 km2. Ciletuh Geopark or Ciletuh Earth Park is one of the areas that has the oldest rocks on the island of Java. These rocks have an age of about 50-65 million years ago. The process of the occurrence of these rocks is predicted to result from the collision of the Eurasian Plate (Continental Plate), which is composed of granite (acid), and the Indo-Australian Plate (Ocean Plate), which is composed of basalt (alkaline) then appears to the surface of the earth.

3.2. Ciletuh Geo Area Bioregion Characteristic
The Ciletuh Geo Area's bioregion characteristics, which consist of four districts, namely Ciemas District, Ciracap District, Surade District, and Waluran District, are the basis for the classification of bioregional space units. These characteristics consist of bio-physical and geological characteristics.

3.2.1. Topography and Land Slope
The slope of the land in the Ciletuh Geo Area is very diverse, divided into five classes of steepness, i.e., flat (about 39% or an area of 223.41 km^2), gently sloping (by 32% or covering an area of 180.39 km^2), moderate steep (by 20% or an area of 114.13 km^2), steep (by 7% or an area of 39.09 km^2), and very steep (by 2% or an area of 11.41 km^2). The very steep slope classes are dominating the fault around the Ciletuh Bay Amphitheatre.

3.2.2. Soil
Ciletuh Geo Area is composed of 7 (seven) types of soil, namely alluvial, laterite, latosol, lithosol, podzolic, Mediterranean, and regosol. Ciletuh Geo area consists of eight groups of soil types, namely alluvial gray-brown, alluvial hydro morph, yellowish red and podsolic lateritic complexes, reddish-brown and lithosol latosol complexes, yellowish red and brown latosol complexes, yellowish-red latosol brown latosol complexes and yellowish red podsolic complexes, mediteran reddish-brown, and lithosol, and regosol gray.
The results of data processing on soil types in Ciletuh Geo area show that the dominant soil types are lateritic complexes of yellowish red and podsolic with a percentage value of 53%

or an area of 299.19 km^2 and a complex of yellowish red and brown latosol with a percentage of 18% or an area of 100.9 km^2. This type of yellowish-red lateric soil has an in-depth soil profile, quickly absorbs water, has a moderate organic matter with a neutral to acidic pH, and contains lots of iron and aluminum, so it is good to be used as a building foundation. The red soil texture is relatively dense and sturdy to support the buildings on it and can be used for plantations for maize, oil palm, rubber, cloves, cocoa, and coffee. Brown-gray alluvial soil types have the smallest percentage, namely 1% or an area of 3.84 km^2. This type of soil contains a lot of sand and clay, does not contains many nutrient elements, and has a high organic matter content. The pH content in this soil is low (5.3 - 5.8), and the dissolved aluminum content is very high.

3.2.3. Climate
The climatic data of Sukabumi Regency for ten years from 2006-2015 was used average annual rainfall in Ciletuh Geo Area uses in. There are two classes of rainfall dept in Ciletuh Geo Area, i.e. 2000-2500 mm / year and 2500-3000 mm / year.

3.2.4. Hydrology
The Ciletuh Geo Area's hydrological conditions include groundwater in the form of springs and surface water in rivers and tributaries. Surface water primarily consists of rivers and tributaries that form 5 (five) watersheds (DAS), namely the Cikaso River Basin (922 km2), the Cikarang Watershed (309.44 km2), the Cimarinjung Watershed (133.89 km2), Ciletuh Watershed (279.88 km2), and Cipamaranga Watershed (81.24 km2). The Ciletuh watershed is a distinctive and unique area because there are rocks that are the oldest (pre-tertiary) in West Java. The river in Ciletuh Geo Area consists of the main river, namely the Ciletuh River with a length of 230.72 km), Cikarang River (259.62 km length), and several tributaries.

3.2.5. Land Cover
Based on satellite imagery's visual interpretation, the land cover class of the Ciletuh Geo Area consists of water bodies (lake and ponds), natural forest, plantations, settlements, rice fields, shrubland, bare land, and upland cultivated land. Land cover is one factor that influences the preservation of the geodiversity area in the Ciletuh Geo Area.

3.2.6. Geological Formation Process
The Ciletuh area has unique structural characteristics. In general, it cannot be separated from the activities of the West Java regional tectonic processes, which are tectonic fossils (subduction during the Cretaceous era). So that makes the geological outcrop in the Ciletuh area very unique and rare. The formation of the geological structure of West Java coincided under the influence of the Indian-Australian plate collision activity with the Eurasian plate that has lasted since the Cretaceous Period until now. The position of the subduction zone during this period has changed several times.
In the Late Eocene, the Java subduction route's position shifted to the south for the first time. The Ciletuh area, including its Massive Area, has experienced several tectonic activities, the first time in Pre-Tertiary tectonic activity, the second after the Eocene, maybe even the Oligocene, the third at the end of the Middle Miocene, and the last rapture in the Pliocene era. Tectonic activities after Eocene (Oligocene) affect massive rocks and Eocene rocks, during which time massive rocks experience shifts to form fractures. The Massive Ciletuh includes the western end of the Southern Mountain Range or the Southern Mountain Line, with approximately 50 km in West Java. In the east, this line narrows down to several kilometers, ends near Cilacap, and appears again south of Jogyakarta. Morphologically Massive Ciletuh is located outside Plateau-Jampang, which is bordered by the surrounding escarpment.

3.2.7. Geological Distribution Pattern
Based on the data obtained, [7] has assessed 25 geological sites found in the Ciletuh Geopark. There are differences between the geological sites at the Ciletuh Geopark that have been determined by UNESCO and the sites that have been assessed due to the absence of data related to the assessment of the 61 sites found in the Ciletuh Geopark. However, based on the results obtained, an assessment has been carried out on the entire geological site but has not been published. Among the 25 sites, 18 geological sites were selected in the Ciletuh Geo Area. The 18 sites have moderate to prominent geological heritage values based on scientific records and unique geological phenomena for educational and research purposes [7].

3.2.8. Bioregional Classification
The bioregional classification is arranged hierarchically based on the biophysical characteristics of the Area. There are 3 (three) classes of biophysical and geological characteristics, namely: Bioregion Unit, Landscape Unit, and Place Unit.

3.2.8.1. Bioregion Unit
The bioregion unit classification is prepared by initiating the existing watershed boundaries in the Ciletuh Geo Area. The natural characteristics contained within a watershed boundary have similarities but differ from one watershed to another. The Ciletuh Geo Area is divided into 5 (five) watersheds, namely Cikarang Watershed Cikaso Watershed, Ciletuh Watershed, Cimarinjung Watershed, and Cimaparangan Watershed. The five watersheds are the basis for determining the bioregion unit in the Ciletuh Geo Area.

3.2.8.2. Landscape Unit
The biophysical characteristics used are sub-watershed, soil type, climate, and land slope. Overlay the four parameters, resulting in 294 landscape units within the Ciletuh Geo Area.

3.2.8.3. Place Unit
The Place Unit is the lowest unit in the bioregion classification. The type of geological formation is used as the basis for determining the unit of Place in this study. The analysis of 294 landscape units with geological formations found in the Ciletuh Geo Area resulted in 711 Place Units. Each class of Place Units is a homogeneous area distinguished from other classes of Place Units.

4. Synthesis
Ciletuh Geopark is a National Geopark area that UNESCO has recognized as a Global Geopark Network (GGN). The uniqueness of geodiversity, which is formed through a process that lasts hundreds to millions of years, is a resource that must be conserved. In the synthesis stage, the compilation is carried out at each place unit with a significance value for 18 geological sites, making each place unit's intrinsic value. Based on the 18 selected geological heritage sites, there are 6 (six) geological sites classified as aquatic units, and 12 sites are terrestrial place unit (Figure 2). The locations of the 18 sites are in different Place Units and landscape units. After that, 12 sites located in terrestrial areas will become objects in the Ciletuh Geo Area landscape's conservation planning. Based on these 12 sites, there are 2 (two) geological sites are in the same Place and landscape unit, namely the Puncakmanik Curug and the Middle Curug. Besides, there are 2 (two) geological sites in the same landscape unit but different Place Units, namely Batununggul Beach and Batu Batik.

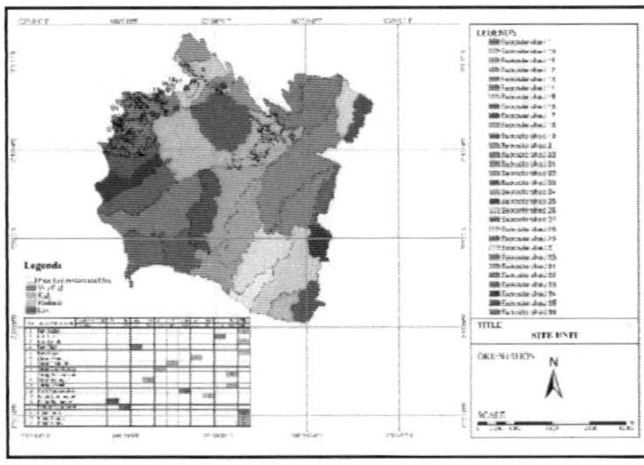

No	GEOLOGICAL SITE	Landscape Unit	39	78	114	123	131	139	230	245	267	267	292	Other
		Site Unit	53	222	248	253	258	309	589	672	561	601	693	Other
1	Batu Badak													1
2	Batu Batik											2		
3	Batu Kodok													3
4	Batu Naga				4									
5	Batu Pagar													5
6	Curug Awang									6				
7	Curug Cikanteh							7						
8	Curug Cimarinjung						8							
9	Curug Puncakmanik												9	
10	Curug Sodong					10								
11	Curug Tengah												11	
12	Fosil Numumulites								12					
13	Pantai Batunungggul										13			
14	Pantai Palampang		14											
15	Peridotit Serpentinit			15										
16	Pulau Kunti													16
17	Pulau Mandra													17
18	Pulau Manuk													18

Very High   High   Moderate   Low

Figure 2 Distribution map of Ciletuh geo area geological sites as place units and distribution matrix of Ciletuh geo area sites in place units

5. Concept And Development

5.1. Planning Concept

Based on the bioregion approach, Ciletuh Geo Area consists of 711 Place Units with homogeneous characteristics in each unit class and can be distinguished from other class unit units. The Place Units are then overlaid with geological heritage sites and selected as many as 18 Place Units with geological heritage sites. The 18 units consist of 12 units in the terrestrial Area, and the other six are in the water unit. Based on 18 geological sites that have been identified, they have a low to leading geological value heritage. The basic concept of planning applied in the Ciletuh Geo Area is to create a conservation area for the Ciletuh Geo Area's geological diversity. The determination of the conservation area is based on 711 Place Units in the Ciletuh Geo Area.

5.2. Concept Devevopement Plan

5.2.1. Block Plan Concept

Based on [8] management blocks in the nature reserve area consist of protection blocks and other blocks. The other block in question includes the rehabilitation block, religion, culture, history, and special block. The protection block is a core part that has a geological object on it. Rehabilitation blocks are blocks that have been damaged and need a rehabilitation process. The rehabilitation block also acts as a buffer for the protection block, a core part of conservation. A particular block is located outside the rehabilitation block and does not directly access the protection block. This conservation area spatial planning is also supported by [9] and [10]. This regulation explains the zoning regulations for areas of the uniqueness of rocks and fossils. In this regulation, it is defined that an area of the uniqueness of rocks and fossils included in a nature reserve area needs to pay attention to several regulations, namely, the construction of buildings is limited to research, education, and nature tourism. The construction of buildings and activities other than the three things mentioned is strictly prohibited and may be subject to sanctions. However, in the regulations regarding the management block of nature reserves, areas in conservation areas with buildings are categorized into special blocks.

5.2.2. Circulation Concept

The circulation concept used to preserve the Ciletuh Geo Area is based on determining blocks and limiting each block's activities. There are 3 (three) circulation concepts applied: primary circulation, secondary circulation, and tertiary circulation. The primary circulation is the main circulation route in the Ciletuh Geo Area. Secondary circulation is a circulation that connects between special blocks with each rehabilitation block in the Area. Tertiary circulation is circulation that is in the core area of the protection and at the same time connects the protection block with the rehabilitation block. Besides, access to the rehabilitation block also needs to be limited because the rehabilitation block activities are ecosystem restoration activities. Meanwhile, for special blocks, circulation is a little loose because the activities in this block are more general. Some settlements existed before the designation of the Ciletuh geological diversity area as a Geopark area.

5.2.3. Vegetation Concept
The concept of vegetation in geodiversity conservation landscape planning is based on vegetation's function on the existing land cover of Ciletuh Geo area. The idea of vegetation is arranged based on its functions, including its conservation and cultivation function. Vegetation with a conservation function is used to protect the ecosystem's richness above it and is found in forest land cover. The cultivation function is found in the agricultural land cover, consisting of rice fields, fields, and plantations. Vegetation with the cultivation function is applied to the entire Ciletuh Geo Area in accordance with existing conditions. Vegetation with a cultivation function can be applied to protection blocks, rehabilitation blocks, and special blocks. In protection blocks, cultivated vegetation can be applied because it allows the use of germplasm for the protection block's cultivation functions.

6. Landscape Planning

6.1. Block Plan
Spatial plans are made based on references from the concept of space. The spatial plan for preserving the geodiversity of Ciletuh Geo Area is divided into three blocks. The three blocks are protection block, rehabilitation block, and special block. Each block has its criteria according to the concept of space. These criteria affect the land cover in each block.

6.1.1. Protection Block
The protection block is a space that is the core of conservation. Protection blocks are based on the presence of a geological site in a Place Unit. The Place Units used as protection blocks are Place Units 53, 222, 248, 253, 258, 309, 589, 672, 561, 601, and 693. In the protection block, the activities carried out are very limited, which affects the land cover. The existing land cover is maintained, including forest land cover, shrubs, open land, and water bodies. Besides, land cover in the form of agricultural lands such as paddy fields, plantations, and fields is also maintained due to the permissibility of utilizing nuftah plasma in the protection block. However, the existence of buildings is prohibited above the protection block. Therefore, the protection block has an existing land cover in the form of a building used as an activity area or a settlement where the location is moved out of the protection block. Palampang beach is located in the protection block in unit 53, where part of the existing land cover is in the form of a newly built lodging building. The building is then moved out of the protection block to preserve the geological site inside.

6.1.2. Rehabilitation block
A rehabilitation block is an area around a protection block that has been damaged, so it needs to be repaired. The general activities that can be carried out in this block are the same as in the protection block but are given additional area restoration activities. The rehabilitation block also functions to maintain the protection block from interference from outside the block.

6.1.3. Special Block
The Special Block is an area in the Ciletuh Geo Area conservation area, which has a building above it, but this building existed before the Ciletuh Bay area was designated as a Geopark area. One Area that becomes a special block is in unit 54 around the Palampang

Beach protection block. In this Area, a building is a center for fish auction activities and a pier for fishing boats fishing in Ciletuh Bay. The building is located downstream on the shore of Ciletuh Bay. The building is inevitable because it is one of the activity centers for fishers in Ciletuh Bay.

6.1.4. Landscape Plan

At the landscape planning stage, the final result is a landscape plan image (Figure 3). Landscape planning is prepared based on spatial plans, circulation plans, and vegetation plans prepared based on planning concepts. In-Place Units with a geological site significance class above them, planning is carried out with the consideration of determining blocks based on [8]. The planning is then outlined in the Ciletuh Geo Area's spatial planning. This geodiversity conservation landscape plan is expected to maintain the geological diversity in this Area in line with the Ciletuh Geo Area's development as a sustainable area.

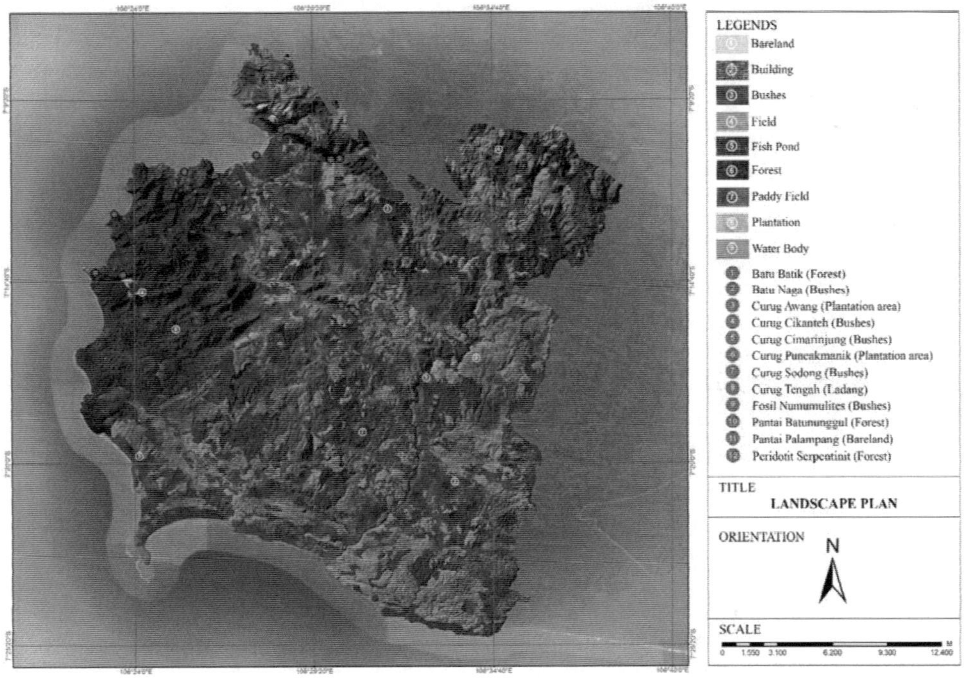

Figure 3 Landscape plan

6.1.5. Circulation Plan

The circulation plan is divided into 3 (three), namely primary circulation, secondary circulation, and tertiary circulation. The primary circulation is only located in special blocks and is a public road traversed by motorized vehicles. This circulation is the main circulation within the Area and a connecting circulation outside the Ciletuh Geo Area. Secondary circulation is located at geological site points. The secondary circulation connects the Area to and from the rehabilitation block of the special block. This secondary circulation is in the form of access that can only be passed by humans and non-motorized vehicles such as bicycles. The third is the tertiary circulation, the circulation within the protection block and the circulation connecting the protection block and the rehabilitation block. The tertiary circulation in this conservation plan cannot be accessed freely due to the limited activities that can be carried out in the Area.

6.1.6. Vegetation Plan

The vegetation plan developed follows the development of the draft vegetation concept. The vegetation used in each space is in accordance with the existing and existing characteristics of the unit. The use of greenery is adjusted to the land cover in the existing Place Units.

7. Conclusions And Suggestions

7.1. Conclusion

The Ciletuh Geo Area has landscape characteristics in biophysics and geology, which form the bioregion units. Based on the analysis results, the Ciletuh Geo Area is included in the five watersheds bioregion unit of 5 watersheds, namely Cikarang watershed, Cikaso watershed, Ciletuh, Cimarinjung watershed, and Cimaparangan watershed. These five bioregion units can be divided into 294 landscape units based on their intrinsic physical value (sub-watershed, soil type, rainfall, and slope). Furthermore, based on the intrinsic value of the geological site significance, there are 711 Place Units.

Landscape planning for geodiversity conservation in the Ciletuh Geo Area can be prepared based on the bioregion approach. Based on the intrinsic value in the form of geological significance in each class of units, a block plan unit can be developed by referring to [8]. Landscape plans are outlined in the form of spatial plans, circulation plans, and vegetation plans.

7.2. Suggestion

Landscape planning for the preservation of geodiversity in the Ciletuh Geo Area with the bioregion approach is expected to be a source of information in developing a sustainable Geopark so that preserving geodiversity in this Area, it can always be maintained. In the future, it is also hoped that based on this landscape plan, regulation can be made by the Sukabumi Regency government related to the preservation of geodiversity in the Ciletuh Geo Area. Besides, to perfect this research, further and more comprehensive research is needed on the properties of the properties so that a protection program can be carried out at each geological site.

References

[1] UNESCO 2004 Guidelines and criteria for National Geoparks seeking UNESCO's assistance to join the Global Geoparks Network (GGN)
[2] Nior S G 2018 Kawasan geopark Ciletuh, Kabupaten Sukabumi sebagai laboratorium alam geologi Bandung (ID): Universitas Padjajaran
[3] Newsome D, Dowling R and Leung Y F 2012 The nature and management of geotourism: a case study of two established iconic geotourism destinations *Tourism Management Perspectives* **2**(3): 19–27
[4] Farsani N T, Coelho C and Costa C 2011 Geotourism and geoparks as novel strategies for socio-economic development in rural areas *International Journal of Tourism Research* p 68-81
[5] Thayer R L Jr 2003 Life place, bioregional thought and practice Berkeley(US): Universitas Of California Press
[6] Pramukanto Q 2004 Pemberdayaan ruang publik: setangkup wilayah biofisik dan domain masyarakatMakalah Semiloka “Pemberdayaan Area Publik di dalam Kota, Area Publik sebagai Tempat Warga Kota Mengekspresikan Diri” Jakarta (ID):IAI
[7] Rosana M F, Yuningsih E T, Haryanto I, Fauzielly L, Zakaria Z, Haryanto A D and Senjaya Y A 2016 Kajian geodiversiti, geoheritage dan geokonservasi daerah ciletuh dan cisolok, Kabupaten Sukabumi, menuju penetapan UNESCO Global Geopark Laporan Bandung (ID): Universitas Padjajaran
[8] Regulation of the Minister of Environment and Forestry of the Republic of Indonesia Number P.76/Menlhk-Setjen/2015 concerning Criteria for National Park Management Zones and Management Blocks of Nature Reserves, Wildlife Reserves, Grand Forest Parks and Nature Tourism Parks
[9] Government Regulation Number 26 of 2008 concerning the National RTRW
[10] Government Regulation Number 13 of 2017 concerning National RTRW

AUTHOR INDEX